Henning Müller · Paul Clough · Thomas Deselaers ·
Barbara Caputo
Editors

ImageCLEF

Experimental Evaluation in Visual Information Retrieval

Springer

Editors
Henning Müller
Business Information Systems
University of Applied Sciences
Western Switzerland (HES–SO)
TechnoArk 3
3960 Sierre
Switzerland
henning.mueller@hevs.ch

Paul Clough
Information School
University of Sheffield
Regent Court
Sheffield S1 4DP
England
p.d.clough@sheffield.ac.uk

Thomas Deselaers
ETH Zürich
Computer Vision Lab/ETF-C 113.2
Zürich
Switzerland
deselaers@vision.ee.ethz.ch

Barbara Caputo
Idiap Research Institute
rue Marconi 19
1920 Martigny
Switzerland
bcaputo@idiap.ch

ISSN 1387-5264 The Information Retrieval Series
ISBN 978-3-642-15180-4 e-ISBN 978-3-642-15181-1
DOI 10.1007/978-3-642-15181-1
Springer Heidelberg Dordrecht London New York

Library of Congress Control Number: 2010934959

ACM Classification (1998): H.3, I.4, C.4, J.3

Cover design: KünkelLopka GmbH, Heidelberg

Printed on acid-free paper

Springer is part of Springer Science+Business Media (www.springer.com)

This book is dedicated to our families and the love, support and encouragement they have given us.

Foreword

The pervasive creation and consumption of content, especially visual content, is ingrained into our modern world. We're constantly consuming visual media content, in printed form and in digital form, in work and in leisure pursuits. Like our cave–man forefathers, we use pictures to record things which are of importance to us as memory cues for the future, but nowadays we also use pictures and images to document processes; we use them in engineering, in art, in science, in medicine, in entertainment and we also use images in advertising. Moreover, when images are in digital format, either scanned from an analogue format or more often than not born digital, we can use the power of our computing and networking to exploit images to great effect.

Most of the technical problems associated with creating, compressing, storing, transmitting, rendering and protecting image data are already solved. We use accepted standards and have tremendous infrastructure and the only outstanding challenges, apart from managing the scale issues associated with growth, are to do with locating images. That involves analysing them to determine their content, classifying them into related groupings, and searching for images. To overcome these challenges we currently rely on image metadata, the description of the images, either captured automatically at creation time or manually added afterwards. Meanwhile we push for developments in the area of *content–based* analysis, indexing and searching of visual media and this is where most of the research in image management is concentrated.

Automatic analysis of the content of images, which in turn would open the door to content–based indexing, classification and retrieval, is an inherently tough problem and because of the difficulty, progress is slow. Like all good science it cannot be rushed yet there is a frustration with the pace of its development because the rollout and development of other related components of image management, components such as capture, storage, transmission, rendering, etc., has been so rapid. We seem to be stuck on the problems of how to effectively find images when we are looking for them. While this is partly caused by the sheer number of images available to us, it is mostly caused by the scientific difficulty of the challenge and so it requires a basic scientific approach to exploring the problem and finding solutions. As in all science, a fundamental aspect is measurement and benchmarking.

In any science, each new development, each approach, algorithm, model, idea or theory has to be measured in order to determine its worth and validity. That is how progress is made, and how fields advance. A theory is put forward and experiments to assess and measure the theory are carried out which may or may not support the theory and we advance the field, either by learning more about what works, or equally important we learn about what does not work. In the technology sector and in the information management area in particular, measuring the validity and worth of new ideas, approaches, etc., now takes place as part of organised benchmarking activities and there are many established examples. The Pascal Visual Object Class recognition challenge addresses recognising objects in images, TRECVid addresses content analysis, retrieval and summarization from video, the KDD competition addresses data mining, and there have been others in machine learning for stock market prediction, shape retrieval, coin classification, text detection and reading, face verification, fingerprint verification, signature verification and of course the well–known NetFlix data mining and recommender competition. All these and many others take place against a backdrop of exploring new ideas, new approaches, and measuring their efficacy in a controlled environment. Which takes us to the present volume which covers ImageCLEF.

Cross–language image retrieval is a niche application domain within the broader area of managing image/visual media. Its importance is huge, though given the directions in which Internet growth is heading with multi–linguality and cross–language resources and processes growing ever more important. Henning Müller and Paul Clough have put together an impressive collection of contributions describing the formation, the growth, the resources, the various tasks and achievements of the ImageCLEF benchmarking activity, covering seven years of development in an annual cycle and involving contributions from hundreds of researchers from across the globe. This book could be described as a capstone volume which brings together all the contributions into one place, but a capstone is a finishing stone or a final achievement, and ImageCLEF continues today, as active as ever. With four parts which address the settings and logistics of ImageCLEF, the various track reports, some reports from participants and finally some external views, the volume is balanced and presents a comprehensive view of the importance and achievements of ImageCLEF towards advancing the field of cross–lingual image retrieval. It will remain an essential reference for anybody interested in how to start up and run a sizeable benchmarking activity, as well as an invaluable source of information on image retrieval in a cross–lingual setting.

Alan F. Smeaton
CLARITY: Centre for Sensor Web Technologies
Dublin City University
Dublin, Ireland, May 2010

Preface

This book contains a collection of texts centred on the evaluation of image retrieval systems. Evaluation, whether it be system–oriented or user–oriented, is an important part of developing effective retrieval systems that meet the actual needs of their end users. To enable reproducible evaluation requires creating standardised benchmarks and evaluation methodologies. This book highlights some of the issues and challenges in evaluating image retrieval systems and describes various initiatives that have sought to provide researchers with the necessary evaluation resources.

In particular the book summarises activities within ImageCLEF, an initiative to evaluate cross–language image retrieval systems that has been running as part of the Cross Language Evaluation Forum (CLEF) since 2003. ImageCLEF has provided resources, such as benchmarks, for evaluating image retrieval systems and complements a number of initiatives within the image retrieval research community, such as TRECvid for video retrieval, PASCAL for object recognition and detection and the many other smaller benchmarks, databases and tools available to researchers.

In addition to providing evaluation resources, ImageCLEF has also run within an annual evaluation cycle culminating in a workshop where participants have been able to present and discuss their ideas and techniques, forming a community with common interests and goals. Over the years ImageCLEF has seen participation from researchers within academic and commercial research groups worldwide, including those from Cross–Language Information Retrieval (CLIR), medical informatics, Content–Based Image Retrieval (CBIR), computer vision and user interaction.

This book comprises contributions from a range of people: those involved directly with ImageCLEF, such as the organisers of specific image retrieval or annotation tasks; participants who have developed techniques to tackle the challenges set forth by the organisers; and people from industry and academia involved with image retrieval and evaluation in general and beyond ImageCLEF. The book is structured into four parts:

- **Part I.** This section describes the context of ImageCLEF and the issues involved with developing evaluation resources, such as test collections and selecting evaluation measures. A focal point throughout ImageCLEF and across many of the tasks has been to investigate how best to combine textual and visualisation information to improve information retrieval. Within the first section we summarise approaches explored within ImageCLEF over the years for this critical step in the retrieval process.

- **Part II.** This section includes seven chapters summarising the activities of each of the main tasks that have run within within ImageCLEF over the years. The track reports are written by those involved in co–ordinating ImageCLEF tasks and provide summaries of individual tasks, describe the participants and their approaches, and discuss some of the findings.
- **Part III.** This section is a selection of chapters by groups participating in various tasks within ImageCLEF 2009. Summaries of the techniques used for various domains such as retrieving diverse sets of photos from a collection of news photographs, multi–modal retrieval from online resources, such as Wikipedia, and retrieval and automatic annotation of medical images are presented. The chapters in this section show the variety and novelty of state–of–the–art techniques used to tackle various ImageCLEF tasks.
- **Part IV.** The final section provides an external perspective on the activities of ImageCLEF. These help to offer insights into the current and emerging needs for image retrieval and evaluation from both a commercial and research perspective. The final chapter helps to put ImageCLEF into the context of existing activities on evaluating multimedia retrieval techniques, providing thoughts on the future directions for evaluation over the coming years.

Sierre, Zürich, Martigny, Switzerland
Sheffield, UK
July 2010

Henning Müller
Paul Clough
Thomas Deselaers
Barbara Caputo

Acknowledgements

It is hard to know where to begin when considering who to thank because ImageCLEF has been very much a collaborative effort and involved a large number of people over the past seven years. We thank the UK Engineering and Physical Sciences Research Council (EPSRC) for funding the Eurovision project (grant number GR/R56778/01): a proposal written by Mark Sanderson that included running a large–scale benchmarking event for cross–language image retrieval that was to become ImageCLEF. In addition to the EPSRC, we thank all of the national and international funding bodies who have supported ImageCLEF in one way or another over the years. In particular we thank the European Union for support through the funding of various EU projects that have supported task organisers. We would like to acknowledge the following projects for supporting ImageCLEF activities:

- UK projects: Eurovision (EPSRC, grant GR/R56778/01)
- EU projects: MUSCLE, Multimatch, SemanticMining, DebutIT, Promise, TrebleCLEF, DELOS;
- Swiss projects: FNS (205321–109304/1), HES–SO for the BeMeVIS project;
- German projects: Theseus (grant 01MQ07017);
- companies such as Google (medical task) and LTU (for the object recognition task);
- American projects: NSF (grant ITR–0325160)

From the start ImageCLEF has been a collaborative activity and involved many people, many bringing new ideas, which is necessary in evaluation as it needs to advance as technology advances. People have given up their time and put tremendous effort into helping co–ordinate and run tasks. This has enabled us to produce re–useable resources for various image retrieval communities. The following is hopefully a complete list of those involved in the organisation of ImageCLEF: Thomas Arni, Peter Dunker, Thomas M. Deserno, Julio Gonzalo, Michael Grubinger, Allan Hanbury, William Hersh, Mark Huiskes, Patric Hensfelt, Charles Kahn, Jayashree Kalpathy–Cramer, Jussi Karlgren, Jana Kludas, Monica Lestari–Paramita, Stefanie Nowak, Adrian Popescu, Andrzej Pronobis, Mark Sanderson, Tatiana Tommasi and Theodora Tsikrika.

Besides those involved in the organisation and actually performing the work, we also need to thank all the data providers who have enabled us to distribute and use their content. In particular we acknowledge the help of Norman Reid from St. Andrews University Library in Scotland for providing us access to the historic set of photographs for the first ImageCLEF evaluation campaign. In addition we thank the following institutions: University of Basel (Pathopic), University of Geneva (Casimage), Mallinckrodt Institute of Radiology (MIR), RSNA, Flickr, viventura, Belga, Wikipedia, LTUtech, PASCAL, UCLA (HEAL, PEIR), OHSU (CORI), MyPACS, KTH, and the IRMA group.

Thanks go to all those involved in carrying out annotations and relevance assessments across the tasks without whom we would have no gold standard to benchmark against. Many people have given up their own time to generate relevance assessments and not been paid for their contributions. We also thank all the participants to ImageCLEF tasks. It is the participants who make an event such as ImageCLEF possible and without whose support and comments there would be no results to report. Over the seven years of ImageCLEF, approximately 200 groups have signed up for at least one for the tasks and over 100 groups have submitted results to one of the tasks in multi–lingual image retrieval.

We thank those who were involved from the start of ImageCLEF. In particular we thank Carol Peters who has provided CLEF with ten years of dedicated service and created a stable environment in which to run not just ImageCLEF, but many important evaluation tasks for comparing and improving multi–lingual information retrieval systems. We thank Carol for allowing us to include ImageCLEF within the CLEF activities and for her continual support and encouragement of our work. We also gratefully acknowledge the support of Donna Harman for her insightful comments and the discussions we had on evaluation methodologies and directions to follow.

Finally, in terms of this book we would like to thank Ralf Gerstner from Springer Verlag for his help and support, David Clough for proofreading the chapters and helping produce the final version of the book and all the contributing authors. We were perhaps a little naïve when fixing the publication deadlines and had to relieve some of the stress on our fellow writers. We apologise for this and thank everyone for staying with us despite the difficulties. We hope that you will enjoy reading your chapters now they are published.

Contents

List of Contributors

Julien Ah-Pine
Xerox Research Centre Europe, Meylan, France

Adil Alpkocak
Dokuz Eylul University, Department of Computer Engineering, Izmir, Turkey

Uri Avni
Tel Aviv University, Tel Aviv, Israel

Steven Bedrick
Oregon Health and Science University, Portland, OR, USA

Tolga Berber
Dokuz Eylul University, Department of Computer Engineering, Izmir, Turkey

Jinbo Bi
Siemens at Malvern, PA, USA

Luca Bogoni
Siemens at Malvern, PA, USA

Barbara Caputo
Idiap Research Institute, Martigny, Switzerland

Jean–Pierre Chevallet
University of Grenoble, Laboratoire d'Informatique de Grenoble, Grenoble, France

Stephane Clinchant,
Xerox Research Centre Europe, Meylan, France

Paul D. Clough
University of Sheffield, Sheffield, UK

Gabriela Csurka,
Xerox Research Centre Europe, Meylan, France

Adrien Depeursinge
University and University Hospitals of Geneva (HUG), Geneva 14, Switzerland

Thomas Deselaers
Computer Vision Laboratory, ETH Zurich, Zurich, Switzerland

Manuel Carlos Díaz Galiano
SINAI Research group, University of Jaén, Jaén, Spain

Charles Florin
Siemens at Malvern, PA, USA

Lluis Garcia
Yahoo! Research, Barcelona, Spain

Miguel Ángel García Cumbreras
SINAI Research group, University of Jaén, Jaén, Spain

Eric Gaussier
University of Grenoble, Laboratoire d'Informatique, Grenoble, France

Theo Gevers
University of Amsterdam, Amsterdam, The Netherlands

Jacob Goldberger
Bar Ilan University, Ramat–Gan, Israel

Julio Gonzalo
E.T.S.I. Informática de la UNED, Madrid, Spain

Hayit Greenspan
Tel Aviv University, Tel Aviv, Israel

Michael Grubinger
Medellín, Colombia

Allan Hanbury
Information Retrieval Facility, Vienna, Austria

William Hersh
Oregon Health and Science University, Portland, OR, USA

Joo Hwee Lim
Institute for Infocom Research, Singapore

Anna K. Jerebko
Siemens at Malvern, PA, USA

Joemon M. Jose
University of Glasgow, Glasgow, UK

Jayashree Kalpathy–Cramer
Oregon Health and Science University, Portland, OR, USA

Jussi Karlgren
SICS, Kista, Sweden

Deniz Kilinc
Dokuz Eylul University, Department of Computer Engineering, Izmir, Turkey

Jana Kludas
CUI, University of Geneva, Switzerland

Arun Krishnan
Siemens at Malvern, PA, USA

Teerapong Leelanupab
University of Glasgow, Glasgow, UK

Monica Lestari Paramita
University of Sheffield, Sheffield, UK

Suzanne Little
KMi, The Open University, UK

Ainhoa Llorente
KMi, The Open University, UK

Loïc Maisonasse
TecKnowMetrix, Voiron, France

María Teresa Martín Valdivia
SINAI Research group, University of Jaén, Jaén, Spain

Arturo Montejo Ráez
SINAI Research group, University of Jaén, Jaén, Spain

Henning Müller
University of Applied Sciences Western Switzerland (HES–SO), Sierre, Switzerland

Vanessa Murdock
Yahoo! Research, Barcelona, Spain

Stefanie Nowak
Fraunhofer IDMT, Ilmenau, Germany

Ximena Olivares
Universitat Pompeu Fabra, Barcelona, Spain

Francesco Orabona
Dipartimento di Scienze dell'Informazione, Universita' degli Studi di Milano, Milano, Italy

Sangmin Park
Siemens at Malvern, PA, USA

Florent Perronnin
Xerox Research Centre Europe, Meylan, France

Andrzej Pronobis
Department of Computer Science, Royal Institute of Technology, Stockholm, Sweden

Saïd Radhouani
Koodya sàrl, Bou Salem, Tunisia

Vikas Raykar
Siemens at Malvern, PA, USA

Jean-Michel Renders
Xerox Research Centre Europe, Meylan, France

Stefan Rüger
KMi, The Open University, Milton Keynes, UK

Marcos Salganicoff
Siemens at Malvern, PA, USA

Koen E. A. van de Sande
University of Amsterdam, Amsterdam, The Netherlands

Mark Sanderson
University of Sheffield, Sheffield, UK

Alan F. Smeaton
CLARITY: Centre for Sensor Web Technologies, Dublin City University, Dublin, Ireland

Martin Stephens
Press Association Images, Nottingham, UK

Dhavalkumar Thakker
Press Association Images, Nottingham, UK

Tatiana Tommasi
Idiap Research Institute, Martigny, Switzerland

Theodora Tsikrika
Centrum Wiskunde & Informatica, Amsterdam, The Netherlands

Guido Zuccon
University of Glasgow, Glasgow, UK

Roelof van Zwol
Yahoo! Research, Barcelona, Spain

Part I
Introduction

The basic concepts of visual information retrieval benchmarks.

Chapter 1
Seven Years of Image Retrieval Evaluation

Paul Clough, Henning Müller, and Mark Sanderson

Abstract In this chapter we discuss evaluation of Information Retrieval (IR) systems and in particular ImageCLEF, a large–scale evaluation campaign that has produced several publicly–accessible resources required for evaluating visual information retrieval systems and is the focus of this book. This chapter sets the scene for the book by describing the purpose of system and user–centred evaluation, the purpose of test collections, the role of evaluation campaigns such as TREC and CLEF, our motivations for starting ImageCLEF and then a summary of the tracks run over the seven years (data, tasks and participants). The chapter will also provide an insight into lessons learned and experiences gained over the years spent organising ImageCLEF, and a summary of the main highlights.

1.1 Introduction

The contents of this book describe ImageCLEF, an initiative for evaluating cross–language image retrieval systems in a standardised manner thereby allowing comparison between the various approaches. ImageCLEF ran for the first time in 2003 as a part of the Cross–Language Evaluation Forum (CLEF), leading to seven years of activities which are summarised in this book. As of 2010, however, the ImageCLEF evaluation campaign is still running evaluation tasks. A major outcome of ImageCLEF has been the creation of a number of publicly–accessible evaluation

Paul Clough
University of Sheffield, Sheffield, United Kingdom e-mail: p.d.clough@sheffield.ac.uk

Henning Müller
Business Information Systems, University of Applied Sciences Western Switzerland (HES–SO), TechnoArk 3, 3960 Sierre, Switzerland e-mail: henning.mueller@hevs.ch

Mark Sanderson
University of Sheffield, Sheffield, United Kingdom
e-mail: m.sanderson@sheffield.ac.uk

H. Müller et al. (eds.), *ImageCLEF*, The Information Retrieval Series 32,
DOI 10.1007/978-3-642-15181-1_1, © Springer-Verlag Berlin Heidelberg 2010

resources. These benchmarks have helped researchers develop new approaches to visual information retrieval and automatic annotation by enabling the performance of various approaches to be assessed. A further outcome, arguably less tangible but just as important, has been to encourage collaboration and interaction between members of various research communities, including image retrieval, computer vision, Cross–Language Information Retrieval (CLIR) and user interaction.

The possibility of creating a publicly available benchmark or test collection for evaluating cross–lingual image retrieval systems was a key objective of the Eurovision project[1]. This included dissemination through an international body, such as CLEF, and in 2002 a new multimedia evaluation task for CLEF was proposed (Sanderson and Clough, 2002). At the same time the CLEF community were looking for new avenues of research to complement the existing multi–lingual document retrieval tasks being offered to participants. Image retrieval was seen as a natural extension to existing CLEF tasks given the language neutrality of visual media, and motivated by wanting to enable multi–lingual users from a global community access to a growing body of multimedia information.

In addition the image retrieval community was calling for a standardised benchmark. Despite the many advances in areas such as visual information retrieval, computer vision, image analysis and pattern recognition over 20 or so years, far less effort has been placed on comparing and evaluating system performance (Müller et al, 2004). Although evaluation was conducted by some researchers, the availability of often only small and copyrighted databases made it hard to compare between systems and provide conclusive results. Calls for a systematic evaluation for image retrieval systems were suggested as a way to make further advances in the field and generate publicly–accessible evaluation resources (Smith, 1998; Goodrum, 2000; Müller et al, 2001), similar to evaluation exercises being carried out in text retrieval such as the U.S. Text REtrieval Conference or TREC[2] (Voorhees and Harman, 2005).

Although Forsyth (2002) argued that such an evaluation of content–based retrieval systems was not productive because the performance of such techniques was too low, the impact of having evaluation resources available for comparative evaluation could clearly be seen in events such as TREC in the text retrieval community and could equally be assumed to advance visual retrieval systems in a similar manner. Over the years, evaluation events such as Benchathlon[3], TRECVID[4], ImagEval[5] and ImageCLEF have helped to foster collaboration between members of the visual retrieval community and provide the frameworks and resources required for systematic and standardised evaluation of image and video retrieval systems. Chapter 27 discusses in more detail various evaluation campaigns for multimedia retrieval.

---

[1] The Eurovision project was funded by the UK Engineering and Physical Sciences Research Council (`http://www.epsrc.ac.uk`) grant number GR/R56778/01

[2] `http://trec.nist.gov/`

[3] `http://www.benchathlon.net/`

[4] `http://trecvid.nist.gov/`

[5] `http://www.imageval.org/`

1.2 Evaluation of IR Systems

Evaluation is the process of assessing the 'worth' of something and evaluating the performance of IR systems is an important part of the development process (Saracevic, 1995; Robertson, 2008). For example, it is necessary to establish to what extent the system being developed meets the needs of the end user, to show the effects of changing the underlying system or its functionality on system performance, and enable quantitative comparison between different systems and approaches. However, although most agree that evaluation is important in IR, much debate exists on exactly how this evaluation should be carried out. Evaluation of retrieval systems tends to focus on either the system or the user. Saracevic (1995) distinguishes six levels of evaluation objectives, not mutually exclusive, for information systems, including IR systems:

1. The *engineering level* deals with aspects of technology, such as computer hardware and networks to assess issues such as reliability, errors, failures and faults.
2. The *input level* deals with assessing the inputs and contents of the system to assess aspects such as coverage of the document collection.
3. The *processing level* deals with how the inputs are processed to assess aspects such as the performance of algorithms for indexing and retrieval.
4. The *output level* deals with interactions with the system and output(s) obtained to assess aspects such as search interactions, feedback and outputs. This could include assessing usability for example.
5. The *use and user level* assesses how well the IR system supports people with their searching tasks in the wider context of information seeking behaviour (e.g. the user's specific seeking and work tasks). This could include, for example, assessing the quality of the information returned from the IR system for work tasks.
6. The *social level* deals with issues of impact on the environment (e.g. within an organisation) and could include assessing aspects such as productivity, effects on decision–making and socio–cognitive relevance.

The first three levels (1–3) are typically considered part of system–centred evaluation; the latter three (4–6) part of user–centred evaluation. For many years evaluation in IR has tended to focus on the first three levels, predominately through the use of standardised benchmarks (or test/reference collections) in a laboratory–style setting. The design of a standardised resource for IR evaluation was first proposed over 50 years ago by Cleverdon (1959) and has since been used in major information retrieval evaluation campaigns, such as TREC (Voorhees and Harman, 2005), CLEF (Peters and Braschler, 2001) and the NII Test Collection for IR Systems or NTCIR (Kando, 2003).

Over the years the creation of a standard test environment has proven invaluable for the design and evaluation of practical retrieval systems by enabling researchers to assess in an objective and systematic way the ability of retrieval systems to locate documents relevant to a specific user need. Although this type of evaluation has met with criticism, such as whether the performance of a system on a benchmark reflects how a system will perform in an operational setting and the limited involvement of end users in evaluating systems, it cannot be denied that this kind of organised large–scale evaluation has done the field tremendous good, both within and outside the environment of evaluation campaigns (Chapter 27 describes the strengths and weaknesses of evaluation campaigns). However, it is important to acknowledge that IR systems are increasingly used in an interactive way and within social contexts. This has motivated evaluation from a user–centred evaluation perspective to assess performance at the latter three levels: output, use and user, and social (Borland, 2000; Dunlop, 2000; Ingwersen and Järvelin, 2005; Petrelli, 2008; Kelly, 2010). Projects such as MIRA (an evaluation framework for interactive and multimedia information retrieval applications) started to address this for visual information from 1996 (Dunlop, 2000).

The contents of this book are mainly related to system–centred evaluation of visual information retrieval systems: the resources generated to support evaluation and advances in image retrieval and annotation that have resulted from experiments within ImageCLEF. This is not to imply that user–centred evaluation has been ignored. In fact, from the very beginning ImageCLEF ran an interactive image retrieval task (described in Chapter 7) that was later subsumed by the interactive CLEF track (iCLEF). In addition, where possible, evaluation resources that are described in the following chapters, were designed with realistic operational settings in mind. However, our primary aim has been to first create the necessary resources and framework in which researchers could develop and compare underlying techniques for visual retrieval across multiple domains and tasks.

1.2.1 IR Test Collections

A core activity of evaluation campaigns such as TREC and CLEF has been to create reusable benchmarks for various tasks and domains in IR (Robertson, 2008; Sanderson, 2010 – to appear). Similar to other fields in science a benchmark provides a standard by which something can be measured. The design of a standardised resource for evaluation of document retrieval systems (a *test collection* was first proposed in the late 1950s in the Cranfield I and II projects (Cleverdon, 1959, 1991), and has since become the standard model for comparative evaluation of IR systems. In this approach to testing IR systems, commonly referred to as the Cranfield paradigm, the focus is on assessing the performance of how well a system can find documents of interest given a specification of the user's information need in a way that is abstracted from an operational environment. Laboratory–based evaluation is popular because user–based evaluation is costly and complex and it is often difficult to interpret results obtained with end users.

The main components of a typical IR test collection are:

1. A *collection of documents* representative of a given domain (each document is given a unique identifier *docid*). Collections created for and used in ImageCLEF are discussed in Chapter 2.

2. A set of *topics* or *queries* (each given a unique identifier *qid*) describing a user's information needs expressed as narrative text or sets of keywords. For image retrieval, topics may also include example relevant images. Topic creation within ImageCLEF is discussed further in Chapter 3.
3. A set of *relevance judgments (qrels)*, or ground truths, provide a representative sample of which documents in the collection are relevant to each topic (a list of qid/docid pairs). Although relevance judgments are commonly binary (relevant/not relevant) the use of *graded* relevance judgments is also commonly utilised in IR evaluation (e.g. highly relevant/partially relevant/not relevant). This has implications for which performance measures can be used to evaluate IR systems. The topic of gathering relevance assessments for ImageCLEF is discussed in Chapter 4.

Performance measures, such as precision and recall, are used to provide absolute measures of retrieval effectiveness, e.g. what proportion of relevant documents are returned by the IR system (see Chapter 5 for further details on IR evaluation measures). Together, the test collection and evaluation measures simulate the users of a search system in an operational setting. In evaluations such as CLEF, the focus is not on absolute values but on relative performance: system outputs can be compared and systems ranked according to scores obtained with the evaluation measures (i.e. comparative testing). Although test collections were originally used to evaluate ad hoc[6] retrieval, evaluation campaigns, such as TREC and CLEF, have extended the use of test collections to other tasks (e.g. document filtering and routing, document classification and automatic annotation).

Evaluation campaigns, such as TREC and CLEF, are founded upon the Cranfield paradigm and make use of test collections to evaluate various aspects of information access. However, a 'TREC–style' evaluation not only includes producing evaluation resources, such as test collections, but also community building through holding organised annual workshops to present and discuss findings with other researchers. Figure 1.1 shows activities commonly undertaken in the evaluation 'cycle' of TREC (although applicable to other campaigns such as CLEF and NTCIR). For TREC and CLEF this cycle operates runs during one year; some evaluation campaigns operate over a longer period (e.g. NTCIR runs the cycle over 18 months). The cycle begins with a call for participation followed by an expression of interest from participating groups and registration. Evaluation tasks are centred on tracks (e.g. ImageCLEF is a track of CLEF) that may involve one or many tasks. The track organisers must define their tasks for prospective participants in addition to preparing the document collection and topics. This may also involve preparing and releasing training data beforehand. The participants run their IR experiments according to a variety of parameters to produce system outputs in standard format (called *runs*) and will submit what they consider their *n* best runs to the evaluation campaign. Typically the runs

[6] Ad hoc retrieval as defined by TREC simulates the situation in which a system knows the set of documents to be searched, but the search topics are not known to the system in advance. It is also characterised by a detailed specification of the user's query (title, narrative description and keywords) and searches are required to achieve high recall.

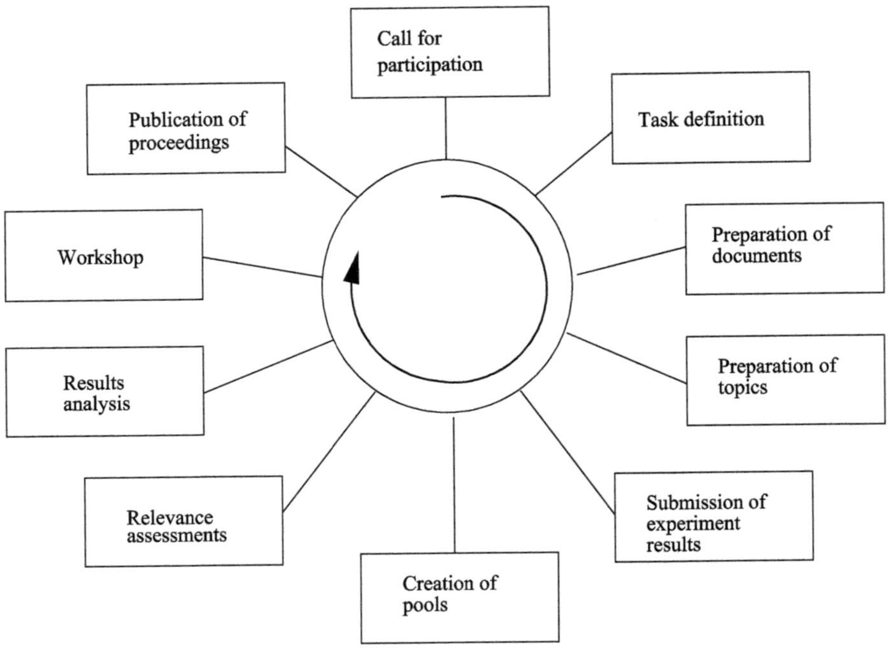

Fig. 1.1: Annual cycle of activities in a TREC–style evaluation (adapted from `http://trec.nist.gov/presentations/TREC2004/04intro.pdf`).

will be based on varying search parameters such as the use of relevance feedback or various combinations of visual and textual modalities.

A sub–set of runs, chosen by the organisers, is used to create *document pools*, one for each topic (Kuriyama et al, 2002). Domain experts (the assessors) are then asked to judge which documents in the pool are relevant or not. Document pools are created because in large collections it is infeasible to judge every single document for relevance. These assessments (qrels) are then used to assess the performance of submitted runs. Evaluation measures are used to assess run performance based on the number of relevant documents found. Although relevance is subjective and can vary between assessors, investigations have shown that relevance assessments can provide consistent evaluation results when ranking runs relative to one another (Voorhees, 2000). Results are released and analysed prior to holding a workshop event to share and discuss findings. Finally, the activities and results are written up in some kind of formal publication, such as workshop proceedings.

1.2.2 Cross–Language Evaluation Forum (CLEF)

CLEF began in 2000 to promote the development of multi–lingual information access systems (Peters and Braschler, 2001). CLEF grew out of the Cross–Language IR track of TREC that ran from 1997–1999. The aims of CLEF are[7] (i) developing an infrastructure for the testing, tuning and evaluation of information retrieval systems operating on European languages in both monolingual and cross–language contexts, and (ii) creating test–suites of reusable data which can be employed by system developers for benchmarking purposes. In the 2009 CLEF campaign the following main tracks were run:

- Ad hoc track, which deals with multi–lingual textual document retrieval;
- ImageCLEF track, which concerns cross–language retrieval in image collections;
- iCLEF track, which addresses interactive cross–language retrieval;
- QA@CLEF track, which covers multiple language question answering;
- INFILE track, which concentrates on multi–lingual information filtering;
- LogCLEF track, which copes with log analysis from search engine and digital library logs;
- CLEF–IP track, which studies multi–lingual access and retrieval in the area of patent retrieval;
- Grid@CLEF track, which performs systematic experiments on individual components of multi–lingual IR systems.

In total there have been 10 CLEF campaigns to date, involving around 200 different participating groups from around the world. Several hundred different research papers have been generated by CLEF participants over the years describing their evaluation experiments and the state of the art contributions to multi–lingual information access.

1.3 ImageCLEF

1.3.1 Aim and Objectives

ImageCLEF first ran in 2003 with the aim of investigating cross–language image retrieval in multiple domains. Retrieval from an image collection offers distinct characteristics and challenges with respect to one in which the document to be retrieved is text (Clough and Sanderson, 2006). For example, the way in which a query is formulated, the methods used for retrieval (e.g. based on low–level features derived from an image, or based on associated textual information such as a caption), the types of query, how relevance is assessed, the involvement of the user during the search process, and fundamental cognitive differences between the interpretation of visual versus textual media. For cross–lingual IR the problem is further complicated

[7] These aims have been taken from the CLEF website: `http://www.clef-campaign.org/`

by user queries being expressed in a language different to that of the document collection or by multi–lingual collections. This requires crossing the language barrier by translating the collection, the queries, or both into the same language. Although the tasks and data sets used in ImageCLEF changed over the years the objectives broadly remained the same:

- To investigate the effectiveness of combining textual and visual features for cross–lingual image retrieval. The combination of modalities is the subject of Chapter 6.
- To collect and provide resources for benchmarking image retrieval systems. These resources include data sets, topics and relevance assessments, which are discussed in Chapters 2–4 and in the track overviews (Chapters 7–12).
- To promote the exchange of ideas to help improve the performance of future image retrieval systems. Work from selected participants from ImageCLEF 2009 is found in Chapters 14–24.

To meet these objectives a number of tasks have been organised by ImageCLEF within two main domains: (1) medical image retrieval and (2) non–medical image retrieval, including historical archives, news photographic collections and Wikipedia pages. Broadly speaking the tasks fell within the following categories: ad hoc retrieval, object and concept recognition, and interactive image retrieval.

Ad hoc retrieval. This simulates a classic document retrieval task: given a statement describing a user's information need, find as many relevant documents as possible and rank the results by relevance. In the case of cross–lingual retrieval the language of the query is different from the language of the metadata used to describe the image. Ad hoc tasks have been run by ImageCLEF from 2003 to 2009 for medical retrieval and non–medical retrieval scenarios, see Chapters 7 and 12 respectively.

Object and concept recognition. Although ad hoc retrieval is a core image retrieval task, a common precursor is to identify whether certain objects from a pre–defined set of classes are contained in an image (object class recognition), assign textual labels or descriptions to an image (automatic image annotation) or classify images into one or many classes (automatic image classification). Chapters 11 and 12 summarise the ImageCLEF object and concept recognition tasks, including medical image classification.

Interactive image retrieval. Image retrieval systems are commonly used by people interacting with them. From 2003 a user–centred task was run as a part of ImageCLEF and eventually subsumed by the interactive CLEF (iCLEF) track in 2005. Interaction in image retrieval can be studied with respect to how effectively the system supports users with query formulation, query translation (in the case of cross–lingual IR), document selection and document examination. See Chapter 7 for further details on the interactive image retrieval tasks of CLEF.

Table 1.1: Participation in the ImageCLEF tasks 2002–2009, distinct number of participants by year and chapter references for further details.

Task	2003	2004	2005	2006	2007	2008	2009	See Chapter
General images								
Photographic retrieval	4	12	11	12	20	24	19	8
Interactive image retrieval	1	2	*2*	*3*	–	*6*	*6*	7
Object and concept recognition				4	7	11	19	11
Wikipedia image retrieval						12	8	9
Robot vision task							7	10
Medical images								
Medical image retrieval		12	13	12	13	15	17	13
Medical image classification			12	12	10	6	7	12
Total (distinct)	4	17	24	30	35	45	65	

1.3.2 Tasks and Participants

Table 1.1 summarise the tasks run during ImageCLEF between 2003 and 2009 and shows the number of participants for each task along with the distinct number of participants in each year. The number of participants and tasks offered by Image-CLEF has continued to grow steadily throughout the years from four participants and two tasks in 2003 to 65 participants and seven tasks in 2009. Participants have come from around the world to participate in ImageCLEF from both academic and commercial institutions. It is difficult to summarise all of the ImageCLEF activities between 2003 and 2009 and we have not provided an exhaustive account, but in brief these are some of the key events year by year:

- In *2003* the first ImageCLEF task was run at the 4th CLEF workshop by Mark Sanderson and Paul Clough involving two tasks and four participants.
- For *2004* a medical image retrieval task organised by Henning Müller was added to ImageCLEF giving a total of three different tasks. This attracted submissions from 17 participating groups and began the focus for us on medical images.
- In *2005* a new medical image annotation task was introduced bringing the total number of tasks offered to four. William Hersh, Thomas Deserno, Michael Grubinger and Thomas Deselaers joined the organisers and we received approximately 300 runs from 24 participants. The interactive task moved to iCLEF in collaboration with Julio Gonzalo and Jussi Karlgren.
- In *2006* 30 participants submitted runs to four tasks that included a new non–medical object annotation task organised by Allan Hanbury and Thomas Deselaers. A new data set (IAPR–TC12) was also developed for the ad hoc retrieval task (referred to as ImageCLEFphoto).
- In *2007* a total of 35 participants submitted runs to four tasks: multi–lingual ad hoc retrieval, medical image retrieval, hierarchical automatic image annotation for medical images and photographic annotation through detection of objects, a purely visual task. Jayashree Kalpathy–Cramer joined the organising team.

- In *2008* we included a new task for cross–lingual image retrieval from Wikipedia (called WikipediaMM) where participants could exploit the structure of Wikipedia for retrieval. This attracted submissions from 12 participants and overall a total of 45 groups submitted over 1,000 runs to ImageCLEF tasks. The photographic retrieval task experimented with promoting diversity in image retrieval and the interactive task, now a part of iCLEF, created a novel evaluation utilising data from Flickr and undertaking log analysis. Thomas Arni, Theodora Tsikrika and Jana Kludas joined the organisers.
- The *2009* ImageCLEF track was run at the 10th and final CLEF workshop. We had the largest number of participants to ImageCLEF (65 groups) across six tasks which included a new robot vision task organised by Andrzej Pronobis and Barbara Caputo that attracted seven participants. Monica Lestari Paramita also joined the organising team of the ImageCLEFphoto task that used a new data set from Belga, a news agency from Belgium, containing over 500,000 images.

1.3.3 Data sets

A major contribution of ImageCLEF has been to collect a variety of data sets for use in different tasks. Table 1.2 shows all 16 data sets used in ImageCLEF over the seven years, which are further discussed in Chapter 2. The table shows the data set, year added to the ImageCLEF campaign, the total number of images contained in the data set and languages used to annotate the image metadata. For data sets where the same data set has been used but added to in subsequent years, such as the Radiological Society of North America (RSNA), the final number of images has been reported in the table. Clearly noticeable is that many collections are annotated in English. As a cross–language track of CLEF the focus has been primarily on translating user's queries (*query translation*) for bilingual retrieval from a query in a non–English language into English. Other CLEF tracks have focused on other cross–language issues such as bilingual retrieval between other language pairs and multi–lingual retrieval: searching document collections that contain texts in multiple languages.

1.3.4 Contributions

Each of the overview chapters in this book (Chapters 7–13) provides a description of activities conducted in ImageCLEF and summarises contributions made in each of the areas covered. This includes a summary of test collections and ground truths produced for each task that have been used within various research communities. It is clear from the participant's reports (Chapters 14–24) that many novel and interesting techniques have been developed as a part of the experiments carried out for ImageCLEF. This highlights the benefits of TREC–style evaluation for IR sys-

Table 1.2: A summary of data sets used in ImageCLEF 2003–2009

Data set	Year Added	#Images	Annotation Languages
General images			
St Andrews (SAC)	2003	28,133	English
IAPR–TC12	2006	20,000	English, Spanish, German
Belga	2009	498,920	English
LTU	2006	1,100	–
PASCAL VOC	2007	2,600	–
Flickr MIR	2009	25,000	–
INEX MM	2008	150,000	English
KTH–IDOL2	2009		
Medical images			
IRMA	2005	14,410	–
Casimage	2004	8,725	English, French
MIR	2005	1,177	English
PEIR	2005	32,319	English
PathoPIC	2005	7,805	English, German
MyPACS	2007	15,140	English
CORI	2007	1,496	English
RSNA	2008	75,000	English

tems. Chapter 27 highlights the benefits (and limitations) of evaluation campaigns for multimedia retrieval researchers, but overall we believe that ImageCLEF has made a number of contributions including the following:

Reuseable benchmarks: one of the largest obstacles in creating a test collection for public use is securing a suitable collection of images for which copyright permission is agreed. This has been a major factor influencing the data sets used in the ImageCLEF campaigns. The ImageCLEF test collections provide a unique contribution to publicly available test collections and complement existing evaluation resources for a range of retrieval tasks and scenarios. These resources include the IAPR–TC12 photographic collection (Grubinger et al, 2006), a segmented version of the IAPR–TC12 data set (Escalante et al, 2010) and Casimage (Müller et al, 2004).

Evaluation measures: a range of performance measures have been experimented with or developed for ImageCLEF including Geometric Mean Average Precision (GMAP), Cluster Recall (for assessing diversity) and a new evaluation metric based on ontology scoring for the 2009 image annotation task (Nowak et al, 2010).

Open forum for exchange of research: ImageCLEF has actively promoted discussion at the CLEF workshops about approaches to ImageCLEF tasks. In addition, a number of activities[8] have been organised in conjunction with the CLEF workshop and a number of European projects: the First, Second and Third MUSCLE/ImageCLEF Workshops on Image and Video Retrieval Evaluation in

[8] See http://www.imageclef.org/events/ for further details and access to workshop proceedings.

2005–2007, the QUAERO/ImageCLEF Workshop on Multimedia Information Retrieval Evaluation in 2008 and the Theseus/ImageCLEF Workshop on Multimedia Information Retrieval Evaluation.

Publications: the CLEF workshop proceedings provide a published set of formal papers that describe ImageCLEF activities over the years. In addition, the organisers of ImageCLEF co–ordinated a Special Issue on Image and Video Retrieval Evaluation (Hanbury et al, 2010) in the journal Computer Vision and Image Understanding (CVIU) and a Special Issue on Medical Image Annotation in ImageCLEF 2007 (Deselaers et al, 2009) for Pattern Recognition Letters (PRL).

Advances in state of the art: ImageCLEF has run various tasks in different image retrieval settings. For example the medical image retrieval task has provided a set of resources for assessing the performance of medical retrieval systems based upon realistic tasks and topics. The organisers have involved medical professionals in creating realistic tasks and carrying out relevance assessments. Chapter 6 on fusion techniques for combining textual and visual information demonstrates a positive contribution in exploring the use of multiple modalities for image retrieval.

1.3.5 Organisational Challenges

Based on our experiences with ImageCLEF over the past seven years we have encountered a number of challenges with running a TREC–style multimedia retrieval evaluation benchmark. The main organisational challenges are detailed below with suggested solutions (adapted from Müller et al (2007)).

One of the greatest challenges facing the organisation of ImageCLEF has been *funding*. Organising a successful event requires a certain level of commitment from the organisers and their host institutions, e.g. to create suitable data sets, organise and pay for relevance assessments, to maintain regular communication with participants and assist with producing publications from the evaluation event (e.g. workshop proceedings). The ImageCLEF organisers have relied on the support of national and international funding bodies in addition to voluntary effort. Running an evaluation campaign over several years requires thinking about funding beyond the lifetime of a single research project. A strength of ImageCLEF has been to involve several different people to distribute the workload and costs.

To produce reusable evaluation resources for multimedia retrieval systems requires *obtaining access* to data sets and *permission* from the owners to distribute the content to participating groups. This is a significant challenge for high–quality multimedia data sets that are often copyrighted and subject to limited distribution. ImageCLEF has been able to gain access to a number of data sets, some with little or no copyright restrictions. Availability of data sets has a direct impact on what can be evaluated in the evaluation campaign and on reusability of the data set after the lifetime of the evaluation campaign.

A difficult task is often *advertising* the evaluation campaign and *motivating participation*. This is particularly relevant to multimedia retrieval where it is often time–consuming to develop systems for specific tasks and submit runs. This is clearly seen by comparing the number of groups that register for the task (to obtain the data sets) compared to the number who eventually submit results: commonly lower than 50%. ImageCLEF has also had to actively advertise the event across multiple domains because of the cross–disciplinary nature of the tasks. ImageCLEF has benefitted from being part of CLEF that already had a following of participants, was well–known in the IR field and offered participants the chance to publish their results in a good quality publication: the Springer Lecture Notes in Computer Science, after the workshop.

An often difficult task has been to encourage *input from commercial organisations*: both collaborating with organisers (e.g. to suggest suitable search tasks) and participating in the evaluation event itself. Ideally having commercial input enables participants to tackle current real–world challenges and offer businesses an opportunity to investigate what state of the art approaches can achieve on their data sets. The 2010 CLEF campaign has been organised around themes that both academics and businesses have identified as important areas of research requiring investigation.

Creating *realistic tasks and user models* is important in estimating the effectiveness of systems in an operational setting based on results obtained in a laboratory–setting using the benchmarks provided. In ImageCLEF, for example, we have developed realistic search tasks and queries based on the knowledge of experts (e.g. discussions with medical professionals in the case of the medical image retrieval tasks) and analysing query logs generated by existing search systems.

A further challenge in ImageCLEF has been to *efficiently create the ground truths*. This is linked with funding as it is often an extensive and time–consuming task. Approaches such as pooling and interactive search and judge are often used to reduce the amount of assessor time required for judging the relevance of documents, but completeness of relevance judgments and variations amongst assessors must be taken into account. A further issue is that criteria for assessing relevance in multimedia retrieval is often different from assessing the results of text retrieval systems, particularly for medical images (Sedghi et al, 2009). This may require the use of domain experts to make the judgments which relies on access to such people and their availability to make judgments.

1.4 Conclusions

To improve multimedia retrieval systems we need to have appropriate evaluation resources, such as test collections, that offer researchers access to visual data sets, example queries and relevance judgments. Over the past seven years ImageCLEF has provided such resources, together with providing a forum in which researchers have been able to interact and discuss their findings. ImageCLEF has provided mainly resources for system–centred evaluation of image retrieval systems, but has also maintained a relationship with user–centred evaluation of image retrieval systems, mainly through its relationship with the CLEF interactive track (iCLEF).

However, there are still many issues to address with regards to evaluation and the results of ImageCLEF by no means provide a 'silver bullet' solution to evaluating image retrieval systems. There is still a tension between running system–centred and user–centred evaluation on a large scale for image retrieval (e.g. (Forsyth, 2002)). Most image retrieval in practice is interactive and should be seen as a priority for future image retrieval evaluation campaigns. Attempts have been made to run interactive tasks, but participation continued to be low across the years. This is not just a problem with image retrieval but an issue with IR evaluation in general.

Specific areas that are still ripe for exploration include: investigating which performance measures best reflect user's satisfaction with image retrieval systems and incorporating measures such as system response time; further investigation of the information seeking behaviours of users searching for images, such as their goals and motivations, search contexts, the queries issued and their reformulation strategies, and especially criteria shaping a user's notion of relevance; assessing user behaviours such as browsing, an important search strategy for image retrieval; continuing to develop publicly–accessible data sets covering multiple domains, tasks and varying in size; investigating the utility of test collections in image retrieval evaluation, especially with respect to the user to generate realistic test resources. Only by doing this can we start to address some of the concerns expressed by researchers such as Saracevic (1995), Forsyth (2002) and Smith (1998).

Acknowledgements We gratefully acknowledge the help of Norman Reid from St. Andrews University Library in Scotland for providing us access to the historic set of photographs for the first ImageCLEF evaluation campaign. Our thanks also extend to all other data providers who have enabled us to distribute and use their content. We thank the national and international funding bodies who have supported ImageCLEF over the years in a variety of ways. In particular we thank the European Union for support through the funding of various EU projects that have supported task organisers. Thanks go to all those involved in carrying out relevance assessments across the tasks without which whom we would have no gold standard to benchmark against. Finally a huge thank you goes to Carol Peters, the co–ordinator of CLEF, who supported ImageCLEF from the very beginning and through seven years of activities.

References

Borland P (2000) Experimental components for the evaluation of interactive information retrieval systems. Journal of Documentation 56(1):71–90

Cleverdon C (1959) The evaluation of systems used in information retrieval. In: Proceedings of the International Conference on Scientific Information — Two Volumes. Washington: National Academy of Sciences, National Research Council, pp 687–698

Cleverdon CW (1991) The significance of the Cranfield tests on index languages. In: 14th annual international ACM SIGIR conference on research and development in information retrieval. ACM, Chicago, IL, USA, pp 3–12

Clough PD, Sanderson M (2006) User experiments with the eurovision cross–language image retrieval system. Journal of the American Society for Information Science and Technology

57(5):679–708
Deselaers T, Müller H, Deserno TM (2009) Editorial to the special issue on medical image annotation in ImageCLEF 2007. Pattern Recognition Letters 29(15):1987
Dunlop M (2000) Reflections on MIRA: Interactive evaluation in information retrieval. Journal of the American Society for Information Science 51(14):1269–1274
Escalante HJ, Hernández CA, Gonzalez JA, López-López A, Montes M, Morales EF, Sucar LE, Villaseñor L, Grubinger M (2010) The segmented and annotated IAPR TC–12 benchmark. Computer Vision and Image Understanding 114(4):419–428
Forsyth D (2002) Benchmarks for storage and retrieval in multimedia databases. In: Proceedings of storage and retrieval for media databases, pp 240–247. SPIE Photonics West Conference
Goodrum A (2000) Image information retrieval: An overview of current research. Informing Science 3(2):63–66
Grubinger M, Clough PD, Müller H, Deselaers T (2006) The IAPR TC–12 benchmark — a new evaluation resource for visual information systems. In: Proceedings of the International Workshop OntoImage 2006 Language Resources for Content–Based Image Retrieval, held in conjunction with LREC 2006, pp 13–23
Hanbury A, Clough PD, Müller H (2010) Special issue on image and video retrieval evaluation. Computer Vision and Image Understanding 114:409–410
Ingwersen P, Järvelin K (2005) The turn: Integration of information seeking and retrieval in context. The information retrieval series, Springer, Secaucus, NJ, USA. 140203850X
Kando N (2003) Evaluation of information access technologies at the NTCIR workshop. In: Peters C, Gonzalo J, Braschler M, Kluck M (eds) Comparative Evaluation of Multilingual Information Access Systems Fourth Workshop of the Cross–Language Evaluation Forum, CLEF 2003. Lecture Notes in Computer Science (LNCS), vol 3237, Trondheim, Norway, pp 29–43
Kelly D (2010) Methods for evaluating interactive information retrieval systems with users. Foundations and Trends in Information Retrieval 3(1–2):1–224
Kuriyama K, Kando N, Nozue T, Eguchi K (2002) Pooling for a large–scale test collection: An analysis of the search results from the first NTCIR workshop. Information Retrieval 5(1):41–59
Müller H, Müller W, McG Squire D, Marchand-Mailet S, Pun T (2001) Performance evaluation in content–based image retrieval: Overview and proposals. Pattern Recognition Letters 22(5):593–601
Müller H, Geissbuhler G, Marchand-Maillet S, Clough PD (2004) Benchmarking image retrieval applications. In: Proceedings of the tenth international conference on distributed multimedia systems (DMS'2004), workshop on visual information systems (VIS 2004), pp 334–337
Müller H, Deselaers T, Grubinger M, Clough PD, Hanbury A, Hersh W (2007) Problems with running a successful multimedia retrieval benchmark. In: Proceedings of the third MUSCLE / ImageCLEF workshop on image and video retrieval evaluation
Nowak S, Lukashevich H, Dunker P, Rüger S (2010) Performance measures for multilabel classification — a case study in the area of image classification. In: ACM SIGMM International conference on multimedia information retrieval (ACM MIR). ACM press, Philadelphia, Pennsylvania
Peters C, Braschler M (2001) Cross–language system evaluation: The CLEF campaigns. Journal of the American Society for Information Science and Technology 52(12):1067–1072
Petrelli D (2008) On the role of user–centred evaluation in the advancement of interactive information retrieval. Information Processing and Management 44(1):22–38
Robertson S (2008) On the history of evaluation in ir. Journal of Information Science 34:439–456
Sanderson M (2010 – to appear) Test Collection Evaluation of Ad–hoc Retrieval Systems. Foundations and Trends in Information Retrieval
Sanderson M, Clough PD (2002) Eurovision — an image–based CLIR system. In: Workshop held at the 25th annual international ACM SIGIR conference on research and development in information retrieval, Workshop 1: Cross–Language Information Retrieval: A Research Roadmap. ACM press, Philadelphia, Pennsylvania, pp 56–59

Saracevic T (1995) Evaluation of evaluation in information retrieval. In: 18th annual international ACM SIGIR conference on research and development in information retrieval. ACM, Seattle, OR, USA, pp 138–146

Sedghi S, Sanderson M, Clough PD (2009) A study on the relevance criteria for medical images. Pattern Recognition Letters 29(15):2046–2057

Smith JR (1998) Image retrieval evaluation. In: Proceedings of the IEEE Workshop on Content–Based Access of Image and Video Libraries (CBAIVL 1998). IEEE Computer Society, Washington, DC, USA, pp 112–113

Voorhees EM (2000) Variations in relevance judgments and the measurement of retrieval effectiveness. Information Processing & Management 36(5):697–716

Voorhees EM, Harman DKe (2005) TREC: Experiments and evaluation in information retrieval. MIT Press, Cambridge, MA

Chapter 2
Data Sets Created in ImageCLEF

Michael Grubinger, Stefanie Nowak, and Paul Clough

Abstract One of the main components of any Text REtrieval Conference (TREC)–style information retrieval benchmark is a collection of documents, such as images, texts, sounds or videos that is representative of a particular domain. Although many image collections exist both on–line and off–line, finding visual resources suitable for evaluation benchmarks such as ImageCLEF is challenging. For example, these resources are often expensive to purchase and subject to specific copyright licenses, restricting both the distribution and future access of such data for evaluation purposes. However, the various ImageCLEF evaluation tasks have managed to create and/or acquire almost a dozen document collections since 2003. This chapter begins by discussing the requirements and specifications for creating a suitable document collection for evaluating multi–modal and cross–lingual image retrieval systems. It then describes each of the eleven document collections created and used for ImageCLEF tasks between 2003 and 2009. The description includes the origins of each document collection, a summary of its content, as well as details regarding the distribution, benefits and limitations of each resource.

2.1 Introduction

A core component of any TREC–style benchmark is a set of documents (e.g. texts, images, sounds, videos) that is representative of a particular domain (Markkula et al, 2001). Although there are hundreds of different collections available, finding such resources for general use is often difficult, not least because of copyright issues,

Michael Grubinger
Carrera 83 Calle 33–93, Medellín, Colombia, e-mail: `michael.grubinger@gmx.at`

Stefanie Nowak
Fraunhofer IDMT, Ilmenau, Germany, e-mail: `stefanie.nowak@idmt.fraunhofer.de`

Paul Clough
University of Sheffield, United Kingdom, e-mail: `p.d.clough@sheffield.ac.uk`

H. Müller et al. (eds.), *ImageCLEF*, The Information Retrieval Series 32,
DOI 10.1007/978-3-642-15181-1_2, © Springer-Verlag Berlin Heidelberg 2010

which restrict the large–scale distribution and future accessibility of data. This is especially true for visual resources as these are, in general, often more valuable than written texts and hence subject to limited availability and access for the research community (Grubinger et al, 2006).

2.1.1 Collection Creation

The organizers of an evaluation event for Visual Information Retrieval (VIR) have, in principle, two different choices with respect to the acquisition of benchmark resources: they can (1) custom–build a document collection from scratch, or (2) gain the distribution rights of an existing collection and adapt it to the research objectives of the particular evaluation event.

2.1.1.1 Custom–built Document Collections

The advantages of building a document collection for the evaluation of VIR from scratch are manifold:

- The contents of the document collection can be created and brought to perfection with respect to the main objective of the evaluation event.
- The image selection process can be pre–defined to ensure that the collection is representative of a particular domain.
- The type, quality and quantity of images and annotations can be created bearing in mind the current state–of–the–art of retrieval algorithms.
- The resources can be made available to participants (and other researchers) freely and without copyright restrictions. This also ensures the reproducibility of research results after the evaluation event.

However, there is the danger that the document collection becomes too contrived unless pre–defined collection creation processes, which are based on real–world studies, are strictly obeyed while building the collection. Also, the manpower and resources required to custom–build such a collection should not be underestimated.

2.1.1.2 Existing Document Collections

Due to lack of manpower and resources, it is often not feasible for benchmark organizers to build their own test collection from scratch, especially in large–scale retrieval evaluation events. In that case, the acquisition of an already existing document collection is often the only choice.

While this option is faster and easier than custom–building the evaluation resources from scratch, one needs to take into consideration that most collections were originally created for purposes other than retrieval evaluation. Thus, a detailed analysis of the collection's content should be undertaken prior to its acquisition and adaptation. In particular, the following questions need to be addressed:

- Is the collection representative for the particular domain in question?
- Can the retrieval evaluation objectives be achieved using this collection? If not, is it feasible to alter the data accordingly?
- Are the data in the collection suitable to create meaningful query topics (i.e. neither too easy nor too difficult) to evaluate the state–of–the–art retrieval methods?
- Can the resources be made available to the benchmark participants royalty–free and without (too many) copyright restrictions?

The acquisition of an existing document collection should only be considered if all these questions can be answered positively.

2.1.2 Requirements and Specification

Regardless of the collection creation approach, it is vital to create (or select and/or adapt) the potential document collection according to pre–defined collection requirements. For the evaluation of multi–modal cross–language VIR in general (and ImageCLEF in particular), the following specifications (Grubinger, 2007) should thereby be taken into account:

2.1.2.1 Evaluation Scope

While many benchmarks in other areas of computing are preoccupied with speed and response time (e.g. TPC–Transaction Processing Performance Council, SPEC–Standard Performance Evaluation Corporation), these measures do not play a central role in ImageCLEF, although they are extremely important for image retrieval. Following the methodology of TREC, the main focus lies in the evaluation of an algorithm's ability to identify relevant images (i.e. retrieval precision) rather than in its ability to carry out efficient search (i.e. retrieval speed). Although retrieval speed is generally considered as an essential factor for the usability of a system, it often depends on extraneous factors such as network connection, disk bandwidth or processor speed, which can hinder the objective comparison of retrieval methods. Within ImageCLEF, only the interactive evaluation had a scope on retrieval speed.

2.1.2.2 Collection Size

Similar to retrieval and processing speed, the collection size is not of primary importance. Of course, it should not be too small in order to produce significant and robust evaluation results. Having too large databases, on the other hand, would be impractical: retrieval by image content requires some degree of indexing and, as a consequence, the costs of indexing the database could be considerable. However, retrieval problems in small and very large collections are quite different and retrieval systems need to scale up to millions or billions of images. Retrieval in extremely large databases is a specific research field currently not fully covered by ImageCLEF. Obtaining such very–large scale collections therefore must be one of the future directions of ImageCLEF.

2.1.2.3 Collection Parameters

Ideally, any benchmark collection would be parametric, thereby allowing the specification of parameters that may be adjusted according to different requirements. Only by these means can the benchmark be geared to meet a variety of needs and be adapted to changing evaluation goals. Reasons for such changes can be due to the development of more powerful retrieval systems or due to changing interests in the research community (e.g. expressed by participants' feedback at evaluation events).

2.1.2.4 Collection Image Quality

Since most Content–Based Image Retrieval (CBIR) approaches are based on the analysis and processing of color, texture and shape, images in the collection should meet the following quality requirements to allow for a meaningful evaluation:

Resolution. Images should exhibit a minimum resolution of 256 x 256 pixels to allow for meaningful application of CBIR methods.

Clarity. Blurry photos due to camera movement or any other reason at the time of capturing the image should be avoided.

Contrast. Only photos with a reasonable level of contrast should be selected for the benchmark collection.

One might argue that not all images in real–world collections are of high quality and a benchmark collection should be as close to reality as possible. While this is certainly true, it would be easy to lower the image quality in a collection (e.g. by decreasing the resolution, blurring, etc.) if this was required for a specific evaluation — yet the converse is not always possible (Leung and Ip, 2000). On the other hand, this could create artificial collections that do not correspond to reality.

2.1.2.5 Image Annotations

Semantically rich image annotations are obviously a benefit to any document collection as they can facilitate the categorization of, and the search for, particular images and also make the query topic creation process easier.

However, they are not only an advantage for collection management purposes, but still a vital part of image retrieval algorithms as well and therefore an indispensable component of any test collection. Currently, the state–of–the–art methods in purely visual CBIR deliver a limited retrieval quality for semantic queries. Research is a long way from bridging the semantic gap using CBIR approaches alone and semantic search requests can still only be successfully processed by the inclusion of textual representations.

Hence, any image retrieval evaluation set–up that follows the traditional TREC methodology should be based on a document collection with quality annotations or with realistic annotations based on a specific user model (e.g. Web search exhibits very limited annotations). In a cross–language evaluation environment such as ImageCLEF, multi–lingual annotations are certainly a benefit as monolingual captions provide little challenge for the participating systems. Yet, it is expected that, as CBIR methods evolve and improve, the importance of text representations will decrease (Grubinger, 2007).

2.1.2.6 Copyright

Ideally, all evaluation resources would be made available to the participants royalty–free and without copyright restrictions (also after the event, so that non–participating researchers can reproduce the retrieval and evaluation results). Unfortunately, in reality this is not always possible. In order to be suitable for benchmarks, the original copyright owners of the collections need to agree to at least the following:

- All evaluation resources are royalty–free. Not many research groups would participate if they had to pay for the evaluation resources.
- The data can be distributed to all participants electronically, whereby all participants are allowed to use the collection data for research purposes at least in the context of the benchmark.
- Participants are allowed to use and illustrate parts of the resources in their corresponding publications at least directly linked to the benchmark.

2.1.3 Collection Overview

Since 2003, ImageCLEF has created (and/or acquired and adapted) almost a dozen document collections to support its various evaluation tasks. Figure 2.1 provides a time–related overview of these collections.

Some collections are freely available for download from the ImageCLEF website[1], while others are subject to signing an end–user agreement with the task organizers and/or original copyright holders. We now describe each of these collections.

[1] http://www.imageclef.org/

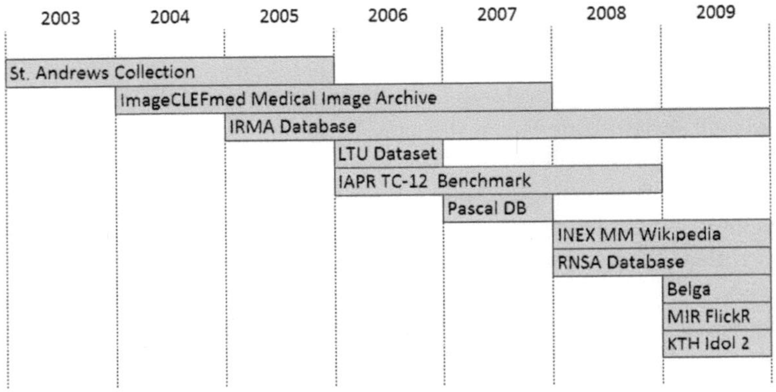

Fig. 2.1: Time–related overview of ImageCLEF document collections.

2.2 Image Collections for Photographic Retrieval

Three databases have been acquired for the evaluation of visual information ad hoc retrieval from generic photographic collections (see Chapter 8): The St. Andrews collection of historic photographs from 2003 to 2005 (Section 2.2.1); the IAPR TC–12 database from 2006 to 2008 (Section 2.2.2); and the Belga news agency photographic collection (Section 2.2.3) in 2009.

2.2.1 The St. Andrews Collection of Historic Photographs

The St. Andrews Collection (SAC) of historic photographs is a subset of one of Scotland's most important archives of historic photography, which was made available to the public via a Web interface[2] in a large–scale digitalization project by St. Andrews University Library (Reid, 1999). This collection of 28,133 photographs from well–known Scottish photographers and photographic companies was a core component of the ImageCLEF Ad hoc Retrieval Task from 2003 to 2005.

2.2.1.1 Collection Content

Most photos in the SAC are monochrome or black-and-white (89.0%), due to the historic nature of the collection, and are specific to Scotland (67.1%) or the UK (95.0%) between 1840 and 1940 (see (Reid, 1999) for detailed statistics). The collection includes photos and postcards of towns and villages, nature (e.g. landscapes,

[2] `http://www.st-andrews.ac.uk/specialcollections/`

(a) Golfers (b) Portraits (c) Landscapes (d) Animals (e) Buildings

Fig. 2.2: Sample images from the SAC.

```
<DOC><DOCNO>stand03_1099/stand03_21287.txt</DOCNO>
  <HEADLINE>Frome, Somerset. Catherine Hill.</HEADLINE>
  <TEXT>
    <RECORD_ID>JV-.094276</RECORD_ID>
    Catherine Hill, Frome.
    Steep road lined with stone buildings with shops at ground
    level; awnings on shop windows; goods in doorway, left.
    Registered 1925
    J Valentine & Co
    Somerset, England
    JV-94276 pc/mb/jf/mb DETAIL: Woman in long skirt and summer
    blouse and hat, carrying shopping basket. PCARD:
    Handwritten date and postmark 1912.
    <CATEGORIES>[shops], [buildings - stone], [Someset all
    views], [Collection - J Valentine & Co] </CATEGORIES>
    <SMALL_IMG>stand03_1099/stand03_21287.jpg</SMALL_IMG>
    <LARGE_IMG>stand03_1099/stand03_21287_big.jpg</LARGE_IMG>
  </TEXT>
</DOC>
```

Fig. 2.3: Sample SGML image caption and image.

animals), architecture (e.g. buildings, statues, monuments), events (e.g. war–related, royal visits), transport (e.g. ships, carriages), family and individual portraits, and sports (especially golf). Figure 2.2 displays sample images from a selection of these categories.

Not all the images in the SAC exhibit exactly the same size: the large versions of the images show an average resolution of 368 x 234 pixels; the corresponding thumbnails exhibit 120 x 76 pixels (see (Reid, 1999) for detailed statistics).

2.2.1.2 Image Captions

Each photograph has an annotation that consists of the following nine fields: (1) a unique record number, (2) a full title, (3) a short title, (4) a textual description of the image content, (5) the date when the photograph was taken, (6) the originator, (7) the location where the photograph was taken, (8) notes for additional information, and (9) its corresponding categories. These captions have been encapsulated in a Standard Generalized Markup Language (SGML) format to be compatible with existing TREC collections (see Figure 2.3 for an example).

The `<DOCNO>` tag contains the pathname of the image as a unique document identifier, and the title and categories are indicated by the `<HEADLINE>` and `<CATEGORIES>` tags, respectively. The remaining caption fields are enclosed by the <TEXT> tag and are not structured. In addition, the <SMALL_IMG> and <LARGE_IMG> tags contain the path of the thumbnail and of the large version. Further examples and information about the SAC can be found in (Clough et al, 2006; Reid, 1999) and the St. Andrews University Library[3].

2.2.1.3 Benefits and Limitations

The SAC was used as the basis for ImageCLEF because of the following advantages: it (1) represented a reasonably–sized collection of images, (2) offered high quality, semi–structured annotations to support Text–Based Image Retrieval (TBIR) methods, and (3) permission was granted by St. Andrews Library to download and distribute the collection for the photographic retrieval task. All this facilitated the birth of ImageCLEF — a very valuable contribution to the evaluation event, indeed.

However, there were also a few limitations in its use. For example, most of the images in the collection are monochrome or black-and-white photographs; they do not contain many clearly separated objects and a few are also of very poor quality (e.g. too dark, too blurry). All this makes the SAC a very difficult collection for purely visual analysis. Furthermore, the domain of the SAC is restricted to mainly photographs specific to life in Scotland and England from 100 years ago, which together with the excessive use of colloquial and domain-specific language affects both its use and effectiveness as a generic evaluation resource.

As a consequence, after three years of image retrieval evaluation using the SAC, it was replaced by a new resource: the IAPR TC–12 database.

2.2.2 The IAPR TC–12 Database

The photographic collection of the IAPR TC–12 database was used in the ImageCLEF General Photographic Retrieval Task (ImageCLEFphoto) from 2006 to 2008, for the 2007 GeoCLEF Geographic Retrieval Task and the 2007 and 2008 ImageCLEF Visual Concept Detection Task (see also Section 2.4).

While most other collections were originally created for purposes other than retrieval evaluation, the goal for the development of the IAPR TC–12 database was to provide a generic photographic collection which could be used for a variety of research and evaluation purposes in general, and for ImageCLEF in particular. More information on the design and implementation of the IAPR TC–12 database, created under Technical Committee 12 (TC–12) of the IAPR[4], can be found in (Grubinger, 2007).

[3] `http://www-library.st-andrews.ac.uk/`

[4] `http://www.iapr.org/`

(a) Landscapes (b) Sports (c) Sunsets (d) Animals (e) People

Fig. 2.4: Sample images from the IAPR TC–12 database.

2.2.2.1 Collection Content

The IAPR TC–12 database contains 20,000 photos taken from locations around the world and comprises a varying cross–section of naturalistic images. Figure 2.4 illustrates a number of sample images from a selection of categories.

The majority of the images in the collection have been provided by viventura[5], an independent travel organization that offers adventure trips to South America. Travel guides accompany the tourists and maintain a daily on–line diary including photographs of trips made as well as general pictures of each location including accommodation facilities and ongoing social projects. The remainder of the images has been collected by the first author from personal experiences (e.g. holidays, sports events) to systematically add to the diversity of the collection. The IAPR TC–12 database therefore contains many images of similar visual content, but varying illumination, viewing angle and background, which provides an additional challenge for the successful application of CBIR methods (Grubinger et al, 2006).

2.2.2.2 Image Captions

Each image in the collection has a corresponding semi–structured annotation consisting of the following seven fields: (1) a unique identifier, (2) a title, (3) a free–text description of the semantic and visual contents of the image, (4) notes for additional information, (5) the provider of the photo and fields describing (6) where and (7) when the photo was taken. These annotations are available in three languages for each image: English, German and Spanish.

Figure 2.5 shows a sample image with its corresponding English annotation. All images, metadata and multi–lingual annotations are stored in a database, allowing the creation of collection subsets with respect to a variety of parameters (e.g. which images or caption fields to use, or which annotation language to select).

Consequently, the ImageCLEF organizers made use of the parametric nature of the IAPR TC–12 database and created a different subset of the test collection each

[5] `http://www.viventura.net/`

```
<DOC>
<DOCNO>annotations/16/16019.eng</DOCNO>
<TITLE>Flamingo Beach</TITLE>
<DESCRIPTION>a photo of a brown sandy beach; the dark
  blue sea with small breaking waves behind it; a dark
  green palm tree in the foreground on the left; a blue
  sky with clouds on the horizon in the background;
</DESCRIPTION>
<NOTES>Original name in Portuguese: "Praia do Flamengo";
  Flamingo Beach is considered as one of the most
  beautiful beaches of Brazil;</NOTES>
<LOCATION>Salvador, Brazil</LOCATION>
<DATE>2 October 2004</DATE>
<IMAGE>images/16/16019.jpg</IMAGE>
<THUMBNAIL>thumbnails/16/16019.jpg</THUMBNAIL>
</DOC>
```

Fig. 2.5: Sample image caption from the IAPR TC–12 database.

year (see also Chapter 8). Moreover, additional subsets of this test collection were used for the 2007 GeoCLEF Geographic Retrieval Task and the 2007 and 2008 ImageCLEF Visual Concept Detection Tasks (see also Section 2.4).

2.2.2.3 Benefits and Limitations

The photographic collection of the IAPR TC–12 database exhibits the following benefits: all photos are high–quality color photographs with excellent levels of resolution and contrast; the generic collection contains a variety of real–life photographs from a range of subjects and settings; high–quality multi–lingual annotations make the collection suitable for the evaluation of a range of retrieval tasks; the parametric nature of the benchmark allows for the fast adaptation to changed retrieval requirements or new evaluation needs; and the collection is available freely and without copyright restrictions that would hinder its redistribution for evaluation purposes.

However, ImageCLEF participants felt in 2008 that the time had come to move on to a bigger image archive for evaluation. Other criticisms were that the collection seemed a bit too contrived: the annotations were created by a single person and in extremely high quality. Hence, in 2009 the IAPR TC–12 database was replaced by the Belga news agency photographic collection.

2.2.3 The Belga News Agency Photographic Collection

In 2009, the ImageCLEF organizers offered a new challenge to the ImageCLEF-photo participants by providing a database that was nearly 25 times larger than the ones used in the years before: the photographic collection of Belga[6], a Belgian news agency covering all aspects of life and current affairs: politics, economics, finance, social affairs, sports, culture and personalities. The news content is thereby supplied in text, pictures, audio and video formats (Lestari Paramita et al, 2009).

[6] http://www.belga.be/

922331|AD001 - 20020424 - MANCHESTER, UNITED
KINGDOM : Manchester United's Laurent Blanc
(L) tackles Bayer Leverkusen's Dimitar
Berbatov during the Champions League Semi
Final first leg match at Old Trafford in
Manchester 24 April 2002.
EPA PHOTO EPA-ADRIAN DENNIS

Fig. 2.6: Sample image and caption of the Belga database.

Table 2.1: Collection overview of the ImageCLEFmed teaching files.

Collection	Year added	Image type(s)	Cases	Images	Annotations
Casimage	2004	Radiology, Pathology	2,076	8,725	2,076
MIR	2005	Nuclear Medicine	407	1,177	407
PEIR	2005	Pathology, Radiology	32,319	32,319	32,319
PathoPIC	2005	Pathology	7,805	7,805	15,610
MyPACS	2007	Radiology	3,577	15,140	3,577
CORI	2007	Endoscopy	1,496	1,496	1,496
TOTAL			47,680	66,662	55,485

ImageCLEFphoto 2009 was provided with a collection of 498,920 photos with unstructured, English–only annotations describing the contents of the image, such as: people shown in the photo, the event taking place, and the location where the image was captured. Figure 2.6 shows an image example with its caption.

The Belga database offered new challenges to the participants in comparison to the SAC and IAPR TC–12 collections. For example, the unstructured nature of the image captions requires the automatic extraction of information about, for example the location, date or photographic source of the image as a part of the indexing and retrieval process. In addition, it contains many cases where pictures were orientated correctly, thereby making CBIR more difficult (Lestari Paramita et al, 2010).

However, one of the few limitations of the collection can be found in the fact that the English–only annotations provide little challenge to Cross–Language Information Retrieval (CLIR) systems other than having a query in a language different from English.

2.3 Image Collections for Medical Retrieval

Two major image archives have been used for the evaluation of ad hoc retrieval from medical collections (see also Chapter 13): The ImageCLEFmed teaching files from 2004 to 2007 (Section 2.3.1) and the collection of the Radiological Society of North America (RSNA) since 2008 (Section 2.3.2).

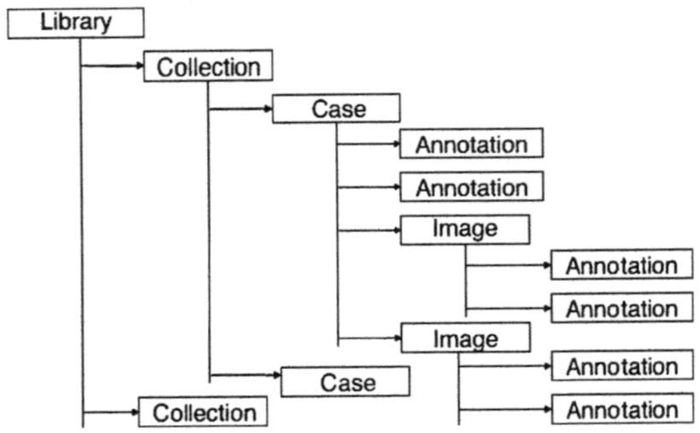

Fig. 2.7: Structure of the ImageCLEFmed teaching files.

2.3.1 *The ImageCLEFmed Teaching Files*

The ImageCLEFmed teaching files are a collection of domain–specific photographs for the medical field, which was used in the medical ad hoc retrieval tasks of ImageCLEF (ImageCLEFmed) from 2004 to 2007. This medical archive comprises in total 66,662 images and is, in fact, a composite of several medical subcollections provided by independent medical institutions and hospitals that granted ImageCLEF permission to use their data sets in its evaluation campaign (Hersh et al, 2007, 2009). Figure 2.7 provides an overview of the conceptual structure of the ImageCLEFmed teaching files. The individual collections are partly organized into cases that represent a group of related images and annotations (some collections have an organization based on single images). Each image is part of a case and has optional associated annotations, which consist of metadata and/or a textual annotation. All images and annotations are stored in separate files, whereby the connections between the collections, cases, images and annotations are established in an XML file. Table 2.1 provides an overview of the collections in the ImageCLEFmed teaching files. In total, it contains 66,662 images, 47,680 medical cases and 55,485 annotations in English, French and/or German. The individual subcollections are briefly introduced below.

2.3.1.1 Casimage

The first collection used by ImageCLEFmed in 2004 is the Casimage collection[7] (Rosset et al, 2004). Most of its 8,725 images are from radiology (but it also contains photographs, presentation slides and illustrations) belonging to 2,075 medical cases (see Figure 2.8). The majority (95%) of these medical cases have corresponding notes which are written in XML, with 75% being annotated in French and 20% in English. These quite elaborate case notes can comprise several images and in-

[7] `http://www.casimage.com/`

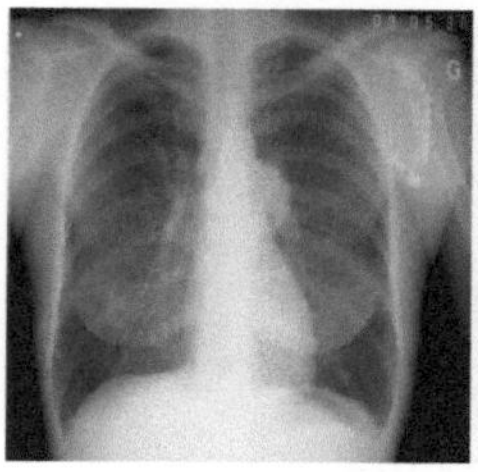

ID: 3179
Description: Sur la radiographie du thorax de face, présence d'un élargissement hilaire gauche. Les structures vasculaires hilaires sont visibles à travers l'effet de masse [...]
Diagnosis: Schwannome
ClinicalPresentation: Douleur dorsale.
Commentary: Le schwannome est une des tumeurs des nerfs périphériques, développées à partir des nerfs intercostaux, pneumogastriques, phréniques et récurrent gauche [...]
KeyWords: Schwannome; médiastin [...]

Fig. 2.8: Sample image and caption of the Casimage collection.

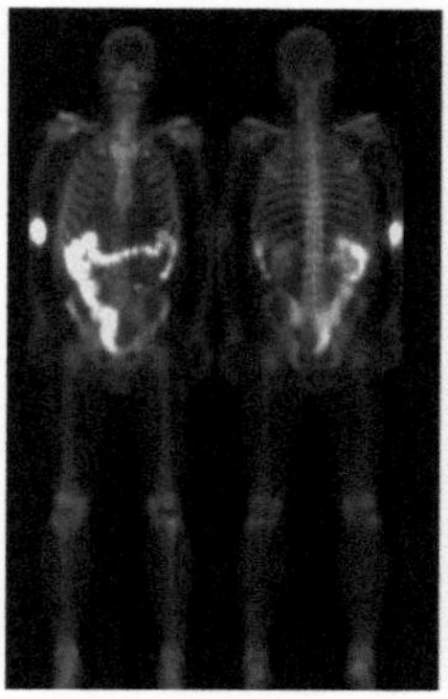

CASE: MIR Teaching file case bs089 Case Author(s): M. Quinn, MD and M. Mintun, MD , 01/30/98 . Rating: #D2, #Q4 Diagnosis: GU-GI fistula Brief history: 61 yo male with back pain Images: Anterior and posterior images are shown View main image(bs) in a separate image viewer Full history/Diagnosis is available below Diagnosis: GU-GI fistula Full history: 61 year old male with prior cystectomy and diverting colostomy for transitional cell carcinoma of the bladder with colonic invasion and obstruction [...] Radiopharmaceutical: Tc99m-MDP Findings: There is no evidence for osseous metastases. However, there is abnormal radiopharmaceutical accumulation throughout the large bowel from the right colon to the site of the diverting colostomy. Discussion: Obviously, the presence of radiopharmaceutical in the colon is abnormal on delayed images from a bone scan. [...] Followup: A subsequent IVP revealed a fistulous connection from high in the ilieal conduit to the right colon at the site of surgical clips in the right lower quadrant, suggesting breakdown of a prior anastomosis [...]

Fig. 2.9: Sample image and caption of the MIR database.

clude a field for the title, diagnosis, free–text description, clinical presentation, hospital, department and keywords; 207 case notes are empty (Müller et al, 2004).

2.3.1.2 MIR

In 2005, ImageCLEFmed was given permission to use the nuclear medicine database of the Mallinckrodt Institute of Radiology[8] (MIR) with 1,177 images mainly from the field of nuclear medicine (Wallis et al, 1995).

Similar to Casimage, the images are assigned to medical cases which are described in English XML files. These rather extensive descriptions are only encapsulated by one CASE tag, as illustrated in Figure 2.9. Yet, some kind of semi–structured information still exists within the text as there are sections for, for example, diagnosis, findings, discussion and follow–up.

[8] http://www.mir.wustl.edu/

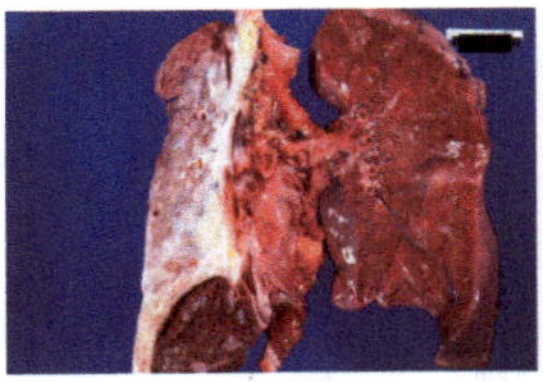

MetadataID: 6435
GlobalID: 1503
ContributeDate: 02/06/2004
Title: RESPIRATORY
Description: RESPIRATORY: Lung: Mesothelioma: Gross good contrast uninvolved left lung with tumor in right lung
ApproveDate: 02/06/2004
CatalogDate: 02/06/2004
SourceCollection: PEIR - University of Alabama at Birmingham Department of Pathology

Fig. 2.10: Sample image and caption of the PEIR data set.

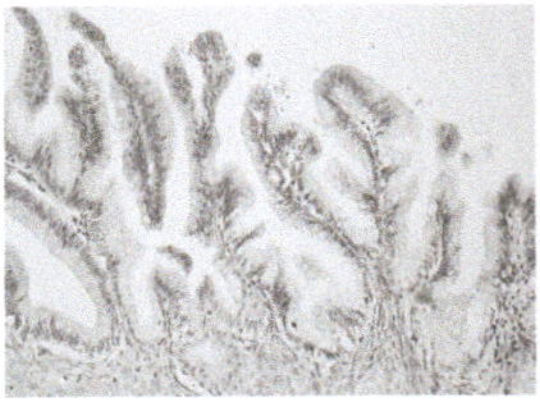

ID: 000154
Diagnose: Muzinöses Zystadenom des Ovars, Borderline Malignität --- (154)
Synonyme: -
Beschreibung: Papilläre muzinöse Epithelproliferate. Kerne leicht vergrössert und hyperchromatisch. Keine Invasion des Stromas.
Zusatzbefund: 6 kg schwerer multilokulärer zystischer Tumor mit fadenziehendem klarem Inhalt. Oberfläche glatt. Zystenauskleidung teils glatt, teils mit kleinen blumenkohlartigen Auflagerungen.

Fig. 2.11: Sample image and caption of the PathoPic collection.

2.3.1.3 PEIR

Another database that was made available in 2005 was the Pathology Educational Instructional Resource (PEIR) Data set[9]; this collection contains 32,319 mainly pathology images (see Figure 2.10).

In contrast to MIR, each image has a corresponding English caption based on the Health Education Assets Library (HEAL) project[10] and is thus not organized in cases but in terms of images. The PEIR Data set also shows a very detailed annotation structure, depicting information such as the filename, a title, a description, a date of contribution, archiving and cataloguing, and the image source. More information on the HEAL project can be found in (Candler et al, 2003).

2.3.1.4 PathoPic

The PathoPic[11] collection (Glatz-Krieger et al, 2003) was also included in 2005 and comprises 7,805 pathology images. Similar to MIR, this collection also comprises structured captions on a per image basis in English and German. However, its captions are not as detailed as those of PEIR, and some English captions are especially very short (see Figure 2.11 for examples).

---

[9] `http://peir.path.uab.edu/`

[10] `http://www.healcentral.com/`

[11] `http://alf3.urz.unibas.ch/pathopic/`

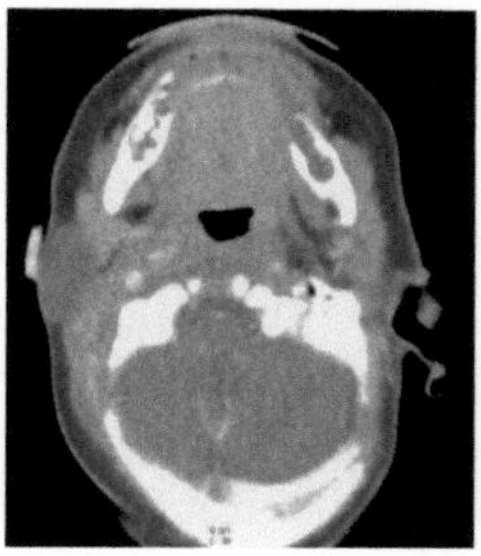

title: BRANCHIAL CLEFT CYST
abstract: Abstract: Newborn with a right neck mass at birth in a rural hospital.
author: Dr Phillip Silberberg, Children's Hospital Omaha, Radiologist, Creighton University, USA
modified_date: 2004-09-21
section: Newborn with a right neck mass at birth in a rural hospital.
section: Dahnert, Wolfgang. Radiology Review Manual 4th Edition. Williams and Wilkins, 1999
document-type: radiologic teaching file
pathology: Congenital **anatomy**: Face and Neck
modality: CT **access**: public

Fig. 2.12: Sample image and caption of the MyPACS data set.

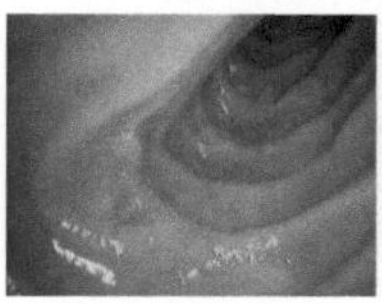

ID: 139000004933
Title: Upper Gastrointestinal Ulcers
Series: Images from the National Endoscopic Database
Subject: Endoscopy, Gastrointestinal; Peptic Ulcer
Description: Location: Duodenal Bulb.

Fig. 2.13: Sample image and caption of the CORI Database.

2.3.1.5 MyPACS

In 2007, two additional databases were added to the ImageCLEFmed teaching files. The first was the MyPACS data set[12] comprising 15,140 radiology images.

These images belong to 3,577 cases, which are described in English only. Figure 2.12 provides a sample image with its annotation.

2.3.1.6 CORI

The second collection that was added to the ImageCLEFmed teaching files in 2007 is the image database of the Clinical Outcomes Research Initiative[13] (CORI), containing 1,496 endoscopic images.

Figure 2.13 illustrates a sample image and its corresponding caption. Each image contains one English annotation, comprising the ID, title, series, subject and description (annotations are hence per image and not per case). The CORI database extends the spectrum of the ImageCLEFmed database in so far as there were very few endoscopic images in the data set.

[12] http://www.mypacs.net/

[13] http://www.cori.org/

2.3.1.7 Benefits and Limitations

The benefits of the ImageCLEFmed teaching files are obvious: the six medical data sets together build a large image set from a variety of medical fields; the high resolution and also the nature of the images are almost predestined for CBIR evaluation, while the multi–lingual captions in English, German and French create a realistic, comprehensive and versatile data set for the evaluation of TBIR as well.

Unfortunately, some of these medical images (especially those from the MyPACS collection) are under copyright restrictions and their redistribution to the participating research groups is only possible through a special agreement with the original copyright holders. In many cases, the captions do not describe the image content itself but rather the context in which the image was taken; they further contain many spelling errors (especially Casimage), many abbreviations, which are used in a non–standardized way, and many terms which are very specific to the medical domain and unlikely to be found within most general purpose dictionaries or stemmers (Müller et al, 2006).

2.3.2 The RSNA Database

In 2008, the ImageCLEFmed organizers managed to obtain the rights to use a subset of a large database of medical images that is also accessible via the Goldminer image search engine[14] and replaced the ImageCLEFmed teaching files as an evaluation resource for the ImageCLEF medical retrieval tasks. This subset was made available by the Radiological Society of North America[15] (RSNA).

2.3.2.1 Collection Content

The content of the RSNA database represents a broad and significant body of medical knowledge and includes high quality images, which are, in fact, original figures used in articles taken from the radiological journals *Radiology* and *Radiographics*; these images are associated with journal articles and can be part of a figure[16].

The collection further provides the corresponding figure captions and links to the full text articles via the PubMed Identifier (PMID), which can further be used to obtain the Medical Subject Heading (MeSH) terms assigned by the National Library of Medicine (NLM) for PubMed[17]. All captions are in English only.

[14] `http://goldminer.arrs.org/`

[15] `http://www.rsna.org/`

[16] Due to copyright restrictions, it is not possible to depict sample images and annotations.

[17] `http://www.pubmed.gov/`

Table 2.2: Overview of the IRMA database at ImageCLEF (2005–2009).

Year	Images (Training)	Classes (Training)	Images (Validation)	Images (Test)	Classes (Test)
2005	9,000	057	-	1,000	057
2006	9,000	116	1,000	1,000	116
2007	10,000	116	1,000	1,000	116
2008	12,076	193	1,000	1,000	187
2009	12,677	193	-	1,733	169

2.3.2.2 Data Distribution

The database that was eventually distributed to the ImageCLEFmed participants included an XML file with the image ID, the captions of the images, the titles of the journal articles in which the image had appeared, and the PMID of the journal article. In addition, a compressed file containing approximately 66,000 images was provided in 2008, and one containing nearly 75,000 images in 2009, respectively.

2.4 Automatic Image Annotation and Object Recognition

While the ImageCLEF medical annotation and classification tasks (see also Chapter 12) have used the IRMA database exclusively from 2005 to 2009 (Section 2.4.1), the data collections have changed quite frequently for the generic visual object/concept recognition and annotation tasks (see also Chapter 11): the LookThatUp (LTU) collection was used in 2006 (Section 2.4.2), the PASCAL VOC collection in 2007 (Section 2.4.3), a subset of the IAPR TC–12 database in 2008 (Section 2.2.2) and a subset of the Flickr MIR data set in 2009 (Section 2.4.4).

2.4.1 The IRMA Database

The Image Retrieval in Medical Applications (IRMA) database[18] is a collection of 15,000 medical radiographs that have been randomly collected from daily routine work at the Department of Diagnostic Radiology of the RWTH Aachen University[19] (Lehmann et al, 2003). Sub sets of this archive were used in the Automatic Medical Image Classification/Annotation Task at ImageCLEF from 2005 to 2009 (see Table 2.2). All images in the IRMA database are provided as PNG files using 256 grey values fitting into a bounding box of the same size. Each image is thereby classified by, and annotated with, its complete IRMA code (see Figure 2.14 for an example).

[18] `http://irma-project.org/`

[19] `http://www.rad.rwth-aachen.de/`

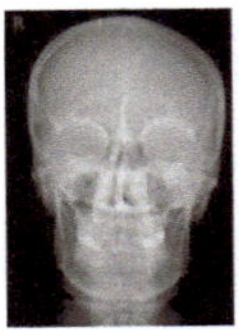

```
IRMA: 1121-120-200-700
T: x-ray, plain radiography, analog, overview image
D: coronal, anteroposterior (AP coronal), unspecified
A: cranium, unspecified, unspecified
B: musculoskeletal system, unspecified, unspecified
```

Fig. 2.14: Sample image with corresponding IRMA code.

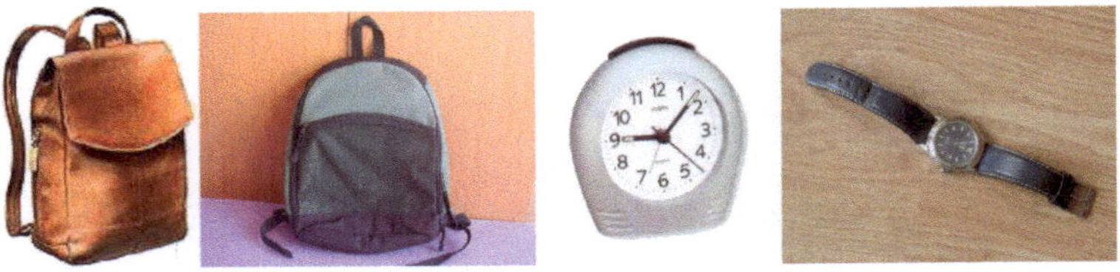

(a) Bag–training (b) Bag–test (c) Clock–training (d) Clock–test

Fig. 2.15: Sample training and test images of the LTU data set.

While many other existing medical terminologies such as the MeSH thesaurus are poly–hierarchical (i.e. several paths can lead to a code entity), the IRMA code relies on class–subclass relations to avoid ambiguities in textual classification (Lehmann et al, 2006).

In particular, the IRMA code comprises four mono–hierarchical axes with three to four positions each: the technical code (T) describes the imaging modality, the directional code (D) models body orientations, the anatomical code (A) refers to the body region examined, and the biological code (B) depicts the biological system examined. The complete IRMA code subsequently exhibits a string of 13 characters, each in {0...9, a...z}: TTTT-DDD-AAA-BBB. More information on the IRMA database and code can be found in (Lehmann et al, 2003, 2006).

2.4.2 *The LookThatUp (LTU) Data set*

LTU Technologies[20] provided their hand–collected data set of mono–object images of 268 classes to the Automatic Object Annotation Task at ImageCLEF 2006. The image collection subset that was used in that event had been reduced to 21 classes (and 13,963 images respectively) to make the task at least somewhat realistic for existing techniques. All the images are in PNG format, and most of them exhibit a resolution of 640 x 480 pixels.

[20] `http://www.ltutech.com/`

Fig. 2.16: Sample images from the PASCAL VOC 2006 collection.

The LTU data set consists of images containing only one object in a rather clean environment, i.e. the images show the object and some mostly homogeneous background. These images were used for training in ImageCLEF 2006. The test collection comprises 1,100 images that show the 21 object classes in a more natural setting, i.e. there is more background clutter than in the training images. Figure 2.15 depicts examples for training data and test images. The task proved to be too difficult for the state–of–the–art retrieval technology of 2006; hence, the LTU data set was replaced by the PASCAL object recognition database one year later.

2.4.3 *The PASCAL Object Recognition Database*

The PASCAL object recognition database is a compilation of image databases with the goal of providing a standardized collection for the evaluation of the current generation of object recognition algorithms (Everingham et al, 2006). A collection subset of approximately 2,600 PNG images belonging to ten classes was used as training data at the ImageCLEF 2007 Object Retrieval Task, while 20,000 images from the IAPR TC–12 database (see Section 2.2.2) were used as test data.

Figure 2.16 provides some sample images. The corresponding annotations denote which of the ten object classes is visible in which area of the image. The images in the collection are PNG images and show objects from a number of classes in mostly realistic scenes (i.e. no pre–segmented objects).

The subset used at ImageCLEF 2007 is available on the Web page of the PASCAL VOC challenge[21], which facilitates the reproduction of evaluation results outside the PASCAL VOC and ImageCLEF campaigns. Yet, some images are of very poor quality, and providing training and test data from different collections, albeit being realistic, was too hard for most retrieval algorithms.

Hence in 2008, 1,827 new training and 1,000 test images were taken from a subset of the IAPR TC–12 database (see Section 2.2.2) that had not been included in the retrieval and annotation tasks of ImageCLEF 2007. Participants felt that the number of training and test images was a bit too low to provide significant results. Therefore, the ImageCLEF organizers decided to move on to use a new and much larger collection from 2009: the MIR Flickr image data set.

---

[21] http://www.pascal-network.org/challenges/VOC/databases.html

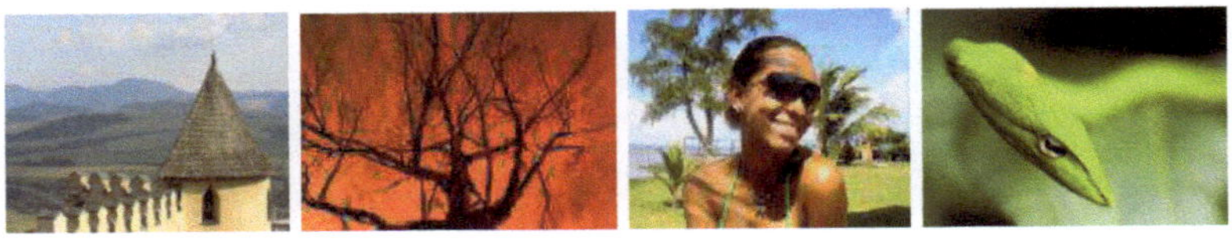

Fig. 2.17: Sample images of the MIR Flickr image data set.

{3DBCA9A2-C56E-4958-A8D7-7832B6783F65}.jpg
Day, Sunset_Sunrise, Citylife, Outdoor, No_Blur,
Overall_Quality, Aesthetic_Impression, Fancy, Lake,
No_Persons, Neutral_Illumination, Water, Clouds,
Sky, Building_Sights, Landscape, No_Visual_Season

Fig. 2.18: Sample image caption from the MIR Flickr image data set.

2.4.4 The MIR Flickr Image Data Set

The ImageCLEF 2009 Large–Scale Visual Concept Detection and Annotation Tasks used a subset of the MIR Flickr image data set (Huiskes and Lew, 2008). This database contains 25,000 consumer photos from Flickr[22] made available under a creative common license (see Figure 2.17 for examples). Most photos contain Exchangeable Image File Format (EXIF) data, which are stored in separate text files.

In 2009, 18,000 photos — 5,000 were used for training, 13,000 as test images — were annotated manually with 53 pre–defined visual concepts and provided to the participants. These multi–label annotations mostly refer to holistic visual concepts, which are organized in a small ontology.

Figure 2.18 shows an example for an image with its corresponding visual concepts. More information on the annotation process and concept ontology can be found in (Nowak and Dunker, 2010).

2.5 Image Collections in Other Tasks

Image collections used in further ImageCLEF tasks include the INEX MM collection in the WikipediaMM Retrieval Tasks 2008 and 2009 (Section 2.5.1) and the KTH–IDOL2 database in the Robot Vision Task in 2009 (Section 2.5.2).

[22] `http://www.flickr.com/`

```
<?xml version = "1.0" ?>
<article><name id="205995">African_Buffalo.JPG</name>
  <text> <h2>Summary</h2>
    An African Buffalo Bull. Photographed at Mabula
    Game Reserve, South Africa, 2004 by Paul M Rae.<p>
    <h2>Licensing</h2>
    <wikitemplate parameters="1">
      <wikiparameter number="0" last="1">
        <value>cc-by-2.5</value>
      </wikiparameter>
    <wikitemplate>
  <text>
<article>
```

Fig. 2.19: Sample Wikipedia image with its corresponding caption.

2.5.1 The INEX MM Wikipedia Collection

The INEX MM Wikipedia collection is a subset of the Wikipedia XML corpus (Denoyer and Gallinari, 2006), which comprises XML collections based on Wikipedia[23], an on–line encyclopedia that is collaboratively written by contributors from all over the world. This multimedia collection had previously been used at the INEX 2006 and 2007 multimedia tasks (Westerveld and van Zwol, 2007) and was also made available to the WikipediaMM Retrieval Task at ImageCLEF in 2008 and 2009 (see also Chapter 9).

2.5.1.1 Collection Contents

The subset of the Wikipedia XML corpus that was eventually provided to the ImageCLEF participants comprises approximately 150,000 JPG and PNG Wikipedia images. The image collection thereby does not only contain photographs, but also maps, satellite images, x–rays, graphs, drawings, sketches, illustrations, and figures.

These come in all dimensions and sizes, ranging from 30 x 30 pixels and 1 KB to 4,800 x 3,600 pixels and 4.7 MB; some exhibit rather extreme dimensions, like 11,880 x 1,683 pixels. Each image belongs to one text file containing user–generated XML captions in English. A sample image caption is shown in Figure 2.19. The image annotations are highly heterogeneous and can be of varying length, but usually contain at least a brief description of the image contents, the Wikipedia user who uploaded the photo, and copyright information. More information on the collection can be found in (Westerveld and van Zwol, 2007).

[23] http://www.wikipedia.org/

2.5.1.2 Additional Resources

Additional resources were made available to the ImageCLEF participants to support their investigations of multi–modal approaches (i.e. combining CBIR with TBIR):

Image similarity matrix. For each image, this matrix contains the list of the top $K = 1.000$ most similar images in the collection with their similarity scores.
Image classification scores. The classification scores for 101 conceptional classes are provided for each image.
Image features. For each image, the set of the 120–dimensional feature vectors that has been used to derive the classification scores above was provided.

These resources are beneficial to researchers who wish to exploit visual evidence without having to pre–process the entire image collection first.

2.5.1.3 Benefits and Limitations

The main benefit of the INEX MM Wikipedia collection lies in the large number of royalty–free images as well as in the extensive annotation text that is associated with them. This allows for close examination of both CBIR and TBIR.

The varying image dimensions and highly heterogeneous and often extremely short captions can be seen as one of the drawbacks of the collection. Since anyone can edit the text files, the annotation quality inherently varies within the collection as well. Furthermore, the English–only captions provide little challenge for the participants in a CLIR evaluation environment such as CLEF.

2.5.2 The KTH–IDOL2 Database

The KTH–IDOL2 database (Luo et al, 2006) is a database made available by the Royal Institute of Technology[24]. A subset of this image collection was used in the ImageCLEF 2009 robot vision task (see also Chapter 10).

2.5.2.1 Collection Content

The database contains 24 image sequences acquired by two mobile robot platforms with a perspective camera using a resolution of 320 x 240 pixels per image. These two robots were manually driven through a five room subsection of a larger office environment while continuously taking images, whereby each of the five rooms represented a different functional area: a one–person office, a two–person office, a kitchen, a corridor, and a printer area (see Figure 2.20).

[24] `http://www.kth.se/`

(a) Corridor (b) Office (1) (c) Office (2) (d) Kitchen (e) Printer Area

Fig. 2.20: Sample images from the KTH–IDOL2 database.

The appearance of the rooms was captured under three different illumination conditions (i.e. cloudy weather, sunny weather, and night) and across a time span of six months. Thus, the sequences exhibit variability that occurs in real–world environments introduced not only by illumination but also by human activity (e.g. presence/absence of people, furniture/objects relocated). Each image was subsequently labeled as belonging to one of the rooms according to the position of the robot during acquisition (Luo et al, 2007).

2.5.2.2 Data Distribution

The training and validation set for the ImageCLEF 2009 Robot Vision Task consisted of a subset of the KTH–IDOL2 database. An additional, previously unreleased image sequence was used for testing, whereby the test sequences were recorded in the same five–office environment 20 months after the acquisition of the original KTH–IDOL2 data.

2.6 Conclusions

This chapter first introduced the requirements and specifications for test collection creation for multi–modal cross–language image retrieval evaluation in general, and then described each of the collections created and used for the ImageCLEF tasks between 2003 and 2009 in particular. This includes the collection origins and contents as well as distribution details, benefits and limitations of each resource.

It is recognized that benchmarks are not static as the field of VIR might (and will) develop, mature and/or even change. Consequently, benchmarks will have to evolve and be augmented with additional features or characteristics depending on the researchers' needs. Hence, ImageCLEF will continue to create and acquire document collections for its evaluation tasks in the future.

Acknowledgements This work has been partly supported by grant No. 01MQ07017 of the German research program THESEUS funded by the Ministry of Economics.

References

Candler CS, Uijtdehaage SHJ, Dennis SE (2003) Introducing HEAL: The Health Education Assets Library. Academic Medicine 78(3):249–253

Clough PD, Sanderson M, Reid N (2006) The Eurovision St. Andrews Collection of Photographs. SIGIR Forum 40(1):21–30

Denoyer L, Gallinari P (2006) The Wikipedia XML Corpus. SIGIR Forum 40(1):64–69

Everingham M, Zisserman A, Williams CKI, van Gool L (2006) The PASCAL Visual Object Classes Challenge 2006 (VOC2006) Results. Tech. rep., University of Oxford, Oxford, UK

Glatz-Krieger K, Glatz D, Gysel M, Dittler M, Mihatsch MJ (2003) Webbasierte Lernwerkzeuge für die Pathologie. Der Pathologe 24(5):394–399

Grubinger M (2007) Analysis and evaluation of visual information systems performance. PhD thesis, School of Computer Science and Mathematics. Faculty of Health, Engineering and Science. Victoria University, Melbourne, Australia

Grubinger M, Clough PD, Müller H, Deselaers T (2006) The IAPR TC–12 Benchmark: A New Evaluation Resource for Visual Information Systems. In: International Workshop OntoImage 2006 Language Resources for Content–Based Image Retrieval, held in conjunction with LREC 2006, Genoa, Italy, pp 13–23

Hersh W, Müller H, Kalpathy-Cramer J, Kim E (2007) Consolidating the ImageCLEF Medical Task Test Collection: 2005–2007. In: Proceedings of the Third Workshop on Image and Video Retrieval Evaluation. MUSCLE, Budapest, Hungary, pp 31–39

Hersh W, Müller H, Kalpathy-Cramer J (2009) The ImageCLEFmed Medical Image Retrieval Task Test Collection. Digital Imaging 22(6):648–655

Huiskes MJ, Lew MS (2008) The MIR FlickR Retrieval Evaluation. In: Proceedings of the 2008 ACM international conference on multimedia information retrieval. ACM press, New York, NY, USA, pp 39–43

Lehmann TM, Deselaers T, Schubert H, Güld MO, Thies C, Fischer B, Spitzer K (2003) The IRMA Code for Unique Classification of Medical Images. In: Huang HK, Ratib OM (eds) Medical Imaging 2003: PACS and Integrated Medical Information Systems: Design and Evaluation. SPIE Proceedings, vol 5033, San Diego, CA, USA, pp 440–451

Lehmann TM, Deselaers T, Schubert H, Güld MO, Thies C, Fischer B, Spitzer K (2006) IRMA — a Content–Based Approach to Image Retrieval in Medical Applications. In: IRMA International Conference 2006, Washington, DC, USA, pp 911–912

Lestari Paramita M, Sanderson M, Clough PD (2009) Developing a Test Collection to Support Diversity Analysis. In: Proceedings of the ACM SIGIR 2009 Workshop: Redundancy, Diversity, and Interdependence Document Relevance. ACM press, Boston, MA, USA, pp 39–45

Lestari Paramita M, Sanderson M, Clough PD (2010) Diversity in Photo Retrieval: Overview of the ImageCLEFphoto Task 2009. In: Peters C, Tsikrika T, Müller H, Kalpathy-Cramer J, Jones JFG, Gonzalo J, Caputo B (eds) Multilingual Information Access Evaluation Vol. II Multimedia Experiments: Proceedings of the 10th Workshop of the Cross–Language Evaluation Forum (CLEF 2009), Revised Selected Papers. Lecture Notes in Computer Science (LNCS), Corfu, Greece

Leung CHC, Ip H (2000) Benchmarking for Content–Based Visual Information Search. In: Laurini R (ed) Fourth International Conference On Visual Information Systems (VISUAL 2000). Lecture Notes in Computer Science (LNCS), vol 1929. Springer, Lyon, France, pp 442–456

Luo J, Pronobis A, Caputo B, Jensfelt P (2006) The KTH–IDOL2 Database. Tech. Rep. CVAP304, Kungliga Tekniska Hoegskolan, Stockholm, Sweden

Luo J, Pronobis A, Caputo B, Jensfelt P (2007) Incremental Learning for Place Recognition in Dynamic Environments. In: Proceedings of the 2007 IEEE/RSJ International Conference on Intelligent Robots and Systems (IROS07). IEEE, San Diego, CA, USA, pp 721–728

Markkula M, Tico M, Sepponen B, Nirkkonen K, Sormunen E (2001) A Test Collection for the Evaluation of Content–Based Image Retrieval Algorithms — A User and Task–Based Approach. Information Retrieval 4(3–4):275–293

Müller H, Rosset A, Vallée JP, Terrier F, Geissbuhler A (2004) A reference data set for the evaluation of medical image retrieval systems. Journal of Computerized Medical Imaging and Graphics 28:65–77

Müller H, Clough PD, Hersh W, Deselaers T, Lehmann T, Geissbuhler A (2006) Using Heterogeneous Annotation and Visual Information for the Benchmarking of Image Retrieval Systems. In: Santini S, Schettini R, Gevers T (eds) Internet Imaging VII. SPIE Proceedings, vol 6061, San José, CA, USA

Nowak S, Dunker P (2010) Overview of the CLEF 2009 Large–Scale Visual Concept Detection and Annotation Task. In: Peters C, Tsikrika T, Müller H, Kalpathy-Cramer J, Jones JFG, Gonzalo J, Caputo B (eds) Multilingual Information Access Evaluation Vol. II Multimedia Experiments: Proceedings of the 10th Workshop of the Cross–Language Evaluation Forum (CLEF 2009), Revised Selected Papers. Lecture Notes in Computer Science (LNCS), Corfu, Greece

Reid N (1999) The Photographic Collections in St Andrews University Library. Scottish Archives 5:83–90

Rosset A, Müller H, Martins M, Vallée JP, Ratib O (2004) Casimage Project—A Digital Teaching Files Authoring Environment. Journal of Thoracic Imaging 19(2):103–108

Wallis JM, Miller MM, Miller TR, Vreeland TH (1995) An Internet–based Nuclear Medicine Teaching File. The Journal of Nuclear Medicine 36(8):1520–1527

Westerveld T, van Zwol R (2007) The INEX 2006 Multimedia Track. In: Fuhr N, Lalmas M, Trotman A (eds) Advances in XML Information Retrieval: Fifth International Workshop of the Initiative for the Evaluation of XML Retrieval. INEX 2006. Revised Selected Papers. Lecture Notes in Computer Science/Lecture Notes in Artificial Intelligence (LNCS/LNAI), vol 4518. Springer, Schloss Dagstuhl, Germany, pp 331–344

Chapter 3
Creating Realistic Topics for Image Retrieval Evaluation

Henning Müller

Abstract This chapter describes the various ways for creating realistic query topics in the context of image retrieval evaluation campaigns such as ImageCLEF. A short overview describes general ways of creating topics, from complete laboratory style evaluations based on the technical capabilities of systems to real–world applications with real end users. The chapter offers help to those planning to evaluate systems on how to develop challenging and realistic topics based on knowledge of the users and of the capabilities of systems. Information sources for created topics are detailed. The main analysis will be the ImageCLEF tasks, and especially the medical retrieval tasks, where many different ways for creating topics have been analyzed over the years.

3.1 Introduction

Evaluation has always been an important aspect of systems development and demonstrating technical progress in all fields of research, including information retrieval. Creating formalised statements of user's information needs (topics) is a core part of IR evaluation using test collections. Topics are used to compare techniques in a particular field of research; however, creating realistic and effective topics is far from trivial. In information retrieval, the first systematic evaluation of research systems were the Cranfield tests in 1962 (Cleverdon, 1962). These tests mention the following as requirements for evaluation: the existence of a data set; the creation of query tasks and detailed topics that correspond to a user's information need; and a judgement of relevance for all documents/images in the collection with respect to the created topics. Almost all current evaluation campaigns such as TREC[1] and

Henning Müller
Business Information Systems, University of Applied Sciences Western Switzerland (HES–SO), TechnoArk 3, 3960 Sierre, Switzerland, e-mail: henning.mueller@hevs.ch

[1] Text REtrieval Conference, http://trec.nist.gov/

H. Müller et al. (eds.), *ImageCLEF*, The Information Retrieval Series 32,
DOI 10.1007/978-3-642-15181-1_3, © Springer-Verlag Berlin Heidelberg 2010

CLEF[2] are still based on this paradigm (Harman, 1992; Savoy, 2002), although with increasing database size judging all items in a database for relevance is not possible and pooling is usually used to limit the amount of work required for the judgments (Sparck Jones and van Rijsbergen, 1975). (See Chapter 4 for more details regarding relevance assessments.) Thus topic creation has been an integral part of the evaluation process in information retrieval.

This chapter focuses on the evaluation of image retrieval, however, rather than textual information retrieval. Image retrieval has been a very active domain over the past 25 years (Smeulders et al, 2000) but evaluation of image retrieval has rather been neglected (Müller et al, 2001) over much of this period. Over the last ten years, this has slowly changed and a large number of evaluation campaigns and more systematic evaluation approaches have also started in visual information retrieval. After initial proposals from Gunther and Beretta (2001) with general ideas, TRECVid[3] has been the first campaign to systematically evaluate video retrieval from large–scale archives with news footage (Smeaton et al, 2003). Other campaigns more focused on image retrieval, such as ImageCLEF[4] or ImageEval[5], followed only a little later.

In terms of topic creation, only very limited systematic analysis has taken place and one of the few papers really describing the process of topic generation for ImageCLEF is by Grubinger and Clough (2007). For most other evaluation campaigns, available data sources such as user log files have been used from a variety of sources such as Web log files (Müller et al, 2007), or library log files (Clough et al, 2006). Another approach is to integrate the participants into the creation of topics (Tsikrika and Kludas, 2009). The goal of topic development is usually to create topics that:

- correspond to a specific user model, i.e. a person searching for information in a particular context;
- correspond to real needs of operational image retrieval systems;
- are at least partly solvable with the existing technology;
- are diverse to allow a good part of the retrieval functionality to be tested and a large part of the data set to be explored;
- differ in coverage from rather broad to very specific needs;
- are solvable with documents from the given collection.

Another problem when considering analyzing visual information retrieval is how to express the information need of a potential user precisely. Information needs can generally be described in words, but for topic generation they can be represented with either text or visual examples, which determines which types of system can be evaluated. Most often, text is used for expressing the topic and textual information retrieval is much further advanced than visual retrieval in this respect. If the goal

[2] Cross Language Evaluation Forum, `http://www.clef-campaign.org/`

[3] `http://trecvid.nist.gov/`

[4] `http://www.imageclef.org/`

[5] `http://www.imageval.org/`

of a benchmark is to evaluate both visual and textual retrieval systems (and also combined retrieval), both media need to be represented in the query formulation. Whereas text can in this case easily be taken from usage log files, image examples are only very rarely available directly from such log files, as there are only very few visual systems in daily use. The choice of images for a query thus becomes an important part of the process and this is most often not analyzed further. Combined visual and textual retrieval really has the potential to improve current information access systems, but the results of evaluation campaigns to date also show how difficult these combinations are to work with.

In several evaluation tasks (Grubinger and Clough, 2007; Müller et al, 2009) the topics are classified into whether they mainly correspond to visual search tasks, where image analysis can be of use; to semantic search tasks, where mainly text retrieval can be useful; or to mixed tasks where the two can be expected to be useful. This classification is usually performed manually by experienced researchers and the results show that this classification is possible when being at least partly familiat with the database. This also means that systems could automatically determine the necessary resources for optimizing retrieval results if this knowledge can be formalized.

Another axis to take into account when developing topics is the topic difficulty, which needs to be challenging for existing systems employed and so rather difficult, but still correspond to the capabilities of the techniques. Particularly when pooling is used, the expected number of relevant images is also important as an excessively large number of relevant images can result in a large number of relevant documents remaining un–judged. On the other hand, a very small number of relevant documents can result in distorted performance measures if only one or two documents are relevant. Topic quantity is another important question that has been analyzed over many years. This is particularly important for getting stable/robust results and avoiding systems being ranked in a random order. Experiences in TREC suggest that at least 25 query topics are necessary for obtaining relatively stable results (Voorhees and Harmann, 2000), whereas others estimate this number to be much higher and near to 200–300 topics (Sparck Jones and van Rijsbergen, 1975). In general 25–50 query topics are recommended for relatively stable results.

An important link exists between the topic development and the relevance judgement process. TREC generally proposes that the topic creator should judge the relevant images themselves so the exact reasoning behind creating the topic can be taken into account for the judgment and means that this corresponds to one clear information need of a particular person. On the other hand, relevance of images has been shown to depend on the person, the situation and is not stable over time even for the same person. Thus, it was often proposed to have several judgments from different people so that the variability and subjectivity of the topics can be measured, e.g. using a kappa score (Müller et al, 2009). In general, results in ImageCLEF suggest that the judgments for image–based topics have less variation than for text–based query topics.

3.2 User Models and Information Sources

This section describes the underlying user models for image retrieval evaluation. Many purely image analysis benchmarks such as PASCAL[6] (Everingham et al, 2006) lack a concrete user model and involve rather basic scientific research tasks without any clearly visible application in mind. Examples for such topics can be detecting dogs or cats in images, which can then be used for future automatic annotation of images.

In general, when specific applications are identified, an appropriate user model is chosen such as journalists searching for images (Markkula and Sormunen, 1998) or Web users operating an image search engine (Goodrum, 2000). This can subsequently be taken into account for the definition of relevance in the evaluation. Relevance in itself is a rather poorly defined concept subject to much interpretation (Mizzaro, 1997) and having a clear user model and goal in mind can reduce this subjectivity. More on relevance judgments can be found in Chapter 4.

3.2.1 Machine–Oriented Evaluation

In image processing and many pattern recognition tasks involving images, the tasks for evaluation tools are more oriented towards advancing the current capabilities of techniques rather than towards real applications involving end users. This does not mean that these tasks cannot be useful, but care needs to be taken that tasks and databases are not too much oriented towards the capabilities of particular algorithms.

In the large majority of evaluation settings in image analysis, objects are to be detected in images such as in the PASCAL network of excellence (Everingham et al, 2006), or images are to be classified into a set of categories (Deselaers et al, 2007). This might currently not deliver results for real applications but it can be a preliminary step to developing tools that can subsequently help in such applications. Many other tasks have a user model in mind, such as clinicians searching for images but then use an outline that does not correspond to any realistic scenario. The risk in pure image classification or too machine–oriented tasks is to first create technologies and then create a data set for which the technology works well. This should really be the other way around and technology should adapt to the tasks (Müller et al, 2002), as otherwise the performance of a system is basically defined through the creation of the database.

One machine–oriented task that has a clear user model in mind is, for example, copy detection (Law-To et al, 2007), where distorted and modified images need to be traced back to their original. This scenario simulates a person or organization searching for copyright infringements, and similar techniques are used when uploading, for example, a video on YouTube, where Google needs to determine ex-

[6] `http://pascallin.ecs.soton.ac.uk/challenges/VOC/`

tremely quickly whether copyrighted material had been used. The ImageEval benchmark had an extensive task on this topic for images and TRECVid for videos. The quality of the current techniques for copy detection tasks is generally very high.

3.2.2 User Models

For general image retrieval, a very large number of applications have been proposed (Smeulders et al, 2000; Enser, 1995) and all application domains can be used to create user models. The first domains used as user models for image retrieval are domains with a wealth of visual data available, such as journalists (Markkula and Sormunen, 1998) and librarians (Clough et al, 2005).

In terms of the application of these user models for visual information retrieval benchmarks, TRECVid first used journalists (Smeaton et al, 2003). ImageCLEF on the other hand started on the photographic retrieval task with librarians searching for images (Clough et al, 2005), then used the general public having personal photo collections (Grubinger et al, 2008), before using journalists in 2010 (Lestari Paramita et al, 2010). The choice of user model basically corresponded to the databases used. For the Wikipedia topics, general Web users of Wikipedia were taken as the user model (Tsikrika and Kludas, 2009). By having the users create the topics, while there can be influence from the researchers based on the knowledge of their own techniques, the topics created should still correspond relatively well to the user model.

ImageCLEFmed always had clinicians in mind, first with an image example, then with a clear information need regarding single images (Müller et al, 2008), and later with a specific clinical task, where similar cases were searched for (Müller et al, 2009).

For all these user models, axes can be found along which topics can be created, and along which many of the information needs can be classified. For personal photo collections, the following axes have been identified for the retrieval (Grubinger and Clough, 2007):

- temporal constraints of the retrieval, so for example during a certain period or in a certain year;
- geographical constraints such as particular places or countries;
- actions defined by the use of verbs in the queries;
- search for particular objects or persons with general nouns and proper names;
- search with adjectives that specify a characteristic of a place, object or person.

In a similar way, the following axes were found for visual information needs in the medical field:

- anatomic region (i.e. lung, liver, leg);
- imaging modality (i.e. x–ray, CT, MRI);
- pathology (i.e. fracture, cancer);
- abnormal observation (i.e. enlarged heart).

Usually much of the topic development was along these axes and normally it was checked that the information needs were not too broad and that they covered at least two of these axes.

3.2.3 Information Sources for Topic Creation

To obtain knowledge for a particular user model it is important to have access to data that underlie such information needs. In the following subsections such information sources are explained that allow for creating realistic topics, though these are mainly textual resources. This means that there is a problem in finding visual examples for realistic search topics for these user models, mainly linked to the fact that very few visual retrieval systems are in routine use. This means that the example images for the topics have to be found from other sources in addition to the textual formulation of such a user need. Such examples should of course not be part of the collection itself as otherwise the corresponding descriptions can easily be used for query expansion with a potential bias of the results

Another problem in the topic generation process is to ensure that there are relevant images in the collection for the information need. Even when the information sources for generating topics were taken into account based on the collection used, the request can still be outside of the actual content of the databases. It is thus important to develop candidate topics first, and then restrict the benchmark to a subset of these candidate topics where a sufficiently high number of relevant images can be found in the collection. The exact number of relevant images or documents is most often not important but at least a few should be findable with example search systems.

3.2.3.1 Classification Tasks

For most classification tasks within ImageCLEF such as the medical image classification task (Deselaers et al, 2007), the photo annotation task (Nowak and Dunker, 2009) and the robot vision task (Caputo et al, 2010) no dedicated topic creation is necessary as the knowledge and the type of topics are contained within the databases or the annotations of the databases. Databases are divided into training and test data and the test data are basically the topics. The exact annotation process of the databases is outside of the scope of this chapter.

These topics can still be based on user models and in the context of ImageCLEF they most often are. For the medical classification task, the user model is clinicians and the situation is that many images have either no annotation or in the case of DICOM files, the annotations are not very detailed and contain errors (Güld et al, 2002). Thus, the collection was annotated by clinicians and new images have to be annotated automatically with a chosen annotation schema based on the the training data. For the photo annotation task, several schemes were tested over the years. In general, a collection of photographs had to be annotated with the objects or concepts contained in the images (concepts can be dogs, cars, outdoor images or night pictures, for example). Usually, a reasonably small number of concepts were chosen, typically in the range of 10–120, as current visual classification techniques often do not work very well when having to deal with a very large number of classes. Slightly different is the situation for the robot vision task, where the goal is to develop robots who can detect their own location based on the pictures they take, using training data from the same locations but under different lighting conditions and potentially with changes in the rooms such as moved furniture or modified objects. The ground truth is the location of the robot that is known and stored when recording the images.

3.2.3.2 Inherent Knowledge

The easiest way of generating topics is often to have a domain expert generate topics that correspond to a particular domain, that are challenging and at the same time useful for the chosen user model. In ImageCLEF, such an approach was taken for the first medical retrieval task (Clough et al, 2005), where a clinician very familiar with the chosen document collection selected a set of relevant images as query topics. This assured that the topics were useful and covered the collection well. On the other hand they represented the view of a single clinician and were thus not representative in this respect.

For the Wikipedia task, the inherent knowledge of the participating research groups was used (Tsikrika and Kludas, 2009), as all participants were asked to provide example topics and the topics for the evaluation were chosen from among this pool. This has an inherent risk that researchers develop topics that work well for their own system, but this risk does not bias results if all participants take part in the process. On the other hand, topics can be based too much on the technical possibilities and not on a real application of a Wikipedia user who searches for images.

3.2.3.3 Surveys and Interviews

Surveys among user groups are an important way to find out how images are being used and how visual image retrieval can help in the information retrieval process. One of the earlier studies analyzing the behavior of journalists in searching for images is described in (Markkula and Sormunen, 1998).

Within ImageCLEF, only the medical retrieval tasks used such surveys to create topics. To create the topics for the 2005 task, two surveys were performed among several groups of medical professionals in Portland (Oregon), USA and Geneva, Switzerland, (Hersh et al, 2005; Müller et al, 2006), located in medical teaching hospitals. The results of the surveys and the examples given by the experts were both used for the topic generation. The surveys also allowed definition of the differences in tasks depending on the roles of the health professionals (teaching, research, clinical work). In 2010 (Radhouani et al, 2009), another survey was performed in

Portland, OR, USA for the topic generation of ImageCLEF 2010. This time, the clinicians had access to a visual and textual retrieval system for executing example queries and analyzing the results during the interview, which can potentially give much more interesting topics and also provide image examples for the query formulation.

3.2.3.4 Usage Log Files

Log files are clearly the resource most often used as a basis for generating topics. The advantage is that they are usually available without requiring additional work and topics can thus be created just by cleaning the text in the logs. A problem with logs, particularly when they are on Web search engines, is the fact that they contain usually extremely short queries of often only one to two words, and creating a well–defined search topic from one or two terms is often hard. Library logs, such as the one used in (Clough et al, 2005) have the advantage that they contain not just a few quick terms formulated for a Web search engine, but rather well–thought–out information needs. They can thus be used more directly than Web search logs containing fewer terms. One solution to this is to add terms to make search requests more specific, or to reformulate them to reduce ambiguity and also potentially the number of relevant images. In specialized domains such as the medical field, log file terms can also be very specific with only a few search terms.

The frequency of the same search request is often used as a criterion for selection, as the most representative information needs should be used for evaluation if possible, or frequent terms should at least have a higher probability of being selected.

Concrete examples of log file use within ImageCLEF are the use of library log files of the St. Andrews library (Clough et al, 2005) for the photographic retrieval task. Other log files used for the photographic task are the Web logs of the Viventura travel agency (Grubinger et al, 2008), where the search requests were only slightly modified to be more specific and thus limit the number of relevant images. Also in the photographic task, the logs of the Belga news agency were used for topic development (Lestari Paramita et al, 2010). In all these cases, the logs corresponded to the database that was used for the retrieval.

For the medical tasks, no log files were available that correspond to the collection used for retrieval. Other information sources thus had to be found. With the health on the net media search engine[7] such a source exists and was used for ImageCLEFmed in 2006 (Müller et al, 2007). In general, some cleaning of the topics was necessary to make them more specific as most search requests were extremely general, e.g. 'heart' or 'lung'. For 2007 a log file of the PubMed[8] literature search engine was used (Müller et al, 2008). This makes the selecting process more difficult as queries with visual information needs had to be found. All imaging modalities were used to pre–filter the search request and only the remaining search requests that included

[7] http://www.hon.ch/HONmedia/

[8] http://www.pubmed.gov/

Table 3.1: Sources used for generating the query topics in ImageCLEF (not including the interactive and geographic query tasks).

Year	2003	2004	2005	2006	2007	2008	2009
Photo retrieval	St. Andrews logs	St. Andrews logs	St. Andrews logs	viventura web logs	viventura web logs	viventura web logs	Belga logs
Photo Annot.			annotated data	annotated data	annotated data	annotated data	annotated data
Medical retrieval		expert knowledge	expert survey	web logfile HON	Medline queries	from previous years	expert survey
Medical Annot.			annotated data	annotated data	annotated data	annotated data	annotated data
Nodule detection							expert annotations
Wikipedia						user generated	user generated
Robot vision						places known	places known

a modality were taken into consideration for the topic development based on the frequency of their occurrence.

3.3 Concrete Examples for Generated Visual Topics in Several Domains

This chapter gives a few examples for topics created in the context of ImageCLEF tracks using the various sources described. Table 3.1 also gives an overview of the ImageCLEF tasks and their way of generating the topics over the seven years of ImageCLEF. It can be seen that all purely visual tasks used only annotated data for generating topics and relevance judgments. This means that the tasks are really classification and not retrieval tasks, and the separation of the data into test data and training data was usually done in a more or less random fashion that took into account a certain distribution among training and test data.

By contrast the Wikipedia task used participant–generated topics, and the photographic retrieval task used three different types of log files. The medical retrieval task changed the topic generation almost every year using first expert knowledge, then user surveys and then two different types of log files for the topic generation. It is not possible to give examples for all tasks in this chapter and the corresponding Chapters 7, 8, 9, 10, 11, 12, and 13 can be used to find further details about each of the tasks.

3.3.1 Photographic Retrieval

In the Wikipedia task the topics were generated by the participants of the task as described by Tsikrika and Kludas (2009). In Figure 3.1, an example for such a topic

```
<topic>
<number> 1 </number>
<title> cities by night <title>
<image> hksky2.jpg </image>
<narrative> I am decorating my flat and as I like photos
  of cities at night, I would like to find some that I could
  possibly print into posters. I would like to find photos of
  skylines or photos that contain parts of a city at night
  (including streets and buildings). Photos of cities
  (or the earth) from space are not relevant.
</narrative>
</topic>
```

Fig. 3.1: Example topic for the Wikipedia task including a visual example, a title and a narrative describing the detailed information need.

can be seen. For the retrieval, the participating research groups could decide to use only the title, or to include the narrative as well. An image was supplied for almost all topics in the first year as can be seen in Figure 3.1, whereas in subsequent years several images were supplied for each topic.

The practice of using task participants for generating the topics was taken from the INEX[9] multimedia track (Westerveld and van Zwol, 2007) and has worked well over the years.

For the ImageCLEF photo retrieval retrieval task, various log files have been used over the years for generating the topics. An example for a topic using the Viventura log file can be seen in Figure 3.2. Several example images were supplied with each of the topics. In addition to the title and the narrative, the language of the topics can vary between German, English and Spanish. The user model is a person having a large personal collection of holiday pictures.

3.3.2 Medical Retrieval

An overview for medical image retrieval and its applications is given by Müller et al (2004). The topic developments for ImageCLEFmed generally modeled a clinician working on a particular case and who had a specific information need. Other roles of clinicians such as teacher and researcher were also considered. Figure 3.3 shows

[9] INitiative for the Evaluation of XML retrieval, `http://www.inex.otago.ac.nz/`

```
<top>
<num> Number: 14 </num>
<title> scenes of footballers in action </title>
<narr> Relevant images will show football (soccer)
players in a game situation during a match. Images with
footballers that are not playing (e.g. players posing for
a group photo, warming up before the game, celebrating
after a game, sitting on the bench, and during the half-
time break) are not relevant. Images with people not
playing football (soccer) but a different code (American
Football, Australian Football, Rugby Union, Rugby League,
Gaelic Football, Canadian Football, International Rules
Football, etc.) or some other sport are not relevant.
</narr>
<image> images/31/31609.jpg </image>
<image> images/31/31673.jpg </image>
<image> images/32/32467.jpg </image>
</top>
```

Fig. 3.2: Examples topic from the photographic retrieval task.

an example topic. Topics were always supplied in three languages (English, French, German) and with several example images. Topics were also developed along the axes anatomy, modality, pathology and abnormality. In the case of the topic shown, the two axes modality (x–ray) and pathology (fractures) are covered.

Due to the large variety of potential results of all anatomic regions in this case, the query can not be considered a visual query as it cannot be solved with visual features alone. It is thus regarded as a mixed query as visual features can help to distinguish x–ray images from other modalities.

3.4 The Influence of Topics on the Results of Evaluation

The various examples and ways of creating topics have shown that topic development is not an easy process. This raises the question of why invest a large amount of time and effort into creating such topics? The answer is that the entire evaluation that follows in an evaluation campaign or a single system evaluation is based on the topics developed. The topics have a much stronger influence on the comparative evaluation than the database itself and the relevance judgments have. Thus,

```
Show me all x-ray images showing fractures.
Zeige mir Roentgenbilder mit Bruechen.
Montres-moi des radiographies avec des fractures.
```

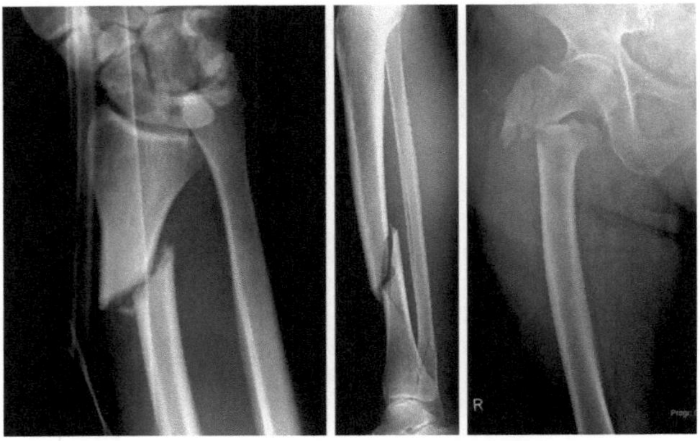

Fig. 3.3: A query requiring more than visual retrieval but where visual features can deliver hints to good results.

the importance of the topic development should not be taken lightly and it needs to be made clear what the main goal in the topic development is. It has to be clearly stated whether the topic development is based on any real application, or whether the capabilities of a certain technique are to be tested mainly in a laboratory style evaluation. Very often topics pretend to be modeling real–world applications when they are really not doing so.

3.4.1 Classifying Topics Into Categories

To further analyze information retrieval techniques, the topics can be classified into groups that can subsequently be used for analyzing techniques separately. Within several ImageCLEF tasks, the topics are classified into visual, textual and mixed topics by an experienced researcher in the field. This allows us to separately measure the best techniques for each of these categories.

Grubinger and Clough (2007) surveyed several of the ImageCLEFphoto topics for their level of 'visualness' (very bad, bad, average, good, very good). Several researchers judged the topics with respect to the visualness and then compared the performance results using a visual system for retrieval, showing that visualness can be estimated very well.

Topics can also be classified into other categories, allowing us to separately analyze the influences of certain techniques for particular tasks (e.g. tasks with a geographical orientation, topics with actions, topics of particular persons or topics with temporal constraints).

3.4.2 Links Between Topics and the Relevance Judgments

As the concept of relevance in retrieval tasks is not very stable, there are several approaches for linking the topic creation process with the relevance judgement process. In TREC, the people creating the topics are usually the people who also judge the pools for relevance. This has the advantage that the topic creator knows what he had in mind with the task creation, but on the other hand this can be very different if another person is judging the same topic. In the Wikipedia task, part of the topic creation and relevance judgement process is also performed by the participants and thus potentially by the topic creators. In the medical tasks of ImageCLEF, domain experts judge the topics but have not created the topics themselves. In general, several people judge the same topics, which allows us to analyze the level of ambiguity in the topic. This also allows us to find out whether the topic was well formulated for the system, and potentially ambiguous topics can still be removed at this point.

An extremely important step when developing topics with judgment in mind is to have a very detailed description or narrative of the task. Particularly if the relevance judges have not created the topics themselves it is important to detail exactly what is to be regarded as relevant. A description of exactly what is regarded as non–relevant is also extremely important as this can help define the border between relevant and non–relevant documents or images. The descriptions for the relevance judgements of the medical task have grown to over five pages, meaning they detail the entire process and define where the border between relevant and non–relevant is.

3.4.3 What Can Be Evaluated and What Can Not?

One of the questions is also with respect to what the limit of system capabilities is that can be evaluated. Jörgensen (1999) details the limits of image retrieval systems with respect to emotions, feelings and impressions but also shows ways how this can at least partially be reached. It is clear that query topics in image retrieval benchmarks need to correspond to current system capabilities and need to propose challenging search problems for the research community. To continue proposing challenging problems it is extremely important to have the topics evolve regularly over time, for example making them more challenging. If the topics of the benchmarks do not evolve sufficiently, the participating teams can be over–optimized for a particular scenario and this has to be avoided. The photo retrieval task has in this context evolved in several directions from evaluating very large databases to evaluating diversity. For the medical task this has been the creation of much larger data sets and also the development from image retrieval to case–based retrieval including images. This evolution has to be retained although it usually means additional work for the participants and also reduces the number of research groups participating as participation means increased work.

Another concept that can be important for generating topics is the concept of diversity. This was used in ImageCLEF for the photographic retrieval task in 2008 and 2009 (Lestari Paramita et al, 2010). In this case not only the topics need to be created but also the clusters of images for each topic that correspond to different representations of a particular search topic.

3.5 Conclusions

Topic creation is an important part of the evaluation of information retrieval systems, especially for visual information retrieval. As systems start to reach a quality where they can be used in real applications, mainly when used in combination with text retrieval, it is important to prove the quality of the tools. For this it is important to direct research efforts towards real problems and scenarios where image retrieval can deliver an added value. For this it seems necessary to have clear user models in mind, then create databases and topics based on the user models and then optimize techniques for these topics and databases. This avoids optimizing the data set to deliver good results for a particular technique (Müller et al, 2002), and so advances the technology.

Topic development is important for the creation of information retrieval tasks and more effort is necessary to control all the variables in this process. Parameters such as topic difficulty, topic variety and particularly the orientation towards real problems has to be taken into account to advance image retrieval through using good evaluation practices.

In the context of cross–language information retrieval it also needs to be stated that image retrieval offers a valuable contribution to language–independent information retrieval, as annotations with concepts can generate annotations in any language. Visual image analysis can also find similar images independent of the language. Within ImageCLEF several tasks are totally language–independent whereas others use collections in English and then propose topics in several languages. Starting from 2010 Wikipedia will have images annotated in various languages, which is the norm in the context of Wikipedia where content is created in many languages. Such a scenario can actually increase the importance of visual retrieval that currently has poorer performance than textual image retrieval.

Acknowledgements I would like to thank all funding institutions who have made this work possible, notably the European Union through the 6th and 7th framework program through the MultiMatch, TrebleCLEF, Chorus+, Promise and Khresmoi projects, the American National Science Foundation (NSF), the Swiss National Science Foundation (SNF), Google, the University of Geneva and the University of Applied Sciences Western Switzerland (HES–SO). Thanks are also due to all those who made their data available to ImageCLEF.

References

Caputo B, Pronobis A, Jensfelt P (2010) Overview of the CLEF 2009 robot vision task. In: Multilingual Information Access Evaluation Vol. II Multimedia Experiments: Proceedings of the 10th Workshop of the Cross–Language Evaluation Forum (CLEF 2009), Revised Selected Papers. Lecture Notes in Computer Science (LNCS). Springer

Cleverdon CW (1962) Report on the testing and analysis of an investigation into the comparative efficiency of indexing systems. Tech. rep., Aslib Cranfield Research Project, Cranfield, USA

Clough PD, Müller H, Sanderson M (2005) The CLEF cross–language image retrieval track (ImageCLEF) 2004. In: Peters C, Clough PD, Gonzalo J, Jones GJF, Kluck M, Magnini B (eds) Multilingual Information Access for Text, Speech and Images: Result of the fifth CLEF evaluation campaign. Lecture Notes in Computer Science (LNCS), vol 3491. Springer, Bath, UK, pp 597–613

Clough PD, Müller H, Deselaers T, Grubinger M, Lehmann TM, Jensen J, Hersh W (2006) The CLEF 2005 cross–language image retrieval track. In: Cross–Language Evaluation Forum (CLEF 2005). Lecture Notes in Computer Science (LNCS). Springer, pp 535–557

Deselaers T, Müller H, Clough PD, Ney H, Lehmann TM (2007) The CLEF 2005 automatic medical image annotation task. International Journal in Computer Vision 74(1):51–58

Enser PGB (1995) Pictorial information retrieval. Journal of Documentation 51(2):126–170

Everingham M, Zisserman A, Williams CKI, van Gool L, Allan M, Bishop CM, Chapelle O, Dalal N, Deselaers T, Dorko G, Duffner S, Eichhorn J, Farquhar JDR, Fritz M, Garcia C, Griffiths T, Jurie F, Keysers D, Koskela M, Laaksonen J, Larlus D, Leibe B, Meng H, Ney H, Schiele B, Schmid C, Seemann E, Shawe-Taylor J, Storkey A, Szedmak S, Triggs B, Ulusoy I, Viitaniemi V, Zhang J (2006) The 2005 PASCAL visual object classes challenge. In: Machine Learning Challenges. Evaluating Predictive Uncertainty, Visual Object Classification, and Recognising Tectual Entailment (PASCAL Workshop 05). Lecture Notes in Artificial Intelligence (LNAI), vol. 3944. Southampton, UK, pp 117–176

Goodrum A (2000) Image information retrieval: An overview of current research. Informing Science 3(2):63–66

Grubinger M, Clough PD (2007) On the creation of query topics for ImageCLEFphoto. In: MUSCLE/ImageCLEF workshop 2007, Budapest, Hungary

Grubinger M, Clough P, Hanbury A, Müller H (2008) Overview of the ImageCLEF 2007 photographic retrieval task. In: CLEF 2007 Proceedings. Lecture Notes in Computer Science (LNCS), vol 5152. Springer, Budapest, Hungary, pp 433–444

Güld MO, Kohnen M, Keysers D, Schubert H, Wein BB, Bredno J, Lehmann TM (2002) Quality of DICOM header information for image categorization. In: International Symposium on Medical Imaging. SPIE Procweedings, vol 4685, San Diego, CA, USA, pp 280–287

Gunther NJ, Beretta G (2001) A benchmark for image retrieval using distributed systems over the Internet: BIRDS–I. Tech. rep., HP Labs, Palo Alto, Tech. Rep. HPL–2000–162, San Jose

Harman D (1992) Overview of the first Text REtrieval Conference (TREC–1). In: Proceedings of the first Text REtrieval Conference (TREC–1), Washington DC, USA, pp 1–20

Hersh W, Jensen J, Müller H, Gorman P, Ruch P (2005) A qualitative task analysis for developing an image retrieval test collection. In: ImageCLEF/MUSCLE workshop on image retrieval evaluation, Vienna, Austria, pp 11–16

Jörgensen C (1999) Retrieving the unretrievable in electronic imaging systems: emotions, themes and stories. In: Rogowitz B, Pappas TN (eds) Human Vision and Electronic Imaging IV. SPIEProc, vol 3644, San Jose, California, USA. (SPIE Photonics West Conference)

Law-To J, Joly LCA, Laptev I, Buisson O, Nozha Boujemaa VGB, Stentifordl F (2007) Video copy detection: a comparative study. In: Proceedings of the 6th ACM international conference on Image and video retrieval. ACM press, pp 371–378

Lestari Paramita M, Sanderson M, Clough P (2010) Diversity in Photo Retrieval: Overview of the ImageCLEFphoto Task 2009. In: Peters C, Tsikrika T, Müller H, Kalpathy-Cramer J, Jones JFG, Gonzalo J, Caputo B (eds) Multilingual Information Access Evaluation Vol. II Multimedia

Experiments: Proceedings of the 10th Workshop of the Cross–Language Evaluation Forum (CLEF 2009), Revised Selected Papers. Lecture Notes in Computer Science (LNCS), Corfu, Greece

Markkula M, Sormunen E (1998) Searching for photos — journalists' practices in pictorial IR. In: Eakins JP, Harper DJ, Jose J (eds) The Challenge of Image Retrieval, A Workshop and Symposium on Image Retrieval. Electronic Workshops in Computing. The British Computer Society, Newcastle upon Tyne

Mizzaro S (1997) Relevance: The whole (hi)story. Journal of the American Society for Information Science 48(9):810–832

Müller H, Müller W, Squire DM, Marchand-Maillet S, Pun T (2001) Performance evaluation in content–based image retrieval: Overview and proposals. PRL 22(5):593–601. Special Issue on Image and Video Indexing

Müller H, Marchand-Maillet S, Pun T (2002) The truth about Corel – Evaluation in image retrieval. In: Lew MS, Sebe N, Eakins JP (eds) Proceedings of the International Conference on the Challenge of Image and Video Retrieval (CIVR 2002). Lecture Notes in Computer Science (LNCS), vol 2383. Springer, London, England, pp 38–49. 3-540-43899-8

Müller H, Michoux N, Bandon D, Geissbuhler A (2004) A review of content–based image retrieval systems in medicine—clinical benefits and future directions. International Journal of Medical Informatics 73(1):1–23

Müller H, Despont-Gros C, Hersh W, Jensen J, Lovis C, Geissbuhler A (2006) Health care professionals' image use and search behaviour. In: Proceedings of the Medical Informatics Europe Conference (MIE 2006). IOS Press, Studies in Health Technology and Informatics, Maastricht, The Netherlands, pp 24–32

Müller H, Boyer C, Gaudinat A, Hersh W, Geissbuhler A (2007) Analyzing web log files of the Health On the Net HONmedia search engine to define typical image search tasks for image retrieval evaluation. In: MedInfo 2007. IOS press, Studies in Health Technology and Informatics, vol 12, Brisbane, Australia, pp 1319–1323

Müller H, Deselaers T, Kim E, Kalpathy-Cramer J, Deserno TM, Clough PD, Hersh W (2008) Overview of the ImageCLEFmed 2007 medical retrieval and annotation tasks. In: CLEF 2007 Proceedings. Lecture Notes in Computer Science (LNCS), vol 5152. Springer, Budapest, Hungary, pp 473–491

Müller H, Kalpathy-Cramer J, Hersh W, Geissbuhler A (2008) Using Medline queries to generate image retrieval tasks for benchmarking. In: Medical Informatics Europe (MIE2008). IOS press, Gothenburg, Sweden, pp 523–528

Müller H, Kalpathy-Cramer J, Eggel I, Bedrick S, Said R, Bakke B, Kahn Jr. CE, Hersh W (2009) Overview of the CLEF 2009 medical image retrieval track. In: Working Notes of CLEF 2009, Corfu, Greece

Nowak S, Dunker P (2009) Overview of the CLEF 2009 Large–Scale Visual Concept Detection and Annotation Task. In: Peters C, Tsikrika T, Müller H, Kalpathy-Cramer J, Jones J, Gonzalo J, Caputo B (eds) Multilingual Information Access Evaluation Vol. II Multimedia Experiments: Proceedings of the 10th Workshop of the Cross–Language Evaluation Forum (CLEF 2009), Revised Selected Papers. LNCS, Corfu, Greece

Radhouani S, Kalpathy-Cramer J, Bedrick S, Hersh W (2009) Medical image retrieval, a user study. Tech. rep., Medical Inforamtics and Outcome Research, OHSU, Portland, OR, USA

Savoy J (2002) Report on CLEF–2001 experiments. In: Report on the CLEF Conference 2001 (Cross Language Evaluation Forum). Lecture Notes in Computer Science (LNCS), vol 2406. Springer, Darmstadt, Germany, pp 27–43

Smeaton AF, Kraaij W, Over P (2003) TRECVID 2003: An overview. In: Proceedings of the TRECVID 2003 conference

Smeulders AWM, Worring M, Santini S, Gupta A, Jain R (2000) Content–based image retrieval at the end of the early years. IEEE Transactions on Pattern Analysis and Machine Intelligence 22(12):1349–1380

Sparck Jones K, van Rijsbergen C (1975) Report on the need for and provision of an ideal information retrieval test collection. British Library Research and Development Report 5266, Computer Laboratory, University of Cambridge

Tsikrika T, Kludas J (2009) Overview of the wikipediaMM task at ImageCLEF 2008. In: Peters C, Giampiccolo D, Ferro N, Petras V, Gonzalo J, Peñas A, Deselaers T, Mandl T, Jones G, Kurimo M (eds) Evaluating Systems for Multilingual and Multimodal Information Access — 9th Workshop of the Cross–Language Evaluation Forum. Lecture Notes in Computer Science (LNCS), Aarhus, Denmark

Voorhees EM, Harmann D (2000) Overview of the ninth Text REtrieval Conference (TREC–9). In: The Ninth Text Retrieval Conference, Gaithersburg, MD, USA, pp 1–13

Westerveld T, van Zwol R (2007) The INEX 2006 Multimedia track. In: Fuhr N, Lalmas M, Trotman A (eds) Comparative Evaluation of XML Information Retrieval Systems, Proceedings of the 5th International Workshop of the Initiative for the Evaluation of XML Retrieval (INEX 2006), Revised Selected Papers. Lecture Notes in Computer Science (LNCS), vol 4518. Springer, pp 331–344

Chapter 4
Relevance Judgments for Image Retrieval Evaluation

Jayashree Kalpathy–Cramer, Steven Bedrick, and William Hersh

Abstract In this chapter, we review our experiences with the relevance judging process at ImageCLEF, using the medical retrieval task as a primary example. We begin with a historic perspective of the precursor to most modern retrieval evaluation campaigns, the Cranfield paradigm, as most modern system–based evaluation campaigns including ImageCLEF are modeled after it. We then briefly describe the stages in an evaluation campaign and provide details of the different aspects of the relevance judgment process. We summarize the recruitment process and describe the various systems used for judgment at ImageCLEF. We discuss the advantages and limitations of creating pools that are then judged by human experts. Finally, we discuss our experiences with the subjectivity of the relevance process and the relative robustness of the performance measures to variability in relevance judging.

4.1 Introduction

The goal of evaluation in information retrieval is to characterize the ability of search engines to meet the information needs of the users. Systematic evaluations of information retrieval evaluations began nearly 50 years ago with the Cranfield tests (Cleverdon, 1962). These experiments defined the necessity for a document collection, query tasks and ground truth for evaluation. They set the stage for much of what was to follow in the evaluation of the performance of search engines. The role model for most current evaluation campaigns is clearly TREC[1] (Text REtrieval

Jayashree Kalpathy–Cramer
Oregon Health & Science University, Portland, OR, USA e-mail: kalpathy@ohsu.edu

Steven Bedrick
Oregon Health & Science University, Portland, OR, USA e-mail: bedricks@ohsu.edu

William Hersh
Oregon Health & Science University, Portland, OR, USA e-mail: hersh@ohsu.edu

[1] http://trec.nist.gov/

H. Müller et al. (eds.), *ImageCLEF*, The Information Retrieval Series 32,
DOI 10.1007/978-3-642-15181-1_4, © Springer-Verlag Berlin Heidelberg 2010

Conference): (Voorhees and Harmann, 1998), a series of conferences that began in 1992 and continues to this day in organizing evaluation campaigns in a diverse set of areas.

Retrieval systems, whether text or multimedia, typically supply the user with an ordered set of results. Users, including experts for highly domain–specific tasks, are then recruited to provide judgments on the relevance of the items in this ordered list for the prescribed search topics. These relevance judgments are then used to compare the performance of the runs from the different systems that participated in the campaign.

In this chapter, we review our experiences with the relevance judging process in ImageCLEF, using the medical retrieval task as a primary example. We begin with a brief description of the stages in an evaluation campaign and provide details of the aspects of the relevance judgment process. We briefly summarize the recruitment process and describe the various systems used for judgment at ImageCLEF. We discuss the techniques typically used to pool documents in order to present the judges with a more manageable subset of documents (i.e. one that is presumed to contain a substantial number of relevant documents). The relevance of a document to a user's information need can be highly subjective, and depends on the user, the user's level of expertise with the subject matter, and the context of the search. We discuss our experiences with the subjectivity of the relevance process and the relative robustness of the performance measures to variability in relevance judging. Although in the past we have typically recruited clinicians as domain experts for the medical retrieval task, we have found interesting differences in terms of judging depending on the topic and the judge's individual level of expertise with it.

4.2 Overview of Relevance Judgments in Information Retrieval

Information retrieval campaigns strive to quantify the differences in the ability of search systems to meet the information needs of potential users. These evaluations can be system–based or user–based. However, since effective user–based evaluations are costly, difficult to conduct and hard to scale, evaluation campaigns such as TREC and CLEF (Cross Language Evaluation Forum) typically rely on system–based evaluations. This is in spite of observations made by (among others) Hersh et al. regarding the lack of correspondence in the performance observed between user–based and system–based evaluations of the same system (Hersh et al, 2000).

4.2.1 Test Collections

The goal of the Cranfield methodology, a precursor to modern system–based evaluations, was to create 'a laboratory type situation where, freed as far as possible from the contamination of operational variables, the performance of index languages could be considered in isolation' (Cleverdon, 1991; Voorhees, 2002). Although there have been concerns about the extensibility of these types of abstracted evaluations to real users, it is generally accepted that there is value in conducting these provided the set of topics and collections is sufficiently large and diverse (Voorhees, 2002).

Thus, a critical component of these system–based evaluations is the compilation of a fairly large and diverse collection of documents. For an image retrieval collection, this collection typically consists of images and associated metadata, possibly in the form of figure captions, associated articles, user–generated tags or text from a Web page found near the image in question.

In addition to the documents (images), a set of information needs (called topics) must be provided in order to characterize the performance of the various search systems. TREC and CLEF distinguish between an information need (the topic) and the data that is given to the retrieval system (the query) (Voorhees, 2002). Again, a large and diverse set of topics is an important part of the test collection. Typically, 25–50 topics have been used at the various TREC and CLEF campaigns.

In addition to the documents in the collection and the topics, the final part of the test collection is the relevance judgments. Real users are recruited to help provide their assessment of the relevance of the returned document for the information need provided in the topic.

This step is typically the most time and resource intensive part of building test collections. Additionally, a lot of subjectivity can be introduced in the evaluation due to potential differences in opinion among judges about the potential relevance of documents. However, it has been reported (Voorhees, 2002) that, although these differences can result in absolute difference in evaluation measures such as precision and recall, overall, the relative rankings remain consistent.

In the batch–mode (Hersh et al, 2000) described above, typical for the system–based evaluations used in the Cranfield and associated tests, a static set of topics and relevance judgments are used to simulate real users. However, in user–based evaluations, real users are typically set in front of retrieval systems during the evaluation and asked to list their own information needs, or are given vague descriptions of information needs and are asked to formulate queries or otherwise attempt to express those needs to the search system.

As these evaluations tend to be highly resource intensive and not easily scalable, there have been relatively few user–based evaluations reported in the literature, especially compared to the ubiquity of system–based evaluations such as TREC and CLEF.

4.2.2 Relevance Judgments

In many classification tasks, including image classification of radiographs into the IRMA (Image Retrieval in Medical Applications) classes (Müller et al, 2008), the ground truth — the status of each image with regard to its relevance or non–relevance — is known ahead of time and can be used for the evaluation. For example, in a collection of images intended to be used for an evaluation of tumor classification programs, it is possible to know a priori whether a given image contains a tumor or not. In the case of multi–class classification in a flat structure, classification accuracy is typically used as the metric. In the case of hierarchical classification, a score that takes into account the level at which the misclassification occurred can be used for the evaluation. Additionally, misclassifications can be penalized more than incomplete classifications.

However, in the case of evaluation for information retrieval, the 'relevance' of each retrieved document for each information need has to be independently assessed, as a document's relevance depends on the topic, the searcher and the retrieved result.

Once a collection and set of topics has been provided to the participants of a retrieval evaluation, the users provide runs that contain an ordered list of documents retrieved for each topic using their systems. Typically, users provide 10–20 runs per system with roughly 1,000 documents for each topic. These runs are obtained using a variety of algorithms or settings within their retrieval systems. The next step in the process is to evaluate the performance of these runs. In order to do that, users must evaluate the relevance of the returned documents to the particular topic.

4.2.2.1 Pooling

The original Cranfield approach was to perform complete relevance judgments on a small collection as it was felt that: 'Experience had shown that a large collection was not essential, but it was vital that there should be a complete set of relevance decisions for every question against every document, and, for this to be practical, the collection had to be limited in size' (Cleverdon, 1991; Voorhees, 2002). However, in the modern era of evaluation campaigns, the size of collections has grown dramatically, and there has been a shift from Cranfield–style complete judgments to the use of pooling as introduced by Sparck Jones and van Rijsbergen (1975). They believed that complete judgments would become impractical with larger collections and proposed pooling as a technique to assist in the generation of the final relevance judgments[2].

To pool submitted runs, a subset of documents, aggregated on a per–topic basis from the full runs submitted by users, is judged during the relevance judgment process. Typically, the most relevant (documents from the top of the ordered list) 30–100 documents from each run for a given topic are aggregated to create the pool for that topic. This approach is taken to try to maximize the number of relevant documents in the collection that are judged. Every document in the pool is judged for relevance for the topic, and documents not in the pool are assumed to not be relevant for the topic.

However, as discussed in the literature (Voorhees, 2002; Zobel, 1998; Buckley et al, 2006), when the number of documents whose relevance is assessed is small

[2] Called 'qrels' in TREC and CLEF.

relative to the number of documents in the collection, there are concerns about a potential bias in the evaluation. One needs to consider whether the manner in which the documents are selected for the relevance judgment process somehow biases the results, either towards groups who have submitted many runs, or to systems which might have good early precision but not necessarily high recall. Consequently, evaluation campaign organizers typically provide a limit for the number of runs a group can submit in order to minimize the biasing of the pool. However, concerns remain whether runs that were not part of the pooling process would be disadvantaged in the evaluation, as potentially relevant runs that were unique to that system would not be judged and would therefore be considered to be not relevant for the purposes of the evaluation.

Modifications suggested to the pooling process in the literature include judging more documents from topics that have a larger number of relevant documents by adding documents to be judged after an initial round (Zobel, 1998).

Other suggestions such as from Cormack et al (1998) include Interactive Search and Judging (ISJ) and performing local or global move–to–front (MTF) techniques where the next target for judging comes from runs that had a recent relevant document. Consequently, in the MTF approach, more documents are judged from runs that have many relevant documents. In the ISJ approach, extensive interactive searches by multiple assessors are used to identify all possible relevant documents. These judgments can be created independently and ahead of the participants' runs. However, since only one or a few select systems are used, the relevance judgments may be biased by the search system used to create the relevance judgments.

Most tasks at ImageCLEF typically use pooling with the top 30–100 images from each run being used to create the set of documents of which the relevance is then assessed. However, in select tasks such as the photo tasks, these pools were supplemented with manual interactive searches (interactive search and judge or ISJ) (Clough et al, 2004) to ensure sufficient quality. The ISJ approach did find relevant images that the standard pooling had missed.

It has been argued that by limiting the number of runs in the pool from each group/technique, having a large number of topics and maintaining a diversity in the types of approaches to the retrieval (visual and textual, language modeling vs. Boolean) as well as in the collection of documents, the bias can be minimized. Even so, it needs to be stressed that system evaluations that depend on relevance judgments typically must be considered in comparative, not absolute terms. Although the raw metrics might change depending on the pooling process and number of documents judged, it has been demonstrated that the overall ranking of the systems participating in the campaigns is relatively robust (Voorhees, 2002).

Binary preference (bpref) (Buckley and Voorhees, 2004) has been proposed as a more robust metric for evaluating retrieval performance in the case of incomplete judgments. As pool sizes grow small when compared to total collection sizes many campaigns have moved to using bpref instead of mean average precision (MAP) as the primary metric of interest. He et al (2008) have argued that bpref, normalized discounted cumulative gain (nDCG) (Järvelin and Kekäläinen, 2002), and inferred average precision (infAP) (Yilmaz and Aslam, 2006) are more stable metrics than MAP.

4.2.2.2 Levels of Relevance

The relevance of documents to the information need, as assessed by humans, can be on a variety of scales. Most modern evaluations using test collections, including the various TREC and CLEF tracks, have assumed that relevance is a binary characteristic: a document is either considered relevant or not relevant. Traditional information retrieval metrics including precision and recall are easily calculated using binary classification of relevance.

Evaluation measures created using trec_eval [3], the software used most commonly at TREC and CLEF, use binary relevance levels. However, over the years, some tasks at ImageCLEF (including the photo and medical tasks) have asked assessors to judge the relevance of images using a ternary scheme: relevant, partially relevant and not relevant, to deal with potential uncertainty in the assessor's judgment (Müller et al, 2008; Clough et al, 2004).

However, as mentioned above, in order to calculate the typical metrics of precision and recall, these ternary judgments need to be converted to binary judgments. This is accomplished by either counting partially relevant documents as relevant ('lenient') or counting partially relevant documents as not relevant ('strict'). The 'strict' relevance set is geared towards a task that desires precision while the 'lenient' qrel is geared towards recall. For some of the tasks, results using both the strict and lenient sets were provided to the participants (Müller et al, 2008; Clough et al, 2004).

Similarly, in cases where more than one judge assessed a topic, the qrels can be created using a variety of schemes. In some instances of the medical task relevance judgements from the first judge to finish the judging were used to create the qrels file. In the photo task (Clough et al, 2004), where two assessors were used per topic, qrels were generated using both *intersection* (where an image is considered to be relevant only if both assessors agree) and *union* (an image is considered to be relevant if either assessor labels it as being relevant). Thus, the photo task created four qrels files: union–strict, union–lenient, intersection–strict and, intersection–lenient.

Again, it has been observed that the rankings of the various participants' systems are typically robust to using any of the above–mentioned qrels (Hersh et al, 2006a; Müller et al, 2007, 2009). Although the absolute values of precision and recall change with changes in the number of images deemed relevant by the various judgement schemes, the relative rankings of the participating systems tends to remain constant.

Graded relevance assessments can also be used where the relevance of each document is stated on an ordinal scale. In this case, nDCG is often used for evaluation.

---

[3] `http://trec.nist.gov/`

This metric considers both the level of relevance of the document as well as its position in the ordered list of results.

Relevance judges are typically provided with instructions on how to estimate the relevance of a particular document for a given topic. These instructions can be highly task–specific. For instance, in some of the original TREC tracks, the judges were asked to determine if each document presented would be of value in preparing a report on the topic (Voorhees, 2002). Under this approach, the relevance of each document is meant to be evaluated without considering any of the other documents in the pool for that topic. In the medical image retrieval tasks of CLEF (Müller et al, 2008), a very domain specific–task, clinicians are typically recruited for the relevance judgment process, and are asked to make judgments of clinical relevance using fairly stringent criteria. Judges of the photo retrieval task of ImageCLEF were instructed to label any image as relevant if any part of the image was deemed relevant (Clough et al, 2004).

Many researchers have studied the variations in relevance judging resulting from different judges, as well as the impact that these variations might have on the overall results of an evaluation. The inter–rater agreement, as measured by variations of the kappa metric, is typically in the moderate to good range, and as has been mentioned previously, the overall ranking resulting from the differing relevance judgements is relatively stable (Hersh et al, 2006a; Müller et al, 2007, 2009).

The concept of relevance as applied to images is particularly problematic, as the relevance of a retrieved image can depend heavily on the context in which the search is being performed and it is often difficult to verbalise a visual information need. An additional source of difficulty with making relevance judgments is that domain experts tend to be stricter than novices (Müller et al, 2009); thus the validity of their judgments for a particular task may depend on the nature of the intended users. These challenges are discussed further in Section 4.3.3.1.

4.2.2.3 Judging Process

In order to perform relevance judgments on the large number of retrieved results for the diverse set of topics, as is common in these evaluation campaigns, judges need to be recruited. A variety of options for recruiting judges have been employed over the years. In many tasks, participants have been asked to judge, especially if they have also been involved in the topic creation process (Tsikrika and Kludas, 2009). In some of the original TREC tasks, participants were asked to create the topics and judge the relevance of the retrieved documents as if they were evaluating the relevance of the document in preparing a report on the subject. However, evaluations focusing on highly domain–specific tasks (such as medical image retrieval) typically recruit domain experts (e.g. clinicians) to carry out relevance judgments. More recently, in order to get a large set of documents evaluated, services such as Mechanical Turk (Nowak and Rüger, 2010) provided by Amazon have been explored. In other examples, community–based judging using games such as LabelMe (Russell et al, 2008) have been used with some success.

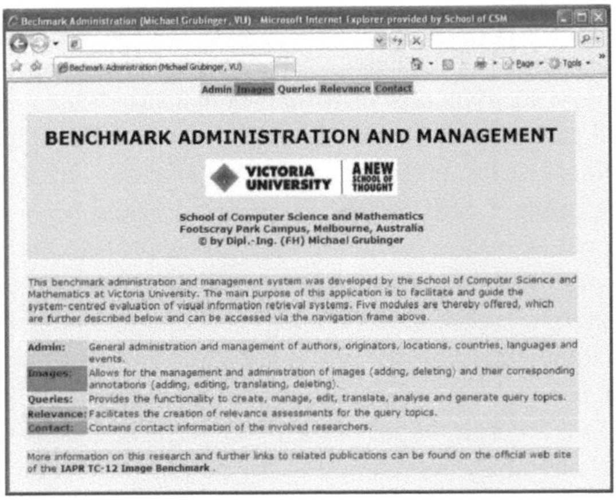

Fig. 4.1: Relevance judging system for the photo task — administration screen.

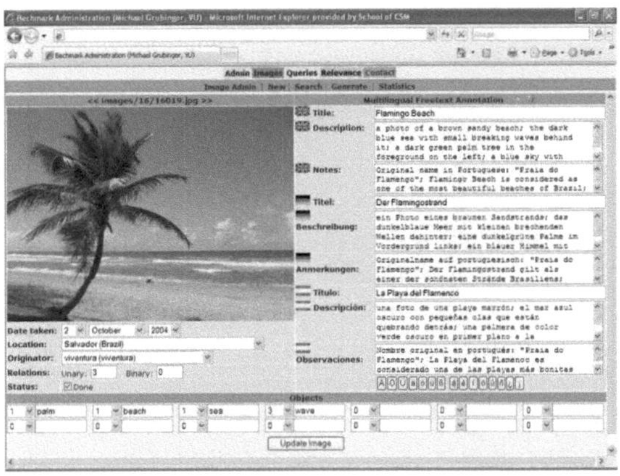

Fig. 4.2: Relevance judging system for the photo task— annotation edit screen.

4.2.2.4 Relevance Judgment Systems

A variety of systems have been built for the task of relevance judgment for image retrieval evaluations. Distributed Information Retrieval Evaluation Campaign Tool (DIRECT) (Di Nunzio and Ferro, 2004), a system built for the task of managing information retrieval campaigns by the organizers of CLEF, has been adapted for use by some the sub–tasks of ImageCLEF. Figures 4.1, 4.2, 4.3, 4.4, 4.5, and 4.6 provide screen shots of the various systems used within the ImageCLEF tasks.

Most systems have an administration page as shown in Figure 4.1. This enables the organizers to manage the creation of users, pools, assignment of topics and other administrative tasks. In some tasks such as the photo annotation task, participants create tags for the various images in the pools. These could be completely ad hoc

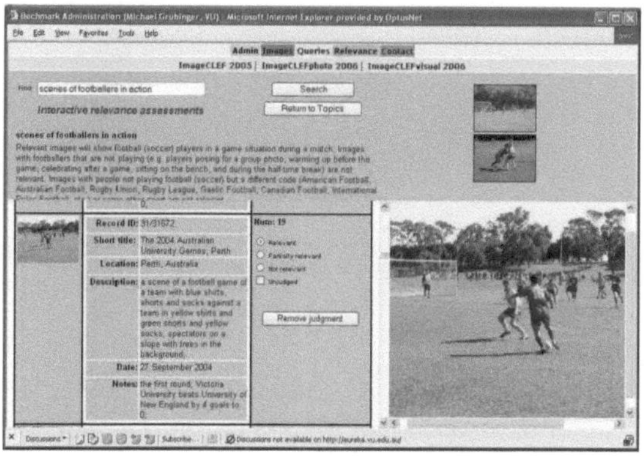

Fig. 4.3: Relevance Judging system for the photo task — assessment screen.

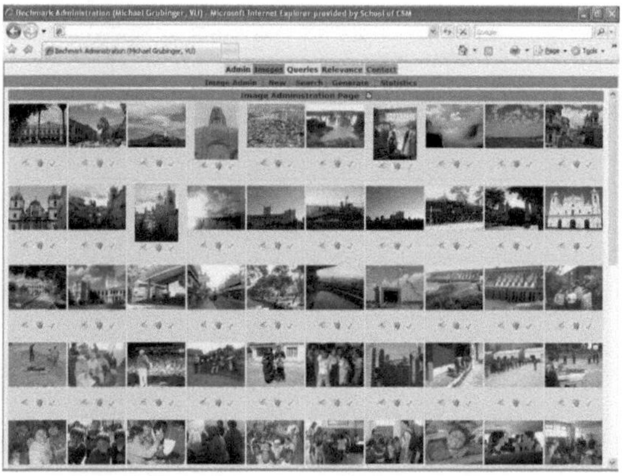

Fig. 4.4: Relevance Judging system for the photo task — image administration screen.

tags and captions (see Figure 4.2), or be part of a small vocabulary created by the organizers for a specific task (see Figure 4.6). Most importantly, there are a set of pages that contain the image to be judged, set of options that the judge can use to indicate relevance (see Figure 4.3), and a way to review the judgments (see Figure 4.4).

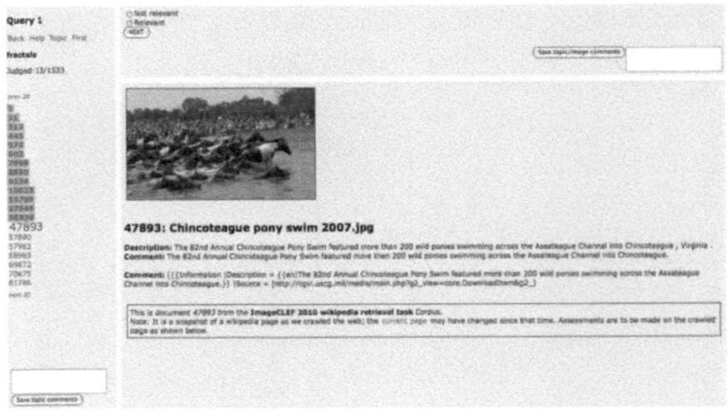

Fig. 4.5: Relevance judging system for the wiki task.

Fig. 4.6: Relevance judging system for the photo annotation task.

4.3 Relevance Judging for the ImageCLEF Medical Retrieval Task

4.3.1 Topics and Collection

Image retrieval is a growing area of research in medical informatics (Hersh et al, 2006a). Effective image annotation and retrieval can be useful in the clinical care of patients, education and research. The medical retrieval task of ImageCLEF was inaugurated in 2004 and has been repeated each year since. The medical image retrieval track's test collection began with a teaching database of 8,000 images in 2004.

For the first several years, the ImageCLEF medical retrieval test collection was an amalgamation of several teaching case files in English, French, and German. By 2007, it had grown to a collection of over 66,000 images from several teaching collections, as well as a set of topics that were known to be well–suited for textual, visual or mixed retrieval methods.

In 2008, images from the medical literature were used for the first time, moving the task one step closer towards applications that could be of interest in clinical scenarios. Both in 2008 and 2009, the Radiological Society of North America (RSNA) made a subset of its journals' image collections available for use by participants in the ImageCLEF campaign. The 2009 database contained a total of 74,902 images, the largest collection yet. All images were taken from the journals *Radiology* and *Radiographics*, both published by the RSNA. This collection constitutes an important body of medical knowledge from the peer–reviewed scientific literature, and includes high quality images with textual annotations.

4.3.2 Judges

Since its inception, relevance judgments for the medical image retrieval task at ImageCLEF were performed by physicians as well as clinician students in Oregon Health and Science University's biomedical informatics graduate program. All were paid an hourly rate for their work. Judges were typically responsible for judging three to five topics, each of about 1,000 images. Many judges continued to participate year after year, even after they had graduated from the program.

We have encountered a number of interesting observations and challenges over the course of the years in which we have been the organizers of the ImageCLEF medical retrieval task.

In the earlier years, the ImageCLEF medical image collection contained annotations in English, French, and German, and the topics were similarly linguistically diverse. Our judges, on the other hand, were almost all monolingual English speakers, and therefore experienced difficulty in judging images whose relevance depended on the content of a non–English annotation. This could have created a bias towards relevance for images with English–language annotations. Although our collection in 2008–2010 was monolingual (English), relevance judging of multi–lingual collections continues to be a challenging problem.

In 2008 (Hersh et al, 2009) we created, and made available, a consolidated collection of images, topics and qrels from years 2005–2007. During the process of creating this consolidated collection, we reviewed some of our older qrels and discovered significant discrepancies in how the judges performed the relevance assessments.

One important discovery we have made is that by providing judges with very explicit directions, the quality and consistency of the judging process improved significantly. This observation holds for all relevance judging tasks; however, we have found that it is particularly important when judging the relevance of images. After several years' worth of experience with this matter, our instructions to judges have become quite detailed. For example, we are now careful to make clear to judges that when a topic specifying an information need includes criteria along a variety of image axes (modality, pathology, anatomical location and view), an image should only be considered relevant if it meets all of the explicitly mentioned terms (i.e. 'ANDing' and not 'ORing').

For instance, in the topic 'CT liver abscess', only actual CT scans showing a liver abscess should be considered relevant. Pathology or MRI images of liver abscesses would not be considered relevant. Images of other abscesses would not be considered relevant. An x–ray image associated with an annotation that refers to a need for a CT scan in the future should not be considered relevant. On the other hand, if the topic had only specified 'liver abscess', a judge would have been correct to include x–ray or photographic images of liver abscesses together with CT images.

Another area in which we have learned to be more precise involves synonyms. Medical vocabulary is rich and there are often many ways to describe any given disease, organ, or other such medical topic. We now instruct our judges to consider annotations featuring synonyms of topic words to be relevant. For example, 'cholangiocarcinoma' is a synonym of 'bile duct cancer'. If a topic were to specify 'bile duct cancer', we would expect our judges to judge images whose annotations mention 'cholangiocarcinoma' to be relevant. This represents a prime example of the value of using domain experts as relevance judges.

After making these, and other similar criteria more explicit in our judging instructions, we observed a marked decrease in the number of images judged to be 'partially relevant', which as a category is meant to be used when exact relevance or irrelevance could not be determined.

We have also begun attempting to characterize variation among our judges in their relevance assessments and are studying the impact of that variation on the ImageCLEF evaluation results and overall rankings. Towards that end, we now make sure that we recruit a sufficient number of judges to be able evaluate inter–rater agreement on at least a subset of the topics. We will describe some of our finding in Section 4.3.3.1.

4.3.3 Relevance Judgment Systems and the Process of Judging

We introduced a new system to manage the process of collecting relevance judgments for the medical retrieval task in 2008, built using the Ruby on Rails framework and a PostgreSQL database. This system enables judges to record simple ternary judgments (e.g., 'relevant', 'partially relevant', and 'not relevant') via a Web interface and enables the administrators to manage which judges are assigned to which topics. The system's architecture is modular and flexible and can be used for all of the various tasks associated with the relevance judgment process. Runs from the track participants are loaded directly into the system's database, after which the system can automatically generate pools (as described in Section 4.2.2.1) using a user–configurable number of top-ranked documents (images) from each topic. Typically, we use the first 35-50 images from each run, and end up with an average pool size of around 900–1,000 images per topic.

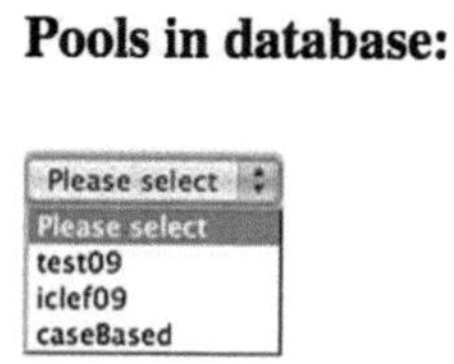

Fig. 4.7: Selecting the pools for relevance judgments.

A variety of pools can be maintained and made available if necessary as shown in Figure 4.7. A screen shot of the interface is provided in Figure 4.8. As seen in the figure, for the image–based topics, each judge is presented with the topic's text, as well as several sample images. Images clearly corresponding to all criteria are judged as 'relevant', images whose relevance cannot be safely confirmed but could possibly be relevant are marked as 'partly relevant', and images for which one or more criteria of the topic are not met were marked as 'not relevant'. Our judges are instructed in these criteria (as described above), and typically are quite proactive about contacting us with questions.

Judgments are saved as soon as the judge selects a radio button. The system provides visual feedback to judges, in that the rows of images marked as relevant turn green after being judged, whereas those marked as non–relevant turn red. Judges are able to log on and off multiple times during the process and all their judgments are saved across multiple sessions. When judges log back on, they are provided with the option of viewing only their remaining un–judged images (as opposed to having to scan through all of their already–judged images to get back to the point at which they left off). In addition to saving time, this feature also enables judges to easily ascertain whether or not they have completed judging all the images in the pool for the topic at hand.

Once all topics in a pool have been completely judged, the system allows the organizers to easily create qrel files after first selecting the judges to be used for each topic. Since TREC–style qrel files assume a binary definition of relevance, the system allows organizers to choose either strict (partially relevant images are considered non–relevant) or lenient (partially relevant images are considered relevant) qrel modes. Additionally, the system facilitates the creation of more complex qrels, by combining judgments from two or more judges (OR or AND of the Boolean relevances). The system also allows the calculation of kappa scores as measure of inter–rater agreement. Once the judgments have been completed, track participants are able to evaluate the performance of their runs and visualize the given relevance of the images in their runs.

In 2009, we introduced a case–based retrieval task. For the case–based topics, the judge was shown the original case description and several images appearing in the original article's text. Along with a short description for the judgments, a full document was prepared to describe the judging process, including what should be

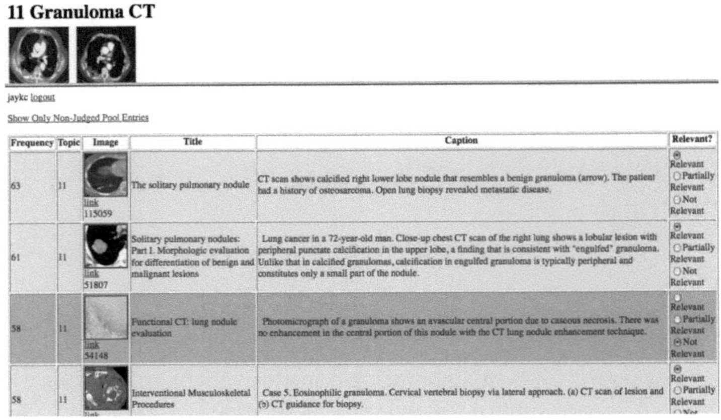

Fig. 4.8: Screenshot of the relevance judgment used for the medical retrieval task.

regarded as relevant versus non–relevant. As described above, this increased clarity was helpful in improving the performance of judgments.

4.3.3.1 Multiple Judges

Although the reliability of judging obtained during ImageCLEF has been slightly better than that obtained from relevance judgments of textual documents in clinical (Hersh et al, 1994) and genomics (Hersh et al, 2006b) tasks, we have found instances of incorrectly judged images, especially with regards to image modality (a vitally important factor in image retrieval). In order to characterize the inter–rater variability, in recent years we have increased the number of topics that are judged by more than one judge. We have also received feedback from the judges about our system and the process of judging. Based on the feedback from the judges, it was apparent that the concept of relevance is highly dependent on the expertise of the judge.

For example, a resident, in a state of naïveté, might consider images of a broad scope to be relevant to a given topic. These images might include those of normal conditions as well as differential diagnosis, essentially anything that might help them gain information about the query topic. On the other hand, a specialist very familiar with a topic might only mark as relevant those images that are exceedingly specific and on–topic. One proposed modification to the judging instructions is to include some language about the expected level of expertise of the simulated information seeker (radiologist, specialist, internist, resident, etc.)

We analyzed the pattern of overlap between pairs of judges and found that, in some cases, they were symmetric: i.e. for each topic, an equal number of discordant images were judged to be relevant by one judge or the other. However, in a large

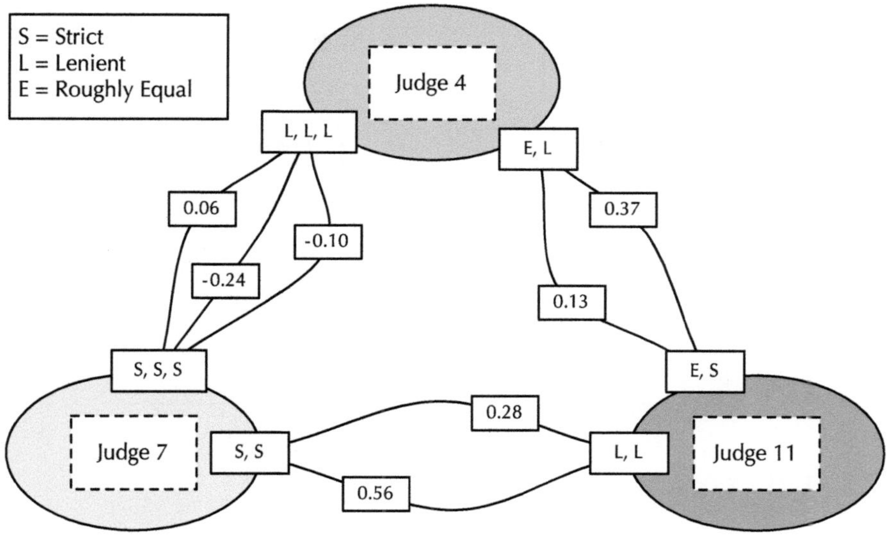

Fig. 4.9: Visualization of leniency difference among judges.

number of cases, the overlap was non–symmetric: i.e. one judge marked a substantially larger number of images as relevant than the other, and thus the relevant images in one case were a subset of the relevant images of the other judge. Anecdotally, the specialists were more stringent than the non–specialists, who in turn were more stringent than the non–clinicians. We compared the metrics of the runs using the different qrels generated using the different relevance judgments, and found that although the raw numbers were different comparative trends held. In other words, the overall rankings were relatively robust — individual systems' scores varied between qrels, but the overall patterns of performance (rankings) were the same. We also observed, somewhat surprisingly at first, that the actual MAP of the best runs *decreased* with the more lenient runs. Although the trends hold across qrels, the difference between the best and the worst runs actually *decreases* if we have very lenient (and potentially more random) judges.

In 2009's medical retrieval campaign, we had two judges for each of the case–based topics. The kappa scores for these topics and judges were lower than for the ad hoc retrieval topics. Additionally, as seen in Figure 4.9, the discordance in relevance judgments was not symmetric. Judge 7 was considerably stricter than judge 11 who was somewhat more strict than judge 4.

We attempted to simulate the effect of having extremely lenient and strict judges by conducting an experiment using the 2009 runs. We believe that extremely lenient judges add noise to the system. To test this hypothesis, we started with the qrels based on the the strict judgments, and added random noise by selecting 50–300 random images from each topic to be relevant. We compared the MAP of the original qrel versus the MAP from the noisy qrel. As the noise increased, the curve

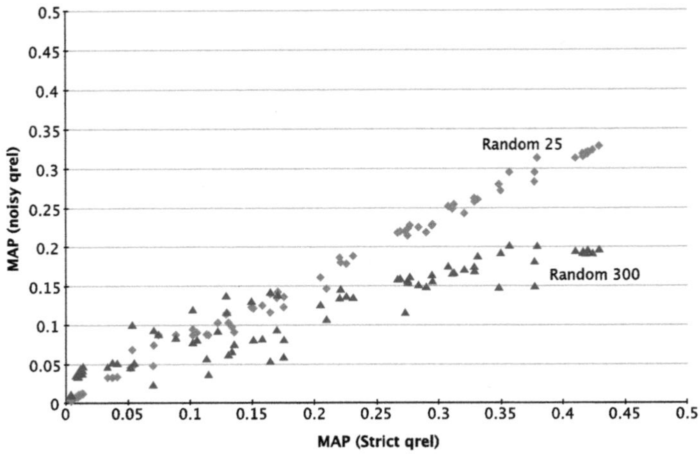

Fig. 4.10: MAP for noisy vs. original qrel.

flattened out, decreasing the difference between a good and a poor system, as seen in Figure 4.10. This is very similar to the effect we saw above between the strict and lenient judges.

4.4 Conclusions and Future Work

In this chapter, we reviewed the process of generating relevance judgments for retrieval evaluation campaigns. We described the different systems used within ImageCLEF with emphasis on the medical retrieval task. Inter–rater agreement amongst judges can be estimated using the kappa measure. Although there can be variations in kappa for different sets of judges, the evaluation metrics are somewhat robust to some noise in the judgment process caused by these differences. Anecdotally, more naive judges tend to be more lenient in the judgment for highly domain–specific tasks. We plan to further examine the impact of the judge's expertise and the context of their search on the evaluation process. Additionally, we will continue to study if and how these system-oriented metrics translate to user satisfaction with systems.

Acknowledgements Jayashree Kalpathy–Cramer was supported by the following grants from the National Library of Medicine: 2T15LM007088 and 1K99LM009889 and Steven Bedrick by grant NLM 2T15LM007088. We would like to thank RSNA for contributing images to ImageCLEF medical retrieval tasks in 2008, 2009 and 2010. We also would like to thank Charles Kahn, MD, Medical College of Wisconsin for his help with the ImageCLEF project.

References

Buckley C, Voorhees EM (2004) Retrieval evaluation with incomplete information. In: Proceedings of the 27th annual international ACM SIGIR conference on research and development in information retrieval. ACM press, New York, NY, USA, pp 25–32

Buckley C, Dimmick D, Soboroff I, Voorhees E (2006) Bias and the limits of pooling. In: Proceedings of the 29th annual international ACM SIGIR conference on research and development in information retrieval. ACM press, New York, NY, USA, pp 619–620

Cleverdon CW (1962) Report on the testing and analysis of an investigation into the comparative efficiency of indexing systems. Tech. rep., Aslib Cranfield Research Project, Cranfield, USA

Cleverdon CW (1991) The significance of the cranfield tests on index languages. In: Proceedings of the 14th annual international ACM SIGIR conference on research and development in information retrieval. ACM press, pp 3–12

Clough P, Sanderson M, Müller H (2004) The CLEF Cross Language Image Retrieval Track (ImageCLEF) 2004. In: Image and Video Retrieval (CIVR 2004). Lecture Notes in Computer Science (LNCS), vol 3115. Springer, pp 243–251

Cormack GV, Palmer CR, Clarke CLA (1998) Efficient construction of large test collections. In: Proceedings of the 21st annual international ACM SIGIR conference on research and development in information retrieval. ACM press, pp 282–289

Di Nunzio G, Ferro N (2004) DIRECT: a system for evaluating information access components of digital libraries. In: Research and Advanced Technology for Digital Libraries. Lecture Notes in Computer Science (LNCS), vol 3652. Springer, pp 483–484

He B, Macdonald C, Ounis I (2008) Retrieval sensitivity under training using different measures. In: Proceedings of the 31st annual international ACM SIGIR conference on research and development in information retrieval. ACM press, New York, NY, USA, pp 67–74

Hersh W, Buckley C, Leone TJ, Hickam D (1994) OHSUMED: an interactive retrieval evaluation and new large test collection for research. In: Proceedings of the 17th annual international ACM SIGIR conference on research and development in information retrieval. ACM press, pp 192–201

Hersh W, Turpin A, Price S, Chan B, Kramer D, Sacherek L, Olson D (2000) Do batch and user evaluations give the same results? In: Proceedings of the 23rd annual international ACM SIGIR conference on research and development in information retrieval. ACM press, pp 17–24

Hersh W, Müller H, Jensen J, Yang J, Gorman P, Ruch P (2006a) Advancing biomedical image retrieval: Development and analysis of a test collection. Journal of the American Medical Informatics Association 13(5):488–496

Hersh W, Müller H, Kalpathy-Cramer J, Kim E, Zhou X (2009) The consolidated ImageCLEFmed medical image retrieval task test collection. Journal of Digital Imaging 22(6):648–655

Hersh WR, Bhupatiraju RT, Ross L, Roberts P, Cohen AM, Kraemer DF (2006b) Enhancing access to the bibliome: the trec 2004 genomics track. Journal of Biomedical Discovery and Collaboration 1:3

Järvelin K, Kekäläinen J (2002) Cumulated gain–based evaluation of ir techniques. ACM Transactions of Information Systems 20(4):422–446

Müller H, Clough P, Hersh W, Geissbuhler A (2007) Variations of relevance assessments for medical image retrieval. In: Adaptive Multimedia Retrieval (AMR). Lecture Notes in Computer Science (LNCS), vol 4398. Springer, Geneva, Switzerland, pp 233–247

Müller H, Deselaers T, Kim E, Kalpathy-Cramer J, Deserno TM, Clough PD, Hersh W (2008) Overview of the ImageCLEFmed 2007 medical retrieval and annotation tasks. In: CLEF 2007 Proceedings. Lecture Notes in Computer Science (LNCS), vol 5152. Springer, Budapest, Hungary, pp 473–491

Müller H, Kalpathy-Cramer J, Eggel I, Bedrick S, Saïd R, Bakke B, Kahn Jr CE, Hersh W (2009) Overview of the CLEF 2009 medical image retrieval track. In: Working Notes of CLEF 2009, Corfu, Greece

Nowak S, Rüger S (2010) How reliable are annotations via crowdsourcing: a study about inter-annotator agreement for multi-label image annotation. In: Proceedings of the international conference on multimedia information retrieval (MIR 2010). ACM press, New York, NY, USA, pp 557–566

Russell B, Torralba A, Murphy K, Freeman W (2008) LabelMe: a database and web–based tool for image annotation. International Journal of Computer Vision 77(1–3):157–173

Sparck Jones K, van Rijsbergen C (1975) Report on the need for and provision of an ideal information retrieval test collection. British Library Research and Development Report 5266, Computer Laboratory, University of Cambridge

Tsikrika T, Kludas J (2009) Overview of the wikipediaMM task at ImageCLEF 2008. In: Peters C, Giampiccolo D, Ferro N, Petras V, Gonzalo J, Peñas A, Deselaers T, Mandl T, Jones G, Kurimo M (eds) Evaluating Systems for Multilingual and Multimodal Information Access — 9th Workshop of the Cross-Language Evaluation Forum. Lecture Notes in Computer Science (LNCS). Springer, Aarhus, Denmark

Voorhees EM (2002) The philosophy of information retrieval evaluation. In: Revised Papers from the Second Workshop of the Cross–Language Evaluation Forum on Evaluation of Cross–Language Information Retrieval Systems — CLEF 2001. Lecture Notes in Computer Science (LNCS). Springer, London, UK, pp 355–370

Voorhees EM, Harmann D (1998) Overview of the seventh Text REtrieval Conference (TREC–7). In: The Seventh Text Retrieval Conference, Gaithersburg, MD, USA, pp 1–23

Yilmaz E, Aslam JA (2006) Estimating average precision with incomplete and imperfect judgments. In: Proceedings of the 15th ACM international conference on Information and knowledge management. ACM press, New York, NY, USA, pp 102–111

Zobel J (1998) How reliable are the results of large–scale information retrieval experiments? In: Croft WB, Moffat A, van Rijsbergen CJ, Wilkinson R, Zobel J (eds) Proceedings of the 21st Annual International ACM SIGIR conference on research and development in information retrieval. ACM press, Melbourne, Australia, pp 307–314

Chapter 5
Performance Measures Used in Image Information Retrieval

Mark Sanderson

Abstract Although during the running of the ImageCLEF tracks there was no explicit co–ordination on the types of evaluation measures employed, the same statistics were often used across ImageCLEF. Therefore, in this chapter, the range of measures used in the evaluation exercise is described. The original research defining a measure, together with their formulations and the relative pros and cons of the measures, are also detailed. Research that both compares the measures and attempts to determine the best is also outlined. Finally, the use of measures in the different tracks and years of ImageCLEF is tabulated.

5.1 Evaluation Measures Used in ImageCLEF

At its most general, measurement has been described as the assignment of numerals to things so as to represent facts and conventions about them (Stevens, 1946). ImageCLEF is a collaborative evaluation exercise examining image searching or image analysis systems. In this context, the main purpose of measurement is as a means of comparison between different systems. More specifically, the evaluation acts as a simulation of a potential operational setting into which the systems may be deployed. The measures provide a way of determining which system submitted to a particular track of ImageCLEF would be best in the setting simulated by the test collection. Different measures reflect different priorities in the simulation of the operational setting. For example, in retrieval some evaluation measures stress the importance of locating as many relevant images as possible, while others focus on a small fixed size output. The choice of evaluation measure employed by different ImageCLEF track co–ordinators reflects varying priorities over the years of ImageCLEF. In this chapter, the descriptions of the measures are organized so as to

Mark Sanderson
University of Sheffield, Sheffield, United Kingdom
e-mail: m.sanderson@sheffield.ac.uk

H. Müller et al. (eds.), *ImageCLEF*, The Information Retrieval Series 32,
DOI 10.1007/978-3-642-15181-1_5, © Springer-Verlag Berlin Heidelberg 2010

Table 5.1: Contingency table of retrieval results.

	Relevant	Non–relevant	
Retrieved	a	b	a+b
Not retrieved	c	d	c+d
	a+c	b+d	a+b+c+d

illustrate these priorities. The measures for image retrieval are described first, followed by those used for image annotation and other tracks of ImageCLEF. Finally, the use of measures across the years of ImageCLEF is described.

5.2 Measures for Retrieval

The earliest work in evaluating Information Retrieval (IR) systems dates back to the 1950s. Although a wide range of properties could be assessed in a searching system — Cleverdon for example described a number (Cleverdon and Keen, 1966) — the main quality of an IR system that is measured is search effectiveness. This is an assessment of the number of relevant documents that are retrieved by a query. In the early days of IR, almost all the IR systems built were so–called Boolean retrieval systems, which partitioned the collection into two sets, those documents that were retrieved by a user's query and those that were not. Combining the sets with information from a test collection on the relevance of documents to queries, the contingency table shown in Table 5.1 can be created.

Early evaluation measures were constructed from combinations of the table cells. Kent et al (1955) described a number of such measures and was the first to describe precision and recall together (though the researchers called them by a different name). They are defined as follows

$$Precision = \frac{a}{a+b} \tag{5.1}$$

$$Recall = \frac{a}{a+c} \tag{5.2}$$

Precision assesses the fraction of retrieved documents that are relevant, recall assesses the fraction of relevant documents retrieved. While, ideal for set based retrieval, the measures as defined in this form, needed to be adapted to the increasingly common IR systems producing ranked output, ordering retrieved documents based on a score detailing the similarity of each document to the user's query. As pointed out by Swets (1963), the density of relevant documents typically reduces as one moves from the top of a ranking downwards. Hull (1993) stated that effectiveness measures for ranked retrieval approached the measurement of this ranking in one of two ways: measuring the density of relevant documents at a commonly chosen recall value, or at a fixed rank number. Most evaluation measures that ImageCLEF organizers used can be grouped into Hull's two categories and they are so organised in this chapter. The remaining statistics are placed into a third miscellaneous category, which is described next. Finally, research comparing different measures is detailed.

5.2.1 Measuring at Fixed Recall

As detailed in the evaluation chapters of any IR book from the 1960s onwards (Salton, 1968; van Rijsbergen, 1979), a common approach for assessing effectiveness was graphing the density of relevant documents across a ranking. Here precision was measured at a set of standard recall points for each topic in a test collection; the mean of precision values at each point was taken, plotted on a graph and a line between the points drawn. Although a range of different recall levels were proposed, the de facto standard became 11 points starting at recall=0% incrementing by 10% up to recall=100%. Interpolation functions were used to ensure that precision was measurable at these fixed points. See Van Rijsbergen for more detail on their construction (van Rijsbergen, 1979, chap. 7). In the first year of ImageCLEF, a Recall/Precision (R/P) graph was used to compare different retrieval systems. However, a typical R/P graph showed similar characteristics of each plotted system. Perhaps for this reason, there was a growing trend in many parts of the IR research community to move from these visual presentations to a single value measure. Here, we describe such measures that are commonly used in ImageCLEF. The first three, Mean Average Precision (MAP), Geometric Mean Average Precision (GMAP) and Binary Preference (BPref) measure the density of relevant documents at the point in the ranking where all known relevant documents are retrieved (recall=100%); the final one, R Precision, measures at the point in the ranking where a perfect IR system would have retrieved all known relevant documents.

5.2.1.1 MAP — Mean Average Precision

For almost two decades the most popular evaluation measure has been MAP. The measure calculated on a single topic is defined as follows:

$$AP = \frac{\sum_{rn=1}^{N}(P(rn) \times rel(rn))}{R} \tag{5.3}$$

Here, N is the number of documents retrieved, rn is the rank number; $rel(rn)$ returns either 1 or 0 depending on the relevance of the document at rn; $P(rn)$ is the precision measured at rank rn and R is the total number of relevant documents for this particular topic. The measure calculates precision at the rank position of each relevant document and takes the average. Note, by summing over N and dividing by R, in effect, precision is measured as being zero for any un–retrieved relevant document. The mean of the AP scores for a set of topics is taken to produce Mean AP. The measure was developed by organisers of the Text REtrieval Conference or TREC (Harman, 1995) for use in the ad hoc track of TREC, which addressed the searching task of an information analyst: someone wishing to locate every possible relevant item. One might question the value of this measure for tasks where users might only be interested in finding a few relevant items near the top of a ranking. See, however, the research below on comparing evaluation measures.

5.2.1.2 GMAP — Geometric MAP

The MAP measure appears to be the first to explicitly state the way that scores across the topics of a test collection were to be averaged. Up to that point, it was simply assumed that the arithmetic mean would be used. There appears to be little or no evidence that this is the appropriate way to summarise the effectiveness of a searching system. Cooper (1968) proposed alternatives to the arithmetic mean, but chose to stick with it. As part of the interest in developing a test collection focused on poorly performing topics, Voorhees and later Robertson described GMAP, which used the geometric mean of AP scores (Voorhees, 2005; Robertson, 2006). As Robertson stated:

> GMAP treats a change in AP from 0.05 to 0.1 as having the same value as a change from 0.25 to 0.5. MAP would equate the former with a change from 0.25 to 0.3, regarding a change from 0.25 to 0.5 as five times larger.

The geometric mean of average precision (AP) values computed over a set of topics Q are as follows:

$$GMAP(Q) = \sqrt[|Q|]{\prod_{k=1}^{|Q|} [AP(Q_k) + \varepsilon]} \tag{5.4}$$

Adding ε avoids GMAP going to zero if $AP = 0$. Robertson discussed this measure in some detail pointing out that using geometric mean emphasized improvements in topics that had a low AP score. Whether the method is a more effective averaging approach than the arithmetic mean is yet to be determined.

5.2.1.3 BPref — Binary Preference

As test collections have grown, in addition to containing a set of documents judged relevant and not relevant for each topic, they also have a large set of documents that have not been judged at all. Commonly, if any such documents were retrieved, evaluation measures considered them as not relevant. Buckley and Voorhees were concerned that with increasing sizes of test collections, the number of unjudged documents was growing and devised an evaluation measure, BPref, so called as it uses binary relevance judgments to define a preference relation (Buckley and Voorhees, 2004). It is defined as follows:

$$BPref = \frac{1}{R}\sum_{r}\left(1 - \frac{N\ ranked\ higher\ than\ R}{min(R,N)}\right) \tag{5.5}$$

Where R is the number of documents judged relevant for a particular topic; N is the number of documents judged not relevant; r is a relevant retrieved document, and n is a member of the first R irrelevant retrieved documents. Several versions of BPref were defined in the literature, the version shown here is considered to be the definitive one (Soboroff, 2006). Since Buckley and Voorhees's work, a number of alternatives to BPref have been created; of particular note is infAP (Yilmaz and Aslam, 2006).

5.2.1.4 R–Precision — Recall Precision

A simple approach to measuring effectiveness is to calculate precision at R, the total number of known relevant documents for a particular topic (Harman, 1993). Note that at rank R, the number of relevant documents ranked below R equals the number of non–relevant documents ranked above R, some refer to R as the equivalence number and call R–precision missed@equivalent (Pearson, 1995).

$$RP = \frac{r(R)}{R} \tag{5.6}$$

Here, R is the number of known relevant documents and $r(R)$ is the number of relevant documents retrieved in the top R.

5.2.2 Measuring at Fixed Rank

A common approach to measuring precision over a ranked document list is to measure at a fixed rank position. Ignoring all documents retrieved below the fixed position is thought to be justified as search systems commonly return a page containing a fixed number of documents and users rarely examine more than the first returned page. There are a number of variants used in ImageCLEF.

5.2.2.1 P(n) — Precision Measured at a Fixed Rank

Precision at a fixed rank is simply defined as:

$$P(n) = \frac{r(n)}{n} \tag{5.7}$$

Where $r(n)$ is the number of relevant documents in the top n. The value of n varies from one evaluation to another. In ImageCLEF, the values of 5, 10, 20, 30 and 100 were used. The measure is commonly written either as $P10$, $P(10)$ or $P@10$. Along with MAP, this has probably been the most popular evaluation measure in the past decade of IR research. However, the measure is simple: ignoring the position of relevant documents within the top n and ignoring all documents retrieved below n.

5.2.2.2 DCG — Discounted Cumulative Gain; Grades of Relevance

An assumption common to all the measures described up to now is that the relevance judgements of test collections are binary: relevant or not relevant. A few of the test collections in ImageCLEF had ternary relevance judgments: highly relevant, partially relevant, not relevant. The simplest way to use the existing measures with such judgements is to map the multiple levels to binary values. A number of measures were developed to exploit the grades of relevance. Although they were not widely used in ImageCLEF, they are mentioned here due to their increasing importance and use in IR research. One of the best known measures is Discounted Cumulative Gain (DCG) (Järvelin and Kekäläinen, 2000). It is defined as follows:

$$DCG = rel(1) + \sum_{i=2}^{n} \frac{rel(i)}{log_b(i)} \tag{5.8}$$

Where $rel(i)$ returns a numerical value corresponding to the relevance grade assigned to the document at rank i; n is the rank that DCG is calculated up to. Järvelin and Kekäläinen assumed that the likelihood of users examining a document reduced as its rank increased. Therefore, they introduced a log–based discount function to model that drop in user interest. The degree of discount could be varied by changing the base b of the log function; Järvelin and Kekäläinen suggested setting $b = 2$. There is a great deal of anecdotal evidence that this measure is commonly used by Web search companies, although the so–called Burges variant of the measure is more used (Burges et al, 2005).

$$DCG = \sum_{i=1}^{n} \frac{2^{rel(i)} - 1}{log(1+i)} \tag{5.9}$$

Both these measures produce values that are not constrained within a particular range. Järvelin and Kekäläinen also created nDCG (Järvelin and Kekäläinen, 2002), which compares the DCG value with the score gained from a perfect ranking of the relevant documents up to rank n: $IDCG(n)$.

$$nDCG(n) = \frac{DCG(n)}{IDCG(n)} \tag{5.10}$$

5.2.3 Measures for Diversity

A core assumption behind many IR test collections is that users of IR systems would provide a detailed unambiguous specification of what they were looking for and that the relevance of a document could be judged independent of other documents. Early on, researchers challenged this view: Verhoeff, Goffman, and Belzer (1961); Fairthorne (1963) and Goffman (1964) respectively pointed out that users' definitions of relevance were diverse, queries were commonly ambiguous, and that the relevance of a document to a query could be strongly influenced by the documents already retrieved. It was not until much later that these concerns were addressed. As detailed in Chapter 8, ImageCLEF started to create test collections with diverse relevance judgments and therefore needed means of measuring the effectiveness of systems searching on those collections. The initial approach used was proposed by Zhai et al (2003) who among other measures created sub–topic recall (S–recall). Sub–topics were the name the researchers gave to the different aspects to which a topic might be relevant. An examination of the literature reveals that the naming of the facets of a topic varies greatly: some refer to nuggets, others to aspects, while others talk of clusters. Considering a topic with n_A sub–topics and a ranking of documents, $d_1..d_m$, S–recall calculates the percentage of sub–topics retrieved by the documents up to rank position K:

$$S-recall(K) = \frac{\bigcup_{i=1}^{K} s(d_i)}{n_A} \tag{5.11}$$

Here, $s(d_i)$ is the set of sub–topics covered in d_i. The measure gave a higher score to runs that covered the largest number of sub–topics. Note in some ImageCLEF documentation, this measure is referred to as cluster recall (CR). This measure simply measures the spread of diversity in a ranking. Clarke et al (2010) proposed an adaptation of nDCG called α–nDCG, which also considered the rank of relevant documents. The researchers re–defined the function $rel(i)$ from nDCG as:

$$rel(i) = \sum_{k=1}^{m} J(d_i,k)(1-\alpha)^{r_{k,i-1}} \tag{5.12}$$

where m is the number of distinct nuggets (the researchers' term for sub–topics), $n_1..n_m$, relevant to a particular topic; $J(d_i,k) = 1$ if an assessor judged that document d_i contained nugget n_k;

$$r_{k,i-1} = \sum_{j=1}^{i-1} J(d_j,k) \tag{5.13}$$

is the number of documents ranked before document d_i that were judged to contain nugget n_k; the constant α represents the probability that the user of the retrieval system observed prior relevant documents.

Evaluation research continues to develop, some of the more recent work includes the following:

- Rank Biased Precision (RBP): a measure with a different model of a user's propensity to examine top ranked documents from that used by DCG, (Moffat and Zobel, 2008);
- Intent aware measures, which allow for the different aspects of a diverse query to be preferred by different proportions of users (Agrawal et al, 2009);
- NRBP: Clarke et al.'s drawing together of ideas drawn from RBP, intent aware measures, and α–nDCG. The derivation and formulation of this measure is left for the reader to pursue (Clarke et al, 2010);
- ERR: a measure that considers the likelihood of a user examining a document at rank position r given that other documents ranked higher than r were also relevant (Chapelle et al, 2009).

5.2.4 Collating Two Measures Into One

At times, the ImageCLEF organisers wished to combine two evaluation measures into a single value. This is not new: starting with precision and recall, researchers often found that two evaluation measures examined different qualities of search output, but wished to have both values summarised into a single score. Rijsbergen (1974) surveyed methods for doing this in 1974. He later proposed using the weighted harmonic mean, which is commonly referred to as f, and is defined as follows (summarising precision and recall).

$$f = \frac{1}{\alpha(\frac{1}{P}) + (1-\alpha)(\frac{1}{R})} \tag{5.14}$$

Here α indicates a preference for how much influence precision or recall has on the value of f; commonly, $\alpha = 0.5$, so f is then:

$$f = \frac{1}{\frac{1}{2}(\frac{1}{R} + \frac{1}{P})} \tag{5.15}$$

or the equivalent form:

$$f = \frac{2 \times R \times P}{P+R} \tag{5.16}$$

5.2.5 Miscellaneous Measures

As well as standard IR evaluation measures, the organisers of the search part of ImageCLEF also used a range of other more ad hoc statistics created by the organisers themselves. They are mentioned here:

- Failed queries — as mentioned for GMAP, there was a concern that a user's overall appreciation of an IR system across a number of topics was not well approximated by the arithmetic mean of scores for the topics. There was a suspicion that a query that returned no relevant documents would be viewed very negatively by users, therefore simply judging the quality of runs by counting the number of these failed queries has been a statistic used in ImageCLEF.
- Relevant retrieved — another simple statistic used was a count of the number of relevant items that were retrieved per topic.

5.2.6 Considering Multiple Measures

With such a range of evaluation measures, it is tempting to ask if some measures are better than others. Among a series of methods devised to examine this question, the common approach was to use a measure to rank the runs submitted to an evaluation campaign (e.g. TREC, NTCIR, CLEF, etc.) and to correlate the run rankings produced by different measures. Early work examining correlations between such rankings showed little difference between measures such as *RP*, $P(n)$, or average precision (Tague–Sutcliffe and Blustein, 1994). More recent work found that high precision measures such as P(1) or P(10), correlated less well with *RP* or *MAP*, see Buckley and Voorhees (2005); Thom and Scholer (2007). Correlation does not address which measure might be better. For this question, the research does not appear to be definitive; here we describe a series of distinct conclusions on this matter:

- It might be tempting to think that the measure that most closely represents one's expectations of how a user might interact with a searching system is the best. For example if searchers are known to only examine the top few documents, a measure such as P(10) might be preferred. However, Yilmaz and Robertson (2009) showed that evaluation measures that make the greatest use of all available relevance information, like MAP, can in fact be the better measure to employ even if the planned use of a retrieval system is to find a few relevant items.
- If comparing evaluation measures where the number of relevance judgments is kept constant, there is some evidence that measures that use a more refined notion of the rank position of relevant documents are better. Sanderson et al (2010) showed that NDCG(10) was better than P(10).
- When considering test collections with few relevant documents, Soboroff (2004) showed that statistically simpler evaluation measures appeared to be more stable. Unlike a lot of research, in this paper P(10) was shown to be better than a number of other popular measures, including MAP.
- Sanderson and Zobel (2005) pointed out that when comparing evaluation measures, it was important to consider the amount of assessor effort required to enable the measure to be calculated. They compared measures after normalizing for assessor effort and showed that this was an important factor.

The early results showing strong correlations between evaluation measures point out that the measures largely assess similar qualities of a ranked document list. However, the more recent work shows a more contradictory picture indicating that one cannot assume that a single measure, such as MAP, will always provide the definitive ordering of runs relative to each other. Use of multiple measures appears to be the best approach; this is supported by a number of examples where unusual outlier runs were spotted only through use of multiple measures. See for example Buckley et al (2007) who by measuring the number of unique relevant documents in a run, identified an important outlier that led to a detailed study of test collection formation. Hawking and Robertson described another research study conducted after finding an unusual retrieval situation where MAP and P(10) showed substantially different results (Hawking and Robertson, 2003).

5.2.7 Measures for Image Annotation and Concept Detection

In general, IR systems rank documents relative to a query and the decision on which is relevant is left to the user. Evaluation measures for IR can be viewed as a form of simulation of a searcher finding the relevant items in the ranking and stopping at some point. In the image annotation, robot vision and concept detection tasks of ImageCLEF, the decision on whether an image is assigned to a particular category or code is done by the system. Consequently, the evaluation measures in the image processing tasks focus on the success of the decisions made by the systems. They are described here.

5.2.7.1 Error Rate

In the initial evaluations of the medical and object annotation tasks, the mean error rate was measured across the possible codes/categories an image could be assigned to. The code assigned to an image was viewed as being either correct or not. The rate was compared to a pair of baseline approaches: a system that always picked the commonest code in the training set; and a basic image similarity system. For the medical annotation task, the codes images were assigned to were made up of four components addressing different aspects (called axes) of a medical image: modality, body orientation, body region, and biological system examined. Each axis code was a number, the digits of which represented a hierarchical scheme which specified progressively more detail about each axis. In later years of the annotation task, participating systems could opt to only code the more significant digits of an axis, leaving the rest unspecified. In order to assign credit for a partial match on an axis code a customised hierarchical error rate was created (Müller et al, 2008). It was defined as follows:

$$error\ rate = \sum_{i=1}^{I} \frac{1}{b_i} \frac{1}{i} \delta(l_i, \hat{l}_i) \tag{5.17}$$

where I is the number of hierarchical levels of the axis code (i.e. number of digits); b is the number of possible values at a particular level; and δ is a function returning the degree of match between the predicted and actual value for that particular level. For every axis, the maximum error rate was calculated and the rates of participating systems were normalised against this. Each axis contributed a quarter of the total error rate, which meant the overall rate ranged between zero and one.

5.2.7.2 Confusion Matrix

In the annotation tasks of ImageCLEF, an image could potentially be assigned to any one of the N possible codes. A confusion matrix was used to understand where common mistaken assignments took place. The matrix was $N \times N$: the rows of the matrix correspond to the correct code for an image, the columns to the assigned code. Each cell totalled up the number of images assigned to a particular correct/assigned code pair. If the annotation was perfect, the matrix equalled N times the identity matrix. A confusion matrix was commonly scatter plotted to visualize mistaken assignments.

5.2.7.3 ROC— Receiver Operating Characteristic

Usually the binary decision of which code to assign an image to, was determined by setting a threshold on a calculated likelihood for the image code pair. In general, the higher the threshold, the more accurate the assignment (i.e. a high true positive rate), but this occurred at the expense of coverage. If the threshold was reduced, while the coverage of images correctly assigned increased, this was at the expense of growing numbers of images incorrectly assigned to a code (i.e. a high false positive rate). An ROC curve plots the balance between TPs (True Positives) and FPs (False Positives) across all possible values of a threshold. They are widely used in many areas of statistics, (Green and Swets, 1966). In ImageCLEF, certain well known properties of ROC curves were used to assess image annotation systems. The average area under an ROC curve (AUC) as well as the average Equal Error Rate (EER) were calculated. The EER is the point at which errors in assignment to a code equal the errors from not assigning images to a code; it has parallels with the IR measure, RPrec. See Fawcett (2006) for more discussion of ROC curves in computer science.

5.3 Use of Measures in ImageCLEF

In Table 5.2 we tabulate the use of measures over the years of ImageCLEF. In each cell we list the ImageCLEF track using this measure. For the precision measured at fixed rank, the value(s) of n are listed.

Table 5.2: Use of the measures in the years and tasks of ImageCLEF (Photo — photo retrieval; Med_R — medical retrieval; Obj — object retrieval; Wik — Wikipedia MM retrieval; Med_A — medical annotation; Img — image annotation; Vis — visual concept detection.).

Year	2003	2004	2005	2006	2007	2008	2009
MAP	Photo	Photo, Med_R	Photo, Med_R	Photo, Med_R	Photo, Med_R	Obj, Med_R, Wik	Med_R, Wik
GMAP				Photo	Photo		
BPref				Photo, Med_R	Photo, Med_R	Med_R	
R Precision				Med_R	Med_R	Wik	Wik
P(n)			Photo (10, 100)	Photo (20), Med_R (30, 100)	Photo (20), Med_R (10, 30, 100)	Photo (20), Med_R (5, 10, 30), Wik (10, 20)	Photo (10), Med_R (5, 10, 30), Wik (10, 20)
Failed queries	Photo						
Relevant retrieved			Photo			Med_R	Med_R
S-recall						Photo	Photo
F						Photo	Photo
Error rate			Med_A	Med_A, Img	Med_A		
Hier. error rate					Med_A	Med_A	Med_A
Confusion matrix				Med_A			
ROC curve						Vis	Vis

From this table it is clear that *MAP* was ImageCLEF's lead retrieval measure with $P(n)$ also well used. The depth at which $P(n)$ was calculated at has reduced somewhat over the years. In the annotation tasks, different forms of error rate were the lead measure.

5.4 Conclusions

In this chapter, the evaluation measures used in ImageCLEF were described, where it was shown that a wide range of measures were employed over the years of the exercise. The workings of the measures were described and the relative merits of one measure over another were detailed.

References

Agrawal R, Gollapudi S, Halverson A, Leong S (2009) Diversifying search results. In: Proceedings of the Second ACM International Conference on Web Search and Data Mining. ACM, pp 5–14

Buckley C, Voorhees EM (2004) Retrieval evaluation with incomplete information. In: Proceedings of the 27th annual international ACM SIGIR conference on research and development in

information retrieval. ACM press, pp 25–32
Buckley C, Voorhees EM (2005) Retrieval system evaluation. In: TREC: Experiment and Evaluation in Information Retrieval. MIT Press, pp 53–75. 0262220733
Buckley C, Dimmick D, Soboroff I, Voorhees EM (2007) Bias and the limits of pooling for large collections. Information Retrieval 10(6):491–508
Burges C, Shaked T, Renshaw E, Lazier A, Deeds M, Hamilton N, Hullender G (2005) Learning to rank using gradient descent. In: Proceedings of the 22nd international conference on Machine learning, Bonn, Germany, pp 89–96
Chapelle O, Metlzer D, Zhang Y, Grinspan P (2009) Expected reciprocal rank for graded relevance. In: Proceeding of the 18th ACM conference on Information and knowledge management. ACM press, pp 621–630
Clarke CLA, Kolla M, Vechtomova O (2010) An effectiveness measure for ambiguous and underspecified queries. In: Advances in Information Retrieval Theory Lecture Notes in Computer Science (LNCS). Springer, pp 188–199
Cleverdon CW, Keen M (1966) Factors affecting the performance of indexing systems, vol 2. ASLIB, Cranfield Research Project. Bedford, UK 37–59
Cooper WS (1968) Expected search length: A single measure of retrieval effectiveness based on the weak ordering action of retrieval systems. American Documentation 19(1):30–41
Fairthorne RA (1963) Implications of test procedures. In: Information Retrieval in Action. Western Reserve UP, Cleveland, Ohio, USA, pp 109–113
Fawcett T (2006) An introduction to ROC analysis. Pattern Recognition Letters 27(8):861–874
Goffman W (1964) On relevance as a measure. Information Storage and Retrieval 2(3):201–203
Green DM, Swets JA (1966) Signal detection theory and psychophysics. John Wiley & Sons, Inc.
Harman DK (1993) Overview of the second text retrieval conference (TREC–2). In: TREC Proceedings. NIST Special Publication. Department of Commerce, National Institute of Standards and Technology, Gaithersburg, MD, USA
Harman DK (1995) Overview of the second text retrieval conference (TREC–2). Information Processing and Management 31(3):271–289
Hawking D, Robertson SE (2003) On collection size and retrieval effectiveness. Information Retrieval 6(1):99–105
Hull D (1993) Using statistical testing in the evaluation of retrieval experiments. In: Proceedings of the 16th annual international ACM SIGIR conference on research and development in information retrieval. ACM press, pp 329–338
Järvelin K, Kekäläinen J (2000) IR evaluation methods for retrieving highly relevant documents. In: Proceedings of the 23rd annual international ACM SIGIR conference on Research and development in information retrieval. ACM press, pp 41–48
Järvelin K, Kekäläinen J (2002) Cumulated gain–based evaluation of IR techniques. ACM Transactions on Information Systems 20(4):422–446
Kent A, Berry MM, Luehrs Jr FU, Perry JW (1955) Machine literature searching VIII. Operational criteria for designing information retrieval systems. American Documentation 6(2):93–101
Moffat A, Zobel J (2008) Rank–biased precision for measurement of retrieval effectiveness. ACM Transactions on Information Systems 27(1)
Müller H, Deselaers T, Deserno T, Kalpathy-Cramer J, Kim E, Hersh W (2008) Overview of the ImageCLEFmed 2007 medical retrieval and medical annotation tasks. Advances in Multilingual and Multimodal Information Retrieval. pp 472–491
Pearson WR (1995) Comparison of methods for searching protein sequence databases. Protein Science: A Publication of the Protein Society 4(6):1145
van Rijsbergen CJ (1979) Information retrieval. Butterworth-Heinemann Ltd., p 224. 0408709294
Rijsbergen CJV (1974) Foundation of evaluation. Journal of Documentation 30(4):365–373
Robertson SE (2006) On GMAP: and other transformations. In: Proceedings of the 15th ACM international conference on Information and knowledge management. ACM press, pp 78–83
Salton G (1968) Automatic information organization and retrieval. McGraw Hill Text

Sanderson M, Zobel J (2005) Information retrieval system evaluation: effort, sensitivity, and reliability. In: Proceedings of the 28th annual international ACM SIGIR conference on research and development in information retrieval. ACM press, pp 162–169

Sanderson M, Lestari Paramita M, Clough P, Kanoulas E (2010) Do user preferences and perfromance measures line up? In: Proceedings of the 33rd annual international ACM SIGIR conference on research and development in information retrieval. ACM press

Soboroff I (2004) On evaluating web search with very few relevant documents. In: Proceedings of the 27th annual international ACM SIGIR conference on research and development in information retrieval. ACM press, pp 530–531

Soboroff I (2006) Dynamic test collections: measuring search effectiveness on the live web. In: Proceedings of the 29th annual international ACM SIGIR conference on research and development in information retrieval. ACM press, pp 276–283

Stevens SS (1946) On the theory of scales of measurement. Science 103(2684):677–680

Swets JA (1963) Information retrieval systems. Science 141(3577):245–250

Tague–Sutcliffe JM, Blustein J (1994) A statistical analysis of the TREC–3 data. In: TREC Proceedings. NIST Special Publication. Department of Commerce, National Institute of Standards and Technology, pp 385–398

Thom JA, Scholer F (2007) A comparison of evaluation measures given how users perform on search tasks. In: The Twelfth Australasian Document Computing Symposium (ADCS 2007), pp 56–63

Verhoeff J, Goffman W, Belzer J (1961) Inefficiency of the use of boolean functions for information retrieval systems. Communications of the ACM 4(12):557–558

Voorhees EM (2005) Overview of the TREC 2004 robust retrieval track. In: TREC Proceedings. NIST Special Publication. Department of Commerce, National Institute of Standards and Technology, Gaithersburg, MD, USA

Yilmaz E, Aslam JA (2006) Estimating average precision with incomplete and imperfect judgments. In: Proceedings of the 15th ACM international conference on information and knowledge management. ACM press, pp 102–111

Yilmaz E, Robertson SE (2009) Learning to rank for information retrieval. In: Workshop in Conjunction with the ACM SIGIR conference on information retrieval. ACM press, Boston, MA, USA

Zhai CX, Cohen WW, Lafferty J (2003) Beyond independent relevance: methods and evaluation metrics for subtopic retrieval. In: Proceedings of the 26th annual international ACM SIGIR conference on research and development in information retrieval. ACM press, pp 10–17

Chapter 6
Fusion Techniques for Combining Textual and Visual Information Retrieval

Adrien Depeursinge and Henning Müller

Abstract This chapter describes several approaches for information fusion that have been used in ImageCLEF over the past seven years. In this context, the fusion of information is mainly meant to combine textual and visual retrieval. Data fusion techniques from 116 papers (62% of ImageCLEF working notes) are categorized, described and discussed. It was observed that three general approaches were used for retrieval that can be categorized based on the system level chosen for combining modalities: 1) at the input of the system with inter–media query expansion, 2) internally to the system with early fusion and 3) at the output of the system with late fusion which is by far the most widely used fusion strategy.

6.1 Introduction

Any concept with even a low level of semantics is best described by the co–occurrence of several events in multiple sources of information. In medicine for instance, diagnosis is established with confidence if, and only if, the laboratory results, the history of the patient and possibly radiographic examinations are all taken into account and converge to a unique conclusion. In another context, a photograph of a football game can be associated with its corresponding event only when the date and the place are known. Consequently, computerized Information Retrieval (IR) must be able to fuse multiple modalities in order to reach satisfactory performance. Information fusion has the potential of improving retrieval performance by relying on the assumption that the heterogeneity of multiple information sources

Adrien Depeursinge
University and University Hospitals of Geneva (HUG), Rue Gabrielle–Perret–Gentil 4, 1211 Geneva 14, Switzerland, e-mail: adrien.depeursinge@sim.hcuge.ch

Henning Müller
Business Information Systems, University of Applied Sciences Western Switzerland (HES–SO), TechnoArk 3, 3960 Sierre, Switzerland, e-mail: henning.mueller@hevs.ch

H. Müller et al. (eds.), *ImageCLEF*, The Information Retrieval Series 32,
DOI 10.1007/978-3-642-15181-1_6, © Springer-Verlag Berlin Heidelberg 2010

and/or algorithms allow cross–correction of some of the errors, leading to better results. Multiple views of the problem potentially allow a reduction of the semantic gap, which is defined in image retrieval as the discrepancy between the user's intentions when searching for a particular image and the visual information that the features are able to model (Smeulders et al, 2000).

Multi–modal information is often available in digital repositories. For example, videos are constituted by synchronized visual and audio modalities. Frequently, images on the Internet come with textual annotations that are semantically related. Modern health information systems enable access to structured information (e.g. age of the patient, gender, laboratory results), free–text in reports, radiological images and biosignals such as electrocardiograms. This means that the major challenge in information fusion is to find adapted techniques for federating multiple sources of information for either decision–making or information retrieval. Fusing multiple information sources is not devoid of risks. Two aspects require particular attention when performing information fusion in order to avoid degradation of the system performance:

- the relevance of all modalities to be fused must be verified to prevent the introduction of noise into the system;
- the fusion scheme must be able to assess trustworthiness of the modalities towards the query in order to allocate confidence in modalities that have high relevance in the context of the query.

Information fusion has been a lively research topic during the last 20 years (see, e.g. (Saracevic and Kantor, 1988; Belkin et al, 1993, 1994; Shaw and Fox, 1994)). Fusion was carried out at three different levels of an IR system (Frank Hsu and Taksa, 2005):

- at the input of the IR system while using multiple queries or query expansion;
- within the system where several algorithms and/or features can be used to increase the heterogeneity of results (i.e. boosting or multiple classifier systems);
- at the output of the system when combining several lists of documents.

Investigation of the effectiveness of combining text and images for retrieval including medical image retrieval is one of the main goals of the ImageCLEF campaign (Hersh et al (2007)). Since its first year in 2003, the organizers of ImageCLEF provided multimedia databases containing images with associated text thus allowing for multi–modal retrieval. During the past seven years of ImageCLEF, three image retrieval tasks elicited research contributions in fusion techniques for combining textual and visual information retrieval:

- the photo retrieval task proposed since 2003,
- the medical image retrieval task proposed since 2004,
- Wikipedia image retrieval task proposed since 2008.

In total, 116 (62%) out of 187 papers in ImageCLEF submissions from 2003 to 2009 attempted to mix Text–Based Image Retrieval (TBIR) with Content–Based Image Retrieval (CBIR) to investigate the complementarity of the two modalities (see Table 6.1).

Table 6.1: Number of papers per task and per year merging textual and visual information during the past seven years of ImageCLEF.

	2003	2004	2005	2006	2007	2008	2009
photo	0/4 (0%)	6/12 (50%)	6/11 (54%)	4/12 (33%)	14/19 (74%)	18/25 (72%)	12/16 (75%)
medical	–	6/11 (54%)	10/14 (71%)	8/10 (80%)	7/9 (78%)	6/11 (54%)	8/14 (57%)
Wikipedia	–	–	–	–	–	8/11 (73%)	3/8 (38%)

6.1.1 Information Fusion and Orthogonality

From a certain point of view, all systems that are using more than one single feature are carrying out information fusion. However, features within a modality may be strongly correlated among them (e.g. consecutive bins of a color histogram, see Depeursinge et al (2010–to appear)). As a consequence, the rank of the space spanned by the feature vector $\mathbf{v_A} = \{a_1 \ldots a_{N_A}\}$ of the modality A is usually much inferior the number of feature N_a of A. We have:

$$rank(A) \ll N_a. \tag{6.1}$$

While taking into account M modalities $\{A_1 \ldots A_M\}$ defined by their respective feature vectors $\{\mathbf{v_{A_1}} \ldots \mathbf{v_{A_M}}\}$, the linear dependence of multi–modal space is given by the number L of possible solutions $(x_1, x_2, \ldots, x_M)$ over all realizations of $\{\mathbf{v_{A_1}} \ldots \mathbf{v_{A_M}}\}$:

$$x_1\mathbf{v_{A_1}} + x_2\mathbf{v_{A_2}} + \cdots + x_M\mathbf{v_{A_M}} = \mathbf{0}, \tag{6.2}$$

with $x_1, x_2, \ldots, x_M \in \mathbb{R} \setminus 0$. Thereby, the amount of heterogeneity H of a combination of modalities can be measured using the number P of linearly independent vectors divided by the number of modalities M:

$$H = \frac{P}{M}. \tag{6.3}$$

H has values in $[0;1] \setminus 0$ and can be seen as the inverse of redundancy. It is important to note that large values of H would not be desirable as it means that no redundancy occurs in the set of modalities, which means that at least M-1 modalities are not related to any concept (or class). An ideal multi–modal system should be composed of modalities that are correlated for no other reason than that these are all related to a corpus of concepts. This was observed by Lee (1997) who stated that "different modalities might retrieve similar sets of relevant documents but retrieve different sets of non–relevant documents". This means that the information gain I_G (according to Quinlan (1986)) of the features from each modality towards the corpus of concepts must be above a critical threshold. I_G was originally defined by Quinlan to iteratively choose informative attributes to build decision trees. $I_G(Y|X)$ of a given attribute X with respect to the class attribute Y quantifies the change in information entropy when the value of X is revealed:

$$I_G(Y|X) = H(Y) - H(Y|X). \tag{6.4}$$

The information entropy $H(Y)$ measures the uncertainty about the value of Y and the conditional information entropy $H(Y|X)$ measures the uncertainty about the value of Y when the value of X is known:

$$H(Y) = -\sum_{y\in\mathscr{Y}} p(y)\log p(y), \tag{6.5}$$

$$H(Y|X) = -\sum_{x\in\mathscr{X},y\in\mathscr{Y}} p(x,y)\log p(y|x). \tag{6.6}$$

To summarize, an optimal multi–modal system should maximize the degree of heterogeneity H while maximizing the information gain I_G of each modality (taken independently) towards the studied corpus of classes.

6.2 Methods

The techniques used through the seven past years in ImageCLEF for fusing textual and visual image information were reviewed and categorized based on their similarities. Only papers that mixed textual and visual retrieval were studied and papers using multiple classifier systems on one single modality were left aside.

In total, techniques from 116 papers from 2004 to 2009 were categorized in the subsections of Section 6.3. An overview of the techniques and trends is presented. Justifications for the approaches and generally known problems are discussed in Section 6.4.

6.3 Results

The various techniques used for fusing textual and visual information in ImageCLEF are described in this section. When available, comparisons of the performances among techniques are detailed. A global view of the data fusion techniques is proposed in Section 6.3.5.

6.3.1 Early Fusion Approaches

An early fusion consists of mixing modalities before making any decisions. The combination takes place in the feature space where the textual and visual attributes

($\{t_1 \ldots t_k\}$ and $\{v_1 \ldots v_l\}$ respectively) are concatenated into one vector to create one unique feature space $\{t_1 \ldots t_k \quad v_1 \ldots v_l\}$ (see, e.g. (Snoek et al, 2005; Gunes and Piccardi, 2005; Depeursinge et al, 2010–to appear)). It enables a true multimedia representation where one decision rule is based on all information sources. The major drawback of this method is that it is confronted with the curse of dimensionality as the the dimension of the resulting feature space is equal to the sum of the dimensions of the subspaces $\mathbf{t}$ and $\mathbf{v}$. High–dimensional spaces tend to scatter the homogeneous clusters of instances belonging to the same concepts. This has to be handled using an appropriate feature weighting scheme, which is usually difficult to achieve in practice for complex multi-class problems where the majority of features are important to predict one particular class but introduce noise for all the other classes.

Early fusion is used without any feature weighting in Ferecatu and Sahbi (2008) in the photo retrieval task where text and visual features are simply normalized before being concatenated. A comparison with a late fusion method based on the combMIN rule (see Section 6.3.2.6) shows that the early fusion performs slightly better but without statistical significance.

Early fusion using various feature weighting schemes for medical image retrieval is investigated in (van Zaanen and de Croon, 2004; Deselaers et al, 2005; Cheng et al, 2005; Deselaers et al, 2006, 2007). Entropy–based feature weighting methods showed to outperform significantly performance obtained using a single modality in (Deselaers et al, 2006, 2007), which is in accordance with our assumptions in Section 6.1.1 as the information gain I_G is based on entropy measures (see Eq. 6.4).

A degradation of the retrieval performance is observed with the Wikipedia task in (Moulin et al, 2008) where a visual vocabulary is first created from basic image features, which is then fused with text features using a TF–IDF weighting (see (Salton and Buckley, 1988)).

In 2009, the best automatic mixed run of the medical task was based on early fusion of text features with very basic image features modeling color information of the whole image (Berber and Alpkoçak, 2009).

6.3.2 *Late Fusion Approaches*

Late fusion approaches concern every technique for combining outputs of distinct systems. The diversity among late fusion strategies is much broader than the early fusion approach and many techniques for combining lists of documents (runs) were used in ImageCLEF and are detailed in this section.

6.3.2.1 Rank–based Fusion vs. Score–based Fusion

When combining runs from different systems there are two main approaches. The relevance of a document d can be measured by either its rank $R_j(d)$ in the list $L_j(d)$

given by an IR system j or by its score $S_j(d)$ (or relevance, similarity, distance to the query). The score–based strategies, although more common, require a normalization among all systems in order to balance the importance of each of them, which is not the case of the rank–based strategies.

Several approaches are found in the literature for normalizing scores. A commonly used technique called MinMax was proposed by Lee (1997, 1995) where the normalized score $\overline{S}$ is computed as follows:

$$\overline{S} = \frac{S - S_{min}}{S_{max} - S_{min}}, \tag{6.7}$$

with S_{min} and S_{max} the lowest and highest scores found among all runs, systems or topics. Montague and Aslam (2001) also proposed two linear transformations for the normalization of scores: Sum and zero–mean and unit–variance ZMUV. Sum maps S_{min} to 0 and the sum of all scores to 1. In ZMUV, the average of all scores is mapped to 0 and their variance to 1. Sum and ZMUV are mostly intended to be used with the combination techniques combSUM and combMNZ respectively (see Sections 6.3.2.4 and 6.3.2.5).

6.3.2.2 Intersection of Runs

The most straightforward combination rule for multiple runs L_j is to intersect each other. The four combination operators used in ImageCLEF are defined as follows (see Villena-Román et al (2007b,a)):

$$\text{OR} \quad L_1 \cup L_2, \tag{6.8}$$

$$\text{AND} \quad L_1 \cap L_2, \tag{6.9}$$

$$\text{LEFT} \quad (L_1 \cup L_2) \cup (L_1 \setminus L_2), \tag{6.10}$$

$$\text{RIGHT} \quad (L_1 \cup L_2) \cup (L_2 \setminus L_1). \tag{6.11}$$

Usually these combination operators were associated with reordering rules (see Sections 6.3.2.3, 6.3.2.4, 6.3.2.5, 6.3.2.6 and 6.3.2.7). In Müller et al (2005), the union of runs (OR) is performed by adding various percentages of top textually– and visually–retrieved documents.

6.3.2.3 Reordering

When documents of various lists are gathered, a rule for reordering the documents is required to obtain a final ranking. In (Hoi et al, 2005; Florea et al, 2006; Gobeill et al, 2006; Fakeri-Tabrizi et al, 2008; Simpson et al, 2009; Mulhem et al, 2009; Besançon and Millet, 2005; Zhou et al, 2008a), the textually–retrieved documents are reordered based on their visual score. Inversely, visually–retrieved documents are reordered with their corresponding textual scores in (Villena-Román et al, 2005;

Gobeill et al, 2006; Clinchant et al, 2007; Chang and Chen, 2007; Jensen and Hersh, 2005; Daumke et al, 2006; Hersh et al, 2006; Granados et al, 2008; Ah-Pine et al, 2008, 2009). In Hare et al (2009); Gao and Lim (2009), a text run is reordered to maximize content–based distance among top images to favor the diversity of top–retrieved images.

6.3.2.4 Linear Combinations

In order to reorder documents based on both textual and visual scores S_t and S_v, a commonly used technique for obtaining the final score $S_{mixed}(d)$ of the document d is to perform a linear combination of scores as follows:

$$S_{mixed}(d) = \alpha S_t(d) + (1-\alpha)S_v(d), \tag{6.12}$$

where S_t and S_v are usually normalized and $\alpha \in [0;1]$. Linear combination of scores was used as defined by Equation 6.12 in a large number of papers (37% of the papers dealing with information fusion in ImageCLEF, (Cheng et al, 2004b,a; Müller et al, 2004; Alvarez et al, 2004; Besançon et al, 2004; Lin et al, 2004; Lim and Chevallet, 2005; Chang et al, 2005; Müller et al, 2005; Adriani and Framadhan, 2005; Ruiz and Southwick, 2005; Besançon and Millet, 2005; Díaz-Galiano et al, 2006; Rahman et al, 2006; Lacoste et al, 2006; Gobeill et al, 2006; Wilhelm and Eibl, 2006; Wilhelm et al, 2007; Maillot et al, 2006; Villena-Román et al, 2007b,a; Clinchant et al, 2007; Jair Escalante et al, 2007; Gao et al, 2007; Díaz-Galiano et al, 2007; Zhou et al, 2007; Hoi, 2007; Kalpathy-Cramer and Hersh, 2007; Yamauchi et al, 2008; Zhou et al, 2008a; Díaz-Galiano et al, 2008; Zhou et al, 2008b; Zhao and Glotin, 2008; Navarro et al, 2008c,b; O'Hare et al, 2008; Ah-Pine et al, 2008; Torjmen et al, 2008; Navarro et al, 2008d,a; Rácz et al, 2008; Ye et al, 2009; Ruiz, 2009; Torjmen et al, 2009; Boutsis and Kalamboukis, 2009; Daróczy et al, 2009; Mulhem et al, 2009; Zhou et al, 2009; Jair Escalante et al, 2009)).

Most often, arbitrary values are used for the weight α with usually more weight on textual scores as textual retrieval performs better than content–based retrieval, at least in terms of recall whereas CBIR tends to have higher early precision (see Müller et al (2008); Belkin et al (1994); Shaw and Fox (1994)). An exception was observed by Douze et al (2009) who obtained best results when applying a strong weight for the visual score.

Some groups used data from the previous year to learn weights (Ruiz, 2009). Järvelin et al (2007) computed the weights based on the variation of the modality towards the corpus of classes. In Rahman et al (2007), the weights are updated dynamically based on the user's relevance feedback. Document–specific weighting is used in Granados et al (2008, 2009) where weight of a document in the 'support' modality is divided by its rank.

In order to foster the modality with higher confidence, a linear combination of the scores is used only if both scores S_t and S_v are above a given threshold in Mulhem (2008); Broda et al (2009). The score of only one of the modalities is used otherwise.

In Zuccon et al (2009), text runs are reordered with a linear combination of text score and visual score based on factor analysis and bi–clustering to favor diversity among the retrieved images.

Linear combinations of ranks are much less frequently used, and were tried by Magalhães et al (2007); Jair Escalante et al (2008). Arithmetic and harmonic means of ranks are employed in Glotin and Zhao (2008). Linear combinations based on ranks have the advantage of not requiring a prior normalization. However, the assessment of confidence of the modalities is lost as two images having the same rank in both textual and visual modalities can have very different relevance towards the query.

CombSUM

A particular case of the linear combination is the combSUM rule where the scores of each modality j are summed to obtain the final score:

$$S_{mixed}(d) = \sum_{j=1}^{N_j} S_j(d), \tag{6.13}$$

with N_j the number of modalities to be combined. CombSUM is equivalent to a linear comb with $\alpha = 0.5$ if the scores are normalized. If not, the influence of each modality is strongly dependent on its scores.

CombSUM with scores was used in Jones et al (2004); Chevallet et al (2005); Martín-Valdivia et al (2005) and was used only once based on rank in El Demerdash et al (2007). Similarly to Mulhem (2008); Broda et al (2009), combSUM is applied if and only if the visual score is above a given threshold based on TF–IDF value for images annotations in Navarro et al (2008c,b,d,a, 2009).

Borda Count

The Borda count election method was developed in the political context in 1770 to create a ranked list of candidates. Each voter ranks all candidates and the sum of the ranks for all voters determines the score of each candidate from which a final ranking can be derived. This method was applied in information fusion in Ho et al (1994); van Erp and Schomaker (2000) and in ImageCLEF in Overell et al (2008). Borda count is strictly equivalent to combSUM on ranks.

6.3.2.5 CombMNZ

A variant of the combSUM method is the combMNZ combination rule which aims at giving more importance to the documents retrieved by several systems as follows (Shaw and Fox, 1994):

$$S_{mixed}(d) = F(d) \sum_{j=1}^{N_j} S_j(d), \tag{6.14}$$

where $F(d)$ is equal to the number of systems that retrieved d. CombMNZ was slightly modified by Inkpen et al (2008) for the photo retrieval task where a weight was applied to the normalized scores of each modality in order to control their respective influences.

6.3.2.6 CombMAX and CombMIN

Contrary to combSUM, the combMAX and combMIN rules put all their confidence in one single modality as follows:

$$\text{combMAX:} \qquad S_{mixed}(d) = \arg \max_{j=1:N_j} (S_j(d)), \tag{6.15}$$

$$\text{combMIN:} \qquad S_{mixed}(d) = \arg \min_{j=1:N_j} (S_j(d)). \tag{6.16}$$

CombMAX and combMIN were used both for photo and medical image retrieval by Besançon and Millet (2005); Chevallet et al (2005); Villena-Román et al (2007b,a) using normalized scores. CombMIN based on ranks was used in Ferecatu and Sahbi (2008) and is similar to combMAX based on score.

A hybrid rule based both combMAX and combMIN is proposed by Villena-Román et al (2007b,a):

$$S_{mixed}(d) = \text{combMAX}(S_j(d)) + \frac{\text{combMIN}^2(S_j(d))}{\text{combMAX}(S_j(d)) + \text{combMIN}(S_j(d))}. \tag{6.17}$$

It allows importance to be given to the minimum scores only if the latter has sufficiently high values.

6.3.2.7 CombPROD

The combPROD combination rule uses the product of scores to compute S_{mixed}:

$$S_{mixed}(d) = \prod_{j=1}^{N_j} (S_j(d)). \tag{6.18}$$

CombPROD favors documents with high scores in all modalities and was used for both photo and medical image retrieval by Martínez-Fernández et al (2004).

6.3.3 Inter–media Feedback with Query Expansion

The idea of query expansion is to modify the original query based on either available documents in the database or given rules (i.e. use of synonyms of query terms) with an aim of guessing the user's intentions. It was successfully applied to TREC[1] test collections in Belkin et al (1993), and Saracevic and Kantor (1988) states explicitly that taking into account the different results of the formulations could lead to retrieval performance better than that of any of the individual query formulations.

Query expansion was widely used in ImageCLEF and particularly for fusing textual and visual information where one modality provides a feedback to the other by means of query expansion, which is commonly called inter–media feedback in ImageCLEF (El Demerdash et al, 2009b).

6.3.3.1 Textual Query Expansion

Inter–media feedback query expansion is based on textual query expansion in most of the papers. Typically textual annotations from the top visually–ranked images (or from a mixed run) are used to expand a textual query (Ruiz and Srikanth, 2004; Müller et al, 2004; Besançon et al, 2004; Jones and McDonald, 2005; Chang et al, 2005; Maillot et al, 2006; Jair Escalante et al, 2007; Chang and Chen, 2007; Torjmen et al, 2007; Gao et al, 2007; Yamauchi et al, 2008; Gao et al, 2008; El Demerdash et al, 2008; Navarro et al, 2008c,b; Chang and Chen, 2008; El Demerdash et al, 2009a; Navarro et al, 2009).

Alternatively, text–based queries are built based on the automatically detected concepts present in the query image in Jair Escalante et al (2007); Tollari et al (2008); Inoue and Grover (2008); Popescu et al (2008).

In Kalpathy-Cramer et al (2008), the medical image modality (x–ray, computed tomography, etc.) is automatically detected from visual features and used as query expansion for text–based retrieval.

6.3.3.2 Visual Query Expansion

A less common approach for inter–media query expansion is proposed by Benczúr et al (2007), where the regions of images that are correlated with the title of the topic are used as visual queries with a CBIR engine.

[1] Text REtrieval Conference (TREC, `http://trec.nist.gov/`)

6.3.4 Other Approaches

Some of the techniques used in ImageCLEF for fusing textual and visual information do not correspond to any of the above–mentioned categories and proposed innovative approaches for merging information sources.

A simple approach is proposed by Radhouani et al (2009) who use visual features to detect the imaging modality in a first step. Then, images returned by a TBIR engine are filtered according to the modality of the query image.

A word–image ontology based on images retrieved by Google images using all nouns contained in the WordNet ontology is used by Chang and Chen (2006); Lacoste et al (2006). The textual query is mapped to a visual query based on the word–image ontology, which is then submitted to a CBIR system to obtain a final list of images.

Two innovative reordering methods based on ranks and applied to subgroups of documents are proposed by Myoupo et al (2009). In the first approach, the combSUM rule is iteratively applied on groups of documents within the lists, where groups are created using a sliding window consisting of groups N consecutive documents within each list. The second merging strategy is based on homogeneous blocks as follows: in the list of text retrieved documents, images are clustered according to their visual similarities to create blocks. Then, blocks are reordered among them according to their internal mean scores.

6.3.5 Overview of the Methods from 2004–2009

An overview of the main techniques and their interdependences is proposed in Figure 6.1. The late fusion techniques are most widely used and developed. The distribution of the various fusion approaches is detailed in Figure 6.2. It is important to note that some groups used a combination of the fusion techniques (see Maillot et al (2006)) and often research groups reused their techniques with slight modifications from one year to another and across tasks, which potentially exaggerates the trends in Figure 6.2.

6.4 Justification for the Approaches and Generally Known Problems

In this section, the justification of the methods, identified trends as well as lessons learned from seven years of multi–modal image retrieval are discussed.

Figures 6.1 and 6.2 clearly show three different choices of the system level for combining the modalities: at the input level with query expansion, internally with early fusion and at the output level with late fusion. Merging modalities at the input

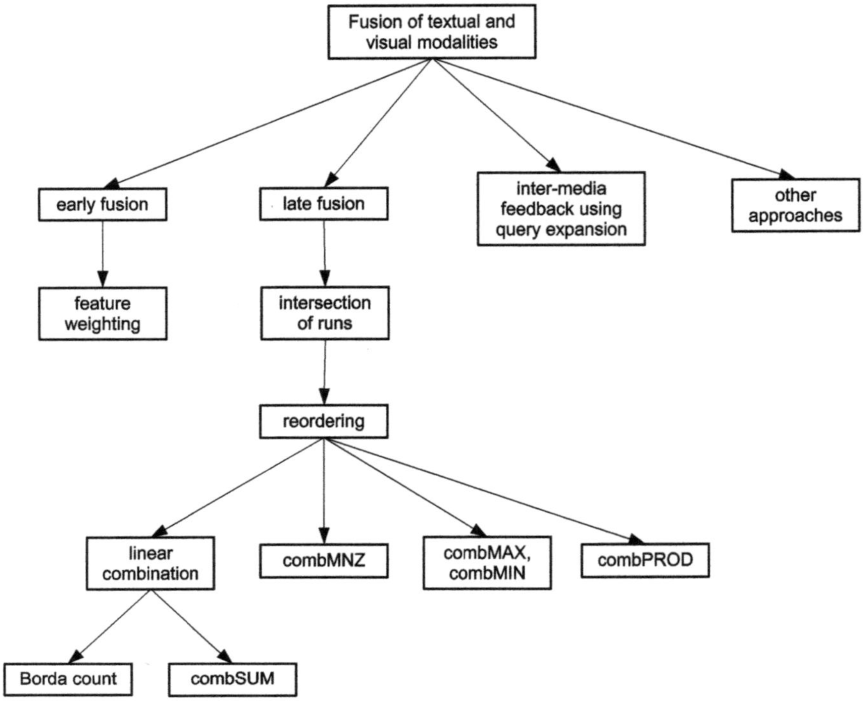

Fig. 6.1: Overview of the techniques.

level with query expansion techniques aims at improving the recall as the additional keywords (or query images) enable it to retrieve more potentially relevant images, but also involve the risk of proposing too many results to the user and thereby decreasing the precision. Early fusion enables a comprehensive overview of the multi–modal information by combining modalities inside the IR system and offers potentially high flexibility for promoting relevant modalities in the context of a particular query. Unfortunately, it is difficult to put into practice because it relies on large and heterogeneous feature spaces that become less distinctive, due to what is called the curse of dimensionality. Moreover, combining binary and categorical variable that are textual attributes with continuous and correlated visual features is not trivial and negative interactions among features can occur (see (Bell, 2003)). Consequently, it was shown to perform very well when textual features are combined with a small number of basic visual features such as in Berber and Alpkoçak (2009), which obtained best performance in last year's (2009) medical image retrieval task. Late fusion techniques are by far the most frequently utilized with more than 60% of the papers dealing with textual and information fusion. This is not surprising as late fusion allows for a straightforward combination of any system

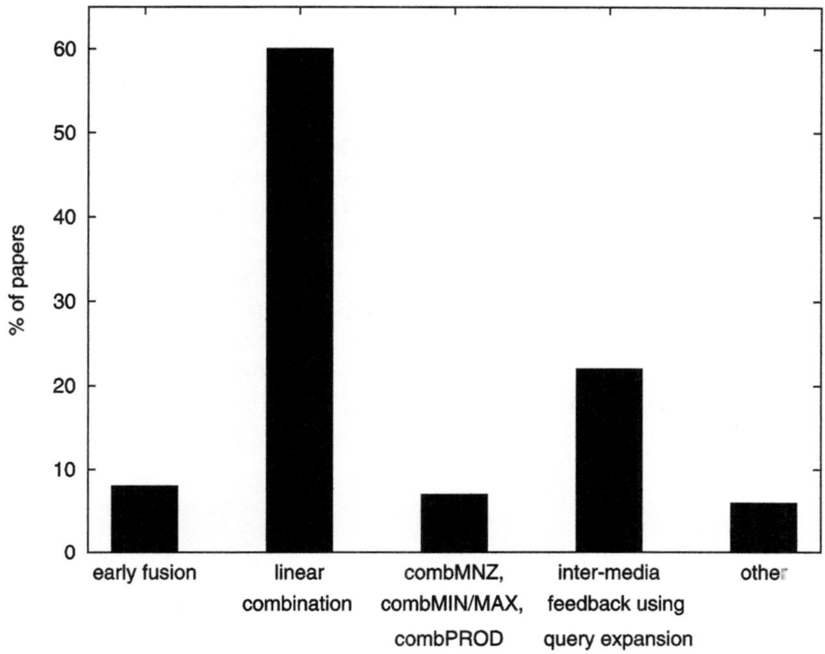

Fig. 6.2: Distribution of fusion approaches.

delivering a ranked list of documents. Most of the research groups focused on the performance of each independent system, which is a necessary condition to achieve high mixed performance.

When both TBIR and CBIR achieve acceptable performance, the choice of the fusion technique should rely on the analysis of the trends of each independent system as well as their complementarity and relevance to the image retrieval task (see (Zhou et al, 2010)). For instance, the combMAX combination rule favors the documents that are highly ranked in one system ('Dark Horse effect', (Vogt and Cottrell, 1999)) and is thus not robust to errors. On the other hand, combSUM and combMNZ favor the documents widely returned to minimize the errors ('Chorus effect') but relevant documents can obtain high ranks even if they are returned by few systems. Nevertheless, some of the approaches have fundamental limitations. This is the case with the linear combination using fixed weight for each document, as it puts blind confidence in one of the modalities and banishes the other one. This is not desirable as each modality usually behaves differently with each query and each set of documents. Consequently, late fusion techniques able to foster the modality with higher confidence are preferable as they allow the selection of the appropriate modality based on the query and the database. The idea of fostering the modality with confidence was found in various approaches such as combPROD or when linear combinations of scores are applied only if the scores of each modality are above a given threshold. Interestingly, Myoupo et al (2009) showed that the reordering of documents was much more adapted when carried out within subgroups of document instead of global reordering.

Several studies tried to enhance the diversity of the retrieved documents using mixed retrieval (see (Chang and Chen, 2008; Ah-Pine et al, 2008; Hare et al, 2009; Zuccon et al, 2009)), which was often based on cross–modality clustering (see (Arni et al, 2008; Lestari Paramita et al, 2009)). This was promoted by the organizers starting from 2008 for the photo retrieval task.

Finally, a quantitative comparison of the various fusion techniques was difficult to perform as the retrieval performance strongly depends on the performance of each independent IR system, which varied significantly among research groups. It was observed that mixed runs achieve better performance than single modalities in most of the cases. Most often, a degradation of performance is observed when the CBIR system achieves poor performances such as in Boutsis and Kalamboukis (2009).

6.5 Conclusions

In this chapter, the various approaches used during the past seven years in the ImageCLEF campaign were reviewed. Clear trends among techniques have been identified and discussed. A major observation is that CBIR systems have become mature enough to extract semantic information that is complementary to textual information, thus allowing enhancement of the quality of retrieval both in terms of precision and recall. However it was observed that combining textual and visual information is not devoid of risks and can degrade the retrieval performance if the fusion technique is not adapted to the information retrieval paradigm as well as to the TBIR and CBIR systems used. The key to using data fusion techniques is making the most of both textual and visual modalities.

Acknowledgements This work was supported by the Swiss National Science Foundation (FNS) with grant 200020–118638/1 and 205321–127355/9.

References

Adriani M, Framadhan A (2005) The University of Indonesia's participation in ImageCLEF 2005. In: Working Notes of CLEF 2005, Vienna, Austria

Ah-Pine J, Cifarelli C, Clinchant S, Renders JM (2008) XRCE's participation to ImageCLEF 2008. In: Working Notes of CLEF 2008, Aarhus, Denmark

Ah-Pine J, Clinchant S, Csurka G, Liu Y (2009) XRCE's participation in ImageCLEF 2009. In: Working Notes of CLEF 20098, Corfu, Greece

Alvarez C, Id Oumohmed A, Mignotte M, Nie JY (2004) Toward cross–language and cross–media image retrieval. In: Working Notes of CLEF 2004, Bath, England

Arni T, Clough PD, Sanderson M, Grubinger M (2008) Overview of the ImageCLEFphoto 2008 photographic retrieval task. In: Working Notes of CLEF 2008, pp 500–511

Belkin NJ, Cool C, Croft WB, Callan JP (1993) The effect of multiple query representations on information retrieval system performance. In: Korfhage R, Rasmussen EM, Willett P (eds) Proceedings of the 16st annual international ACM SIGIR conference on research and development in information retrieval. ACM press, Pittsburgh, PA, USA, pp 339–346

Belkin NJ, Kantor P, Cool C, Quatrain R (1994) Combining evidence for information retrieval. In: TREC–2: The Second Text REtrieval Conference, pp 35–44

Bell AJ (2003) The co–information lattice. In: Proceedings of the 4th international symposium on independent component analysis and blind signal separation (ICA2003). Springer, Nara, Japan, pp 921–926

Benczúr A, Bíró I, Brendel M, Csalogány K, Daróczy B, Siklósi D (2007) Cross–modal retrieval by text and image feature biclustering. In: Working Notes of CLEF 2007, Budapest, Hungary

Berber T, Alpkoçak A (2009) DEU at ImageCLEFmed 2009: Evaluating re–ranking and integrated retrieval model. In: Working Notes of CLEF 2009, Corfu, Greece

Besançon R, Millet C (2005) Merging results from different media: Lic2m experiments at ImageCLEF 2005. In: Working Notes of CLEF 2005, Vienna, Austria

Besançon R, Hède P, Moëllic PA, Fluhr C (2004) LIC2M experiments at ImageCLEF 2004. In: Working Notes of CLEF 2004, Bath, England

Boutsis I, Kalamboukis T (2009) Combined content–based and semantic image retrieval. In: Working Notes of CLEF 2009, Corfu, Greece

Broda B, Paradowski M, Kwanicka H (2009) Multimodal photo retrieval through finding similar documents enhanced with visual clues — a baseline method. In: Working Notes of CLEF 2009, Corfu, Greece

Chang YC, Chen HH (2006) Approaches of using a word–image ontology and an annotated image corpus as intermedia for cross–language image retrieval. In: Working Notes of CLEF 2006, Alicante, Spain

Chang YC, Chen HH (2007) Experiment for using web information to do query and document expansion. In: Working Notes of the CLEF 2007, Budapest, Hungary

Chang YC, Chen HH (2008) Increasing relevance and diversity in photo retrieval by result fusion. In: Working Notes of CLEF 2008, Aarhus, Denmark

Chang YC, Lin WC, Chen HH (2005) Combing text and image queries at ImageCLEF2005. In: Working Notes of CLEF 2005, Vienna, Austria

Cheng PC, Chien BC, Ke HR, Yang WP (2004) KIDS's evaluation in the medical image retrieval task at ImageCLEF 2004. In: Working Notes of CLEF 2004, Bath, England

Cheng PC, Yeh JY, Ke HR, Chien BC, Yang WP (2004) NCTU–ISU's evaluation for the user–centered search task at ImageCLEF 2004. In: Working Notes of CLEF 2004, Bath, England

Cheng PC, Chien BC, Ke HR, Yang WP (2005) NCTU_DBLAB@ImageCLEFmed 2005: Medical image retrieval task. In: Working Notes of CLEF 2005, Vienna, Austria

Chevallet JP, Lim JH, Radhouani S (2005) Using ontology dimensions and negative expansion to solve precise queries in the CLEF medical task. In: Working Notes of the CLEF 2005, Vienna, Austria

Clinchant S, Renders JM, Csurka G (2007) XRCE's participation to ImageCLEFphoto 2007. In: Working Notes of CLEF 2007, Budapest, Hungary

Daróczy B, Petrás I, Benczúr AA, Zsolt Fekete Z, Nemeskey D, Siklósi D, Weiner Z (2009) SZTAKI @ ImageCLEF 2009. In: Working Notes of CLEF 2009, Corfu, Greece

Daumke P, Paetzold J, Markó K (2006) Morphosaurus in ImageCLEF 2006: The effect of subwords on biomedical IR. In: Working Notes of CLEF 2006, Alicante, Spain

Depeursinge A, Racoceanu D, Iavindrasana J, Cohen G, Platon A, Poletti PA, Müller H (2010–to appear) Fusing visual and clinical information for lung tissue classification in HRCT data. Journal of Artificial Intelligence in Medicine

Deselaers T, Weyand T, Keysers D, Macherey W, Ney H (2005) FIRE in ImageCLEF 2005: Combining content–based image retrieval with textual information retrieval. In: Working Notes of the CLEF Workshop, Vienna, Austria

Deselaers T, Weyand T, Ney H (2006) Image retrieval and annotation using maximum entropy. In: Working Notes of CLEF 2006, Alicante, Spain

Deselaers T, Gass T, Weyand T, Ney H (2007) FIRE in ImageCLEF 2007. In: Working Notes of CLEF 2007, Budapest, Hungary

Díaz-Galiano MC, García-Cumbreras MA, Martín-Valdivia MT, Montejo-Raez A, Ureña-López LA (2006) SINAI at ImageCLEF 2006. In: Working Notes of CLEF 2006, Alicante, Spain

Díaz-Galiano MC, García-Cumbreras MA, Martín-Valdivia MT, Montejo-Raez A, Ureña-López LA (2007) SINAI at ImageCLEF 2007. In: Working Notes of CLEF2007, Budapest, Hungary

Díaz-Galiano MC, García-Cumbreras MA, Martín-Valdivia MT, Ureña-López LA, Montejo-Raez A (2008) SINAI at ImageCLEFmed 2008. In: Working Notes of CLEF 2008, Aarhus, Denmark

Douze M, Guillaumin M, Mensink T, Schmid C, Verbeek J (2009) INRIA–LEARs participation to ImageCLEF 2009. In: Working Notes of CLEF 2009, Corfu, Greece

El Demerdash O, Kosseim L, Bergler S (2007) Experiments with clustering the collection at ImageCLEF 2007. In: Working Notes of CLEF 2007, Budapest, Hungary

El Demerdash O, Kosseim L, Bergler S (2008) CLaC at ImageCLEFPhoto 2008. In: Working Notes of CLEF 2008, Aarhus, Denmark

El Demerdash O, Bergler S, Kosseim L (2009a) CLaC at ImageCLEF 2009. In: Working Notes of CLEF 2009, Corfu, Greece

El Demerdash O, Kosseim L, Bergler S (2009b) Image retrieval by inter–media fusion and pseudo–relevance feedback. In: Evaluating systems for multilingual and multimodal information access, pp 605–611

van Erp M, Schomaker L (2000) Variants of the borda count method for combining ranked classifier hypotheses. In: Seventh International Workshop on Frontiers in Handwriting Recognition, pp 443–452

Fakeri-Tabrizi A, Amini MR, Tollari S, Gallinari P (2008) UPMC/LIP6 at ImageCLEF's WikipediaMM: An image–annotation model for an image search–engine. In: Working Notes of CLEF 2008, Aarhus, Denmark

Ferecatu M, Sahbi H (2008) TELECOM ParisTech at ImageClefphoto 2008: Bi–modal text and image retrieval with diversity enhancement. In: Working Notes of CLEF 2008, Aarhus, Denmark

Florea F, Rogozan A, Cornea V, Bensrhair A, Darmoni S (2006) MedIC/CISMeF at ImageCLEF 2006: Image annotation and retrieval tasks. In: Working Notes of CLEF 2006, Alicante, Spain

Frank Hsu D, Taksa I (2005) Comparing rank and score combination methods for data fusion in information retrieval. Information Retrieval 8(3):449–480

Gao S, Lim JH (2009) I2R at ImageCLEF photo retrieval 2009. In: Working Notes of CLEF 2009, Corfu, Greece

Gao S, Chevallet JP, Le THD, Pham TT, Lim JH (2007) IPAL at ImageClef 2007 mixing features, models and knowledge. In: Working Notes of CLEF 2007, Budapest, Hungary

Gao S, Chevallet JP, Lim JH (2008) IPAL at CLEF 2008: Mixed–modality based image search, novelty based re–ranking and extended matching. In: Working Notes of CLEF 2008, Aarhus, Denmark

Glotin H, Zhao Z (2008) Affnity propagation promoting diversity in visuo–entropic and text features for CLEF Photo retrieval 2008 campaign. In: Working Notes of CLEF 2008, Aarhus, Denmark

Gobeill J, Müller H, Ruch P (2006) Query and document translation by automatic text categorization: A simple approach to establish a strong textual baseline for ImageCLEFmed 2006. In: Working Notes of CLEF 2006, Alicante, Spain

Granados R, Benavent X, García-Serrano A, Goñi JM (2008) MIRACLE–FI at ImageCLEFphoto 2008: Experiences in merging text–based and content–based retrievals. In: Working Notes of CLEF 2008, Aarhus, Denmark

Granados R, Benavent X, Agerri R, García-Serrano A, Goñi JM, Gomar J, de Ves E, Domingo J, Ayala G (2009) MIRACLE (FI) at ImageCLEFphoto 2009. In: Working Notes of CLEF 2009, Corfu, Greece

Gunes H, Piccardi M (2005) Affect recognition from face and body: early fusion vs. late fusion. In: 2005 IEEE International Conference on Systems, Man and Cybernetics, vol 4. IEEE Computer Society, Big Island, Hawaii, pp 3437–3443
Hare JS, Dupplaw DP, Lewis PH (2009) IAM@ImageCLEFphoto 2009: Experiments on maximising diversity using image features. In: Working Notes of CLEF 2009, Corfu, Greece
Hersh W, Kalpathy-Cramer J, Jensen J (2006) Medical image retrieval and automated annotation: OHSU at ImageCLEF 2006. In: Working Notes of CLEF 2006, Alicante, Spain
Hersh W, Kalpathy-Cramer J, Jensen J (2007) Medical image retrieval and automated annotation: OHSU at ImageCLEF 2006. In: Peters C, D. PC, Gey FC, Karlgren J, Magnini B, Oard DW, de Rijke M, Stempfhuber M (eds) CLEF. Lecture Notes in Computer Science (LNCS), vol 4730. Springer, pp 660–669
Ho TK, Hull JJ, Srihari SN (1994) Decision combination in multiple classifier systems. IEEE Transactions on Pattern Analysis and Machine Intelligence 16:66–75
Hoi SCH (2007) Cross–language and cross–media image retrieval: An empirical study at ImageCLEF2007. In: Working Notes of CLEF 2007, Budapest, Hungary
Hoi SCH, Zhu J, Lyu MR (2005) CUHK experiments with ImageCLEF 2005. In: Working Notes of CLEF 2005, Vienna, Austria
Inkpen D, Stogaitis M, DeGuire F, Alzghool M (2008) Clustering for photo retrieval at ImageCLEF 2008. In: Working Notes of CLEF 2008, Aarhus, Denmark
Inoue M, Grover P (2008) Effects of visual concept–based post–retrieval clustering in ImageCLEFphoto 2008. In: Working Notes of CLEF 2008, Aarhus, Denmark
Jair Escalante H, Hernández CA, López A, Marín HM, Montes M, Morales E, Sucar LE, Villaseñor L (2007) TIA–INAOE's participation at ImageCLEF 2007. In: Working Notes of CLEF 2007, Budapest, Hungary
Jair Escalante H, Gonzáles JA, Hernández CA, López A, Montes M, Morales E, Sucar LE, Villaseñor L (2008) TIA–INAOE's participation at ImageCLEF 2008. In: Working Notes of CLEF 2008, Aarhus, Denmark
Jair Escalante H, Gonzáles JA, Hernández CA, López A, Montes M, Morales E, Ruiz E, Sucar LE, Villaseñor L (2009) TIA–INAOE's participation at ImageCLEF 2009. In: Working Notes of CLEF 2009, Corfu, Greece
Järvelin A, Wilkins P, Adamek T, Airio E, Jones GJF, Smeaton AF, Sormunen E (2007) DCU and UTA at ImageCLEFphoto 2007. In: Working Notes of CLEF 2007, Budapest, Hungary
Jensen JR, Hersh WR (2005) Manual query modification and automatic translation to improve cross–language medical image retrieval. In: Working Notes of CLEF 2005, Vienna, Austria
Jones GJF, McDonald K (2005) Dublin City University at CLEF 2005: Experiments with the ImageCLEF St Andrew's collection. In: Working Notes of CLEF 2005, Vienna, Austria
Jones GJF, Groves D, Khasin A, Lam-Adesina A, Mellebeek B, Way A (2004) Dublin City University at CLEF 2004: Experiments with the ImageCLEF St Andrew's collection. In: Working Notes of CLEF 2004, Bath, England
Kalpathy-Cramer J, Hersh W (2007) Medical image retrieval and automatic annotation: OHSU at ImageCLEF 2007. In: Working Notes of CLEF 2007, Budapest, Hungary
Kalpathy-Cramer J, Bedrick S, Hatt W, Hersh W (2008) Multimodal medical image retrieval: OHSU at ImageCLEF 2008. In: Working Notes of CLEF 2008, Aarhus, Denmark
Lacoste C, Chevallet JP, Lim JH, Xiong W, Racoceanu D (2006) IPAL knowledge–based medical image retrieval in ImageCLEFmed 2006. In: Working Notes of CLEF 2006, Alicante, Spain
Lee JH (1995) Combining multiple evidence from different properties of weighting schemes. In: Proceedings of the 18th annual international ACM SIGIR conference on research and development in information retrieval. ACM press, pp 180–188
Lee JH (1997) Analyses of multiple evidence combination. In: Proceedings of the 20th annual international ACM SIGIR conference on research and development in information retrieval. ACM press, New York, NY, USA, pp 267–276
Lestari Paramita M, Sanderson M, Clough PD (2009) Diversity in photo retrieval: Overview of the ImageCLEFPhoto task 2009. In: Working Notes of CLEF 2009, Corfu, Greece

Lim JH, Chevallet JP (2005) A structured learning approach for medical image indexing and retrieval. In: Working Notes of CLEF 2005, Vienna, Austria

Lin WC, Chang YC, Chen HH (2004) From text to image: Generating visual query for image retrieval. In: Working Notes of CLEF 2004, Bath, England

Magalhães J, Overell S, Rüger S (2007) Exploring image, text and geographic evidences in ImageCLEF 2007. In: Working Notes of CLEF 2007, Budapest, Hungary

Maillot N, Chevallet JP, Valea V, Lim JH (2006) IPAL inter–media pseudo–relevance feedback approach to ImageCLEF 2006 photo retrieval. In: Working Notes of CLEF 2006, Alicante, Spain

Martín-Valdivia MT, García-Cumbreras MA, Díaz-Galiano MC, Ureña-López LA, Montejo-Raez A (2005) SINAI at ImageCLEF 2005. In: Working Notes of CLEF 2005, Vienna, Austria

Martínez-Fernández JL, Serrano AG, Villena-Román J, Sáenz VDM, Tortosa SG, Castagnone M, Alonso J (2004) MIRACLE at ImageCLEF 2004. In: Working Notes of CLEF 2004, Bath, England

Montague M, Aslam JA (2001) Relevance score normalization for metasearch. In: Proceedings of the tenth international conference on information and knowledge management. ACM press, pp 427–433

Moulin C, Barat C, Géry M, Ducottet C, Largeron C (2008) UJM at ImageCLEFwiki 2008. In: Working Notes of CLEF 2008, Aarhus, Denmark

Moulin C, Barat C, Lemaître C, Géry M, Ducottet C, Largeron C (2009) Combining text/image in WikipediaMM task 2009. In: Working Notes of CLEF 2009, Corfu, Greece

Mulhem P (2008) LIG at ImageCLEFphoto 2008. In: Working Notes of CLEF 2008, Aarhus, Denmark

Mulhem P, Chevallet JP, Quénot G, Al Batal R (2009) MRIM–LIG at ImageCLEF 2009: Photo retrieval and photo annotation tasks. In: Working Notes of CLEF 2009, Corfu, Greece

Müller H, Geissbuhler A, Ruch P (2004) Report on the CLEF experiment: Combining image and multi–lingual search for medical image retrieval. In: Working Notes of CLEF 2004, Bath, England

Müller H, Geissbuhler A, Marty J, Lovis C, Ruch P (2005) Using medGIFT and easyIR for the ImageCLEF 2005 evaluation tasks. In: Working Notes of CLEF 2005, Vienna, Austria

Müller H, Deselaers T, Kim E, Kalpathy-Cramer J, Deserno TM, Clough PD, Hersh W (2008) Overview of the ImageCLEFmed 2007 medical retrieval and annotation tasks. In: CLEF 2007 Proceedings. Lecture Notes in Computer Science (LNCS), vol 5152. Springer, Budapest, Hungary, pp 473–491

Myoupo D, Popescu A, Le Borgne H, Moëllic PA (2009) Visual reranking for image retrieval over the Wikipedia corpus. In: Working Notes of CLEF 2009, Corfu, Greece

Navarro S, Díaz MC, Muñoz R, García MA, Llopis F, Martín MT, Ureña-López LA, Montejo-Raez A (2008a) Text–mess in the medical retrieval ImageCLEF08 task. In: Working Notes of CLEF 2008, Aarhus, Denmark

Navarro S, García MA, Llopis F, Díaz MC, Muñoz R, Martín MT, Ureña-López LA, Montejo-Raez A (2008b) Text–mess in the ImageCLEFphoto task. In: Working Notes of CLEF 2008, Aarhus, Denmark

Navarro S, Llopis F, Muñoz R (2008c) Different multimodal approaches using IR–n in ImageCLEFphoto 2008. In: Working Notes of CLEF 2008, Aarhus, Denmark

Navarro S, Muñoz R, Llopis F (2008d) A multimodal approach to the medical retrieval task using IR–n. In: Working Notes of CLEF 2008, Aarhus, Denmark

Navarro S, Muñoz R, Llopis F (2009) Evaluating fusion techniques at different domains at ImageCLEF subtasks. In: Working Notes of CLEF 2009, Corfu, Greece

O'Hare N, Wilkins P, Gurrin C, Newman E, Jones GJF, Smeaton AF (2008) DCU at ImageCLEFPhoto 2008. In: Working Notes of CLEF 2008, Aarhus, Denmark

Overell S, Llorente A, Liu H, Hu R, Rae A, Zhu J, Song D, Rüger S (2008) MMIS at ImageCLEF 2008: Experiments combining different evidence sources. In: Working Notes of CLEF 2008, Aarhus, Denmark

Popescu A, Le Borgne H, Moëllic PA (2008) Conceptual image retrieval over the Wikipedia corpus. In: Working Notes of CLEF 2008, Aarhus, Denmark

Quinlan RJ (1986) Induction of decision trees. Machine Learning 1(1):81–106

Rácz S, Daróczy B, Siklósi D, Pereszlényi A, Brendel M, Benczúr A (2008) Increasing cluster recall of cross–modal image retrieval. In: Working Notes of CLEF 2008, Aarhus, Denmark

Radhouani S, Kalpathy-Cramer J, Bedrick S, Bakke B, Hersh W (2009) Multimodal medical image retrieval improving precision at ImageCLEF 2009. In: Working Notes of CLEF 2009, Corfu, Greece

Rahman MM, Sood V, Desai BC, Bhattacharya P (2006) CINDI at ImageCLEF 2006: Image retrieval & annotation tasks for the general photographic and medical image collections. In: Working Notes of CLEF 2006, Alicante, Spain

Rahman MM, Desai BC, Bhattacharya P (2007) Multi–modal interactive approach to ImageCLEF 2007 photographic and medical retrieval tasks by CINDI. In: Working Notes of CLEF 2007, Budapest, Hungary

Ruiz ME (2009) UNT at ImageCLEFmed 2009. In: Working Notes of CLEF 2009, Corfu, Greece

Ruiz ME, Southwick SB (2005) UB at CLEF 2005: Medical image retrieval task. In: Working Notes of CLEF 2005, Vienna, Austria

Ruiz ME, Srikanth M (2004) UB at CLEF2004: Part 2 — cross language medical image retrieval. In: Working Notes of CLEF 2004, Bath, England

Salton G, Buckley C (1988) Term weighting approaches in automatic text retrieval. Information Processing and Management 24(5):513–523

Saracevic T, Kantor P (1988) A study of information seeking and retrieving. ii. users, questions, and effectiveness. Journal of the American Society for Information Science 39:177–196

Shaw JA, Fox EA (1994) Combination of multiple searches. In: TREC–2: The Second Text REtrieval Conference, pp 243–252

Simpson M, Rahman MM, Demner-Fushman D, Antani S, Thoma GR (2009) Text– and content–based approaches to image retrieval for the ImageCLEF 2009 medical retrieval track. In: Working Notes of CLEF 2009, Corfu, Greece

Smeulders AWM, Worring M, Santini S, Gupta A, Jain R (2000) Content–based image retrieval at the end of the early years. IEEE Transactions on Pattern Analysis and Machine Intelligence 22(12):1349–1380

Snoek CGM, Worring M, Smeulders AWM (2005) Early versus late fusion in semantic video analysis. In: Proceedings of the 13th annual ACM international conference on Multimedia. ACM press, New York, NY, USA, pp 399–402

Tollari S, Detyniecki M, Fakeri-Tabrizi MR, Ali ad Amini, Gallinari P (2008) UPMC/LIP6 at ImageCLEFphoto 2008: on the exploitation of visual concepts (VCDT). In: Working Notes of CLEF 2008, Aarhus, Denmark

Torjmen M, Pinel-Sauvagnat K, Boughanem M (2007) Using pseudo–relevance feedback to improve image retrieval results. In: Working Notes of CLEF 2007, Budapest, Hungary

Torjmen M, Pinel-Sauvagnat K, Boughanem M (2008) Methods for combining content–based and textual–based approaches in medical image retrieval. In: Working Notes of CLEF 2008, Aarhus, Denmark

Villena-Román J, González-Cristóbal JC, Goñi-Menoyo JM, Martínez-Fernández JL, Fernández JJ (2005) MIRACLE's combination of visual and textual queries for medical images retrieval. In: Working Notes of CLEF 2005, Vienna, Austria

Villena-Román J, Lana-Serrano S, González-Cristóbal JC (2007a) MIRACLE at ImageCLEFmed 2007: Merging textual and visual strategies to improve medical image retrieval. In: Working Notes of CLEF 2007, Budapest, Hungary

Villena-Román J, Lana-Serrano S, Martínez-Fernández JL, González-Cristóbal JC (2007b) MIRACLE at ImageCLEFphoto 2007: Evaluation of merging strategies for multilingual and multimedia information retrieval. In: Working Notes of CLEF 2007, Budapest, Hungary

Vogt CC, Cottrell GW (1999) Fusion via a linear combination of scores. Information Retrieval 1(3):151–173

Wilhelm T, Eibl M (2006) ImageCLEF 2006 experiments at the Chemnitz Technical University. In: Working Notes of CLEF 2006, Alicante, Spain

Wilhelm T, Kürsten J, Eibl M (2007) Experiments for the ImageCLEF 2007 photographic retrieval task. In: Working Notes of CLEF 2007, Budapest, Hungary

Yamauchi K, Nomura T, Usui K, Kamoi Y, Eto M, Takagi T (2008) Meiji University at ImageCLEF2008 photo retrieval task: Evaluation of image retrieval methods integrating different media. In: Working Notes of CLEF 2008, Aarhus, Denmark

Ye Z, Huang X, Lin H (2009) Towards a better performance for medical image retrieval using an integrated approach. In: Working Notes of CLEF 2009, Corfu, Greece

van Zaanen M, de Croon G (2004) FINT: Find Images aNd Text. In: Working Notes of CLEF 2004, Bath, England

Zhao Z, Glotin H (2008) Concept content based Wikipedia WEB image retrieval using CLEF VCDT 2008. In: Working Notes of CLEF 2008, Aarhus, Denmark

Zhou X, Gobeill J, Ruch P, Müller H (2007) University and Hospitals of Geneva at ImageCLEF 2007. In: Working Notes of CLEF 2007, Budapest, Hungary

Zhou X, Gobeill J, Müller H (2008a) MedGIFT at ImageCLEF 2008. In: Working Notes of CLEF 2008, Aarhus, Denmark

Zhou X, Eggel I, Müller H (2009) The MedGIFT group at ImageCLEF 2009. In: Working Notes of CLEF 2009, Corfu, Greece

Zhou X, Depeursinge A, Müller H (2010) Information fusion for combining visual and textual image retrieval. In: Pattern Recognition, International Conference on. IEEE Computer Society, Istambul, Turkey

Zhou Z, Tian Y, Li Y, Liu T, Huang T, Gao W (2008b) PKU at ImageCLEF 2008: Experiments with query extension techniques for text–based and content–based image retrieval. In: Working Notes of CLEF 2008, Aarhus, Denmark

Zuccon G, Leelanupab T, Goyal A, Halvey M, Punitha P, Jose JM (2009) The University of Glasgow at ImageClefPhoto 2009. In: Working Notes of CLEF 2009, Corfu, Greece

Part II
Track Reports

Selected reports of ImageCLEF tracks.

Chapter 7
Interactive Image Retrieval

Jussi Karlgren and Julio Gonzalo

Abstract Information retrieval access research is based on evaluation as the main vehicle of research: benchmarking procedures are regularly pursued by all contributors to the field. But benchmarking is only one half of evaluation: to validate the results the evaluation must include the study of user behaviour while performing tasks for which the system under consideration is intended. Designing and performing such studies systematically on research systems is a challenge, breaking the mould on how benchmarking evaluation can be performed and how results can be perceived. This is the key research question of interactive information retrieval. The question of evaluation has also come to the fore through applications moving from exclusively treating topic–oriented text to including other media, most notably images. This development challenges many of the underlying assumptions of topical text retrieval, and requires new evaluation frameworks, not unrelated to the questions raised by interactive study. This chapter describes how the interactive track of the Cross–Language Evaluation Forum (iCLEF) has addressed some of those theoretical and practical challenges.

7.1 Interactive Studies in Information Retrieval

Information access research in general, whatever the media under consideration, is based on evaluation as the main vehicle of research. Evaluation of information retrieval systems is typically done with a test set of pre–assessed target documents used as a benchmark, under the assumptions that an *information need* can be formulated satisfactorily and appropriately; that documents can be assessed as being

Jussi Karlgren
SICS, Isafjordsgatan 22, 120 64 Kista, Sweden e-mail: jussi@sics.se

Julio Gonzalo
E.T.S.I. Informática de la UNED C/ Juan del Rosal, 16 28040 Madrid, Spain e-mail: julio@lsi.uned.es

H. Müller et al. (eds.), *ImageCLEF*, The Information Retrieval Series 32,
DOI 10.1007/978-3-642-15181-1_7, © Springer-Verlag Berlin Heidelberg 2010

relevant or not (or more or less relevant) for some given information need; that the relevance of a document with respect to that information need is *independent* of other documents in the collection, based solely on the qualities of that document. This abstracts the evaluation away from variation of factors such as task, situation, context, user preferences or characteristics, interaction design, network latency and other such system–external qualities, systematically and intentionally ignoring factors relating to human behaviour and human interaction with information systems.

Early information retrieval research posed questions beyond those concerned with the relation between immediate information need and documents: how the characteristics of the searcher, the task, the feedback, and system qualities all are parameters that information system design needs to take into account (Bennett, 1971, 1972). These considerations have not been put to rest — this discussion is very much still open. All of the basic assumptions of system evaluation can be shown as having problems, and current discussion in the field of interactive retrieval is busily discussing how future evaluation might proceed without relying on overly simple operationalisations of those assumptions (Ingwersen, 1992; Hearst, 1999; Järvelin and Ingwersen, 2005; Fuhr et al, 2009; Belkin et al, 2009; Kamps et al, 2009) that do not directly serve the goals of the underlying top–level objective, that of improving human access to information.

Many of these basic tenets of information retrieval and, more generally, information access, change when moving from the standard model of information retrieval, of retrieving topically focused text documents in an information access session focused on retrieval of documents in a timely fashion to address some specific and well–formulated information need. A major difficulty is understanding how language, which at first glance would seem to be a fairly precise representation of topical content, in fact is situation–specific and dynamic, and that this characteristic is pervasive and necessary for human communication and not something that in general can be avoided through judicious standardisation schemes. Another major difficulty is tracking and understanding usage over time, learning and adaptation on part of the user, and the specific characters of real–life tasks as factors influencing success or failure of interaction with a system.

Specifically, moving from text to other media will entail a necessary change and challenge with respect to formulation of information need; similarly, moving from monolingual to multi–lingual or cross–lingual information retrieval will change the way the system is able to match expressed information need to document content. Evaluating multi–lingual and cross–lingual information retrieval is a serious challenge in its own right, and has been a major topic both in the annual Text REtrieval Conference (TREC) evaluations for several years, in the annual Cross Language Evaluation Forum (CLEF) evaluation cycle, as well as in the related NII Test Collection for IR Systems (NTCIR) and the Forum for Information Retrieval Evaluation (FIRE) initiatives. In the CLEF evaluation campaigns, the interactive track has sought to address questions related to interactive access in multi–lingual target collections.

In general, studies of human behaviour are cumbersome to set up and administer — instructing test subjects and ensuring adequate volume, reliability, and repeatability of results has been a challenge for any interactive study. Ideally, human behaviour should be studied in the field, observing human behaviour in the wild; in practice, to be able to study some specific facet of behaviour, other variables must be fixed and some variables kept under test control — which usually is most convenient in a laboratory environment. Specifically, for information systems the challenge is twofold. Firstly, the coverage and breadth of the data source is one of the obviously important user satisfaction factors, the overhead effort of setting up a realistic test environment is a challenge in itself, and can seldom be practicably done for real–life tasks. This reduces most studies to mock–up or scaled–down environments. Secondly, the variation and breadth of information needs in the user population is immense, and variation in task characteristics may be greater than the variation caused by the variable under study. This tends to require most information retrieval studies to study simulated tasks — where test subjects are given a task to perform, seldom anchored in any real–life context.

7.2 iCLEF Experiments on Interactive Image Retrieval

CLEF has been devoted to the study of multi–lingual information access problems since its foundation in 2000. Before iCLEF 2001, the vast majority of multi–lingual information access research had focused on the automatic components of a system. From 2001 to 2009, the interactive track, iCLEF, focused on the problem of multi–lingual search assistance, i.e. on the interaction design aspects of multi–lingual retrieval. During these years, iCLEF has moved from the study of cross–lingual and multi–lingual retrieval of text to the study of retrieval of images, retaining the connection to the study of multi–linguality. This development was initiated for several reasons, but largely to ensure task realism. The objective of the track is to provide insights in realistic multi–lingual and cross–media information retrieval simultaneously — with the hope of bringing some results back to the field of text retrieval in the process.

The iCLEF track has during its years of operation addressed two main aspects of the multi–lingual access problem: (i) document selection and results exploration; and (ii) query formulation, refinement and translation. Both aspects have been addressed applied to various information access tasks: document retrieval, question answering and image retrieval; using various methodological perspectives: hypothesis–driven, observational studies, search log analysis; and considering different language competencies, i.e. different degrees of familiarity of the user with the target languages.

In the Encyclopedia of Library and Information Sciences, Douglas W. Oard points out that:

> 'Whether people can learn to formulate effective queries is at this point the [question about MLIA] we know the least about.'

(Oard, 2009). In the context of iCLEF, however, we have found some evidence that can be of help for the task of designing multi–lingual information access systems. One of the most basic outcomes of the iCLEF experiments is that support for user–assisted translation of the query improves search results. But the fact that user–assisted translation improves search results does not imply that this feature must be shown to the user by default. On the contrary, the results of most observational studies in iCLEF indicate that this can be annoying to users. Following the minimal cognitive effort principle, users are only interested in checking or modifying the system's query translation when things go wrong. Petrelli specifically addresses this question (Should the user check the query translation first?) and the answer is a clear 'no' (Petrelli et al, 2003). There seems to be, however, intermediate solutions between assisted query translation and full automatic translation that lead to better search results without imposing too much extra effort on the user.

7.2.1 iCLEF Image Retrieval Experiments: The Latin Square Phase

The first years of iCLEF experimentation were executed through a hypothesis–driven, within–subjects Latin–square based experimental design where a reference and a contrastive system are compared using prescribed combinations of system/user/topic to find system effects avoiding other types of dependencies (user, topic, user/system, etc.).

The first interactive image retrieval experiments were conducted jointly at the image retrieval track of CLEF and iCLEF in 2004 and 2005, following the same evaluation paradigm. The case of image retrieval is particularly strong for cross–language search, because the retrieved objects can often be used without the need for further translation, and yet it is often the case that search is — at least partially — based on matching between query words and the image textual metadata.

Some essential questions on the problem of interactive multi–lingual image retrieval are as follows (Gonzalo et al, 2006):

- How well a system supports user query formulation for images with associated texts (e.g. captions or metadata) written in a language different from the native language of the users. This is also an opportunity to study how the images themselves could also be used as part of the query formulation process.
- How well a system supports query reformulation, e.g. the support of positive and negative feedback to improve the user's search experience, and how this affects retrieval. This aims to address issues such as how visual and textual features can be combined for query reformulation and expansion.
- How well a system allows users to browse the image collection. This might include support for summarising results (e.g. grouping images by some pre–assigned categorisation scheme or by visual feature such as shape, colour or tex-

ture). Browsing becomes particularly important in a cross–lingual information retrieval system when query translation fails and returns irrelevant or no results.

- How well a system presents the retrieved results to the user to enable the selection of relevant images. This might include how the system presents the caption to the user (particularly if they are not familiar with the language of the text associated with the images, or some of the specific and colloquial language used in the caption) and investigates the relationship between the image and caption for retrieval purposes.

7.2.1.1 Experimental Procedure

Query reformulation and results presentation were the main focus of the experiments in 2004 and 2005.

Participants were required to compare two interactive cross–language image retrieval systems (one intended as a baseline) that differ in the facilities provided for interactive retrieval. For example, comparing the use of visual versus textual features in query formulation and refinement. As a cross–language image retrieval task, the initial query was required to be in a language different from the collection (i.e. not English) and translated into English for retrieval. Any text displayed to the user was also required to be translated into the user's source language. This might include captions, summaries, pre–defined image categories, etc.

The same search task was used in 2004 and 2005: given an image (not including the caption) from the St. Andrews collection of historic photographs (Reid, 1999), the goal for the searcher is to find the same image again using a cross–language image retrieval system. This models the situation in which a user searches with a specific image in mind (perhaps they have seen it before) but without knowing key information thereby requiring them to describe the image instead, e.g. searches for a familiar painting whose title and artist name are unknown (i.e. a high precision task or target search). While this is not necessarily the most common image search scenario, it is one of the situations where the visual components of the image necessarily play some role in the search; hopefully that emphasises the particular features introduced by the fact that the objects to be retrieved are not textual.

The user–centred search task required groups to recruit a minimum of eight users (native speakers in the source language) to complete 16 search tasks (eight per system). Users were given a maximum of five minutes only to find each image. Topics and systems were presented to the user in combinations following a Latin–square design to ensure minimisation of user/topic and system/topic interactions.

Participants were encouraged to make use of questionnaires to obtain feedback from the user about their level of satisfaction with the system and how useful the interfaces were for retrieval. To measure the effectiveness and efficiency with which a cross–language image retrieval search could be performed, participants were asked to submit the following information: whether or not the user could find the intended image (mandatory), the time taken to find the image (mandatory), the number of steps/iterations required to reach the solution (e.g. the number of clicks or the number of queries — optional), and the number of images displayed to the user (optional).

7.2.1.2 Participation and Results

Overall, four experiments were conducted in 2004 and 2005, coming from NCTU–ISU (Taiwan) and Michigan State University (USA) in 2004, and Miracle (from Madrid, Spain) and the University of Sheffield (UK) in 2005.

NCTU–ISU (Cheng et al, 2005) compared two search interfaces, one implementing text–only relevance feedback, and the other combining textual and visual information to help users find the target image. Their results show that the latter had a better performance, leading to the target images in less iterations.

Michigan State University presented work that was further revised in a SIGIR 2005 paper (Zhang et al, 2005). They compared a system where query refinements were done manually, with a system that provided term suggestions. Although they expected term suggestions to have a positive effect, their results indicated otherwise: success rate was lower (0.27 vs. 0.48) and average search time higher (2:41 vs. 1:41), with a similar average number of iterations. Interestingly, in terms of the ability to help users identify relevant keywords, the term feedback interface works better than the manual refinement interface; but it also provokes a tendency to select irrelevant terms which ultimately damage the average retrieval performance.

Zhang et. al. then performed search simulations, which led them to conclude that retrieval performance using term feedback depends on two factors: the term generation rate, and the term selection rate that measures users' responses to the prompted term list; and both are difficult to improve so that term feedback starts to give better results than manual refinement.

Petrelli and Clough (2006) tested an alternative visualisation of the search results. The proposal was to cluster the results into a hierarchy of text concepts. In spite of the fact that users claimed to prefer this alternative visualisation, the results showed that their performance was slightly more effective (in terms of number of target images successfully found) and more efficient (in terms of average time used) when using the simplest system. Remarkably, user perceptions and satisfaction do not always correlate to actual performance as measured extrinsically.

Miracle compared the same interface but using Spanish (European) versus English versions (Villena et al, 2006). The focus of the experiment was to find whether it is better to use an AND operator to group terms of multi–word queries (in the English system) or combine terms using an OR operator (in the Spanish system). Their aim was to compare whether it is better to use English queries with terms conjuncted (which have to be precise and use the exact vocabulary — may be difficult for a specialised domain such as historical Scottish photographs) or to use the disjunction of terms in Spanish and have the option of relevance feedback (a more fuzzy and noisy search but which does not require precise vocabulary and exact translations). Results were similar for both systems evaluated, although a number of interesting points were made, including: (i) domain–specific terminology causes problems for cross–language searches, and therefore impacts far more on queries with a conjunction of terms; (ii) from questionnaires, users preferred the English version because the conjunction of terms often gave results users could interpret (i.e. a set of documents containing all query terms).

7.2.2 *iCLEF Experiments with Flickr*

7.2.2.1 Exploring the Task Space: iCLEF 2006

Although iCLEF produced a substantial body of knowledge (probably the largest hitherto on the topic of interactive Cross–Language Retrieval) between 2001 and 2005, there were a few limitations in our experimental set up.

- The search task itself was unrealistic: news collections are comparable across languages, and most information tends to be available in the user's native language; therefore why would a user ever want to search for this information in an unknown language? If we translate this problem to the Web, it would be like asking a Spanish speaker to look for information about Norah Jones (the singer) in English, in spite of the fact that there are over 150,000 pages in Spanish about her[1].
- Relevance does not cover all aspects that make an interactive search session successful. Our observational studies indicated that one can get higher user satisfaction even when this does not correspond with higher search success.
- The Latin–square design imposed heavy constraints on the experiments, making them costly and with a limited validity (the number of users was necessary limited, and statistically significant differences were hard to obtain).

In order to overcome these limitations, we decided to propose a new framework for iCLEF with two essential features:

- Using `http://www.flickr.com/` (the popular photo sharing service) as the target collection. Flickr is a large–scale, Web–based image database serving a large social network of Web users. It has the potential to offer both challenging and realistic multi–lingual search tasks for interactive experiments.
- The iCLEF track should provide means to explore alternative evaluation methodologies for interactive information access. For this reason, we decided to fix the search tasks, but to keep the evaluation methodology open. This would allow each participant to contribute with their own ideas about how to study interactive issues in cross–lingual information access.

Since this decision was made, iCLEF has been focused on the image retrieval problem, which is perhaps the most natural and frequent for real–world multi–lingual search needs.

[1] `http://www.google.com/` results as of 11 August 2006.

7.2.3 The Target Collection: Flickr

The majority of Web image search is text–based, and the success of such approaches often depends on reliably identifying relevant text associated with a particular image. Flickr is an on–line tool for managing and sharing personal photographs, and currently contains over four billion freely accessible images[2]. These are updated daily by a large number of registered users and available to all Web users.

Flickr provides both private and public image storage, and photos which are shared can be protected under a Creative Commons (CC) licensing agreement (an alternative to full copyright). Images from a wide variety of topics can be accessed through Flickr, including people, places, landscapes, objects, animals, events, etc. This makes the collection a rich resource for image retrieval research.

There were two possibilities to use the collection: reaching an agreement with Flickr to get a subset of their database, or simply using Flickr's public Application Programming Interface (API) to interact with the full database. The first option is more attractive from the point of view of system design, because it is possible to obtain collection statistics to enhance the search (for instance, tf–idf weights, term suggestions and term translations adapted to the collection) and because it gives total control on the search mechanism. A crucial advantage of having the collection locally stored is enabling the possibility of doing content–based retrieval.

The second option, while more restricted, is also attractive, because users can interact with a larger and more up–to–date database. As the possibility of reaching an agreement with Flickr did not materialise, the second option was finally the only choice available for iCLEF experiments.

One problem of using the full Flickr collection is that it keeps constantly growing, making experiments impossible to replicate. Therefore, we decided to use images uploaded before 21 June 2006 (immediately before the first iCLEF experiments began). The target collection is, then, relatively stable (only removed images may alter the collection, but this happened very rarely).

7.2.4 Annotations

In Flickr, photos are annotated by authors with freely chosen keywords (tags) in a naturally multi–lingual manner: most authors use keywords in their native language; some combine more than one language. User tags may describe anything related to the picture, including themes, places, colours, textures and even technical aspects on how the photograph was taken (camera, lens, etc.). Some tags become naturally standardised among subsets of Flickr users, in a typical process of so–called folksonomies (Mathes, 2004). In addition, photographs have titles, descriptions, collaborative annotations, and comments in many languages. Figure 7.1 provides an example photo with multi–lingual annotations.

---

[2] As of August, 2006.

Fig. 7.1: An example Flickr image, with title, description, classified in three sets (user–defined) and three pools (community shared), and annotated with more than 15 English, Spanish and Portuguese tags.

7.2.5 The Task

As an iCLEF task, searching for images in Flickr presents interesting challenges:

- Different types of associated text, e.g. tags, titles, comments and description fields.
- Collective classification and annotation using freely selected keywords (folksonomies) resulting in non–uniform and subjective categorisation of images.

Fig. 7.2: Visually oriented task: *what is the name of the beach where this crab is resting?*

- Fully multi–lingual image annotation, with all widely–spoken languages represented and mixed up in the collection.

The experiment design in 2006 consisted of three search tasks, where users could employ a maximum of twenty minutes per task. We chose three tasks of a very different nature:

- Topical ad hoc retrieval: *Find as many European parliament buildings as possible, pictures from the assembly hall as well as from the outside.*
- Creative open–ended retrieval: *Find five illustrations to the article "The story of saffron"* (a seven paragraph story about saffron growing in Abruzzo, Italy).
- Visually oriented task: *What is the name of the beach where this crab is resting?* (together with a picture of a crab lying in the sand, see Figure 7.2). The name of the beach is included in the Flickr description of the photograph, so the task is basically finding the photograph, which is annotated in German (a fact that the users ignore).

All tasks could benefit from a multi–lingual search: Flickr has photographs of European parliament buildings described in many languages, photographs about the Abruzzo area and saffron are only annotated in certain languages, and the crab photograph can only be found with German terms. At the same time, the nature of each task is different from the others. The European parliaments topic is biased towards recall, and one can expect users to stop searching only when the twenty minutes expire. The text illustration task only demands five photographs, but it is quite open–ended and very much depending on the taste and subjective approach of each user; we expected the search strategies to be more diverse here. Finally, the 'find the crab' task is more of a known–item retrieval task, where the image is presumed to be annotated in a foreign language, but the user does not know which one; the need for cross–language search and visual description is therefore greater.

Given that part of the experience consisted of proposing new ways of evaluating interactive cross–language search, we did not prescribe any fixed procedure or measure for the task.

To lower the cost of participation, we provided an AJAX–based basic multilingual interface to Flickr, which every participant could use as a basis to build their systems.

7.2.6 Experiments

Fourteen research groups officially signed in for the task, more than in any previous iCLEF edition. However, only the three organising teams (SICS, U. Sheffield and UNED) submitted results (the worst rate in iCLEF campaigns). Perhaps the fact that the task was only partially thought through made it less appropriate as a CLEF event, where teams are usually rushing off to meet deadlines and need crystal clear guidelines. However, the exercise was successful in terms of paving the way for alternative ways of approaching interactive image retrieval evaluation.

UNED (Artiles et al, 2007) measured the attitude of users towards cross–language searching when the search system provides the possibility (as an option) of searching cross–language, using a system which allowed for three search modes: 'no translation', 'automatic translation' (the users choose the source language and the target languages, and the system chooses a translation for every word in the query) and 'assisted translation' (like the previous mode, but now the user can change the translation choices made by the system). Their results over 22 users indicate that users tend to avoid translating their query into unknown languages, even when the results are images that can be judged visually.

U. Sheffield and IBM (Clough et al, 2007) experimented with providing an Arabic interface to Flickr, using an Arabic–English dictionary as an initial translation step, followed by the use of Babelfish to translate into the experiment additional target languages (French, German, Dutch, Italian and Spanish). Users were able to modify the English translation if they had the necessary language skills. With a user group of bilingual Arabic–English users it was found that they: (i) preferred to query in English, although liked having the option of formulating initial queries in Arabic; (ii) found viewing photos with results in multiple languages more helpful than the initial query translation step.

SICS (Karlgren and Olsson, 2007) focused on user satisfaction and confidence as target measures for evaluation. Users were given the tasks, and after some time were given a terminological display of terms they had made use of, together with related terms. This enabled them to broaden their queries: success was not measured in retrieval results but in changes of self–reported satisfaction and confidence as related to the pick–up of displayed terms by the user (Karlgren and Sahlgren, 2005).

7.2.6.1 Search Log Analysis: iCLEF 2008–2009

In 2008 and 2009, iCLEF completed the transition from hypothesis–based experimentation to observational studies, basing the experiments on search log analysis.

The main novelty of the iCLEF 2008/2009 experience has been to focus on the shared analysis of a large search log from a single search interface provided by the iCLEF organisers. The focus was, therefore, on search log analysis rather than on system design. The idea was to study the behaviour of users in an (almost) naturalistic search scenario, having a much larger data set than in previous iCLEF campaigns. The search interface provided by the iCLEF organisers was a basic cross-language retrieval system to access images in Flickr, presented as an on–line game: the user is given an image, and she must find it again without any a–priori knowledge of the language(s) in which the image is annotated. Note that the task is similar to that used in 2004/2005, although the target collection and its degree of multi–linguality is drastically different. Game–like features were intended to engage casual users and therefore increase the chances of achieving a large, representative search log.

7.2.6.2 Search Task Definition

Our primary goal was harvesting a large search log of users performing multi–lingual searches on the Flickr database. Rather than recruiting users (which inevitably leads to small populations), we wanted to publicise the task and attract as many users as possible from all around the world, and engage them in searching. To reach this goal, we needed to observe some restrictions:

- The search task should be clear and simple, requiring no a–priori training or reading for the casual user.
- The search task should be engaging and addictive. Making it an on–line game — with a rank of users — helps achieve that, with the rank providing a clear indication of success.
- It should have an adaptive level of difficulty to prevent novice users from being discouraged, and to prevent advanced users from being unchallenged.
- The task should be naturally multi–lingual.

We decided to adopt a known–item retrieval search task: the user is given a raw (unannotated) image and the goal is to find the image again in the Flickr database, using a multi–lingual search interface provided by the iCLEF organisers. The user does not know in advance in which languages the image is annotated; therefore searching in multiple languages is essential to get optimal results.

The task is organised as an on–line game: the more images found, the higher a user is ranked. Depending on the image, the source and target languages, this can be a very challenging task. To have an adaptive level of difficulty, we implemented a hints mechanism. At any time while searching, the user is allowed to quit the search (skip to next image) or ask for a hint. The first hint is always the target language (and therefore the search becomes mono or bilingual as opposed to multi–lingual). The rest of the hints are keywords used to annotate the image. Each image found scores 25 points, but for every hint requested, there is a penalty of five points.

Initially a five minute time limit per image was considered, but initial testing indicated that such a limitation was not natural and changed users' search behaviour. Therefore we decided to remove time restrictions from the task definition.

7.2.6.3 Search interface

We designed the *Flickling* interface to provide a basic cross-language search front–end to Flickr. Flickling is described in detail in (Peinado et al, 2009a); we will only summarise its basic functionalities here:

- User registration, which records the user's native language and language skills in each of the six European languages considered (EN, ES, IT, DE, NL, FR).
- Localisation of the interface in all six languages[3].
- Two search modes: mono and multi–lingual. The latter takes the query in one language and returns search results in up to six languages, by launching a full Boolean query to the Flickr search API.
- Cross–language search is performed via term–to–term translations between six languages using free dictionaries (taken from: `http://xdxf.revdanica.com/down`).
- A term–to–term automatic translation facility which selects the best target translations according to: (i) string similarity between the source and target words; (ii) presence of the candidate translation in the suggested terms offered by Flickr for the whole query; and (iii) user translation preferences.
- A query translation assistant that allows users to pick/remove translations, and add their own translations (which go into a personal dictionary). We did not provide back–translations to support this process, in order to study correlations between target language abilities (active, passive, none) and selection of translations.
- A query refinement assistant that allows users to refine or modify their query with terms suggested by Flickr and terms extracted from the image rank. When the term is in a foreign language, the assistant tries to display translations in the user's preferred language to facilitate feedback.

[3] Thanks go to the CLEF groups at the U. of Amsterdam, U. of Hildesheim, ELDA and CNR for providing native translations of the interface texts.

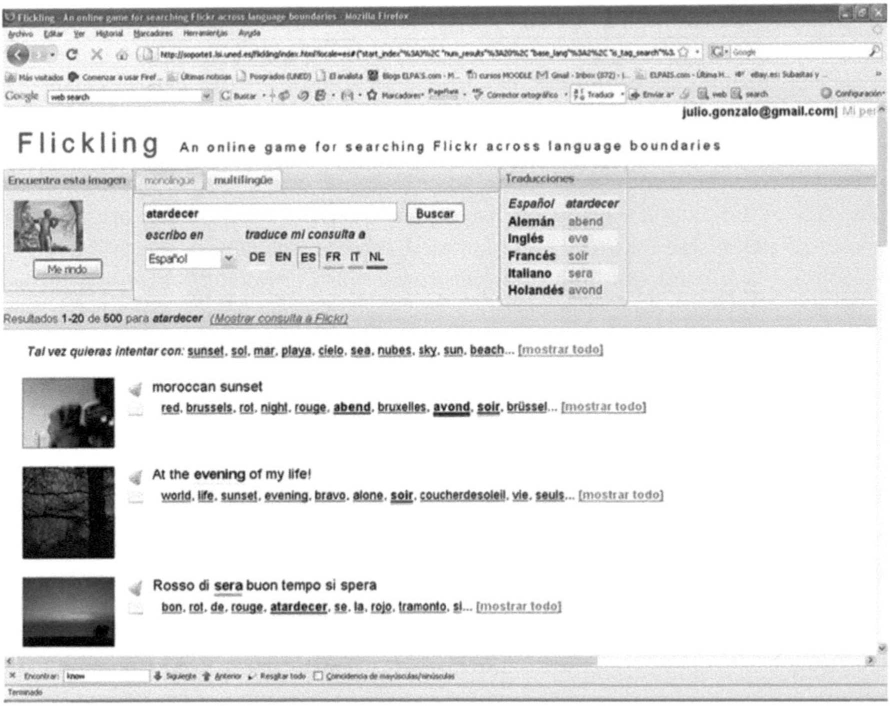

Fig. 7.3: The Flickling search interface used to harvest search logs.

- Control of the game–like features of the task: user registration and user profiles, groups, ordering of images, recording of session logs and access to the hall of fame.
- Post–search questionnaires (launched after each image is found or the search failed) and final questionnaires (launched after the user has searched 15 images, not necessarily at the end of the experience).

Figure 7.3 shows a snapshot of the search interface. Note that we did not intend to provide the best possible cross–language assistance to search the Flickr collection. As we wanted to focus on user behaviour — rather than on hypothesis testing for a particular interactive facility — our intention was to provide a standard, baseline interface that is not dependent on a particular approach to cross-language search assistance.

7.2.6.4 Participation Instructions

Participants in iCLEF 2008/2009 could essentially perform two tasks: (1) analyse log files based on all participating users (which is the default option) and, (2) execute their own interactive experiments with the interface provided by the organisers.

Generation of search logs: Participants can mine data from the search session logs, for example looking for differences in search behaviour according to language skills, or correlations between search success and search strategies.

Interactive experiments: Participants can recruit their own users and conduct their own experiments with the interface. For instance, they could recruit a set of users with passive language abilities and another with active abilities in certain languages and, besides studying the search logs, they could perform observational studies on how they search, conduct interviews, etc. The iCLEF organisers provided assistance with defining appropriate user groups and image lists, e.g. within the common search interface.

7.2.6.5 Data set: Flickling Search Logs

Search logs were harvested from the Flickling search interface in periods of approximately two months: one in May–June 2008 for the first campaign, and a similar one in 2009 for the last campaign.

Dissemination was successful: during the first log harvesting period, the interface was visited by users from 40 different countries from Europe, the Americas, Asia and Oceania. More than 300 people registered (around 230 were active searchers) and 104 performed searches for at least ten different images. Out of them, 18 users attempted all 103 images considered for the task. Apart from general users, the group affiliation revealed at least three user profiles: researchers in information retrieval, linguistics students (most from the University of Padova) and photography fans (many entering from a Spanish blog specialising in photography[4]).

Profiles of user's language skills were very diverse, with a wide range of native and second language abilities. There was a total of 5,101 complete search sessions (i.e. a user starts searching for an image and either finds the image or gives up), out of which the image was annotated in an active language (for the user) in 2,809 cases, in an unknown language in 1,566 cases, and in a passive language (when the user can partially read but cannot write) in 726 cases. Note that, even when the image is annotated in an active language for the user, this is not known by the user a priori, and therefore the search behaviour is equally multi–lingual.

On average each search session included around four queries launched in the monolingual search mode, and four queries in the multi–lingual search mode. Overall, it was possible to collect a large controlled multi–lingual search log, which includes both search behaviour (interactions with the system) and users' subjective

[4] http://dzoom.org.es/

Table 7.1: Statistics of iCLEF 2008/2009 search logs.

	2008	2009
subjects	305	130
log lines	1,483,806	617,947
target images	103	132
valid search sessions	5,101	2,410
successful sessions	4,033	2,149
unsuccessful sessions	1,068	261
hints asked	11,044	5,805
queries in monolingual mode	37,125	13,037
queries in multi-lingual mode	36,504	17,872
manually promoted translations	584	725
manually penalised translations	215	353
image descriptions inspected	418	100

impressions of the system (via questionnaires). This offers a rich source of information for helping to understand multi–lingual search characteristics from a user's perspective. A reusable data source was produced for the first time since iCLEF first began.

One problem of the first search log was that many images were annotated in English (often in addition to a different primary language), and that led to a corpus where in half of the search sessions the user had active language skills in at least one of the target languages used to describe the query. So for 2009 we decided to refine the selection of target images, excluding those which had key information in English. As a result, in most search sessions of the 2009 search log the image is annotated in a language unknown to the user. That makes the 2009 log complementary to the first one.

Overall, the logs collected and released during the iCLEF 2008 and 2009 campaigns contain more than two million lines. Table 7.1 summarises the most relevant statistics of both search logs.

7.2.6.6 Participation and Findings

There were two main types of contributions to the task: (i) groups that analysed the search logs, and (ii) groups that recruited their own set of users. These are some of the main findings:

UNED examined the effects of searcher competence in the target language and system learning effects, studying the logs and examining user responses to the questionnaires given to users at the completion of each completed or aborted task (Peinado et al, 2009b). Analyses showed that when users had competence in the target language, their success at searching was higher; with passive knowledge, user interaction showed similar success to those with active competence, but requiring more interactions with the system. Finally, users with no competence in the target language found less images and with a higher cognitive effort. In 2009,

UNED Peinado et al (2010) focused on discovering successful search strategies when searching in a foreign, unknown language. They found that the usage of cross–language search assistance features has an impact on search success, and that such features are highly appreciated by users.

SICS investigated methods for how to study the confidence and satisfaction of users. In 2009, some preliminary studies of the number of reformulations versus success rate were performed. The SICS team found that the length of query sequences which eventually were successful were longer, indicating persistence when a search appears to be in the right direction. The number of query reformulations also correlate well with success: successful query sequences are a result of active exploration of the query space. However, for users who persist in working with monolingual searches, the SICS team found that queries, firstly tended to be vastly less often reformulated to begin with, and that the successful sequences were more parsimonious than the failed ones; instead, the number of scroll actions were higher. This would seem to indicate that if users are fairly confident of a well put query, they will persist by scrolling through result lists.

Manchester Metropolitan University (MMU) limited their studies to users recruited and observed in the laboratory, instead of considering the whole search log. In 2008, MMU studied how users considered language and cross–linguistic issues during a session and how they switched between the cross–lingual and mono–lingual interfaces. This was done through think–aloud protocols, observation, and interviews of users engaged in search tasks (Vassilakaki et al, 2009). Their main finding is that their users (who were native or near–native English speakers) did not make significant use of the cross–lingual functionalities of the system, nor did they think about language aspects when searching for an image. Comparing with experiments run at other sites, this seems to indicate that English speakers have a different attitude towards multi–lingual search, tending to assume that everything must be findable with English.

The submission from the **University of Westminster** explored user interaction with the facility provided by Flickling to add user–specific translation terms (Tanase and Kapetanios, 2009). By exploring the user's perceived language skills and usage of the personal dictionary feature, experiments demonstrated that even with modest language skills, users were interacting with and using the dictionary–edit feature.

In 2009, the **University of Alicante** (Navarro et al, 2010) investigated whether there is a correlation between lexical ambiguity in queries and search success and, if so, whether explicit word sense disambiguation can potentially solve the problem. To do so, they mined data from the search log distributed by the iCLEF organisation, and found that less ambiguous queries lead to better search results and that coarse-grained Word Sense Disambiguation might be helpful in the process.

The **University of North Texas** (Ruiz and Chin, 2010) aimed at understanding the challenges that users face when searching for images that have multi–lingual annotations, and how they cope with these challenges to find the information they need. Similarly to MMU, instead of using the search log this group recruited their own set of six north American students and studied their search behaviour and subjective impressions using questionnaires, training, interviews and observational analysis. They found that users have considerable difficulties using Flickr tags, particularly when peforming cross–language searches, and that their typical session requires two hints: the target language and a keyword.

7.2.6.7 Directions

The search logs generated by the iCLEF track in 2008 and 2009 together are a reusable resource (the first generated in an interactive cross–lingual retrieval setting) for future user–orientated studies of cross–language search behaviour, and we hope to see new outcomes in the near future coming from in–depth analysis of our logs. The results reported above only scratch the surface of what can be done using log files. Researchers interested in this resource can contact the iCLEF organisation (see [5]) for details on how to obtain the logs for research purposes. In addition, we hope to serve as an inspiration for other similar initiatives – collecting log files for indirect observational studies very conveniently allows results to be compiled and used for multiple purposes.

7.3 Task Space, Technology and Research Questions

7.3.1 Use Cases for Interactive Image Retrieval

It is crucial in any discussion on evaluation of interactive systems to note the difference between benchmarking and validation. Benchmarking is what Cranfield–style studies do, as best represented today by TREC, CLEF, NTCIR and other similar evaluation campaigns. The origin of the metaphor of benchmarking is useful to understand the point of it: bolting a piece of machinery to a workshop bench and running it with various inputs and recording its performance. Benchmarking answers the question 'Is it any good?' Validation is another sort of exercise, investigating if tools and technologies (and the design principles behind them) actually work for the tasks they are envisioned to address. Validation answers the question: 'Is it good for anything?'

Benchmarking can be done from a system perspective, setting a baseline level for some metric and improving the system under consideration with respect to that metric and that baseline. Validation, on the other hand, must be done with respect to some system–external model of intended or assumed usage. Many models are conceivable, and one such model is *use cases*, a fairly informal specification of usage of interactive systems, typically used in the design phase of an industrial project (Jacobson et al, 1992; Cockburn, 2002). Whatever model of usage one adopts, some channel of information from benchmarking to validation and back again is necessary. Usage requirements must be passed to design engineers, measures of variation

[5] `http://nlp.uned.es/iCLEF/`

in system performance must be tested against effects on user behaviour, measures of user satisfaction must be passed back to system engineering. Use cases are one way of bridging the divide between interaction design and system construction and the evaluation of the two.

It is crucial also to note that classic Cranfield–style benchmarking studies are not agnostic with respect to use cases. While the notion of a use case has not been explored to any great extent in information access research, there is an implicit notion of retrieval being task–based, topical, with active, focused and well–spoken users. This implicit use case informs both evaluation and design of systems. The classic target notion of relevance and the classic evaluation measures of recall and precision can be used as a fair proxy for user satisfaction in that usage scenario, even when abstracted to be a relation between query and document rather than between need and fulfilling that need.

The advent of multimedia information access breaks the implicit information retrieval use case, which is a good thing. Multimedia is different, used differently, by different users, and for different reasons than is text. Systems used for entertainment rather than immediate information needs, users that expect the systems to provide information by push actions rather than requesting it by pull actions, in a lean–backward setting rather than a lean–forward setting — all these interaction features have a bearing on evaluation methodologies (Karlgren, 2008, 2009; Boujemaa et al, 2009; Karlgren et al, 2009). Benchmarking must change to capture the most important criteria for success for multimedia information access systems, adding appeal and satisfaction to completeness and precision in the palette of target notions for evaluation.

7.3.2 Challenges: Technology and Interaction

7.3.2.1 Specification of Information Needs

Any interactive information system must have means for users to communicate their needs to the system, whatever level of interactivity it is designed for. Query formulation is the most obvious challenge, but even in other interaction frameworks, the formulation of what the user wishes the system to find, provide, or process needs to be achieved somehow.

The general movement towards multimedia information access from text search takes as its premise that text search technologies can be profitably generalised to cover other media. But text is a very special case, especially as applied to ad hoc, task– and topic–oriented retrieval, as argued above. Text wears its semantics on its sleeve, misleadingly simply, through the words that form one of the building blocks of texts. This is not entirely to the benefit of the field of text retrieval, since some simple questions can be answered through the application of very simple technologies. The formulation of information needs is obviously feasible using the same representational mechanisms of which the text is composed: words are known to all language users and can be expected to be invariant across texts to some extent.

Moving to multimedia content, we now are dealing with a more general case. There are no situation–independent conventional semantic symbols in the signal to bootstrap the representation from. The proposed 'visual words' approach for image retrieval attempts to redress this, but involves considerable and as yet unaddressed challenges. How to proceed towards an acceptable mid–level representation, between the abstract level of understanding the content of the information item on the one hand, and the concrete level of decoding the signal components into pixels, surfaces, textures, and lines on the other is not obvious, as it can be claimed to be for text–based information systems. It is quite likely that any approach which successfully resolves the negotiation between user and system with respect to mid–level representation for image retrieval and other multimedia retrieval tasks, will provide valuable lessons which can be used to improve information access to text as well – solutions which currently are not being pursued in view of the utility of words as an analysis base.

7.3.2.2 Target Notions: is Relevance the Final Word?

Our analysis of usage informs our choice of target notion. Cranfield studies have worked well to establish usefulness of systems with respect to some human activities if the activities in question fit the implicit use case as given above. If they do not, as in most or many multimedia information access cases, the evaluations will fail to establish success criteria. Relevance – the momentary quality of an information object that makes it valuable enough to access and use – is a function of untold numbers of factors, many unimportant in isolation (Mizzaro, 1998). The concept of relevance lies at the convergence of understanding users, information needs, items of information, and interaction. In traditional Cranfield–style information retrieval research efforts the target concept of relevance is based on the everyday notion, but operationally to be a relation between query and document.

In light of the new usage situations opened up by multimedia information access systems, relevance, in its task– and topic–related form, may not be broad enough to cover the range of aspects of user satisfaction that govern the acceptance and take–up of a system under evaluation. We may need to move, for example, to satisfaction– or appeal–related target notions to be able to capture the usefulness of systems. First steps in this direction are being taken in evaluation campaigns such as MediaEval (formerly VideoCLEF).

7.3.2.3 Longitudinal and In–Situ Studies

Traditional user studies expose a set of users to a system for a brief while in a laboratory experiment, often with a pre–set information seeking task given to the test subjects. This sort of study may be useful to evaluate the ergonomics of some specific interface widget, but they certainly are very unlikely to provide a basis to establish the usefulness of a system solution for a new task. The craft of performing longitudinal and field studies is well–established in the general human–computer interaction field, but seldom applied to multimedia information access systems and tasks. That gap between engineering and application oriented research in information retrieval and the craft of designing and building appealing and habitable interfaces, and studying users in action, needs to be closed. In lieu of executing a labour–intensive longitudinal field study, the study of log files provides a cost–effective and efficient alternative.

Acknowledgements We thank all the *N* (a very large *N*) subjects of all the interactive experiments profusely.

References

Artiles J, Gonzalo J, López-Ostenero F, Peinado V (2007) Are users willing to search cross–language? An experiment with the Flickr image sharing repository. In: Evaluation of Multilingual and Multi-modal Information Retrieval: 7th Workshop of the Cross–Language Evaluation Forum, CLEF 2006. Lecture Notes in Computer Science (LNCS), vol 4730. Springer, pp 195–204

Belkin NJ, Cole M, Liu J (2009) A model for evaluation of interactive information retrieval. In: In Proceedings of the SIGIR 2009 Workshop on the Future of IR Evaluation, pp 7–8

Bennett JL (1971) Interactive bibliographic search as a challenge to interface design. In: Walker DE (ed) Interactive bibliographic search: The User/Computer Interface. AFIPS Press, pp 1–16

Bennett JL (1972) The user interface in interactive systems. Annual Review of Information Science and Technology 7:159–196

Boujemaa N, Gouraud H, Compano R, Karlgren J, van der Linden P, King P, Sebe N, Köhler J, Joly A, Geurts J, Dosch C, Ortgies R, Rudström A, Kauber M, Point JC, Moine JYL (2009) CHORUS deliverable 2.3: Final report future of multimedia search engines — findings by the eu project CHORUS. Tech. rep., Chorus Project Consortium

Cheng P, Yeh J, Chien B, Ke H, Yang W (2005) NCTU–ISU's evaluation for the user–centered search task and ImageCLEF 2004. In: Multilingual Information Access for Text, Speech and Images: 5th Workshop of the Cross–Language Evaluation Forum, CLEF 2004. Lecture Notes in Computer Science (LNCS), vol 3491. Springer, pp 793–804

Clough PD, Al-Maskari A, Darwish K (2007) Providing multilingual access to Flickr for Arabic users. In: Evaluation of Multilingual and Multi-modal Information Retrieval: 7th Workshop of the Cross-Language Evaluation Forum, CLEF 2006. Lecture Notes in Computer Science (LNCS). Springer, pp 205–216

Cockburn A (2002) Agile software development. Addison–Wesley

Fuhr N, Belkin NJ, Jose J, van Rijsbergen KCJ (2009) 09101 workshop report — Interactive information retrieval. In: Belkin NJ, Fuhr N, Jose J, van Rijsbergen CJK (eds) Interactive Information Retrieval. Dagstuhl Seminar Proceedings. Schloss Dagstuhl — Leibniz–Zentrum für Informatik, Germany, Dagstuhl, Germany

Gonzalo J, Clough P, Vallin A (2006) Overview of the CLEF 2005 interactive track. In: Accessing Multilingual Information Repositories: 6th Workshop of the Cross–Language Evaluation Forum, CLEF 2005. Lecture Notes in Computer Science (LNCS), vol 4022. Springer, pp 251–262

Hearst M (1999) User interfaces and visualization. In: Baeza-Yates R, Ribeiro-Neto B (eds) Modern Information Retrieval. Addison Wesley

Ingwersen P (1992) Information retrieval interaction. Taylor Graham, London

Jacobson I, Christerson M, Jonsson P, Övergaard G (1992) Object–oriented software engineering: A use case driven approach. Addison–Wesley

Järvelin K, Ingwersen P (2005) The turn: Integration of information seeking and retrieval in context. Springer

Kamps J, Geva S, Peters C, Sakai T, Trotman A, Voorhees E (2009) Report on the SIGIR 2009 workshop on the future of IR evaluation. SIGIR Forum 43(2):13–23

Karlgren J (2008) The CHORUS gap analysis on user–centered methodology for design and evaluation of multimedia information access systems. In: Second International Workshop on Evaluating Information Access (EVIA 2008). NTCIR, Tokyo, Japan

Karlgren J (2009) Affect, appeal, and sentiment as factors influencing interaction with multimedia information. In: Theseus/ImageCLEF workshop on visual information retrieval evaluation. Fraunhofer Society, Corfu, Greece

Karlgren J, Olsson F (2007) Trusting the results in crosslingual keyword–based image retrieval. In: Evaluation of Multilingual and Multi-modal Information Retrieval: 7th Workshop of the Cross-Language Evaluation Forum, CLEF 2006. Lecture Notes in Computer Science (LNCS), vol 4730. Springer, pp 217–222

Karlgren J, Sahlgren M (2005) Automatic bilingual lexicon aquisition using random indexing of parallel corpora. Natural Language Engineering 11(3):327–341

Karlgren J, Kauber M, Boujemaa N, Compano R, Dosch C, Geurts J, Gouraud H, King P, Köhler J, van der Linden P, Ortgies R, Rudström A, Sebe N (2009) CHORUS deliverable 3.4: Vision document. Tech. rep., Chorus Project Consortium

Mathes A (2004) Folksonomies—cooperative classification and communication through shared metadata. Tech. rep., Computer Mediated Communication, Graduate School of Library and Information Science, university of Illinois Urbana–Chanpaign

Mizzaro S (1998) How many relevances in information retrieval? Interacting With Computers 10:305–322

Navarro B, Puchol-Blasco M, Terol R, Vázquez S, Lloret E (2010) Lexical ambiguity in cross–language image retrieval: a preliminary analysis. In: Proceedings of the 10th Workshop of the Cross–Language Evaluation Forum, CLEF 2009. Lecture Notes in Computer Science (LNCS). Springer

Oard D (2009) Multilingual Information Access. Encyclopedia of Library and Information Sciences, 3rd Ed. Taylor and Francis

Peinado V, Artiles J, Gonzalo J, Barker E, López-Ostenero F (2009a) FlickLing: A multilingual search interface for Flickr. In: Evaluating Systems for Multilingual and Multimodal Information Access: 9th Workshop of the Cross–Language Evaluation Forum, CLEF 2008. Lecture Notes in Computer Science (LNCS). Springer

Peinado V, Gonzalo J, Artiles J, López-Ostenero F (2009b) UNED at iCLEF 2008: Analysis of a large log of multilingual image searches in Flickr. In: Evaluating Systems for Multilingual and Multimodal Information Access. 9th Workshop of the Cross–Language Evaluation Forum, CLEF 2008. Lecture Notes in Computer Science (LNCS). Springer

Peinado V, López-Ostenero F, Gonzalo J (2010) UNED at iCLEF 2009: analysis of multilingual image search sessions. In: Proceedings of the 10th Workshop of the Cross–Language Evaluation Forum, CLEF 2009. Lecture Notes in Computer Science (LNCS). Springer

Petrelli D, Clough PD (2006) Concept hierarchy across languages in text–based image retrieval. In: Accessing Multilingual Information Repositories: 6th Workshop of the Cross–Language Evaluation Forum, CLEF 2005. Lecture Notes in Computer Science (LNCS), vol 4022. Springer, pp 297–306

Petrelli D, Demetriou G, Herring P, Beaulieu M, Sanderson M (2003) Exploring the effect of query translation when searching cross–language. In: Peters C (ed) Advances in Cross–Language Information Retrieval: Third Workshop of the Cross–Language Evaluation Forum, CLEF 2002. Lecture Notes in Computer Science (LNCS), vol 2785. Springer, pp 430–445

Reid NH (1999) The photographic collections in St. Andrews university library. Scottish Archives 5:83–90

Ruiz M, Chin P (2010) Users' image seeking behavior in a multilingual tag environment. In: Proceedings of the 10th Workshop of the Cross–Language Evaluation Forum, CLEF 2009. Lecture Notes in Computer Science (LNCS). Springer

Tanase D, Kapetanios E (2009) Evaluating the impact of personal dictionaries for cross–language information retrieval of socially annotated images. In: Evaluating Systems for Multilingual and Multimodal Information Access: 9th Workshop of the Cross–Language Evaluation Forum, CLEF 2008. Lecture Notes in Computer Science (LNCS). Springer

Vassilakaki E, Johnson F, Hartle R, Randall D (2009) A study of users image seeking behavior in Flickling. In: Evaluating Systems for Multilingual and Multimodal Information Access: 9th Workshop of the Cross–Language Evaluation Forum, CLEF 2008. Lecture Notes in Computer Science (LNCS). Springer

Villena J, Crespo-García R, González-Cristóbal JC (2006) Boolean operators in interactive search. In: Accessing Multilingual Information Repositories: 6th Workshop of the Cross–Language Evaluation Forum, CLEF 2005. Lecture Notes in Computer Science (LNCS), vol 4022. Springer, pp 293–296

Zhang C, Chai J, Jin R (2005) User term feedback in interactive text–based image retrieval. In: Proceedings of the 28th annual international ACM SIGIR conference on Research and development in information retrieval. ACM press, pp 51–58

Chapter 8
Photographic Image Retrieval

Monica Lestari Paramita and Michael Grubinger

Abstract CLEF[1] was the first benchmarking campaign that organized an evaluation event for image retrieval: the ImageCLEF photographic ad hoc retrieval task in 2003. Since then, this task has become one of the most popular tasks of ImageCLEF, providing both the resources and a framework necessary to carry out comparative laboratory–style evaluation of multi–lingual visual information retrieval from photographic collections. Running for seven years, several challenges have been given to participants, including: retrieval from a collection of historic photographs; retrieval from a more generic collection with multi–lingual annotations; and retrieval from a large news archive, promoting result diversity. This chapter summarizes each of these tasks, describes the individual test collections and evaluation scenarios, analyzes the retrieval results, and discusses potential findings for a number of research questions.

8.1 Introduction

At the turn of the millennium, several calls (Goodrum, 2000; Leung and Ip, 2000) were made to develop a standardized test collection for Visual Information Retrieval (VIR). In 2003, ImageCLEF[2] was the first evaluation event to answer these calls by providing a benchmark suite comprising an image collection, query topics, relevance assessments and performance measures for cross–language image retrieval, which encompasses two main domains of VIR: (1) image retrieval, and (2) Cross–Language Information Retrieval (CLIR).

Monica Lestari Paramita
University of Sheffield, United Kingdom e-mail: m.paramita@sheffield.ac.uk

Michael Grubinger
Carrera 83 Calle 33-93, Medellín, Colombia e-mail: michael.grubinger@gmx.at

[1] http://www.clef-campaign.org/

[2] http://www.imageclef.org/

H. Müller et al. (eds.), *ImageCLEF*, The Information Retrieval Series 32,
DOI 10.1007/978-3-642-15181-1_8, © Springer-Verlag Berlin Heidelberg 2010

Images by their very nature are language–independent; hence, the language used to express the associated texts or textual queries should not affect retrieval, i.e. an image with a caption written in English should be searchable in languages other than English. The main goals of ImageCLEF thereby include:

- to investigate the effectiveness of combining text and image for retrieval;
- to collect and provide resources for benchmarking image retrieval systems;
- to promote the exchange of ideas which may help improve retrieval performance;
- to evaluate VIR systems in a multi–lingual environment.

To achieve these goals, several tasks have been offered to participating groups between 2003 and 2009, including ad hoc retrieval (hereinafter, but also Chapter 13), object recognition and automatic classification tasks (see Chapters 11 and 12) as well as interactive evaluation of retrieval systems (see Chapter 7). ImageCLEF has provided these tasks within two main areas: retrieval of images from photographic collections and retrieval of images from medical collections.

One of the key tasks of ImageCLEF is concerned with evaluation of system performance for ad hoc image retrieval from photographic collections in a laboratory style setting. This kind of evaluation is system–centered and similar to the classic Text REtrieval Conference or TREC[3] ad hoc retrieval task: simulation of the situation in which a system knows the set of documents to be searched, but the query topics are not known to the system in advance. Evaluation thereby only concentrates on comparing algorithms and systems and not on aspects such as retrieval speed or user interaction, as such evaluation is carried out in other tasks (see Chapter 7).

The specific goal of the photographic ad hoc retrieval task is: given a semantic statement (and/or sample images) describing a user information need, find as many relevant images as possible from a given photographic collection (with the query language either being identical to, or different from, that used to describe the images).

Three major phases can be identified in the history of photographic ad hoc retrieval evaluation at ImageCLEF: From 2003 to 2005, the evaluation was based on retrieval from a historic photographic collection (see Section 8.2). In 2006 and 2007, a generic photographic collection with multi–lingual annotations was used (see Section 8.3). Finally, in 2008 and 2009, the evaluation concentrated not only on retrieval precision, but also on retrieval diversity (see Section 8.4).

8.2 Ad hoc Retrieval of Historic Photographs: ImageCLEF 2003–2005

The ImageCLEF 2003 ad hoc retrieval task was the first evaluation event to finally fulfil the calls for a TREC–style evaluation framework for VIR. The research questions concentrated on the evaluation of retrieval from a collection of historic photographs within the first three years, including:

[3] `http://trec.nist.gov/`

Title: Old Tom Morris, golfer, St Andrews.
Short title: Old Tom Morris, golfer.
Location: Fife, Scotland
Description: Portrait of bearded elderly man in tweed jacket, waistcoat with watch chain and flat cap, hand in pockets; painted backdrop.
Date: ca.1900
Photographer: John Fairweather
Categories: [golf - general], [identified male], [St. Andrews Portraits], [Collection - G M Cowie]
Notes: GMC-F202 pc/BIOG: Tom Morris (1821-1908) Golf ball and clubmaker before turning professional, later Custodian of the Links, St Andrews; golfer and four times winner of the Open Championship; father of Young Tom Morris (1851-1875). DETAIL: Studio portrait.

Fig. 8.1: Sample image and caption from the SAC.

- How can researchers be attracted to participate and submit their results?
- How can representative topics and objective relevance judgments be created?
- What methods can be applied to improve retrieval performance?
- How does monolingual retrieval compare to bilingual image retrieval?
- Is it possible to estimate retrieval difficulty in advance?

This section describes both the pilot task of 2003 as well as the follow–up tasks of 2004 and 2005. Further information can be found in the corresponding overview papers: (Clough and Sanderson, 2004; Clough et al, 2005, 2006).

8.2.1 Test Collection and Distribution

The St. Andrews Collection (SAC) of historic photographs is a subset of one of Scotland's most important archives of historic photography and was provided to ImageCLEF by St. Andrews University Library[4]. This collection of 28,133 photographs was the core component of the ImageCLEF ad hoc retrieval task from 2003 to 2005. Detailed information on the SAC can be found in Section 2.2.1.

Each image contains a semi–structured English annotation, describing the image content in detail (see Figure 8.1 for an example). Participants were provided with these annotations, a 368 x 234 large version and a 120 x 76 thumbnail of each image.

The SAC was chosen as the basis for ImageCLEF because the collection represents a realistic archive of images with high quality captions, and because permission was granted by St. Andrews Library to download and distribute the collection for use in the ad hoc retrieval task.

[4] `http://www-library.st-andrews.ac.uk/`

Table 8.1: Ad hoc query topics at ImageCLEF 2003.

ID	Topic Title	ID	Topic Title
1	Men and women processing fish	26	Portraits of Robert Burns
2	A baby in a pram	27	Children playing on beaches
3	Picture postcard views of St. Andrews	28	Pictures of golfers in the nineteenth century
4	Seating inside a church	29	Wartime aviation
5	Woodland scenes	30	Glasgow before 1920
6	Scottish marching bands	31	Exterior views of Indian temples
7	Home guard on parade during World War II	32	Scottish fishing vessels by the photographer Thompson
8	Tea rooms by the seaside	33	Male portraits
9	Fishermen by the photographer Adamson	34	Dogs rounding-up sheep
10	Ships on the river Clyde	35	The mountain Ben Nevis
11	Portraits of Mary Queen of Scots	36	Churches with tall spires
12	North Street St. Andrews	37	Men holding tennis racquets
13	War memorials in the shape of a cross	38	People using spinning machines
14	Boats on Loch Lomond	39	Men cutting peat
15	Tay bridge rail disaster	40	Welsh national dress
16	City chambers in Dundee or Glasgow	41	A coat of arms
17	Great Yarmouth beach	42	University buildings
18	Metal railway bridges	43	British Windmills
19	Culross abbey	44	Waterfalls in Wales
20	Road bridges	45	Harvesting
21	Animals by the photographer Lady Henrietta Gilmour	46	Postcards by the Valentine photographic company
22	Ruined castles in England	47	People dancing
23	London bridge	48	Museum exhibits
24	Damage due to war	49	Musician and their instruments
25	Golf course bunkers	50	Mountain scenery

8.2.2 Query Topics

In the first year, the topic creation process was based on two different approaches. First, the task organizers browsed the SAC to familiarize themselves with the subjects, which are available throughout the collection. Second, an analysis of the log files taken from the St. Andrews Library Web server that hosted the SAC for several years was carried out to identify popular queries.

Based on the log file analysis, which found that queries are commonly short and specific, modifications were made on some of the original queries to make them more suitable for visual retrieval. For example, the query 'church' was changed to 'churches with tall spires'. A total of 50 English query topics (see Table 8.1) were created to test various aspects of query translation and image retrieval, e.g. pictures of specific objects vs. pictures containing actions, broad vs. narrow concepts, topics containing proper names, compound words, abbreviations, morphological variants and idiomatic expressions.

Each topic consisted of a title (i.e. a short phrase describing the search request), a narrative (i.e. a description of what constitutes a relevant or non–relevant image for

```
<top>
<num> Number: 1 </num>
<EN-title n="1"> Men and women processing fish </EN-title>
<EN-narr> A relevant image will show men and/or women processing fish after catching them.
Processing may include gutting or curing and the picture must show the fish processors at work;
not just mention fish processing, e.g. that fish processing takes place at this port. An example
relevant document is [stand03_2093/stand03_2382]. </EN-narr>
</top>
```

```
<top>
<num> Number: 1 </num>
<IT-title n="1"> Uomini e donne che puliscono il pesce </IT-title>
<IT-title n="2"> Pulizia del pesce al porto </IT-title>
<IT-title n="3"> uomini e donne che lavorano il pesce </IT-title>
</top>
```

Fig. 8.2: ImageCLEF 2003 sample topic.

Table 8.2: Ad hoc query topics at ImageCLEF 2004.

ID	Topic Title	ID	Topic Title
1	Portrait pictures of church ministers by Thomas Rodger	14	Elizabeth the Queen Mother visiting Crail Camp, 1954
2	Photos of Rome taken in April 1908	15	Bomb damage due to World War II
3	St. Andrews cathedral by John Fairweather	16	Pictures of York Minster
4	Men in military uniform, George Middlemass Cowie	17	Pictures of Edinburgh Castle taken before 1900
5	Fishing vessels in Northern Ireland	18	All views of North Street, St. Andrews
6	Views of scenery in British Columbia	19	People marching or parading
7	Exterior views of temples in Egypt	20	River with a viaduct in background
8	College or university buildings, Cambridge	21	Photos showing traditional Scottish dancers
9	Pictures of English lighthouses	22	War memorials in the shape of a cross
10	Busy street scenes in London	23	Photos of swans on a lake
11	Composite postcard views of Bute, Scotland	24	Golfers swinging their clubs
12	Tay Bridge rail disaster, 1879	25	Boats on a canal
13	The Open Championship golf tournament, St. Andrews 1939		

that search request), and an example relevant image to facilitate Content–Based Image Retrieval (CBIR) as well. Moreover, both topic titles and narratives were translated into Italian, German, Dutch, French, Spanish and Chinese to encourage participants to research Cross–Language Information Retrieval (CLIR) methods, too. Each translation was carried out by native speakers, who were also asked to specify alternative translations if appropriate. Figure 8.2 shows one sample topic and its Italian translation.

In 2004, 25 new topics (see Table 8.2) were created using a similar approach. Further, several categories (e.g. queries modified by date/location/photographer) were defined and the topics were modified to be distributed evenly within these categories.

Participants at ImageCLEF 2004 had suggested the creation of more visually–based query topics to allow for a more meaningful application of CBIR methods.

Table 8.3: Ad hoc query topics at ImageCLEF 2005.

ID	Topic Title	ID	Topic Title
1	Aircraft on the ground	15	Golfer putting on green
2	People gathered at bandstand	16	Waves breaking on beach
3	Dog in sitting position	17	Man or woman reading
4	Steam ship docked	18	Woman in white dress
5	Animal statue	19	Composite postcards of Northern Ireland
6	Small sailing boat	20	Royal visit to Scotland (not Fife)
7	Fishermen in boat	21	Monument to poet Robert Burns
8	Building covered in snow	22	Building with waving flag
9	Horse pulling cart or carriage	23	Tomb inside church or cathedral
10	Sun pictures & Scotland	24	Close-up picture of bird
11	Swiss mountain scenery	25	Arched gateway
12	Postcards from Iona, Scotland	26	Portrait pictures of mixed sex group
13	Stone viaduct with several arches	27	Woman or girl carrying basket
14	People at the marketplace	28	Colour pictures of woodland scenes around St Andrews

Table 8.4: Languages researched at ImageCLEF 2003–2005.

Language	2003	2004	2005	Language	2003	2004	2005
Arabic			✓	Hungarian			✓
Bulgarian			✓	Indonesian			✓
Chinese	✓	✓	✓	Italian	✓	✓	✓
Croatian			✓	Japanese		✓	✓
Czech			✓	Norwegian			✓
Danish		✓		Polish			✓
Dutch	✓	✓	✓	Portuguese			✓
Finnish		✓	✓	Romanian			✓
French	✓	✓	✓	Russian		✓	✓
English	✓	✓	✓	Spanish	✓	✓	✓
Filipino			✓	Swedish		✓	✓
German	✓	✓	✓	Turkish			✓
Greek			✓	Visual	✓	✓	✓

Hence, in 2005 the task organizers not only based the topic creation process on the log file analysis and Text–Based Image Retrieval (TBIR) challenges, but also on CBIR baseline runs and provided two sample images (compared to only one in the first two years). These query topics are depicted in Table 8.3.

The number of topic languages increased every year thanks to the help of many participants who contributed translations for the query topics in their native languages. Each translation was double–checked by another native speaker of the same language. By 2005, the topic titles had been translated into up to 31 different languages; yet, not all of them were used by participating groups. The actual use of languages in the retrieval experiments from 2003 to 2005 is summarized in Table 8.4.

Table 8.5: Highest MAP for each query language at ImageCLEF 2003.

Language	Group	MAP	Language	Group	MAP
English	Daedalus	0.5718 (monolingual)			
French	Sheffield	0.4380	Italian	Sheffield	0.4047
Spanish	Daedalus	0.4323	Dutch	Sheffield	0.3904
German	Sheffield	0.4285	Chinese	NTU	0.2888

8.2.3 Relevance Judgments and Performance Measures

The creation of relevance judgment was based on a pooling method and Interactive Search and Judge (ISJ). Both approaches are explained in Chapter 4.

In 2003, the top 100 results from all submitted runs were used to create image pools to be assessed for each topic (and in 2004 and 2005, the top 50 results respectively). To reduce judging subjectivity, each image in the topic pools was assessed by the topic creator and at least two other assessors using a ternary classification scheme: (1) relevant, (2) partially relevant, or (3) not relevant.

Based on these judgments, various combinations could be used to create the final set of relevant images (qrels). In all three years, the qrels were based on the pisac–total set: all images judged as relevant or partially relevant by the topic creator and at least one other assessor. ISJ was also used to supplement the image pools with further relevant images that had not been retrieved by any of the participants.

To evaluate the runs, the retrieval results were computed using the newest version of trec_eval[5]. In 2003 and 2004, only the (arithmetic) mean average precision (MAP) was used, while in 2005 methods were also compared using Precision at 10 and 100 images, P10 and P100 respectively, and the number of relevant images retrieved (RelRet). These and other performance measures are defined in Chapter 5.

8.2.4 Results and Analysis

Four groups participated at ImageCLEF 2003 and experimented with different translation methods, such as dictionary and on–line translation tools, and used Query Expansion (QE) to improve TBIR performance. Monolingual runs thereby consistently achieved higher performance than bilingual runs. Table 8.5 provides an overview of the highest MAP for each topic language.

All runs submitted in 2003 retrieved images based on their captions only. To encourage the use of visual methods, a CBIR system[6] was made available for the participants and query topics were also modified to be more visual in 2004 and 2005. As shown in Table 8.6, these measures taken by the ImageCLEF organizers

[5] `http://trec.nist.gov/trec_eval/`

[6] GIFT system (`http://www.gnu.org/software/gift/`)

Table 8.6: Number and percentage of runs with respect to query dimensions.

Query Dimension	2003	2004	2005
Text only	45 (100%)	106 (56%)	318 (91%)
Combined	-	78 (41%)	27 (8%)
Visual only	-	6 (3%)	4 (1%)
TOTAL	45	190	349

Table 8.7: Highest MAP values for the top six languages at ImageCLEF 2004.

Language	Group	Run ID	MAP	QE	Text	Visual	Title	Narr
English	daedalus	mirobaseen	0.5865		✓		✓	
German	dcu	delsmgimg	0.5327	✓	✓	✓	✓	
Spanish	UNED	unedesent	0.5171	✓	✓		✓	
French	montreal	UMfrTFBTI	0.5125	✓	✓	✓	✓	
Italian	dcu	itlsstimg	0.4379	✓	✓		✓	
Dutch	dcu	nllsstimg	0.4321	✓	✓		✓	
Visual	geneva	GE_andrew4	0.0919	✓		✓		

Table 8.8: Top six languages with highest MAP at ImageCLEF 2005.

Language	Group	Run ID	MAP	QE	Text	Visual	Title	Narr	Image
English	CUHK	ad-eng-tv-kl-jm2	0.4135	✓	✓	✓	✓		✓
Chinese	NTU	CE-TN-WEprf-Ponly	0.3993	✓	✓	✓	✓	✓	
Spanish	Alicante, Jaen	R2D2vot2SpL	0.3447	✓	✓		✓		
Dutch	Alicante, Jaen	R2D2vot2Du	0.3435	✓	✓		✓		
Visual	NTU	NTU-adhoc05-EX-prf	0.3425	✓		✓			✓
German	Alicante, Jaen	R2D2vot2Ge	0.3375	✓	✓		✓		

subsequently proved to be effective as more participating groups submitted runs exploring the use of CBIR, or the combination of CBIR and TBIR, respectively.

Table 8.7 provides an overview of the highest MAP values for the languages in 2004. Popular translation methods included machine translation (73%), bilingual dictionaries and parallel corpora. A number of groups also improved their retrieval results by performing structured and constrained searches in order to identify named entities such as the photographer, date and location.

In most combined approaches, CBIR and TBIR were first performed separately, and then the ranked lists from both searches were merged. However, the combination of visual and textual approaches only managed to improve the performance of some topics. Also, purely visual searches performed poorly. This was (1) due to the fact that the query topics in 2004 did not involve enough visually–related topics, and (2) due to the nature of the images in the SAC which made CBIR difficult.

In 2005, the task organizers created query topics exhibiting more visual features. As a result, the results of visual approaches improved significantly. Table 8.8 provides the MAP scores for the top six highest performing languages. Most of these runs used QE and/or Relevance Feedback (RF). Twenty seven runs combined CBIR

Table 8.9: Average MAP by different modalities for ImageCLEF 2004 and 2005.

Modality	2004	2005
Text only	0.3787	0.2121
Combined text & image	0.4508	0.3086

and TBIR results, including the best monolingual run. On average, combined modality runs outperformed text–only runs in 2004 and 2005, as shown in Table 8.9.

Even though more visual queries were used in 2005, the number of runs using CBIR decreased compared to 2004. CBIR approaches did not seem to benefit from the visual features that could be extracted from the SAC. The evaluation using the SAC had reached a plateau due to several limitations with the collection: mainly black–and–white and grey–scale images (limiting the use of colour, as visual feature playing a vital role in CBIR), domain–specific annotation vocabulary in only one language (English), and restricted retrieval scenarios (i.e. search for historic photographs).

8.3 Ad hoc Retrieval of Generic Photographs: ImageCLEFphoto 2006-2007

At ImageCLEF 2005, participants had called for a test collection with richer visual features and multi–lingual annotations. Hence, in 2006 the SAC was replaced by a more generic photographic collection, the IAPR TC–12 database, created under Technical Committee 12 (TC–12) of the International Association of Pattern Recognition[7] (IAPR). Furthermore, the general photographic ad hoc retrieval task was given a new name (ImageCLEFphoto) in order to avoid confusion with the medical ad hoc retrieval task (ImageCLEFmed). Evaluation objectives and questions included:

- Are evaluation results obtained from the SAC also applicable to generic photos?
- Can combining CBIR and TBIR methods as well as using RF and/or QE improve retrieval performance also with generic photos?
- How does retrieval using short captions compare to using extensive captions?
- Are traditional TBIR methods still applicable for short captions?
- How significant is the choice of the retrieval and/or annotation language?

This section summarizes ImageCLEFphoto 2006 and 2007. More information can be found in the related overview papers: (Clough et al, 2007; Grubinger et al, 2008).

[7] `http://www.iapr.org/`

Title: Flamingo Beach
Description: A photo of a brown sandy beach; the dark blue sea with small breaking waves behind it; a dark green palm tree in the foreground on the left; a blue sky with clouds on the horizon in the background;
Notes: Original name in Portuguese: 'Praia do Flamengo'; Flamingo Beach is considered as one of the most beautiful beaches of Brazil
Location: Salvador, Brazil
Dates: 2 October 2004

Fig. 8.3: Sample image and caption of the IAPR TC–12 database.

8.3.1 Test Collection and Distribution

The photographic collection of the IAPR TC–12 database contains 20,000 colour photos taken from locations around the world and comprises a varying cross–section of still natural images. This test collection, which was specifically built to support the evaluation needs of ImageCLEF, was the core component of ImageCLEFphoto 2006 and 2007. Detailed information on the creation and contents of the IAPR TC–12 database can be found in Section 2.2.2 of Chapter 2 and (Grubinger et al, 2006).

Figure 8.3 illustrates a sample image from the IAPR TC–12 database. Each image in the collection comprises corresponding semi–structured annotations in three different languages: English, German and Spanish. The annotation structure was thereby very similar to that used in the SAC (compare Table 8.1) to provide a smooth transition for returning participants. Only the 'categories' field was missing as it had hardly been used in retrieval from the SAC.

The ImageCLEF organizers used the parametric nature of the test collection and created a different subset of the test collection each year. Consequently, the participants of ImageCLEFphoto 2006 were provided with 20,000 images and the corresponding English and German captions exhibiting a varying degree of annotation 'completeness':

- 70% of the annotations contained title, description, notes, location and date.
- 10% of the annotations contained title, location and date.
- 10% of the annotations contained location and date.
- 10% of the images were not annotated (or had empty tags respectively).

One year later, ImageCLEFphoto 2007 focused on whether TBIR methods would still be suitable to find images with short captions. Thus, the description field was removed from the annotations and participants were provided with annotations only containing title, notes, location and date. The lack of textual information should encourage participants to use CBIR techniques. Four sets of annotations were provided: (1) English, (2) German, (3) Spanish, and (4) one set whereby the annotation language was randomly selected for each of the images.

Table 8.10: Query topics in the IAPR TC–12 database.

ID	Topic Title	ID	Topic Title
1	accommodation with swimming pool	31	volcanos around Quito
2	church with more than two towers	32	photos of female guides
3	religious statue in the foreground	33	people on surfboards
4	group standing in front of mountain landscape in Patagonia	34	group pictures on a beach
5	animal swimming	35	bird flying
6	straight road in the USA	36	photos with Machu Picchu in the background
7	group standing in salt pan	37	sights along the Inka-Trail
8	host families posing for a photo	38	Machu Picchu and Huayna Picchu in bad weather
9	tourist accommodation near Lake Titicaca	39	people in bad weather
10	destination in Venezuela	40	tourist destinations in bad weather
11	black and white photos of Russia	41	winter landscape in South America
12	people observing football match	42	pictures taken on Ayers Rock
13	exterior view of school building	43	sunset over water
14	scenes of footballers in action	44	mountains on mainland Australia
15	night shots of cathedrals	45	South American meat dishes
16	people in San Francisco	46	Asian women and/or girls
17	lighthouses at the sea	47	photos of heavy traffic in Asia
18	sport stadium outside Australia	48	vehicle in South Korea
19	exterior view of sport stadia	49	images of typical Australian animals
20	close-up photograph of an animal	50	indoor photos of churches or cathedrals
21	accommodation provided by host families	51	photos of goddaughters from Brazil
22	tennis player during rally	52	sports people with prizes
23	sport photos from California	53	views of walls with unsymmetric stones
24	snowcapped buildings in Europe	54	famous television (and telecommunication) towers
25	people with a flag	55	drawings in Peruvian deserts
26	godson with baseball cap	56	photos of oxidised vehicles
27	motorcyclists racing at the Australian Motorcycle Grand Prix	57	photos of radio telescopes
28	cathedrals in Ecuador	58	seals near water
29	views of Sydney's world-famous landmarks	59	creative group pictures in Uyuni
30	room with more than two beds	60	salt heaps in salt pan

8.3.2 Query Topics

The participants were given 60 query topics (see Table 8.10) representing typical search requests for the photographic collection of the IAPR TC–12 database.

The creation of these topics had been based on several factors, including: the analysis of a log file from on–line access to the image collection; knowledge of the collection content; various types of linguistic and pictorial attributes; the use of geographic constraints; and the estimated difficulty of the topic.

In particular, 40 topics were directly taken from the log files with slight syntactic modification (e.g. 'lighthouse sea' was changed to 'lighthouse at the sea'). Another ten were derived from the logs (e.g. 'straight roads in Argentina' was changed to

```
<top>
<num> Number: 14 </num>
<title> Scenes of footballers in action </title>
<narr> Relevant images will show football (soccer) players in a game situation during a match.
Images with footballers that are not playing (e.g. players posing for a group photo, warming up
before the game, celebrating after a game, sitting on the bench, and during the half-time break)
are not relevant. Images with people not playing football (soccer) but a different code (American
Football, Australian Football, Rugby Union, Rugby League, Gaelic Football, Canadian Football,
International Rules Football, etc.) or some other sport are not relevant. </narr>
<image> images/31/31609.jpg </image>
<image> images/31/31673.jpg </image>
<image> images/31/32467.jpg </image>
</top>
```

Fig. 8.4: Sample query topic at ImageCLEFphoto 2006.

'straight roads in the USA'). The rest of the queries was not taken from the logs, but created to assess specific aspects of CBIR and TBIR (e.g. 'black and white photos of Russia'). There were 24 queries which contained geographical constraints (e.g. 'tourist accommodation near Lake Titicaca') since these queries were quite common in the log. Half of the topics were classified as 'semantic', one third as 'neutral' and the rest as 'visual'. CBIR approaches were not expected to improve retrieval results in semantic topics, while the visual topics would benefit from the use of visual approaches. More information can be found in (Grubinger, 2007).

The format of the topics (see Figure 8.4) was identical with the one used in previous years, again to provide a smooth transition for returning participants: each topic contained a title, a narrative description and three sample images. The same queries were used in the two year period to allow for a comparison of retrieval from collections of fully annotated (2006) and lightly annotated (2007) photographs.

In both years, the topic titles were provided in 16 languages, including: English, German, Spanish, Italian, French, Portuguese, Chinese, Japanese, Russian, Polish, Swedish, Finnish, Norwegian, Danish and Dutch. All translations were provided by native speakers and verified by at least one other native speaker. Since the annotations were provided in two languages in 2006 (and four sets in 2007), this created 32 potential bilingual retrieval pairs (and even 64 in 2007, respectively).

8.3.3 Relevance Judgments and Performance Measures

Similar to the first three years, the relevance assessments at ImageCLEFphoto 2006 and 2007 were based on the pooling method and ISJ. Image pools were created by taking the top 40 results from all participants' runs, yielding an average 1,045 images to be judged per query topic in 2006 (and 2,299 images in 2007, respectively). ISJ was also being deployed to find more relevant images that were not returned by any methods within the top 40 results, and the resulting pools in 2007 were

Table 8.11: Top results at ImageCLEF 2006.

Topic	Caption	Group	Run ID	MAP	P20	GMAP	bpref
EN	EN	CINDI	Cindi_Exp_RF	0.385	0.530	0.282	0.874
PT	EN	NTU	PT-EN-AUTO-FB-TXTIMG-T-WEprf	0.285	0.403	0.177	0.755
ZH	EN	NTU	ZHS-EN-AUTO-FB-TXTIMG-TOnt-WEprf	0.279	0.464	0.154	0.669
RU	EN	NTU	RU-EN-AUTO-FB-TXTIMG-T-WEprf	0.279	0.408	0.153	0.755
SP	EN	NTU	SP-EN-AUTO-FB-TXTIMG-T-WEprf	0.278	0.407	0.175	0.757
DE	DE	NTU	DE-DE-AUTO-FB-TXTIMG-T-WEprf	0.311	0.335	0.132	0.974
EN	DE	DCU	combTextVisual_ENDEEN	0.122	0.175	0.036	0.524
FR	DE	DCU	combTextVisual_FRDEEN	0.104	0.147	0.002	0.245
Vis.	-	RWTH	RWTHi6-IFHTAM	0.063	0.182	0.022	0.366

complemented with further relevant images found in 2006 to avoid missing out on relevant images not found in 2007 due to the reduced captions. The assessments were, again, based on a ternary classification system, whereby this time, only those images judged relevant by both assessors were considered for the qrels.

The runs were evaluated using MAP and P20. The latter was chosen because most on–line image retrieval search engines display 20 images by default. Other measures used include the GMAP, which tests system robustness, and the binary preference (bpref) to indicate the bias due to incompleteness of relevance judgments.

8.3.4 Results and Analysis

There was an increasing number of participating groups: 12 groups submitted in 2006, and 20 groups in 2007. This was the highest number of participants at ImageCLEF thus far, which was an indication that the need for evaluation of VIR had increased over the years, and that ImageCLEFphoto was considered as a suitable track to explore this field of research. As a consequence, many novel retrieval methods and ideas were investigated. Tables 8.11 and 8.12 show the results for the best performing language pairs (MAP) in both years, but also indicate that the choice of the performance measure does affect system ranking. An overview of all retrieval methods and complete results are available in the ImageCLEF overview papers (Clough et al, 2007; Grubinger et al, 2008).

Comparing the results from both years, it is interesting to see how monolingual results were more affected by the annotation reduction than bilingual results. While monolingual retrieval produced better results than bilingual retrieval in 2006, the results at ImageCLEFphoto 2007 suggested that, on average, bilingual results were as competitive as monolingual results. This might be due to the short image captions provided in 2007, but could also be credited to improved translation resources. Moreover, the choice of the query language was almost negligible in 2007, most likely because many of the short captions contained proper nouns.

Table 8.12: Top results at ImageCLEF 2007.

Topic	Caption	Group	Run ID	MAP	P20	GMAP	bpref
EN	EN	CUT	cut-EN2EN-F50	0.318	0.459	0.298	0.162
PT	EN	NTU	PT-EN-AUTO-FBQE-TXTIMG	0.282	0.388	0.266	0.127
ZH	EN	NTU	ZHT-EN-AUTO-FBQE-TXTIMG	0.257	0.360	0.240	0.089
RU	EN	NTU	RU-EN-AUTO-FBQE-TXTIMG	0.273	0.383	0.256	0.115
ES	EN	NTU	ES-EN-AUTO-FBQE-TXTIMG	0.279	0.383	0.259	0.128
DE	DE	NTU	DE-DE-AUTO-FBQE-TXTIMG	0.245	0.379	0.239	0.108
EN	DE	DCU	combTextVisual_ENDEEN	0.278	0.362	0.250	0.112
FR	DE	DCU	combTextVisual_FRDEEN	0.164	0.237	0.144	0.004
Vis.	-	XRCE	AUTO-NOFB-IMG_COMBFK	0.189	0.352	0.201	0.102

Table 8.13: Results by retrieval modalities at ImageCLEFphoto 2006 and 2007.

Year	2006				2007			
Modality	MAP	P20	bpref	GMAP	MAP	P20	bpref	GMAP
Image	0.041	0.134	0.296	0.014	0.068	0.157	0.080	0.022
Text	0.129	0.173	0.465	0.027	0.120	0.152	0.141	0.018
Combined	0.199	0.281	0.650	0.095	0.149	0.225	0.203	0.050

Table 8.13 shows that combined text and image retrieval outperformed text–only and visual–only retrieval approaches. This trend had already been indicated for retrieval from historic photographic collections and has now continued for retrieval from generic photographic collections as well. The same is true for the use of RF and/or QE, which were also shown to improve retrieval performance in 2006 and 2007.

8.3.5 Visual Sub–task

To attract more visually–orientated groups, a visual sub–task was run in 2006 to investigate CBIR–only techniques: all image captions were deleted, and retrieval had to rely on CBIR techniques only. Thirty queries were selected from the original 60 query topics, with some modifications being made to remove non–visual constraints such as location. For example, the query 'black and white photos from Russia' was changed to 'black and white photos'. Some examples of the visual topics are shown in Table 8.14.

Only two out of 36 registered participants eventually submitted runs to this subtask. The highest performing run, submitted by RWTH University Aachen, Germany, used invariant and Tamura texture feature histograms. The evaluation showed promising results for P20, which was 0.285. However, MAP was very low (0.101 for the best run). This was due to the fact that relevant images found in P20 were quite similar to sample images given in the query (Clough et al, 2007).

Table 8.14: Example of topics in the visual sub–task of ImageCLEFphoto 2006.

ID	Topic Title	Level
82	sunset over water	Easy
78	bird flying	Easy
67	scenes of footballers in action	Medium
84	indoor photos of churches or cathedrals	Medium
83	images of typical Australian animals	Difficult
61	church with more than two towers	Difficult

8.4 Ad hoc Retrieval and Result Diversity: ImageCLEFphoto 2008–2009

The ImageCLEF ad hoc retrieval tasks had followed the evaluation scenario similar to the classical TREC ad hoc retrieval task during the first five years (see Section 8.1). However, in 2008 this scenario was slightly changed: systems were not only expected to return relevant images for a given search request, but also to return these relevant images from as many different sub-topics as possible (to promote retrieval diversity) in the top n results. This novel challenge allowed for the investigation of a number of new research questions, including:

- Is it possible to promote diversity within the top n results?
- Which retrieval approaches work best at promoting diversity?
- Does promoting diversity sacrifice relevance (i.e. precision)?
- How do results compare between bilingual and multi–lingual annotations?
- Do mixed approaches still outperform text or image only methods?
- How much does a priori knowledge about query clusters help to increase diversity?

This section summarizes the ImageCLEFphoto 2008 and 2009 tasks. More information can be found in the respective overview papers: Arni et al (2009); Paramita et al (2010).

8.4.1 Test Collection and Distribution

As in previous years, the IAPR TC–12 database provided the resources for ImageCLEFphoto 2008. Since the evaluation concentrated on diversity within the top retrieval results, a different collection subset to that used in 2006 and 2007 was generated: participants were given two sets of complete annotations (i.e. all caption fields were provided) in (1) English and (2) 'Random', whereby the language for each caption was randomly selected from either English or German.

Reusing the same image collection as in previous years allowed for the investigation of whether precision is affected when diversity is implemented. However, ImageCLEF participants felt in 2008 that the time had come to move on to a bigger

837661 MOS06-20020212-MOSCOW, RUSSIAN FEDERATION: Russian Foreign Minister Igor Ivanov (L) shakes hands with Afghanistan's interim Defence Minister General Mohammad Qasim Fahim (R) to start their talks at the Foreign Ministry in Moscow on Tuesday 12 February 2002. Russia will give the technical and logistical assistance to Afghanistan's army but will not train Afghan military specialists, it was announced. EPA PHOTO EPA-SERGEI CHIRIKOV-vk-fob

Fig. 8.5: Sample image and caption from Belga.

Table 8.15: Examples of different clusters at ImageCLEFphoto 2008.

ID	Topic Title	Cluster
2	church with more than two towers	city
3	religious statue in the foreground	statue
5	animal swimming	animal
12	people observing football match	venue
23	sport photos from Australia	sport
50	indoor photos of a church or cathedral	country

image archive for evaluation. Hence, in 2009 a new challenge was offered by replacing the IAPR TC–12 database with a database that was nearly 25 times larger: the photographic collection of *Belga*[8], a Belgian news agency (see also Section 2.2.3 in Chapter 2).

This data set comprised 498,920 photos with unstructured, English–only annotations (see Figure 8.5). This offered new challenges to the participants in comparison to the SAC and IAPR TC–12 collections. For example, the unstructured nature of the image captions required the automatic extraction of information about, for example the location, date or photographic source of the image as a part of the indexing and retrieval process. In addition, it contained many cases where pictures had not been orientated correctly, thereby making CBIR more difficult (Paramita et al, 2009).

8.4.2 Query Topics

ImageCLEFphoto 2008 used a subset of the previous year's queries: 39 topics were identified that would also be useful for the evaluation of retrieval diversity. The annotation structure was thereby identical to that used in 2006 and 2007, apart from an additional cluster field that was included to represent the diversity need.

For example, the query 'vehicle in South Korea' would benefit from retrieval diversity with respect to 'vehicle types' (see Figure 8.6). A selection of query examples together with their corresponding clusters is illustrated in Table 8.15.

[8] `http://www.belga.be/`

```
<top>
<num> Number: 48 </num>
<title> vehicle in South Korea </title>
<cluster> vehicle type </cluster>
<narr> Relevant images will show vehicles in South Korea, including cars, trains, buses, forklifts,
boats, and so on. Images with vehicles outside of South Korea are not relevant. Images from South
Korea without a single vehicle are not relevant either. </narr>
<image> SampleImages/48/35645.jpg </image>
<image> SampleImages/48/35705.jpg </image>
<image> SampleImages/48/35982.jpg </image>
</top>
```

Fig. 8.6: Sample query topic at ImageCLEFphoto 2008.

```
<top>
<num> 12 </num>
<title> clinton </title>
<clusterTitle> hillary clinton </clusterTitle>
<clusterDesc> Relevant images show photographs of Hillary Clinton. Images of Hillary with
other people are relevant if she is shown in the foreground. Images of her in the background are
not relevant. </clusterDesc>
<image> belga26/05859430.jpg </image>
<clusterTitle> obama clinton </clusterTitle>
<clusterDesc> Relevant images show ... </clusterDesc>
<image> belga... </image>
<clusterTitle> bill clinton </clusterTitle>
<clusterDesc> Relevant images show ... </clusterDesc>
<image> belga... </image>
```

Fig. 8.7: Example of Query Part 1 at ImageCLEFphoto 2009.

The topic creation process for ImageCLEFphoto 2009 was based on search query logs from Belga. In contrast to 2008, where the cluster fields had been estimated based on the query topics, the information on query variations could also be extracted from the log file. For example, 'Victoria Beckham' and 'David Beckham' were variations (and at the same time clusters) for a query looking for 'Beckham'. Eventually, 50 topics (with an average number of four clusters each) were generated, divided in two sets of 25 topics each and released in two different formats: 'Query Part 1' and 'Query Part 2'.

Figure 8.7 provides an example for Query Part 1, which includes a topic title, cluster title, cluster description and an example image. All potential retrieval clusters were provided as a part of the query topic, simulating the situation in which search engines have access to query logs telling the system what variations to expect.

However, in real–life scenarios, often little or no query log information is available to indicate potential clusters. Thus, in the second set of query topics, Query Part 2, little evidence was given for what kind of diversity was expected: the `clusterTitle` and `clusterDesc` fields were hidden, and only the topic title and three example images were provided for CBIR approaches (which, in many

```
<top>
<title> obama </title>
<num> 26 </num>
<image> belga30/06098170.jpg </image>
<image> belga28/06019914.jpg </image>
<image> belga30/06017499.jpg </image>
```

Fig. 8.8: Example of Query Part 2 at ImageCLEFphoto 2009.

cases, would not cover all clusters). Figure 8.8 provides an example for Query Part 2. Further information regarding query and cluster development at ImageCLEFphoto 2009 is available in Paramita et al (2009).

8.4.3 Relevance Judgments and Performance Measures

The relevance assessments from 2007 were reused for ImageCLEFphoto 2008. In addition, the images were assigned one (or more) predefined clusters to enable the quantification of retrieval diversity. Two assessors carried out the classification process, while a third assessor was used to resolve any inconsistent judgments.

In 2009, the relevance assessments were performed using Distributed Information Retrieval Evaluation Campaign Tool[9] (DIRECT) and were carried out in two phases: (i) the relevant images for each query were identified; and (ii) these relevant images were assigned to the clusters. Due to the large collection, the pool sizes rose drastically compared to previous years; thus, each image was only evaluated by one assessor. An average of 700 images were found to be relevant for each query, and around 200 images were relevant for each cluster.

To evaluate the search results, standard IR measures were used: MAP, GMAP and bpref. Retrieval diversity was evaluated using cluster recall CR(n), which represents the percentage of clusters retrieved in the top n documents (Zhai et al, 2003). Moreover, F_1 was used to combine P20 and CR20 in 2008, and P10 and CR10 in 2009 respectively, because the number of clusters had an upper bound of 10 in that year. For a definition of these performance measures, see Chapter 5.

8.4.4 Results and Analysis

ImageCLEFphoto managed to attract more than 40 groups, which registered in both years of the task; 24 submitted results in 2008, and 19 in 2009 respectively. Most participants employed post–processing methods to achieve result diversity. They started the retrieval process by using TBIR baseline runs enhanced by RF/QE to

[9] http://direct.dei.unipd.it/

Table 8.16: Systems with highest F_1 across all 39 topics at ImageCLEFphoto 2008.

Group	Run-ID	Run Type	Modality	P20	CR20	F_1
PTECH	EN-EN-MAN-TXTIMG	MAN	TXT-IMG	0.6885	0.6801	0.6843
PTECH	EN-EN-MAN-TXTIMG-MMBMI	MAN	TXT-IMG	0.6962	0.6718	0.6838
PTECH	EN-EN-MAN-TXT-MTBTN	MAN	TXT	0.5756	0.5814	0.5785
XRCE	xrce_tilo_nbdiv_15	AUTO	TXT-IMG	0.5115	0.4262	0.4650
DCU	EN-EN-AUTO-TXTIMG-QE	AUTO	TXT-IMG	0.4756	0.4542	0.4647
XRCE	xrce_tilo_nbdiv_10	AUTO	TXT-IMG	0.5282	0.4146	0.4646

Table 8.17: Systems with highest F_1 across all 50 topics at ImageCLEFphoto 2009.

Group	Run Name	Topic Fields*	Modality	P10	CR10	F_1
XEROX-SAS	XRCEXKNND	T-CT-I	TXT-IMG	0.794	0.8239	0.8087
XEROX-SAS	XRCECLUST	T-CT-I	TXT-IMG	0.772	0.8177	0.7942
XEROX-SAS	KNND	T-CT-I	TXT-IMG	0.800	0.7273	0.7619
INRIA	LEAR5_TI_TXTIMG	T-I	TXT-IMG	0.798	0.7289	0.7619
INRIA	LEAR1_TI_TXTIMG	T-I	TXT-IMG	0.776	0.7409	0.7580
InfoComm	LRI2R_TI_TXT	T-I	TXT	0.848	0.6710	0.7492

* T = Title, CT = Cluster Title, I = Image

Table 8.18: Performance measures for different query formats.

Queries	Runs	P10	CR10	F_1
Queries part 1 with CT	52	0.6845	0.5939	0.6249
Queries part 1 without CT	32	0.6641	0.5006	0.5581
Queries part 2	84	0.6315	0.5415	0.5693

maximize the number of relevant images in the top n results. Diversity was then promoted by re–ranking the initial run, clustering the top n documents, and selecting the highest ranked document in each cluster to create diverse results.

The top six results across all query topics of ImageCLEFphoto 2008 and 2009 are shown in Tables 8.16 and 8.17. In 2008, the top ten results were all monolingual (English), with the highest bilingual run exhibiting P20 of 0.4397, CR20 of 0.4673 and F_1 of 0.4531. On average, however, the margin between monolingual and bilingual runs was low, continuing the trend of previous years. In 2009, only monolingual runs were evaluated since English was the only language for both annotations and topics. Retrieval results were much higher than in 2008, which was due to less semantic and hence easier topics compared to those used the year before.

Table 8.18 provides the results of the analysis on whether the different query formats influence retrieval effectiveness. Since participants could choose which query fields to use for retrieval, the scores for Query Part 1 were divided into runs which used the cluster title (CT) and runs which did not. The scores between Query Parts 1 and 2 were found to be significantly different.

Table 8.19 shows average scores of the top 20 results across all runs with respect to their retrieval modalities for 2008 and 2009. Mixed CBIR and TBIR methods

Table 8.19: Results by retrieval modalities at ImageCLEFphoto 2008 and 2009.

Year	2008			2009		
Modality	P20	CR20	F_1	P20	CR20	F_1
Image only	0.1625	0.2127	0.1784	0.0787	0.2986	0.1244
Text only	0.2431	0.3915	0.2957	0.6915	0.622	0.6454
Combined	0.2538	0.3998	0.3034	0.6994	0.6883	0.6913

Table 8.20: Participation overview for ImageCLEFphoto 2003-2009.

Queries	2003	2004	2005	2006	2007	2008	2009
Registered groups			19	36	32	43	44
Participating groups	4	12	11	12	20	24	19
Submitted runs	45	190	349	157	616	1042	84

provided the best results also in evaluation scenarios promoting retrieval diversity, although the difference to TBIR–only methods was, on average, only marginal. Yet, looking at the best runs (see Tables 8.16 and 8.17), mixed approaches still outperform TBIR–only approaches. CBIR methods have slightly caught up, but still lag behind.

8.5 Conclusion and Future Prospects

After the image retrieval community had been calling for resources similar to those used by TREC in its ad hoc retrieval tasks for the text retrieval domain, ImageCLEF began in 2003 to also provide similar resources within the context of VIR to facilitate standardised laboratory–style testing of cross–language image retrieval systems. While these resources have predominately been used by systems applying a TBIR approach, there has also been an increasing number of groups using CBIR approaches over the years. Benchmark resources created for ad hoc retrieval from photographic collections include the following:

- historic photographs with extensive semi–structured annotations;
- generic photographs with multi–lingual semi–structured annotations;
- a large press collection containing photos with unstructured annotations.

ImageCLEF ran seven ad hoc cross–language image retrieval tasks for the domain of photographic collections from 2003 to 2009, thereby addressing two main fields of information retrieval research: (1) image retrieval and (2) CLIR. The tasks were modelled on scenarios found in multimedia use at the time and proved to be very popular among researchers as shown by an increasing number of participants over the years (see Table 8.20).

Moreover, each year a large number of participants also registered without eventually submitting results, only to get access to the valuable benchmark resources. In 2009, the much lower number of submitted runs was due to a limitation of five runs for each participating group (before that, an unlimited number of runs could be submitted, all of which were evaluated). In the first four years (2003 to 2006), retrieval from collections with extensive captions suggested the following trends for both historic and generic photographs:

- Using QE and/or RF improves retrieval performance.
- Combining CBIR and TBIR methods improves retrieval performance.
- Monolingual runs outperform bilingual runs.
- Retrieval success does still depend on the annotation language.
- The retrieval difficulty of a topic can be pre–determined.
- The choice of qrels and performance measures can affect system ranking.

At ImageCLEFphoto 2007, most of these trends could be verified also for retrieval from image collections with light annotations, with the following exceptions that indicated that for short captions:

- Bilingual runs perform as well as monolingual runs.
- The choice of query or annotation language hardly affects retrieval success.

The challenge of ImageCLEFphoto in 2008 and 2009 was slightly different to that in previous years and was based on promoting diversity in the search results. Results from both years showed that:

- It is possible to present a diverse result without sacrificing precision.
- A priori information about the cluster title is essential for retrieval diversity.
- A combination of title, cluster title and image maximizes diversity and relevance.
- Mixed runs (CBIR and TBIR) outperform runs based on TBIR or CBIR alone.
- Bilingual retrieval performs nearly as well as monolingual retrieval.

The change of direction in the evaluation objective in 2008 showed that, as the field of VIR develops, test collections and evaluation events need to evolve and react to those changes as well. ImageCLEFphoto is not an exception and will, hence, continue to provide resources to the VIR community in the future to facilitate standardized laboratory–style testing of image retrieval systems.

Acknowledgements We would like to thank the providers of the data sets: St. Andrews University Library for ImageCLEF 2003–2005, viventura for ImageCLEFphoto 2006–2008; Belga Press Agency for ImageCLEFphoto 2009; Michael Grubinger for the IAPR TC–12 database; Theodora Tsikrika for the Belga query log analysis; and Giorgio di Nunzio and the DIRECT team for their support at ImageCLEFphoto 2009. ImageCLEFphoto was funded by the EU under the projects: TrebleCLEF (Grant agreement: 215231), Multimatch (contract IST–2005–2.5.10) and the MUSCLE NoE (FP6–507752).

References

Arni T, Clough PD, Sanderson M, Grubinger M (2009) Overview of the ImageCLEFphoto 2008 Photographic Retrieval Task. In: Peters C, Deselaers T, Ferro N, Gonzalo J, Jones G, Kurimo

M, Mandl T, Peñas A, Petras V (eds) Evaluating Systems for Multilingual and Multimodal Information Access: 9th Workshop of the Cross–Language Evaluation Forum (CLEF 2008). Lecture Notes in Computer Science (LNCS), vol 5706. Springer, Aarhus, Denmark, pp 500–511

Clough PD, Sanderson M (2004) The CLEF 2003 Cross Language Image Retrieval Track. In: Peters C, Gonzalo J, Braschler M, Kluck M (eds) Comparative Evaluation of Multilingual Information Access Systems: Fourth Workshop of the Cross–Language Evaluation Forum (CLEF 2003). Lecture Notes in Computer Science (LNCS), vol 3237. Springer, Trondheim, Norway, pp 581–593

Clough PD, Müller H, Sanderson M (2005) The CLEF Cross Language Image Retrieval Track 2004. In: Peters C, Clough P, Gonzalo J, Jones G, Kluck M, Magnini B (eds) Multilingual Information Access for Text, Speech and Images: Results of the Fifth CLEF Evaluation Campaign. Lecture Notes in Computer Science (LNCS), vol 3491. Springer, Bath, United Kingdom, pp 597–613

Clough PD, Müller H, Deselaers T, Grubinger M, Lehmann TM, Jensen J, Hersh W (2006) The CLEF 2005 Cross–Language Image Retrieval Track. In: Peters C, Gey FC, Gonzalo J, Müller H, Jones GJF, Kluck M, Magnini B, de Rijke M, Giampiccolo D (eds) Accessing Multilingual Information Repositories: Sixth Workshop of the Cross–Language Evaluation Forum (CLEF 2005). Lecture Notes in Computer Science (LNCS), vol 4022. Springer, Vienna, Austria, pp 535–557

Clough PD, Grubinger M, Deselaers T, Hanbury A, Müller H (2007) Overview of the ImageCLEF 2006 Photographic Retrieval and Object Annotation Tasks. In: Peters C, Clough PD, Gey FC, Karlgren J, Magnini B, Oard DW, de Rijke M, Stempfhuber M (eds) Evaluation of Multilingual and Multi-modal Information Retrieval. Lecture Notes in Computer Science (LNCS), vol 4730. Springer, Alicante, Spain, pp 579–594

Goodrum A (2000) Image Information Retrieval: An Overview of Current Research. Informing Science. Special Issue on Information Science Research 3(2):63–66

Grubinger M (2007) Analysis and Evaluation of Visual Information Systems Performance. PhD thesis, School of Computer Science and Mathematics. Faculty of Health, Engineering and Science. Victoria University, Melbourne, Australia

Grubinger M, Clough PD, Müller H, Deselaers T (2006) The IAPR–TC12 Benchmark: A New Evaluation Resource for Visual Information Systems. In: International Workshop OntoImage'2006 Language Resources for Content–Based Image Retrieval, held in conjunction with LREC 2006, Genoa, Italy, pp 13–23

Grubinger M, Clough PD, Hanbury A, Müller H (2008) Overview of the ImageCLEFphoto 2007 Photographic Retrieval Task. In: Peters C, Jijkoun V, Mandl T, Müller H, Oard DW, Peñas A, Petras V, Santos D (eds) Advances in Multilingual and Multimodal Information Retrieval. 8th Workshop of the Cross–Language Evaluation Forum (CLEF 2007). Lecture Notes in Computer Science (LNCS), vol 5152. Springer, Budapest, Hungary, pp 433–444

Leung CHC, Ip H (2000) Benchmarking for Content–Based Visual Information Search. In: Laurini R (ed) Fourth International Conference On Visual Information Systems (VISUAL'2000). Lecture Notes in Computer Science (LNCS), vol 1929. Springer, Lyon, France, pp 442–456

Paramita ML, Sanderson M, Clough PD (2009) Developing a Test Collection to Support Diversity Analysis. In: Proceedings of the ACM SIGIR 2009 Workshop: Redundancy, Diversity, and Interdependence Document Relevance, Boston, MA, USA, pp 39–45

Paramita ML, Sanderson M, Clough PD (2010) Diversity in Photo Retrieval: Overview of the ImageCLEFphoto Task 2009. In: Peters C, Tsikrika T, Müller H, Kalpathy-Cramer J, Jones JFG, Gonzalo J, Caputo B (eds) Multilingual Information Access Evaluation Vol. II Multimedia Experiments: Proceedings of the 10th Workshop of the Cross–Language Evaluation Forum (CLEF 2009), Revised Selected Papers. Lecture Notes in Computer Science (LNCS). Springer

Zhai CX, Cohen WW, Lafferty J (2003) Beyond Independent Relevance: Methods and Evaluation Metrics for Subtopic Retrieval. In: SIGIR 2003: Proceedings of the 26th Annual International ACM SIGIR Conference on Research and Development in Informaion Retrieval. ACM press, Toronto, Canada, pp 10–17

Chapter 9
The Wikipedia Image Retrieval Task

Theodora Tsikrika and Jana Kludas

Abstract The Wikipedia image retrieval task at ImageCLEF provides a test–bed for the system–oriented evaluation of visual information retrieval from a collection of Wikipedia images. The aim is to investigate the effectiveness of retrieval approaches that exploit textual and visual evidence in the context of a large and heterogeneous collection of images that are searched for by users with diverse information needs. This chapter presents an overview of the available test collections, summarises the retrieval approaches employed by the groups that participated in the task during the 2008 and 2009 ImageCLEF campaigns, provides an analysis of the main evaluation results, identifies best practices for effective retrieval, and discusses open issues.

9.1 Introduction

The Wikipedia image retrieval task, also referred to as the WikipediaMM task, is an ad hoc image retrieval task whereby retrieval systems are given access to a collection of images to be searched but cannot anticipate the particular topics that will be investigated. The image collection consists of freely distributable Wikipedia[1] images annotated with user–generated textual descriptions of varying quality and length. Given a user's multimedia information need expressed both as a textual query and also through visual cues in the form of one or more sample images or visual concepts, the aim is to find as many relevant images as possible. Retrieval approaches should exploit the available textual and visual evidence, either in isolation or in combination, in order to achieve the best possible ranking for the user.

Theodora Tsikrika
Centrum Wiskunde & Informatica, Science Park 123, 1098 XG Amsterdam, The Netherlands
e-mail: theodora.tsikrika@acm.org

Jana Kludas
CUI, University of Geneva, Switzerland e-mail: Jana.Kludas@unige.ch

[1] http://www.wikipedia.org/

H. Müller et al. (eds.), *ImageCLEF*, The Information Retrieval Series 32,
DOI 10.1007/978-3-642-15181-1_9, © Springer-Verlag Berlin Heidelberg 2010

The task was set up in 2006 as part of the activities of the INEX Multimedia track (Westerveld and van Zwol, 2007), where it was referred to as the MMimages task. In 2008, the task moved to ImageCLEF, which not only forms a more natural environment for hosting this type of benchmark but also attracts more participants from the content–based image retrieval community. The overall goal of the task is to promote progress in large scale, multi–modal image retrieval via the provision of appropriate test collections that can be used to reliably benchmark the performance of different retrieval approaches using a metrics–based evaluation.

This chapter presents an overview of the Wikipedia image retrieval task in the ImageCLEF 2008 and 2009 evaluation campaigns (Tsikrika and Kludas, 2009, 2010). Section 9.2 presents the evaluation objectives of this task and describes the task's resources, i.e. the Wikipedia image collection and additional resources, the topics, and the relevance assessments. Section 9.3 lists the research groups that participated in these two years of the task under ImageCLEF, outlines the approaches they employed, and presents the results of the evaluation. Section 9.4 examines the results achieved by specific approaches in more detail so as to identify best practices and discuss open issues. Section 9.5 concludes this chapter, provides information on how to access the available resources, and discusses the future of the task.

9.2 Task Overview

9.2.1 Evaluation Objectives

The Wikipedia image retrieval task during the ImageCLEF 2008 and 2009 campaigns aimed to provide appropriate test collections for fostering research towards the following objectives:

Firstly, the task aimed to investigate how well image retrieval approaches, particularly those that exploit visual features, could deal with larger scale image collections. To this end, the goal was to provide a collection of more than 150,000 images; such a collection would be, for instance, much larger than the IAPR TC–12 image collection (Grubinger et al, 2006) that consists of 20,000 photographs and that was, at the time, employed in the ImageCLEF 2008 photo retrieval task (Arni et al, 2009).

Secondly, it aimed to examine how well image retrieval approaches could deal with a collection that contains highly heterogeneous items both in terms of their textual descriptions and their visual content. The textual metadata accompanying the Wikipedia images are user–generated, and thus outside any editorial control and correspond to noisy and unstructured textual descriptions of varying quality and length. Similarly, Wikipedia images cover highly diverse topics and since they are also contributed by Wikipedia users, their quality cannot be guaranteed. Such characteristics pose challenges for both text–based and visual–based retrieval approaches.

Finally, the main aim was to study the effectiveness of retrieval approaches that combine textual and visual evidence in order to satisfy a user's multimedia information need. Textual approaches had proven hard to beat in well–annotated image collections. However, such collections are not the norm in realistic settings, particularly in the Web environment. Therefore, there was a need to develop multi–modal approaches able to leverage all available evidence.

9.2.2 Wikipedia Image Collection

The collection of Wikipedia images used in the Wikipedia image retrieval task during the 2008 and 2009 ImageCLEF campaigns is a cleaned–up version of the image collection created in 2006 in the context of the activities of the INEX Multimedia track, where it was employed for the MMimages task in 2006 (Westerveld and van Zwol, 2007) and 2007 (Tsikrika and Westerveld, 2008). Due to its origins, the collection is referred to as the (INEX MM) Wikipedia image collection.

This image collection was created out of the more than 300,000 images contained within the 659,388 English Wikipedia articles that were downloaded and converted to XML (Denoyer and Gallinari, 2007) so as to form the structured document collection used for the ad hoc and other tasks at INEX 2006 (Malik et al, 2007) The user-generated metadata accompanying these Wikipedia images, usually a brief caption or description of the image, the Wikipedia user who uploaded the image, and the copyright information, were then downloaded and also converted to XML. Due to copyright issues or parsing problems with the downloaded metadata, some images had to be removed leaving a collection of approximately 170,000 images that was used in the INEX Multimedia tracks of 2006 (Westerveld and van Zwol, 2007) and 2007 (Tsikrika and Westerveld, 2008). Once the task became part of ImageCLEF in 2008, the collection was further cleaned up with the aim of keeping only JPEG and PNG images, leading to a collection of 151,519 diverse images with highly heterogeneous and noisy textual descriptions of varying length.

9.2.3 Additional Resources

To encourage participants to investigate multi-modal approaches that combine textual and visual evidence, particularly research groups with expertise only in the field of textual Information Retrieval, a number of additional resources were also provided.

In 2008, the following resources, computed during the INEX 2006 Multimedia track, were made available to support researchers who wished to exploit visual evidence without performing image analysis:

Image classification scores: For each image in the collection, the classification scores for the 101 MediaMill concepts were provided by the University of Amsterdam (Snoek et al, 2006). Their classifiers had been trained on manually annotated TREC Video Retrieval Evaluation (TRECVID) video data for concepts selected for the broadcast news domain.

Visual features: For each image in the collection, the set of the 120D feature vectors that had been used to derive the above image classification scores (van Gemert et al, 2006) were also made available. Participants could use these feature vectors to custom–build a content–based image retrieval system, without having to pre–process the image collection.

In 2009, the following resource was added:

Image similarity matrix: The similarity matrix for the images in the collection was constructed by the IMEDIA group at INRIA. For each image in the collection, this matrix contains the list of the top $K = 1,000$ most similar images in the collection together with their similarity scores. The same was given for each image used as a query example in the topics. The similarity scores are based on the distance between images; therefore, the lower the score, the more similar the images. Further details on the features and distance metric used can be found in Ferecatu (2005).

9.2.4 Topics

Topics are descriptions of multimedia information needs, with each topic containing textual and visual cues that can be used as evidence of the relevance of the images that should be retrieved. A number of factors have to be taken into consideration when creating topics for a test collection since such topics should reflect the real needs of operational retrieval systems, represent the types of services such systems might provide, be diverse, and differ in their coverage.

In 2008, the Wikipedia image retrieval task adopted the topic creation process introduced in INEX, whereby all participating groups were required to submit candidate topics. The participants were provided with topic development guidelines (Kludas and Tsikrika, 2008) which were based on guidelines created earlier in the context of INEX tasks (Larsen and Trotman, 2006). The participating groups submitted 70 topics altogether, which, together with 35 topics previously used in the INEX 2006 and 2007 Multimedia track, formed a pool of 105 candidate topics. Out of these, the task organisers selected a set of 75 topics. In 2009, participation in the topic development process was not mandatory, so only two of the participating groups submitted a total of 11 candidate topics. The rest of the candidate topics were created by the organisers with the help of the log of an image search engine. After a selection process performed by the organisers, a final list of 45 topics was created.

The topics consist of the following parts:

<title> query by keywords,
<image> query by image examples (one or more) — *optional in 2008*,
<concept> query by visual concepts (one or more) — *only in 2008 and optional*,

<narrative> definitive description of relevance and irrelevance.

The topic's <title> simulates a user who does not have (or does not want to use) example images or other visual cues. The query expressed in the <title> is therefore a text–only query. Upon discovering that a text–only query does not produce many relevant results, a user might decide to add visual cues and formulate a multimedia query. The topic's <image> provides visual cues that correspond to example images taken from outside or inside the (INEX MM) Wikipedia image collection and can be of any common format. In 2008, it was optional for topics to contain such image examples, whereas in 2009, each of the topics had at least one, and in many cases several, example images that could help describe the visual diversity of the topic. In 2008, additional visual cues were provided in the <concept> field that contained one or more of the 101 MediaMill concepts for which classification scores were provided.

These textual and visual evidences of relevance can be used in any combination by the retrieval systems; it is up to them how to use, combine or ignore this information. The relevance of a result does not directly depend on these constraints, but is decided by manual assessments based on the <narrative>. This field is not provided to the participants, but only to the assessors, and contains a clear and precise description of the information need in order to unambiguously determine whether or not a given image fulfils the given information need. The <narrative> is the only true and accurate interpretation of a user's needs. Precise recording of the narrative is important for scientific repeatability — there must exist, somewhere, a definitive description of what is and is not relevant to the user. To aid this, the <narrative> should explain not only what information is being sought, but also the context and motivation of the information need, i.e. why the information is being sought and what work–task it might help to solve.

Table 9.1 lists some statistics for the topics that were used during these two years of the task. The titles of these topics can be found in the overview papers of the task (Tsikrika and Kludas, 2009, 2010). The topics range from simple and thus relatively easy (e.g. 'bikes') to semantic and hence highly difficult (e.g. 'aerial photos of non–artificial landscapes'), with the latter forming the bulk of the topics. Semantic topics typically have a complex set of constraints, need world knowledge, and/or contain ambiguous terms; they were created so as to be challenging for current state–of–the–art retrieval algorithms. As mentioned above, in 2008, not all topics contained visual cues since the aim was to represent scenarios where users expressing their multimedia information needs do not necessarily employ visual evidence.

9.2.5 Relevance Assessments

In the Wikipedia image retrieval task, each image was assessed either as being relevant or as being non relevant, i.e. binary relevance was assumed. The retrieved images contained in the runs submitted by the participants were pooled together using a pool depth of 100 in 2008, which resulted in pools that ranged from 753 to

Table 9.1: Statistics for the topics in the ImageCLEF 2008 and 2009 Wikipedia image retrieval.

	2008	2009
Number of topics	75	45
Average number of terms in title	2.64	2.7
Average number of images per topic	0.61	1.9
Number of topics with image(s)	43	45
Number of topics with concept(s)	45	–
Number of topics with both image(s) and concept(s)	28	–
Number of topics with title only	15	–

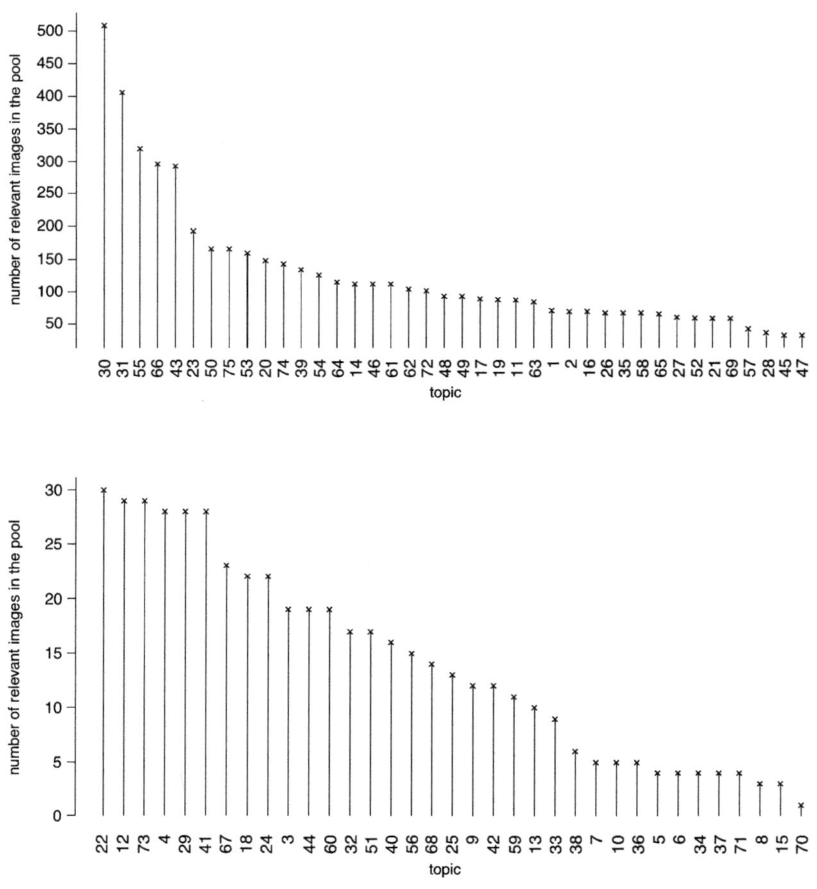

Fig. 9.1: Number of relevant images for each of the 2008 topics; topics are sorted in decreasing order of the number of their relevant images.

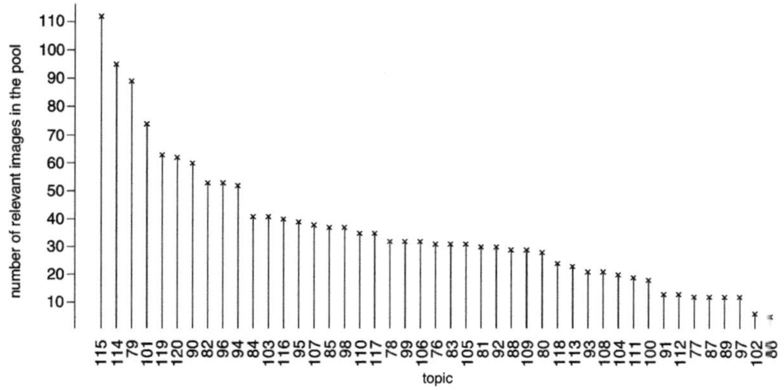

Fig. 9.2: Number of relevant images for each of the 2009 topics; topics are sorted in decreasing order of the number of their relevant images.

1,850 images with a mean and median both around 1,290, and a pool depth of 50 in 2009, which resulted in pools that ranged from 299 to 802 images with a mean and median both around 545. The evaluation was performed by the participants of the task within a period of four weeks after the submission of runs: 13 groups participated in 2008 and seven groups in 2009. The assessors used a Web–based relevance assessment system that had been previously employed in the INEX Multimedia and TREC Enterprise tracks (see Chapter 4 in this volume for further information on this system). In 2008, given that most topics were created by the participants, who were also employed as assessors, an effort was made so as to ensure that most of the topics were assigned to their creators. This was achieved in 76% of the assignments of the topics that were created that year.

Figures 9.1 and 9.2 depict the distribution of relevant images in the judged pools for each of the topics in 2008 and 2009, respectively. The variability in the number of relevant images across topics is evident, with most topics though having less than 100 relevant images. The mean number of relevant images per topic is 74.6 for 2008 and 36 for 2009, while the median is 36 and 31, respectively. Over all 120 topics, the mean number of relevant images per topic is 60.1 and the median 32.

9.3 Evaluation

9.3.1 Participants

Compared to its previous incarnation in the context of the INEX Multimedia track, the Wikipedia image retrieval task attracted more interest once it moved under the

Table 9.2: Groups that participated in the Wikipedia image retrieval task during the 2008 and 2009 ImageCLEF campaigns. Each entry lists the group ID, the academic or research institute hosting the group, the country where it is located, and the number of runs the group submitted in each of the campaigns.

Group ID	Institution	Country	2008	2009
cea	CEA-LIST	France	2	12
chemnitz	Chemnitz University of Technology	Germany	4	–
cwi	Centrum Wiskunde & Informatica	Netherlands	2	–
dcu	Dublin City University	Ireland	–	5
deuceng	Dokuz Eylul University	Turkey	–	6
iiit-h	International Institute of Information Technology, Hyderabad	India	–	1
imperial	Imperial College	UK	6	–
irit	Institut de Recherche en Informatique de Toulouse	France	4	–
lahc	Université Jean Monnet, Saint–Étienne	France	6	13
sinai	University of Jaen	Spain	–	4
sztaki	Hungarian Academy of Science	Hungary	8	7
ualicante	University of Alicante	Spain	24	9
unige	Université de Genève	Switzerland	2	–
upeking	Peking University	China	7	–
upmc/lip6	LIP6, Université Pierre et Marie Curie	France	7	–
utoulon	Université Sud Toulon–Var	France	5	–
		Total runs	77	57

auspices of ImageCLEF. The number of groups that participated by submitting runs was 12 in 2008 and eight in 2009, four of which were returning participants. Table 9.2 lists the participating groups along with the number of runs they submitted for the official evaluation; a total of 77 runs were submitted in 2008, while 57 runs were submitted in 2009. The overwhelming majority of participants are based in Europe, with the exception of only two groups, one from China and one from India.

9.3.2 Approaches

The approaches employed by the participants have been quite diverse. Both textual and visual features have been considered, either in isolation or in combination. Query and document expansion techniques that exploit semantic knowledge bases have been widely applied, as well as query expansion approaches that rely on blind relevance feedback. A short description of the participants' approaches is provided next. Each group is represented by its ID, followed by the year(s) in which the group participated in the task, and the publication(s) where the employed approaches are described in more detail. The groups are listed in alphabetical order of their ID.

cea (2008, 2009) (Popescu et al, 2009; Myoupo et al, 2010) In 2008, they employed Wikipedia and WordNet[2] as knowledge bases for automatically identifying and ranking concepts considered to be semantically related to those in the textual part of the query topics. These concepts were used for expanding the query, which was then submitted against the index of the images' textual descriptions, so as to generate a text–based ranking. In their visual analysis, the images in the collection were classified with respect to several visual concepts using Support Vector Machine (SVM)–based classifiers that exploited colour histogram and texture Local–Edge Pattern (LEP) visual features. Textual concepts in the queries triggered the use of visual concepts (e.g., persons' names triggered the use of the face detector) and the images' classification scores for these concepts were used for re–ranking the text–based results. In 2009, they refined the textual query expansion process by using knowledge extracted only from Wikipedia, whereas for the visual re–ranking they introduced a k–Nearest Neighbour (k–NN) based method. This method builds a visual model of the query using the top–ranked images retrieved by Google[3] and Yahoo![4] for that query and re–ranks the images in the text–based results based on their visual similarity to the query model.

chemnitz (2008) (Wilhelm et al, 2008) They employed their Xtrieval framework, which is based on Lucene[5] and PostgreSQL[6], and considered both textual and visual features, as well as the provided resources (image classification scores and low–level visual features). The text–based retrieval scores were combined with the visual similarity scores and further combined with the concept–based image classification scores. A thesaurus–based query expansion approach was also investigated.

cwi (2008) (Tsikrika et al, 2008) They employed PF/Tijah[7], an XML retrieval framework for investigating a language modelling approach based on purely textual evidence. A length prior was also incorporated so as to bias retrieval towards images with longer descriptions than the ones retrieved by the language model.

deuceng (2009) (Kilinc and Alpkocak, 2009) They applied a two–step approach: 1) text–based retrieval using expanded image descriptions and queries, and 2) re–ranking based on Boolean retrieval and text–based clustering. Terms and term phrases in both image descriptions and queries were expanded using WordNet, through the application of word sense disambiguation and WordNet similarity functions. The text–based results generated in this first step were then re–ranked in a Boolean manner by boosting the scores of the images that contained in their descriptions all the query terms in the exact same order as the query. The vectors of textual features of the results generated in the first step together with the vector of the expanded query were then clustered using the cover coefficient–

---

[2] `http://wordnet.princeton.edu/`

[3] `http://images.google.com/`

[4] `http://images.search.yahoo.com/`

[5] `http://lucene.apache.org/`

[6] `http://www.postgresql.org/`

[7] `http://dbappl.cs.utwente.nl/pftijah/`

based clustering methodology (C^3M). This allowed the calculation of similarity scores between the query vector and the vectors of the retrieved images. The final score was computed as a weighted sum of the Boolean re–ranking and the C^3M re–ranking scores. For further details, see Chapter 14 in this volume.

dcu (2009) (Min et al, 2010) They focused their experimentations on the expansion of the images' textual descriptions and of the textual part of the topics, using the Wikipedia abstracts' collection DBpedia[8] and blind relevance feedback. When DBpedia was employed, the terms from its top–ranked documents retrieved in response to the image description (or textual query) were sorted by their frequency and the top–ranked were selected to expand the images' (or queries') text. The term re–weighting was performed using Rocchio's formula. Query expansion was also performed using blind relevance feedback and BM25 term re–weighting. Lemur[9] was employed as the underlying retrieval framework.

iiit-h (2009) (Vundavalli, 2009) They employed a simple text–based approach that first used Boolean retrieval so as to narrow down the collection to the images accompanied by descriptions that contained all query terms and then ranked these images by applying the vector space model using a *tf.idf* weighting scheme.

imperial (2008) (Overell et al, 2008) They examined textual features, visual features, and their combination. Their text-based approach also took into account evidence derived from a geographic co-occurrence model mined from Wikipedia which aimed at disambiguating geographic references in a context-independent or a context-dependent manner. Their visual-based approach employed Gabor texture features and the City Block distance as a similarity measure. Text-based and visual-based results were combined using a convex combination of ranks. The results of this combination were further merged with results generated from using the top-ranked text-based results as blind relevance feedback in their visual retrieval approach.

irit (2008) (Torjmen et al, 2009) They explored the use of image names as evidence in text-based image retrieval. They first used them in isolation by computing a similarity score between the query and the name of the images in the collection using the vector space model. Then they used them in combination with textual evidence either by linearly combining the ranking of their text-based approach implemented in their XFIRM retrieval system with the ranking produced by the name-based technique or by applying a text-based approach that boosts the weights of terms that also occur in the image name.

lahc (2008, 2009) (Moulin et al, 2009, 2010) In 2008, they used a vector space model to compute similarities between vectors of both textual and visual terms. The textual terms corresponded to textual words and their weights were computed using BM25. The visual terms were obtained through a bag of words approach and corresponded to six–dimensional vectors of clusters of local colour features extracted from the images and quantized by k–means. Both manual

8 `http://dbpedia.org/`

9 `http://www.lemurproject.org/`

and blind relevance feedback were applied to a text–based run so as to expand the query with visual terms. In 2009, their document model was simplified so as to consider textual and visual terms separately and their approach was extended as follows. Additional textual information was extracted from the original Wikipedia articles that contained the images. Several local colour and texture features, including Scale Invariant Feature Transform (SIFT) descriptors, were extracted. Finally, the text–image combination was now performed by linearly combining the text–based and visual–based rankings.

sinai (2009) (Díaz-Galiano et al, 2010) Their approach focused on the expansion of the images' textual descriptions and of the textual part of the topics using WordNet. All nouns and verbs in the image descriptions and text queries were expanded by adding all unique words from all of their WordNet synsets without applying any disambiguation. Lemur was employed as the underlying retrieval framework.

sztaki (2008, 2009) (Racz et al, 2008; Daróczy et al, 2009) In 2008, they used their own retrieval system developed by the Hungarian Academy of Sciences and experimented with a text–based approach that used BM25 and query expansion based on Local Context Analysis (LCA), and its linear combination with a segment–based visual approach. In 2009, they preprocessed the textual image descriptions in order to remove author and copyright information with the aim to reduce the noise in the index. Their text–based approach again used BM25, but query expansion was performed by employing an on–line thesaurus. Their visual runs employed image segmentation and SIFT descriptors. The text–based and visual–based rankings were linearly combined to produce the final score.

ualicante (2008, 2009) (Navarro et al, 2008, 2009) In 2008, they employed their textual passage–based IR–n retrieval system as their baseline approach which was enhanced 1) by a module that decomposed the (compound) image file names in camel case notation into single terms, and 2) by a module that performed geographical query expansion. They also investigated two different term selection strategies for query expansion: probabilistic relevance feedback and local context analysis. In 2009, they further extended their approach by also using the top–ranked images (and their textual descriptions) returned by a content–based visual retrieval system as input for the above term selection strategies performing text–based query expansion.

unige (2008) They employed only textual features and their approach was based on the preference ranking option of the SVM light library developed by Cornell University. One run also applied feature selection to the high dimensional textual feature vector, based on the features relevant to each query.

upeking (2008) (Zhou et al, 2009) They investigated the following approaches: 1) a text–based approach based on the vector space model with *tf.idf* term weights, also using query expansion where the expansion terms were automatically selected from a knowledge base that was (semi–)automatically constructed from Wikipedia, 2) a content–based visual approach, where they first trained 1 vs. all classifiers for all queries by using the training images obtained by Yahoo! image search and then treated the retrieval task as a visual concept detection in

Runs	2008	2009
Textual	35	26
Visual	5	2
Mixed	37	29
All	77	57

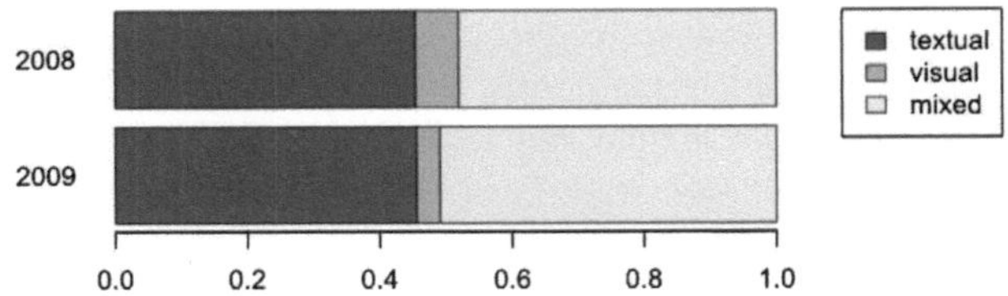

Fig. 9.3: Distribution of runs that employed textual, visual, or a combination of textual and visual low level features over the two years of the Wikipedia image retrieval task.

the given Wikipedia image set, and 3) a cross–media approach that combined the textual and visual rankings using the weighted sum of the retrieval scores.

umpc/lip6 (2008) (Fakeri-Tabrizi et al, 2008) They investigated text–based image retrieval by using a *tf.idf* approach, a language modelling framework, and their combination based on the ranks of retrieved images. They also experimented with the combination of textual and visual evidence by re–ranking the text–based results using visual similarity scores computed by either the Euclidean distance or a manifold–based technique, both on Hue/Saturation/Value (HSV) features.

utoulon (2008) (Zhao and Glotin, 2008) They applied the same techniques they used for the visual concept detection task at ImageCLEF 2008 (see Chapter 11 in this volume for details of that task) by relating each of the topics to one or more visual concepts from that task. These visual–based rankings were also fused with the results of a text–based approach.

All these different approaches can be classified with respect to whether they employ textual or visual low level features or a combination of both; in the latter case, an approach is characterised as mixed. Half of the groups that participated over the two years (eight out of the 16 groups) employed mixed approaches, whereas the other half relied only on textual features. Figure 9.3 shows the distribution of the submitted runs over the types of features they used. In both years of the task, mixed runs had a very slight edge over the textual runs.

The description of the runs submitted by the various groups also reveals that query expansion has been a very popular strategy as it has been applied by 11 of the 16 groups, either through the use of existing or purpose–built semantic knowledge bases (six out of the 11 groups), or through blind relevance feedback that takes into account textual or visual features (three out of the 11 groups), or as a combination of both these techniques (two out of the 11 groups). The application of query expansion aims to deal with the vocabulary mismatch problem, an issue which is particularly prominent in this test collection given both the short textual descriptions accompanying the images and the small number of images provided as query examples. A similar approach that has been applied by three out of the 16 groups with the aim to enrich the available textual descriptions of the Wikipedia images has been document expansion with the use of semantic knowledge bases. Next, the results of the runs submitted by the participating groups over the two years of the task are presented.

9.3.3 Results

The effectiveness of the submitted runs has been evaluated using the following measures: Mean Average Precision (MAP), P@10, P@20, and R–precision, i.e. precision when R (=number of relevant) documents are retrieved; see Chapter 5 in this volume for further details on these evaluation measures.

Figure 9.4 presents the best submitted run for each of the participating groups. Overall, the groups performed slightly better in 2008, an indication perhaps that the 2009 topics were more challenging for the participants. The best performing groups, *upeking* and *cea* in 2008, and *deuceng* in 2009, all employed query expansion, with the latter also performing document expansion, using semantic knowledge bases, such as WordNet and information extracted from Wikipedia. This indicates the usefulness of this approach in this particular setting. Furthermore, the best performing run both in 2008 and in 2009 relied only on textual evidence. This is better illustrated in Table 9.3 that presents a more complete overview of the submitted runs.

Table 9.3 shows for all runs, as well as for only the textual, visual, and mixed runs submitted in a year, the best, worst, median, and mean achieved performance for various evaluation measures. Both in 2008 and in 2009, the best values were achieved by runs that exploit only the available textual evidence. However, the differences between the best textual and the best mixed run for 2008 are not statistically significant for P@10 and P@20 ($p < 0.05$). Furthermore, the differences between the best textual and the best mixed run for 2009 are not statistically significant for all of the reported evaluation measures. On average, the median performance achieved by a mixed run in 2008 is slightly better than the median performance achieved by a textual run in terms of MAP and R–precision, while in 2009 the median values of all reported evaluation measures are higher for the mixed compared to the textual runs. On the other hand, the performance of the visual–only runs is comparatively low.

Given that a number of different evaluation measures were reported, a question that can be raised is whether there are any differences in these measures with respect to how they rank the submitted runs. To investigate this issue, the correlations among these measures were computed using the methodology described by Buckley and Voorhees (2005). For each evaluation measure, the runs are first ranked in order of decreasing performance with respect to that measure. The correlation between any two measures is then defined as the Kendall's τ correlation between the respective rankings. Table 9.4 presents the results of this analysis, where in addition to the evaluation measures previously reported, i.e. MAP, P@10, P@20, and R–precision, the total number of relevant images retrieved (abbreviated as 'Rel ret'), i.e. the sum of the number of relevant images retrieved across all topics for

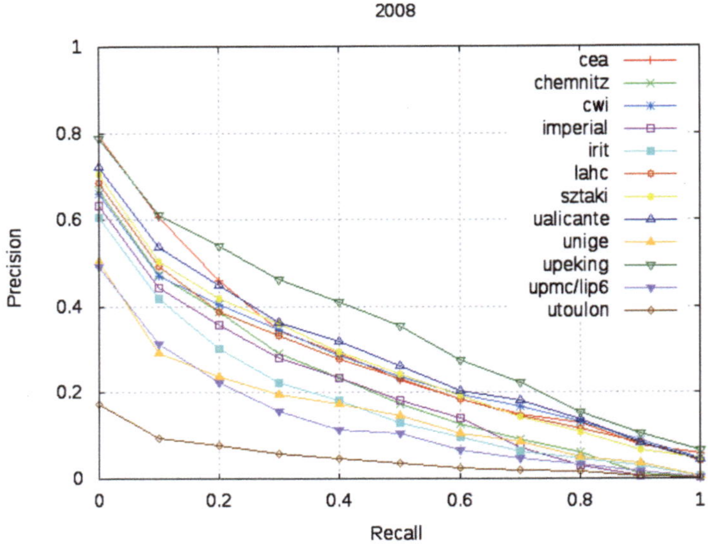

(a) The best retrieval results per group for the 2008 Wikipedia image retrieval task.

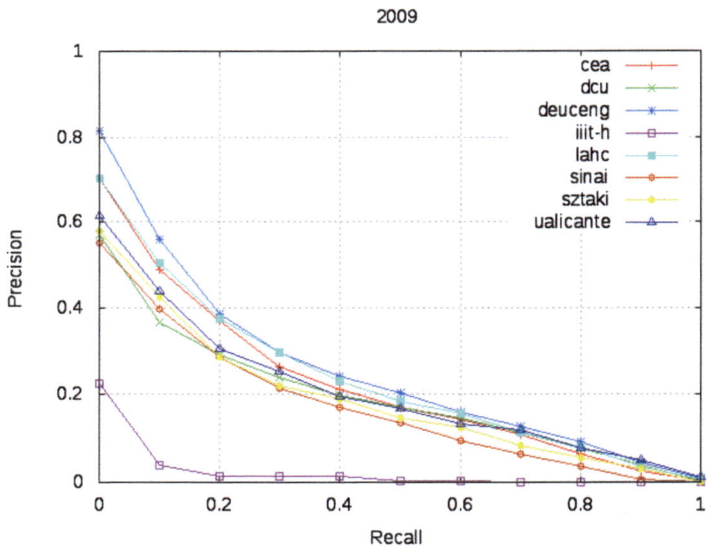

(b) The best retrieval results per group for the 2009 Wikipedia image retrieval task.

Fig. 9.4: The best retrieval results per group.

a year, is also reported. The correlations between the MAP, P@10, P@20, and R–precision measures are all at least 0.67 showing that each pair of measures is corre-

Table 9.3: The best, worst, median and mean performance achieved by all, text only, visual only, and mixed only runs for MAP, P@10, P@20, and R–precision in the 2008 and the 2009 Wikipedia image retrieval tasks. The standard deviation of the performance achieved by the runs in each case is also listed.

		2008				2009			
		MAP	P@10	P@20	R-prec.	MAP	P@10	P@20	R-prec.
		77 runs				57 runs			
All runs	max	0.3444	0.4760	0.3993	0.3794	0.2397	0.4000	0.3189	0.2708
	min	0.0010	0.0027	0.0033	0.0049	0.0068	0.0244	0.0144	0.0130
	median	0.2033	0.3053	0.2560	0.2472	0.1699	0.2644	0.2267	0.2018
	mean	0.1756	0.2761	0.2230	0.2122	0.1578	0.2624	0.2153	0.1880
	stdev	0.0819	0.1169	0.0936	0.0920	0.0571	0.0861	0.0702	0.0631
		35 runs				26 runs			
Textual runs	max	0.3444	0.4760	0.3993	0.3794	0.2397	0.4000	0.3189	0.2708
	min	0.0399	0.0467	0.0673	0.0583	0.0186	0.0689	0.0389	0.0246
	median	0.2033	0.3107	0.2587	0.2472	0.1680	0.2600	0.2178	0.1987
	mean	0.1953	0.2972	0.2453	0.2356	0.1693	0.2717	0.2232	0.1992
	stdev	0.0662	0.0859	0.0690	0.0684	0.0452	0.0717	0.0574	0.0487
		5 runs				2 runs			
Visual runs	max	0.1928	0.4507	0.3227	0.2309	0.0079	0.0222	0.0222	0.0229
	min	0.0010	0.0027	0.0033	0.0049	0.0068	0.0144	0.0144	0.0130
	median	0.0037	0.0147	0.0120	0.0108	0.0074	0.0183	0.0183	0.0179
	mean	0.0781	0.1848	0.1336	0.0962	0.0074	0.0183	0.0183	0.0179
	stdev	0.1039	0.2415	0.1726	0.0122	0.0008	0.0055	0.0055	0.0070
		37 runs				29 runs			
Mixed runs	max	0.2735	0.4653	0.3840	0.3225	0.2178	0.3689	0.2867	0.2538
	min	0.0053	0.0040	0.0047	0.0049	0.0321	0.1044	0.0644	0.0423
	median	0.2083	0.3053	0.2547	0.2536	0.1801	0.2778	0.2389	0.2103
	mean	0.1701	0.2684	0.2139	0.2056	0.1578	0.2706	0.2218	0.1898
	stdev	0.0841	0.1172	0.0949	0.0967	0.0543	0.0776	0.0063	0.0605

Table 9.4: Kendall's τ correlations between pairs of system rankings based on different evaluation measures.

	2008				2009			
	P@10	P@20	R-prec.	Rel ret	P@10	P@20	R-prec.	Rel ret
MAP	0.725	0.797	0.917	0.602	0.808	0.853	0.868	0.538
P@10		0.675	0.715	0.505		0.807	0.777	0.424
P@20			0.779	0.533			0.810	0.489
R-prec.				0.589				0.466

lated, whereas their correlation with the number of relevant images retrieved is relatively low. The highest correlation is between R–precision and MAP; this has also been observed in the analysis of the TREC–7 ad hoc results (Buckley and Voorhees, 2005). Even though R–precision evaluates at exactly one point in a retrieval ranking, while MAP represents the entire area underneath the recall–precision curve, the fact that these two measures rank runs in a similar manner supports the consideration of R–precision as an overall system performance measure.

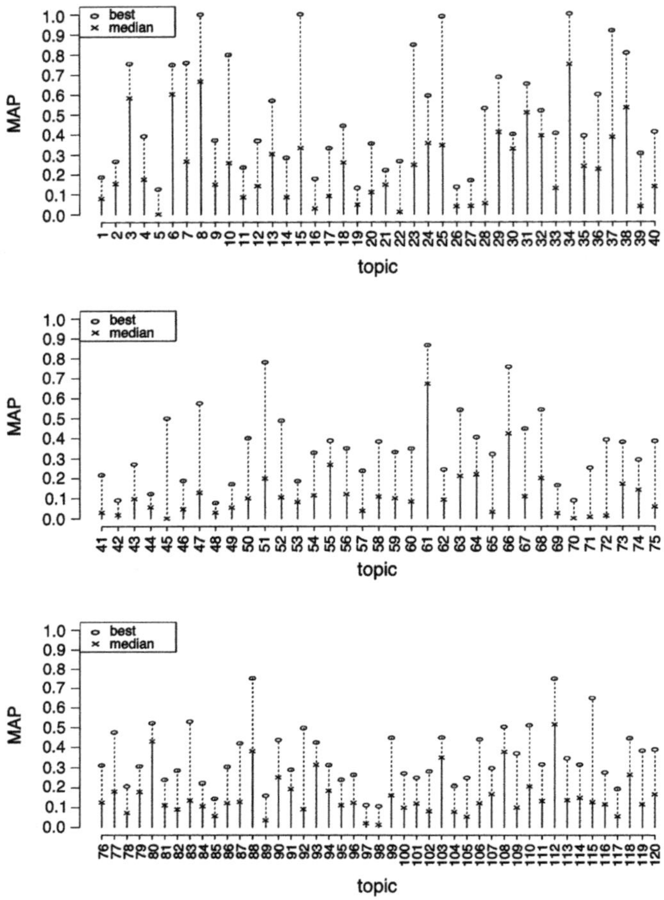

Fig. 9.5: Best and median MAP value achieved for the 2008 topics (top, middle) and for the 2009 topics (bottom).

Apart from the performance achieved over all topics in a year, it is also useful to examine the per topic performance so as to identify the problematic cases. Figure 9.5 presents for each of the topics the best MAP value achieved for that topic by a submitted run, as well as the median performance of all runs for that topic. The variability of the systems' performances over topics indicates the differences in their levels of difficulty, with some topics being very difficult for many of the submitted runs, as illustrated by the low values of the median performance. More detailed per topic analyses can be found in the overview papers of the task (Tsikrika and Kludas, 2009, 2010). Next, the results achieved by specific approaches are further examined so as to identify best practices and discuss open issues.

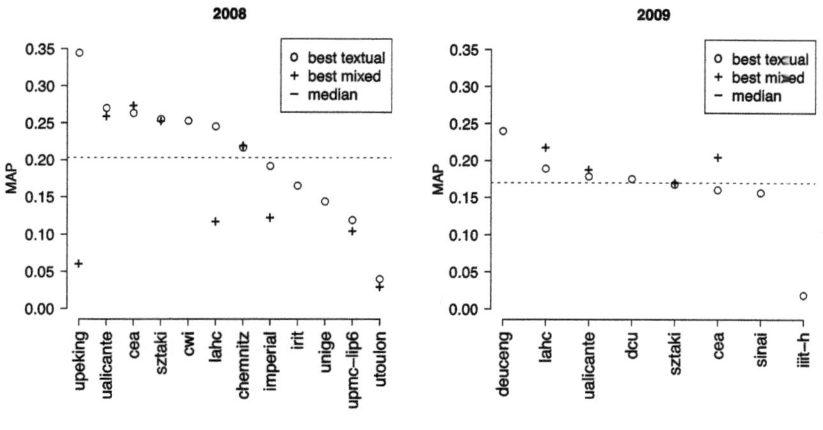

Fig. 9.6: Best textual and best mixed run (if any) for each of the participants in the Wikipedia image retrieval 2008 and 2009 tasks. The groups are ranked in decreasing order of the MAP of their best textual run.

9.4 Discussion

9.4.1 Best Practices

Over the course of these two years, a variety of different approaches have been evaluated using the test collections constructed in the context of the activities of the Wikipedia image retrieval task. To identify some of the best practices among the various techniques that have been applied, the relative performance of the submitted runs is examined.

Figure 9.6 presents for each of the groups that participated in each of the two years, the MAP achieved by its best textual and by its best mixed run (if any), together with the median MAP of all the runs submitted in that year. The group that performed best in each of the two years, *upeking* (Zhou et al, 2009) in 2008 and *deuceng* (Kilinc and Alpkocak, 2009) in 2009, applied textual query expansion using semantic knowledge bases, such a WordNet or knowledge bases extracted from Wikipedia. A similar approach was also applied by the group that achieved the third highest performance of a textual run in 2008, i.e. *cea* (Popescu et al, 2009). Furthermore, the best performing group in 2009, *deuceng* (Kilinc and Alpkocak, 2009), also applied document expansion using semantic knowledge bases. Document and query expansion using DBpedia were also applied by *dcu* (Min et al, 2010) in 2009 and achieved improvements over their textual baseline. All this constitutes strong evidence that such expansion techniques, particularly when applied judiciously so as to deal with the noise that can be potentially added, are particularly effective for such collections of images that are accompanied by short and possibly noisy textual descriptions.

An interesting observation regarding the relative performance of textual and mixed runs is that in 2009 the groups that submitted both textual and mixed runs achieved their best results with their mixed runs. Notable cases are the *lahc* (Moulin et al, 2009, 2010) and *cea* (Popescu et al, 2009; Myoupo et al, 2010) groups that also managed to dramatically improve the performance of their mixed runs in comparison to their 2008 submissions. The improvements achieved by *lahc* were mainly due to the extraction of additional low–level visual features, including SIFT descriptors, and the combination taking place at the post–retrieval stage, as a linear combination of the text-based and visual-based rankings, rather than by considering vectors of both textual and visual terms, as they did in 2008. For *cea*, the major improvement was derived from the employment of a query model that was built using a large number of sample images automatically retrieved from the Web; in their post–submission runs, they managed to further improve the performance of their mixed runs after correcting a bug (Myoupo et al, 2010).

A final source of evidence that has also shown to be useful in this Wikipedia setting corresponds to the image names. Approaches that take them into account have shown improvements over equivalent approaches that do not in three separate cases: *ualicante* (Navarro et al, 2008) and *irit* (Torjmen et al, 2009) in 2008, and *dcu* (Min et al, 2010).

9.4.2 Open Issues

The results presented provide some clear indications on the usefulness of particular textual techniques in the context of this task but do not yet provide sufficient evidence on the best practice to follow when combining multiple modalities; further research is needed in this direction. Furthermore, apart from the encouraging results achieved in 2008 by *cea* (Popescu et al, 2009), the effectiveness of using visual concepts in an ad hoc retrieval task has not been fully explored. To this end, an effort should be made to provide classification scores for the images in the Wikipedia collection. Given the poor generalisation of concept classifiers to domains other than their training domain (Yang and Hauptmann, 2008), it would be best to build classifiers using training samples from Wikipedia. This could potentially be explored in synergy with the image annotation task (see Chapter 11 in this volume). Finally, there should be further efforts in lowering the threshold for the participation in the benchmark by providing resources to support the participants' experiments.

9.5 Conclusions and the Future of the Task

The Wikipedia image retrieval task provides test collections with the aim of supporting the reliable benchmarking of the performance of retrieval approaches that exploit textual and visual evidence for ad hoc image retrieval in the context of a large and heterogeneous collection of freely distributable Wikipedia images that are searched for by users with diverse information needs. Over the course of these two years at ImageCLEF, a variety of retrieval approaches have been been investigated and interesting conclusions have been reached regarding best practices in the field. Nonetheless, much work remains to be done. Future runs of the task will continue to examine the same evaluation objectives using even larger image collections (already the collection provided in 2010 consists of approximately 250,000 Wikipedia images) and exploring their multi–lingual aspects. Further experimentation with the test collections constructed thus far is possible by downloading them from ImageCLEF's resources page[10].

Acknowledgements Theodora Tsikrika was supported by the European Union via the European Commission project VITALAS (EU–IST–IP 045389). Jana Kludas was funded by the European Commission project MultiMATCH (EU–IST–STREP 033104) and the Swiss National Fund (SNF). The authors would also like to thank Thijs Westerveld and Roelof van Zwol for creating the (INEX MM) Wikipedia image collection and setting up the task in the context of the activities of the INEX Multimedia track, Thomas Deselaers for invaluable technical support during ImageCLEF 2008, and all the groups that participated in the task and in the relevance assessment process.

References

Arni T, Clough PD, Sanderson M, Grubinger M (2009) Overview of the ImageCLEFphoto 2008 photographic retrieval task. In: Peters et al (2009), pp 500–511

Buckley C, Voorhees EM (2005) Retrieval system evaluation. In: Voorhees EM, Harman DK (eds) TREC: Experiment and Evaluation in Information Retrieval, Digital Libraries and Electronic Publishing, MIT Press, chap 3, pp 53–75

Daróczy B, Petrás I, Benczúr AA, Fekete Z, Nemeskey D, Siklósi D, Weiner Z (2009) SZTAKI @ ImageCLEF 2009. In: Working Notes of CLEF 2009, Corfu, Greece

Denoyer L, Gallinari P (2007) The Wikipedia XML corpus. In: Fuhr et al (2007), pp 12–19

Díaz-Galiano M, Martín-Valdivia M, Urena-López L, Perea-Ortega J (2010) Using WordNet in multimedia information retrieval. In: Peters et al (2010)

Fakeri-Tabrizi A, Amini MR, Tollari S, Gallinari P (2008) UPMC/LIP6 at ImageCLEF wikipediaMM: an image–annotation model for an image search–engine. In: Working Notes of CLEF 2008, Aarhus, Denmark

Ferecatu M (2005) Image retrieval with active relevance feedback using both visual and keyword-based descriptors. In: Ph.D. Thesis, Université de Versailles, France

Fuhr N, Lalmas M, Trotman A (eds) (2007) Comparative Evaluation of XML Information Retrieval Systems, Proceedings of the 5th International Workshop of the Initiative for the Evaluation of XML Retrieval (INEX 2006), Revised Selected Papers, Lecture Notes in Computer Science (LNCS), vol 4518, Springer

[10] http://www.imageclef.org/datasets/

van Gemert JC, Geusebroek JM, Veenman CJ, Snoek CGM, Smeulders AWM (2006) Robust scene categorization by learning image statistics in context. In: Proceedings of the 2006 Conference on Computer Vision and Pattern Recognition Workshop. IEEE Computer Society, Washington, DC, USA, p 105

Grubinger M, Clough PD, Leung C (2006) The IAPR TC–12 benchmark for visual information search. IAPR Newsletter 28(2):10–12

Kilinc D, Alpkocak A (2009) DEU at ImageCLEF 2009 wikipediaMM task: Experiments with expansion and reranking approaches. In: Working Notes of CLEF 2009, Corfu, Greece

Kludas J, Tsikrika T (2008) ImageCLEF 2008 wikipediaMM task guidelines. Unpublished document distributed to ImageCLEF 2008 wikipediaMM participants

Larsen B, Trotman A (2006) INEX 2006 guidelines for topic development. In: Fuhr N, Lalmas M, Trotman A (eds) Preproceedings of the 5th International Workshop of the Initiative for the Evaluation of XML Retrieval (INEX 2006), pp 373–380

Malik S, Trotman A, Lalmas M, Fuhr N (2007) Overview of INEX 2006. In: Fuhr et al (2007), pp 1–11

Min J, Wilkins P, Leveling J, Jones GJF (2010) Document expansion for text-based image retrieval at CLEF 2009. In: Peters et al (2010)

Moulin C, Barat C, Géry M, Ducottet C, Largeron C (2009) UJM at ImageCLEFwiki 2008. In: Peters et al (2009), pp 779–786

Moulin C, Barat C, Lemaître C, Géry M, Ducottet C, Largeron C (2010) Combining text/image in wikipediaMM task 2009. In: Peters et al (2010)

Myoupo D, Popescu A, Borgne HL, Moëllic PA (2010) Multimodal image retrieval over a large database. In: Peters et al (2010)

Navarro S, Muñoz R, Llopis F (2008) A textual approach based on passages using IR–n in wikipediaMM task 2008. In: Working Notes for the CLEF 2008 Workshop, Aarhus, Denmark

Navarro S, Muñoz R, Llopis F (2009) Evaluating fusion techniques at different domains at ImageCLEF subtasks. In: Working Notes of CLEF 2009, Corfu, Greece

Overell S, Llorente A, Liu H, Hu R, Rae A, Zhu J, Song D, Rüger S (2008) MMIS at ImageCLEF 2008: Experiments combining different evidence sources. In: Working Notes for the CLEF 2008 Workshop, Aarhus, Denmark

Peters C, Deselaers T, Ferro N, Gonzalo J, Jones GJF, Kurimo M, Mandl T, Peñas A, Petras V (eds) (2009) Evaluating Systems for Multilingual and Multimodal Information Access: Proceedings of the 9th Workshop of the Cross–Language Evaluation Forum (CLEF 2008), Revised Selected Papers, Lecture Notes in Computer Science (LNCS), vol 5706, Springer

Peters C, Tsikrika T, Müller H, Kalpathy-Cramer J, Jones GJF, Gonzalo J, Caputo B (eds) (2010) Multilingual Information Access Evaluation II, Multimedia Experiments: Proceedings of the 10th Workshop of the Cross–Language Evaluation Forum (CLEF 2009), Revised Selected Papers, Lecture Notes in Computer Science (LNCS), Springer

Popescu A, Borgne HL, Moëllic PA (2009) Conceptual image retrieval over a large scale database. In: Peters et al (2009), pp 771–778

Racz S, Daróczy B, Siklósi D, Pereszlényi A, Brendel M, Benczúr AA (2008) Increasing cluster recall of cross-modal image retrieval. In: Working Notes for the CLEF 2008 Workshop, Aarhus, Denmark

Snoek CGM, Worring M, van Gemert JC, Geusebroek JM, Smeulders AWM (2006) The challenge problem for automated detection of 101 semantic concepts in multimedia. In: Proceedings of the 14th Annual ACM International Conference on Multimedia. ACM press, New York, NY, USA, pp 421–430

Torjmen M, Pinel-Sauvagnat K, Boughanem M (2009) Evaluating the impact of image names in context-based image retrieval. In: Peters et al (2009), pp 756–762

Tsikrika T, Kludas J (2009) Overview of the wikipediaMM task at ImageCLEF 2008. In: Peters et al (2009), pp 539–550

Tsikrika T, Kludas J (2010) Overview of the wikipediaMM task at ImageCLEF 2009. In: Peters et al (2010)

Tsikrika T, Westerveld T (2008) The INEX 2007 Multimedia track. In: Fuhr N, Lalmas M, Trotman A, Kamps J (eds) Focused access to XML documents, Proceedings of the 6th International Workshop of the Initiative for the Evaluation of XML Retrieval (INEX 2007), Springer, Lecture Notes in Computer Science (LNCS), vol 4862, pp 440–453

Tsikrika T, Rode H, de Vries AP (2008) CWI at ImageCLEF 2008. In: Working Notes for the CLEF 2008 Workshop, Aarhus, Denmark

Vundavalli S (2009) IIIT-H at ImageCLEF Wikipedia MM 2009. In: Working Notes of CLEF 2009, Corfu, Greece

Westerveld T, van Zwol R (2007) The INEX 2006 Multimedia track. In: Fuhr et al (2007), pp 331–344

Wilhelm T, Kürsten J, Eibl M (2008) The Xtrieval framework at CLEF 2008: ImageCLEF wikipediaMM task. In: Working Notes for the CLEF 2008 Workshop, Aarhus, Denmark

Yang J, Hauptmann AG (2008) (Un)Reliability of video concept detection. In: Luo J, Guan L, Hanjalic A, Kankanhalli MS, Lee I (eds) Proceedings of the 7th International Conference on Content–based Image and Video Retrieval (CIVR 2008), ACM press, pp 85–94

Zhao ZQ, Glotin H (2008) Concept content based Wikipedia web image retrieval using CLEF VCDT 2008. In: Working Notes of CLEF 2008, Aarhus, Denmark

Zhou Z, Tian Y, Li Y, Huang T, Gao W (2009) Large–scale cross-media retrieval of wikipediaMM images with textual and visual query expansion. In: Peters et al (2009), pp 763–770

Chapter 10
The Robot Vision Task

Andrzej Pronobis and Barbara Caputo

Abstract In 2009, ImageCLEF expanded its tasks with the introduction of the first robot vision challenge. The overall focus of the challenge is semantic localization of a robot platform using visual place recognition. This is a key topic of research in the robotics community today. This chapter presents the goals and achievements of the first edition of the robot vision task. We describe the task, the method of data collection used and the evaluation procedure. We give an overview of the obtained results and briefly highlight the most promising approaches. We then outline how the task will evolve in the near and distant future.

10.1 Introduction

A fundamental competence for a mobile robot is to know its position in the world. Providing robots with the ability to build an internal representation of the surrounding space, so as to be able to derive robust information about their location therein, can be considered as one of the most relevant research challenges for the robotics community today. The topic has been vastly researched, resulting in a broad range of approaches spanning from the purely metric (Jogan and Leonardis, 2003; Dissanayake et al, 2001; Wolf et al, 2005), to topological (Ulrich and Nourbakhsh, 2000; Ullah et al, 2008; Cummins and Newman, 2008), and hybrid (Thrun, 1998; Brunskill et al, 2007). As robots break down the barriers and start to interact with people (Zender et al, 2008) and operate in large–scale environments (Cummins and Newman, 2008; Ullah et al, 2008), topological models are becoming more popular

Andrzej Pronobis
Department of Computer Science, Royal Institute of Technology, Stockholm, Sweden e-mail: pronobis@nada.kth.se

Barbara Caputo
Idiap Research Institute, Centre du Parc, Rue Marconi 19, PO Box 592, CH-1920 Martigny, Switzerland e-mail: bcaputo@idiap.ch

H. Müller et al. (eds.), *ImageCLEF*, The Information Retrieval Series 32,
DOI 10.1007/978-3-642-15181-1_10,

as a way to augment, or even replace, purely metric space representations. In particular, research on building topological maps has been pushing for methods suitable for place recognition.

Traditionally, sonar and/or laser have been the sensory modalities of choice for place recognition and topological localization (Nourbakhsh et al, 1995; Martínez MozoOs et al, 2007). The assumption that the world can be represented in terms of two dimensional geometrical information proved convenient for many applications. However, the inability to capture important aspects of complex realistic environments leads to the problem of perceptual aliasing (Kuipers and Beeson, 2002), and vastly limits the usability of purely geometrical methods. Recent advances in vision have made this modality emerge as a natural and viable solution. Vision provides richer sensory input allowing for better discrimination. It opens up new possibilities for building cognitive systems, actively relying on the semantic context. Not unimportant is the cost effectiveness, portability and popularity of visual sensors. As a result, this line of research is attracting more and more attention, and several methods have been proposed using vision alone (Torralba et al, 2003; Pronobis and Caputo, 2006; Siagian and Itti, 2007; Cummins and Newman, 2008), or combined with more traditional range sensors (Kortenkamp and Weymouth, 1994; Tapus and Siegwart, 2005; Pronobis et al, 2008).

In spite of significant progress, vision–based localization still represents a major challenge. Firstly, visual information tends to be noisy and difficult to interpret. The visual appearance of places varies with time because of illumination changes (day and night, artificial light either on and off) and because of human activities (furniture moved around, objects being taken out of drawers, and so on). Thus, the solutions must be highly robust, provide good generalization abilities and in general be adaptive. Additionally, the application puts strong constraints on the computational complexity and the increased resolution, while dimensionality of the visual data still constitutes a problem. The fact that so many different parameters influence the accuracy of a vision–based localization system is another challenge in itself, proving especially burdensome at the design stage. As the results depend greatly on the choice of training and test input data, which are unstable over time, it is hard to measure the influence of the different parameters on the overall performance of the system. For the same reason, it becomes nearly impossible to compare solutions in a fair way, as they are usually evaluated in different environments, in different conditions, and under varying assumptions. This is a major obstacle slowing down progress in the field. There is a need for standardized benchmarks and databases, which would allow for fair comparisons, simplify the experimental process and boost development of new solutions. Databases are heavily exploited in the computer vision community, especially for object recognition and categorization (Griffin et al, 2007; MIT-CSAIL, 2006). As the community acknowledges the need for benchmarking, a lot of attention is directed towards designing new data sets, reflecting the increasing capabilities of visual algorithms (Ponce et al, 2006). Also in robotics, research on Simultaneous Localization and Mapping (SLAM) makes use of several publicly available data sets (Howard and Roy, 2003; Nebot, 2006).

However, no database has yet emerged as a standard benchmark for visual place recognition applied to robot localization.

The robot vision task aims at filling this gap, and provides a benchmark to the research community working on the issues described above. The task has been introduced for the first time in 2009 and it has immediately attracted a considerable attention, with seven participating groups and a total of 24 valid runs submitted. These very encouraging first results support us in our vision and make us foresee several future editions of the challenge.

In the rest of the chapter we describe in detail the first edition of the task in Section 10.2, then give a brief overview on how the task is currently evolving in its 2010 implementations in Section 10.3. We conclude with an overall discussion and discussion on future goals.

10.2 The Robot Vision Task at ImageCLEF 2009: Objectives and Overview

The two main objectives of the robot vision task at ImageCLEF are to push forward research on semantic spatial modeling for robot localization, while at the same time making this research field easier to approach also by groups from other research fields, with no previous experience on robotics and robot vision.

To achieve this last objective, we are committed to provide to participants data sequences acquired from mobile robot platforms. This is in contrast with existing benchmark evaluation challenges in robotics, where participants are requested to operate their algorithms on robot platforms (Nebot, 2006; Howard and Roy, 2003). By making this choice, we aim at attracting the attention of researchers from the pattern recognition, computer vision and machine learning fields, who usually test their algorithms on benchmark databases but who would find it daunting to approach a full robotic system for the same task.

The achievement of the first objective requires the definition of a set of subsequent tasks, of increasing complexity over the years, so as to progressively raise the bar and focus on the open challenges that are timely to attack. In the rest of this section we describe in detail the first edition of the robot vision task, which was held in 2009, and where the focus was on topological localization from data acquired by a perspective camera. Here, the challenge was to achieve robustness under varying imaging conditions. We first give a general description of the task (Section 10.2.1). Then, we describe the data set used in more detail (Section 10.2.2). Section 10.2.3 describes how we evaluated the performance of the submitted runs. A thorough description of the outcome of the task is given in Section 10.2.4. The two coming editions of the robot vision task, organized for 2010, shifted the focus onto the place categorization problem (Section 10.3).

10.2.1 The Robot Vision Task 2009

The robot vision task at ImageCLEF 2009 addressed the problem of topological localization of a mobile robot using visual information. We asked participants to determine the topological location of a robot based on images acquired with a perspective camera, mounted on a robot platform. The image sequences were recorded in a five room subsection of an indoor environment, under fixed illumination conditions and at a fixed time. The challenge was to build a system able to answer the question 'where are you?' ('I'm in the kitchen', 'in the corridor', etc) when presented with a test sequence containing images acquired in the previously observed part of the environment, or in additional rooms that were not imaged in the training sequence. The test images were acquired 6–20 months after the training sequence, possibly under different illumination settings. The system had to assign each test image to one of the rooms that were present in the training sequence, or it had to indicate that the image came from a room that was not seen during training.

The overall task was further divided in two separate sub–tasks, one mandatory and one optional. In the mandatory task, the algorithm had to provide information about the location of the robot separately for each test image. In the optional task, the algorithm was allowed to exploit the continuity of the sequences and to rely on the test images already seen.

10.2.2 Robot Vision 2009: The Database

The image sequences consisted of a subset of the publicly available IDOL2 database (Luo et al, 2007) for the training and validation set, and of a previously unreleased sequence for test. All sequences were acquired with a Canon VC–C4 perspective camera, using the resolution of 320 x 240 pixels, mounted on a MobileRobots PowerBot robot platform (see Figure 10.1). The acquisition was performed in a five room subsection of a larger office environment, selected so that each of the five rooms represented a different functional area: a one–person office, a two–person office, a kitchen, a corridor, and a printer area. Figure 10.2 shows the map of the environment.

For the training and validation sequences, the visual appearance of the rooms was captured under three different illumination conditions: in cloudy weather, in sunny weather, and at night. The robot was manually driven through each of the five rooms while continuously acquiring images and laser range scans at a rate of 5 fps. Each data sample was then labeled as belonging to one of the rooms according to the position of the robot during acquisition, rather than according to the content of the images. Examples of images showing the interior of the rooms, variations observed over time and caused by activities in the environments, as well as induced by changes in illumination, are shown in Figure 10.3.

The database was designed to test the robustness of place recognition algorithms to variations that occur over a long period of time. Therefore, the acquisition pro-

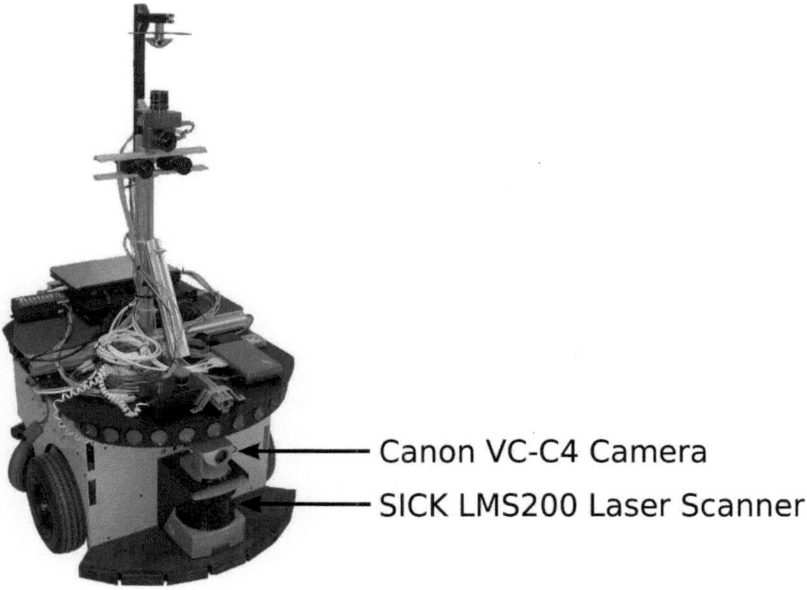

Fig. 10.1: The MobileRobots PowerBot mobile robot platform used for data acquisition in the robot vision task.

cess was conducted in two phases. Two sequences were acquired for each type of illumination conditions over the time span of more than two weeks, and another two sequences for each setting were recorded six months later (12 sequences in total). Thus the sequences captured the variability introduced not only by illumination but also by natural activities in the environment (presence/absence of people, furniture/objects relocated, etc.).

The test sequences were acquired in the same environment, using the same camera set–up. The acquisition was performed 20 months after the training data. The sequences contain additional rooms that were not imaged in the IDOL2 database.

10.2.3 Robot Vision 2009: Performance Evaluation

The image sequences used in the competition were annotated with ground truth. The annotations of the training and validation sequences were available to the participants, while the ground truth for the test sequence was released after the results were announced. Each image in the sequences was labelled according to the position of the robot during acquisition as belonging to one of the rooms used for training or as an unknown room. The ground truth was then used to calculate a score indicating

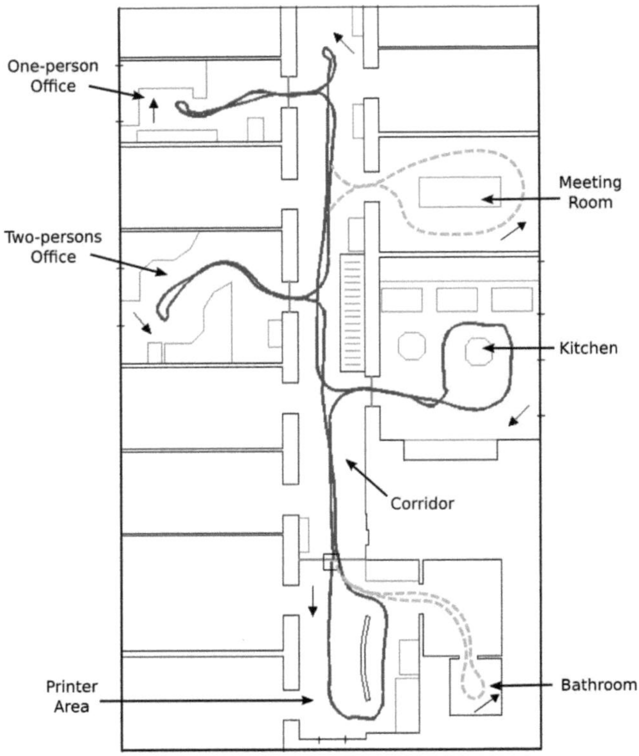

Fig. 10.2: Map of the environment with the approximate path followed by the robot during acquisition of the training, validation and testing data for the 2009 edition of the robot vision task. The dashed segments of the path correspond to the rooms available only in the test set.

the performance of an algorithm on the test sequence. The following rules were used when calculating the overall score for the whole test sequence:

- +1.0 point was given for each image classified correctly;
- +1.0 point was given for each image identified correctly as an unknown room;
- -0.5 points were given for each image misclassified;
- 0.0 points were given for each image where the algorithm did not provide any indication, i.e. for each not classified image.

The sum of all scores obtained for all images in the test sequences gave the overall score for each submitted run.

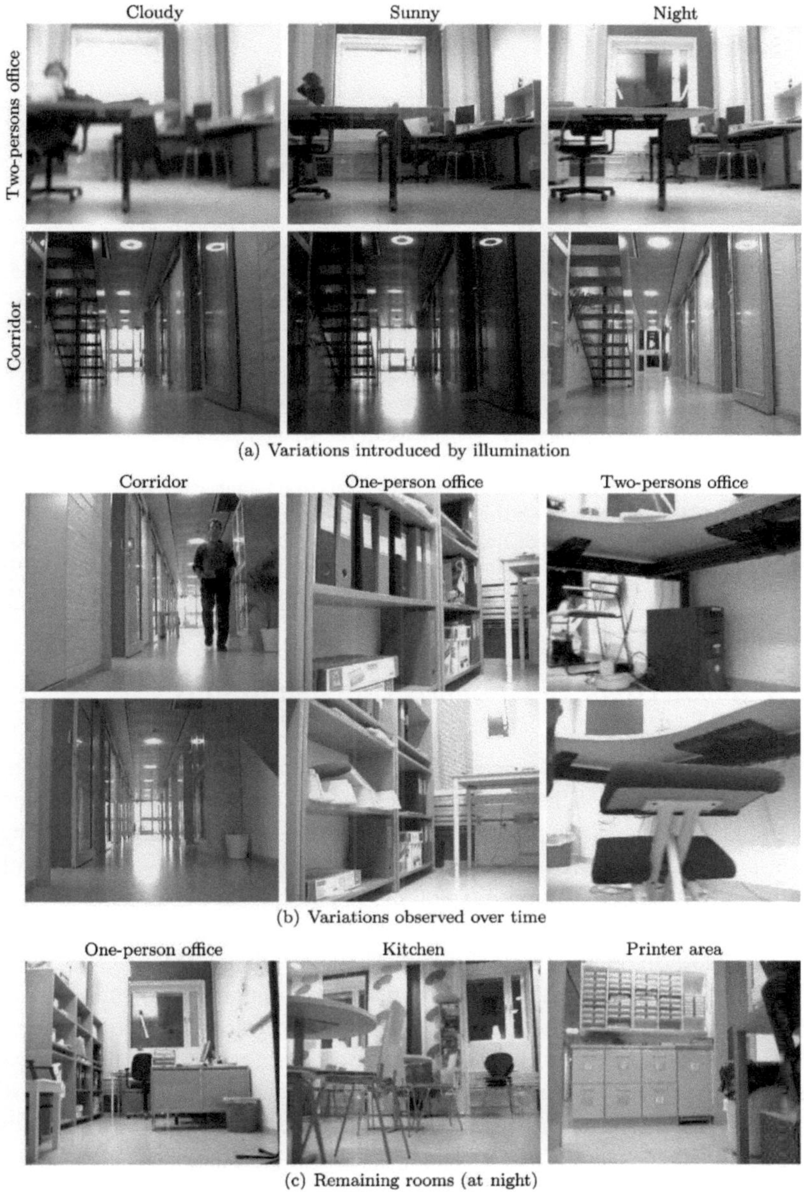

(a) Variations introduced by illumination

(b) Variations observed over time

(c) Remaining rooms (at night)

Fig. 10.3: Examples of pictures taken from the IDOL2 database showing the interior of the rooms, variations observed over time and caused by activity in the environment as well as introduced by changing illumination.

Table 10.1: Results for each run submitted to the mandatory task at the robot vision task 2009.

#	Group	Score
1	Idiap Research Institute, Switzerland	793.0
2	Faculty of Computer Science, The Alexandru Ioan Cuza University, Romania	787.0
3	Faculty of Computer Science, The Alexandru Ioan Cuza University, Romania	787.0
4	Computer Vision and Image Understanding Department, Singapore	784.0
5	Faculty of Computer Science, The Alexandru Ioan Cuza University, Romania	599.5
6	Faculty of Computer Science, The Alexandru Ioan Cuza University, Romania	599.5
7	Laboratoire des Sciences de lInformation et des Systemes	544.0
8	Intelligent Systems and Data Mining Group, Spain	511.0
9	Laboratoire des Sciences de lInformation et des Systemes	509.5
10	Multimedia Information Modeling and Retrieval Group, France	456.5
11	Multimedia Information Modeling and Retrieval Group, France	415.0
12	Multimedia Information Modeling and Retrieval Group, France	328.0
13	Faculty of Computer Science, The Alexandru Ioan Cuza University, Romania	296.5
14	Multimedia Information Modeling and Retrieval Group, France	25.0
15	Laboratoire des Sciences de lInformation et des Systemes	-32.0
16	Laboratoire des Sciences de lInformation et des Systemes	-32.0
17	Laboratoire des Sciences de lInformation et des Systemes	-32.0
18	Laboratoire des Sciences de lInformation et des Systemes	-32.0

10.2.4 Robot Vision 2009: Approaches and Results

The submissions used a wide range of techniques for representing visual information, building models of the appearance of the environment and spatio–temporal integration. It is interesting to note, though, that most of the groups, including the two groups that ranked first in the two tasks, employed approaches based on local features either used as the only image representation or in combination with other visual cues. This confirms a consolidated trend in the robot vision community that treats local descriptors as the off the shelf feature of choice for visual recognition. At the same time, the algorithms used for place recognition spanned from statistical methods to approaches transplanted from the language modeling community.

Table 10.1 shows the results for the mandatory task, while Table 10.2 shows the result for the optional task. Scores are presented for each of the submitted runs that complied with the rules of the contest. We see that the majority of runs were submitted to the mandatory task. A possible explanation is that the optional task requires a higher expertise in robotics that the mandatory task, which therefore represents a very good entry point.

In the following we provide an overview of the approaches used by the participants. The Scale Invariant Feature Transform (SIFT) (Lowe, 2004) was employed most frequently as a local descriptor and the groups winning in both tasks used SIFT in order to represent visual information. The approach used by Idiap (Xing and Pronobis, 2010), which ranked first in the mandatory task, used SIFT combined with several other descriptors including two global image representations: Composed Receptive Field Histograms (CRFH) and PCA Census Transform Histograms

Table 10.2: Results for each run submitted to the optional task of the robot vision task 2009.

#	Group	Score
1	Intelligent Systems and Data Mining Group, Spain	916.5
2	Computer Vision and Image Understanding Department, Singapore	884.5
3	Idiap Research Institute, Switzerland	853.0
4	Intelligent Systems and Data Mining Group, Spain	711.0
5	Intelligent Systems and Data Mining Group, Spain	711.0
6	Intelligent Systems and Data Mining Group, Spain	609.0

(PACT). The algorithm employed by SIMD (Martínez-Gómez et al, 2009) relied mainly on the SIFT descriptor complemented with lines and squares detected using the Hough transform. Other participants also used SIFT (UAIC: (Boroş et al, 2009)); color SIFT (SIFT features extracted from the red, green and blue channels) combined with Hue/Saturation/Value (HSV) color histograms and multi–scale canny edge histograms (MRIM: (Pham et al, 2009)); local features extracted from patches formed around interest points found using the Harris corner detector in images pre-processed using an illumination filter based on the Retinex algorithm (MIRG: (Feng et al, 2009)); or Profile Entropy Features (PEF) encoding RGB color and texture information (LSIS: (Glotin et al, 2009)). Techniques using color descriptors ranked lower in general in the mandatory task, which might suggest that color information was not sufficiently robust to the large variations in illumination captured in the data set.

The participants applied a wide range of techniques to the place recognition problem in the mandatory task. Several variations of a simple image matching strategy were used by SIMD (Martínez-Gómez et al, 2009), UAIC (Boroş et al, 2009) and MIRG (Feng et al, 2009). Idiap (Xing and Pronobis, 2010) built models of places using Support Vector Machines (SVM), separately for several visual cues, and combined the outputs using a Discriminative Accumulation Scheme (DAS). The CVIU group also used Support Vector Machines, while LSIS (Glotin et al, 2009) used Least Squares Support Vector Machines (LS-SVM). Finally, MRIM (Pham et al, 2009) applied a framework based on visual vocabulary and a language model (Conceptual Unigram Model).

Four groups submitted runs to the optional task. The approach used by SIMD (Martínez-Gómez et al, 2009), which ranked first in this track, employed a particle filter to perform Monte Carlo localization. MIRG (Feng et al, 2009) used decision rules to process the results obtained for separate frames. CVIU and Idiap (Xing and Pronobis, 2010) applied simple temporal smoothing techniques, which obtained lower scores than the other approaches.

10.3 Moving Forward: Robot Vision in 2010

In this section we describe how the robot vision task has evolved during 2010. The editions of the challenge maintained the focus on visual place classification for topological localization. Specifically, we organized two editions of the task, one in conjunction with the International Conference on Pattern Recognition (ICPR 2010) and one, ongoing at the time of writing, under the ImageCLEF 2010 umbrella. The level of difficulty of the tasks proposed grew mainly in two directions:

- Image sequences were acquired by a stereo camera, as opposed to a perspective camera as in 2009.
- The number of areas to be recognized grew from five in 2009 (kitchen, corridor, one person office, two person office, printer area) to nine for the robot vision task organized jointly with ICPR2010 (elevator, corridor, kitchen, large office 1, large office 2, student office, laboratory, printer area) up to ten for the robot vision task at ImageCLEF2010 (corridor, elevator, kitchen, large office, meeting room, printer area, recycle area, small office, toilet, large meeting room).

For both editions of the task, the image sequences were acquired using a MobileRobots PowerBot robot platform equipped with a stereo camera system consisting of two Prosilica GC1380C cameras (Figure10.1). We now give a general overview of the robot vision task@ICPR2010 (Section 10.3.1) and look at the ongoing edition of the task at ImageCLEF 2010 (Section 10.3.2).

10.3.1 The Robot Vision Task at ICPR2010

In the second edition of the robot vision task the challenge was again to build a system able to answer the question 'where are you?' ('I'm in the kitchen', 'in the corridor', etc.) when presented with test sequences containing images acquired in the previously observed part of the environment, or in additional rooms that were not imaged in the training sequences. The test images were acquired under different illumination settings than the training data. The system had to assign each test image to one of the rooms that were present in the training sequences, or indicate that the image came from an unknown room. We also allowed the system to abstain from decision in the case of low confidence in the decision.

We considered two separate tasks: task 1 (mandatory) and task 2 (optional) as we did in 2009. The tasks employed two sets of training, validation and testing sequences. The first, easier set contained sequences with constrained viewpoint variability. In this set, training, validation and testing sequences were acquired following a similar path through the environment. The second, more challenging set contained sequences acquired following different paths (e.g. the robot was driven in the opposite direction). The final score for each task was calculated based on the results obtained for both sets. The image sequences used for the contest were taken from

Fig. 10.4: Example pictures of the nine rooms used for the robot vision task at ICPR 2010.

the previously unreleased COLD–Stockholm database (Figure 10.4). The following rules were used for calculating the final score for a run:

- +1.0 point for each correctly classified image;
- correct detection of an unknown room was treated the same way as correct classification;
- -0.5 points for each misclassified image;
- 0.0 points for each image that was not classified (the algorithm refrained from the decision);
- the final score was a sum of points obtained for both sets (easy and hard).

Nine groups participated in the competition, submitting a total of 34 runs. At the time of writing, evaluation and reporting of the results are still ongoing.

Fig. 10.5: Example pictures showing the room categories used for the robot vision task at ImageCLEF 2010.

10.3.2 The Robot Vision Task at ImageCLEF2010

The third edition of the challenge, running at the time of writing, has a special focus on generalization. Participants are being asked to classify rooms and functional areas on the basis of image sequences, captured by a stereo camera mounted on a mobile robot within an office environment (Figure 10.5). The challenge is to build a system able to answer the question 'where are you?' when presented with test sequences containing images acquired in a different environment (different floor of the same building) containing areas belonging to the semantic categories observed previously (present in the training sequence) or to new semantic categories (not imaged in the training sequence). The system should assign each test image to one of the semantic categories of the areas that were present in the training sequence or indicate that the image belongs to an unknown semantic category not included during training. Moreover, the system can refrain from making a decision (e.g. in the case of lack of confidence).

We consider two separate tasks: task 1 (mandatory) and task 2 (optional). The following rules are used when calculating the final score for a run:

- +1.0 point for each correctly classified image belonging to one of the known categories;
- -1.0 point for each misclassified image belonging to one of the known or unknown categories;
- 0.0 points for each image that was not classified (the algorithm refrained from the decision);
- +2.0 points for a correct detection of a sample belonging to an unknown category (true positive);
- -2.0 points for an incorrect detection of a sample belonging to an unknown category (false positive).

10.4 Conclusions

This chapter presents an overview over the newly established robot vision task of ImageCLEF. The overall aim of the task is to push forward research in the field of semantic robot localization using visual information. Therefore, the three editions of the task that have been held so far have addressed the issues of place recognition under varying imaging conditions (robot vision task at ImageCLEF 2009, robot vision task at ICPR 2010) and the visual place categorization problem (robot vision task at ImageCLEF 2010). Participation has been encouraging since its first edition, and it has been growing steadily over the editions.

For the future, we plan to continue posing challenging tasks on the visual place recognition problem for mobile robots, with the aim of attracting contributions to this problem from as many groups outside of the robotics community as possible. A strong focus that we foresee for the near future is the place categorization problem, that is currently one of the most baffling research issues in computer vision as well as robot vision. By introducing stereo data, we will also gently push participants to use more and more 3–D information and temporal continuity in the image sequences.

Acknowledgements We would like to thank the organizers of CLEF and ImageCLEF for their support in organizing the robot vision task.

References

Boroş E, Roşca G, Iftene A (2009) Uaic: Participation in ImageCLEF 2009 robot vision task. In: Working Notes of CLEF 2009, Corfu, Greece. 978-88-88506-84-5

Brunskill E, Kollar T, Roy N (2007) Topological mapping using spectral clustering and classification. In: Proceedings of the IEEE/RSJ International Conference on Intelligent Robots and Systems (IROS)

Cummins M, Newman P (2008) FAB–MAP: Probabilistic localization and mapping in the space of appearance. The International Journal of Robotics Research 27(6):647–665

Dissanayake M, Newman P, Clark S, Durrant-Whyte H, Csorba M (2001) A solution to the simultaneous localization and map building (slam) problem. IEEE Transactions on Robotics and Automation 17(3):229–241

Feng Y, Halvey M, Jose JM (2009) University of glasgow at ImageCLEF 2009 robot vision task. In: Working Notes of CLEF 2009, Corfu, Greece.

Glotin H, Zhao ZQ, Dumont E (2009) Fast LSIS profile entropy features for robot visual self–localization. In: Working Notes of CLEF 2009, Corfu, Greece.

Griffin G, Holub A, Perona P (2007) Caltech–256 Object Category Dataset. Technical Report 7694. Available at `http://authors.library.caltech.edu/7694/`

Howard A, Roy N (2003) The Robotics Data Set Repository (Radish). Available at `http://radish.sourceforge.net/`

Jogan M, Leonardis A (2003) Robust localization using an omnidirectional appearance-based subspace model of environment. Robotics and Autonomous Systems 45(1):51–72

Kortenkamp D, Weymouth T (1994) Topological mapping for mobile robots using a combination of sonar and vision sensing. In: Proceedings of the 12th National Conference on Artificial Intelligence

Kuipers B, Beeson P (2002) Bootstrap learning for place recognition. In: Proceedings of the 18th National Conference on Artificial Intelligence (AAAI'02)

Lowe D (2004) Distinctive image features from scale–invariant keypoints. International Journal of Computer Vision 60(2)

Luo J, Pronobis A, Caputo B, Jensfelt P (2007) Incremental learning for place recognition in dynamic environments. In: Proceedings of the IEEE/RSJ International Conference on Intelligent Robots and Systems (IROS07), San Diego, CA, USA

Martínez-Gómez J, Jiménez-Picazo A, García-Varea I (2009) A particle–filter-based self–localization method using invariant features as visual information. In: Working Notes of CLEF 2009, Corfu, Greece.

Martínez Mozos O, Triebel R, Jensfelt P, Rottmann A, Burgard W (2007) Supervised semantic labeling of places using information extracted from sensor data. Robotics and Autonomous Systems 55(5)

MIT-CSAIL (2006) The MIT–CSAIL database of objects and scenes. Available at `http://web.mit.edu/torralba/www/database.html`.

Nebot E (2006) The Sydney Victoria Park Dataset. Available at `http://www-personal.acfr.usyd.edu.au/nebot/dataset.htm`

Nourbakhsh I, Powers R, Birchfield S (1995) Dervish: An office navigation robot. AI Magazine 16(2):53–60

Pham TT, Maisonnasse L, Mulhem P (2009) Visual language modeling for mobile localization. In: Working Notes of CLEF 2009, Corfu, Greece.

Ponce J, Berg T, Everingham M, Forsyth D, Hebert M, Lazebnik S, Marszalek M, Schmid C, Russell C, Torralba A, Williams C, Zhang J, Zisserman A (2006) Dataset issues in object recognition. In: Towards Category–Level Object Recognition

Pronobis A, Caputo B (2006) A discriminative approach to robust visual place recognition. In: Proceedings of the IEEE/RSJ International Conference on Intelligent Robots and Systems (IROS'06)

Pronobis A, Martínez Mozos O, Caputo B (2008) SVM–based discriminative accumulation scheme for place recognition. In: Proceedings of the IEEE International Conference on Robotics and Automation (ICRA'08)

Siagian C, Itti L (2007) Biologically–inspired robotics vision monte– carlo localization in the outdoor environment. In: Proceedings of the IEEE/RSJ International Conference on Intelligent Robots and Systems (IROS'07)

Tapus A, Siegwart R (2005) Incremental robot mapping with fingerprints of places. In: Proceedings of the IEEE/RSJ International Conference on Intelligent Robots and Systems (IROS'05)

Thrun S (1998) Learning metric–topological maps for indoor mobile robot navigation. Artificial Intelligence 1:30–42

Torralba A, Murphy K, Freeman B, Rubin M (2003) Context–based vision system for place and object recognition. In: Proceedings of the IEEE International Conference on Computer Vision (ICCV'03)

Ullah MM, Pronobis A, Caputo B, Luo J, Jensfelt P, Christensen H (2008) Towards robust place recognition for robot localization. In: Proceedings of the 2008 IEEE International Conference on Robotics and Automation,

Ulrich I, Nourbakhsh I (2000) Appearance–based place recognition for topological localization. In: Proceedings of the IEEE International Conference on Robotics and Automation (ICRA 2000)

Wolf J, Burgard W, Burkhardt H (2005) Robust vision–based localization by combining an image retrieval system with monte carlo localization. IEEE Transactions Transactions on Robotics 21(2):208–216

Xing L, Pronobis A (2010) Multi–cue discriminative place recognition. In: Peters C, Tsikrika T, Müller H, Kalpathy-Cramer J, Jones GJF, Gonzalo J, Caputo B (eds) Multilingual Information Access Evaluation Vol. II Multimedia Experiments, Springer

Zender H, Martinez Mozos O, Jensfelt P, Kruijff GJ, Burgard W (2008) Conceptual spatial representations for indoor mobile robots. Robotics and Autonomous Systems 56(6):493–502

Chapter 11
Object and Concept Recognition for Image Retrieval

Stefanie Nowak, Allan Hanbury, and Thomas Deselaers

Abstract ImageCLEF introduced its first automatic annotation task for photos in 2006. The visual object and concept detection task evolved over the years to become an inherent part of the yearly ImageCLEF evaluation cycle with growing interest and participation from the research community. Although the task can be solved purely visually, the incorporation of multi–modal information such as EXIF (Exchangeable Image File Format) data, concept hierarchies or concept relations is supported. In this chapter, the development, goals and achievements of four cycles of object and concept recognition for image retrieval are presented. This includes the task definitions and the participation of the research community. In addition, the approaches applied to solve the tasks and the lessons learnt are outlined. The results of all years are illustrated, compared and the most promising approaches are highlighted. Finally, the interactions with the photo retrieval task are presented.

11.1 Introduction

In 2006, ImageCLEF added an 'Automatic annotation task for general photographs', which over the years evolved from an image classification task into an object retrieval task (2007), and then into a hierarchical concept annotation task (2008–2009). It has developed into an inherent part of the annual ImageCLEF evaluation cycle with interactions with other tasks. As the task names indicate, the focus of the task changed over the years but the objective has always been to analyze the

Stefanie Nowak
Fraunhofer IDMT, Ilmenau, Germany, e-mail: stefanie.nowak@idmt.fraunhofer.de

Allan Hanbury
Information Retrieval Facility (IRF), Vienna, Austria e-mail: a.hanbury@ir-facility.org

Thomas Deselaers
Computer Vision Laboratory, ETH Zurich, Switzerland e-mail: deselaers@vision.ee.ethz.ch

H. Müller et al. (eds.), *ImageCLEF*, The Information Retrieval Series 32,
DOI 10.1007/978-3-642-15181-1_11, © Springer-Verlag Berlin Heidelberg 2010

content of images based on their visual appearance only. Object class recognition, automatic image annotation, and object retrieval are strongly related tasks. In object class recognition, the aim is to identify whether a certain object is contained in an image; in automatic image annotation, a textual description of a given image is created; and in object retrieval, images containing certain objects or object classes have to be retrieved out of a large set of images. Each of these techniques is important to allow for semantic retrieval from image collections.

Evaluation campaigns for object detection (Everingham et al, 2006, 2010), content–based image retrieval (Clough et al, 2005) and image classification (Moellic and Fluhr, 2006) have been established since 2005. Although these evaluation initiatives have a certain overlap with the tasks described in this chapter, ImageCLEF has always focused on multi–modal analysis and the integration of detection technologies into actual retrieval systems. For example, in 2006 and 2007 the generalization of object recognition algorithms across different databases was tested. This scenario denies the often made assumption that the training and testing images are drawn from the same database and have similar characteristics. In 2008, the participants were provided with a taxonomy and in 2009 with an ontology as additional knowledge sources. These knowledge sources structured the visual concepts into sub and super classes. The ontology also specifies additional relations and restrictions. This textual information was available to enhance the visual analysis algorithms and, for example, to validate the output of the classifiers.

In this chapter, we summarize and analyze the development and the insights gained from four years of object and concept recognition in ImageCLEF. This also allows us to analyze the progress of visual image analysis techniques over these years. This chapter is structured as follows: Section 11.2 introduces the ImageCLEF object and concept retrieval tasks of 2006–2009 in detail and illustrates their aims and the data sets used. Section 11.3 summarizes the approaches of the participants to solve the tasks. Section 11.4 presents the results of the individual years and summarizes the most promising methods. Finally, the combinations of the object retrieval task with the photo retrieval task (Chapter 8) are discussed in Section 11.5, and we conclude in Section 11.6.

11.2 History of the ImageCLEF Object and Concept Recognition Tasks

The first automatic image annotation task was organized in ImageCLEF 2006. A summary of the four cycles of the object and concept recognition tasks from 2006 to 2009 is shown in Table 11.1. The task changed significantly from year to year, which is rather unusual in evaluation campaigns. These changes are manifested in the data sets used (see Chapter 2 for a detailed analysis of the data sets) as well as in the task to be solved by the participants. They reflect the aim to move from a classification task to a full image annotation system that can be combined with other modalities. Every year the task was adapted considering the insights of the

Table 11.1: Summary of the ImageCLEF object and concept recognition tasks characteristics. The table illustrates the type of task, the training and test sets used, the number of images each set contains, the number of visual classes and the number of participants and runs for the years 2006–2009 (OC=Object Categorization, CD=Concept Detection).

Year	Task	Training Set	Num. Images	Test Set	Num. Images	Num. Class	Num. Partic.	Runs
2006	OC	LTU	13,963	Photos	1,100	21	4	10
2007	OC	PASCAL	2,618	IAPR TC–12	20,000	10	7	26
2008	CD	IAPR TC–12 suppl.	1,827	IAPR TC–12 suppl.	1,000	17	11	53
2009	CD	MIR Flickr	5,000	MIR Flickr	13,000	53	19	73

visual tasks of the previous years as well as of the other ImageCLEF tasks. One drawback resulting from these changes is that it is difficult to assess the progress of participating methods over the years.

11.2.1 2006: Object Annotation Task

The *Object Annotation Task* in 2006 (Clough et al, 2007) aimed at the analysis of how purely visual information can be made accessible to text–based searches. The task was designed as a plain classification task to keep the entry barrier low for potential participants. Although the 21 classes were labelled by an object name in English, in fact the task was completely language independent: any other language, or just class numbers, could have been used. A further aim was to investigate how well object categorization algorithms can generalize to images of the same objects that do not necessarily have the same acquisition characteristics. This is a commonly occurring situation in practice, as it is usually not viable to collect a training set large enough to cover all variabilities; however, in other object recognition evaluations this is typically not considered. The training images used were generally clean, containing very little clutter and few obscuring features, while the test images showed objects in a more realistic setting without constraints on acquisition parameters. The training images were taken from a manually collected data set of images in 268 classes kindly provided by LTU technologies[1], from which we selected 21 classes, leading to 13,963 training images. The classes chosen were ashtrays, backpacks, balls, banknotes, benches, books, bottles, calculators, cans, chairs, clocks, coins, computer equipment, cups mugs, hifi equipment, knives forks spoons, mobile phones, plates, sofas, tables, and wallets.

For the test set, 1,100 images of these objects were taken by the organizers. In each test image, at least one object of one of the 21 classes appears, although objects not belonging to any of the classes frequently appear as background clutter.

[1] `http://www.ltutech.com/`

Fig. 11.1: Example images for four of the 21 classes used in the image annotation task in 2006: (left) training set, (right) test set.

The distribution of classes in both training and test sets was non–uniform. Examples of images from the training and test set are shown in Figure 11.1. Along with the training images, 100 of the test images were provided to participants as a development set. The test set was released at a later stage to make training on the testing data difficult. As this was the first time the task was run, and due to its difficulty, only four groups participated, submitting a total of ten runs. For evaluation, the error rate (percentage of incorrectly classified images) was used.

11.2.2 2007: Object Retrieval Task

In the *Object Retrieval Task* in 2007 (Deselaers et al, 2008), the aim was to identify all images showing objects of a certain class. For training, the 'training and validation set' of the PASCAL VOC 2006 data set was used (2,618 images). Objects in these images are annotated with a class label and bounding boxes, having a total of 4,754 objects in ten classes. For testing, the 20,000 images in the IAPR TC–12 database (Grubinger et al, 2006) were used. Examples of images from the training and test sets are shown in Figure 11.2. The task was formulated as a retrieval task with ten queries corresponding to the ten object classes. The relevance assessments on the IAPR TC–12 were obtained in three ways: 1. Pooling: a Web–interface allowed the relevance of the obtained image categorizations to be manually assessed. These categorizations were obtained by pooling all runs (Braschler and Peters, 2003); 2. Additional relevance judgments: the Web interface also offered the assessors the ability to provide additional information on the objects present in the image. The Web interface allowed relevance to be judged rapidly by members of the research groups of the organizers; 3. Manual categorization: Ville Viitaniemi of the Helsinki University of Technology judged all 20,000 images for relevance to the ten queries with stricter definitions of the relevances. Seven groups participated and submitted 26 runs. Performance was measured using Mean Average Precision (MAP).

Fig. 11.2: Example images for four of the ten classes used in the object retrieval task in 2007: bicycle, car, motorbike, person, with (top) PASCAL training set, (bottom) IAPR TC–12 test set. Note the bounding boxes in the training set, and that more than one object can appear in an image.

indoor	outdoor, person, day, vegetation, animal	outdoor, night, water, buildings	outdoor, day, road vegetation, mountains, buildings, sky, overcast

Fig. 11.3: Example images and their concepts from the visual concept detection task in 2008.

11.2.3 2008: Visual Concept Detection Task

In the 2008 *Visual Concept Detection Task* (VCDT) (Deselaers and Hanbury, 2008), the focus was moved from recognizing objects to recognizing concepts, such as indoor/outdoor, day/night, buildings, beach, etc. This is a task that has direct application in annotating images with concepts that are often considered as too obvious to be added to images manually, but have a large potential as useful search terms. 17 hierarchically arranged concepts were chosen. The use of training and test sets with differing characteristics was not continued for the concept detection task. The data set consisted of 2,827 images that were taken from the same pool as those used to create the IAPR TC–12 data set, but were not included in the IAPR TC–12 data set. Example images are shown in Figure 11.3. The data set was divided into 1,827 training images and 1,000 test images. As in 2006, a Web interface was used to annotate the images. Eleven groups participated and submitted 53 runs. The Equal Error Rate (EER) and Area Under Curve (AUC) evaluation measures were used.

Family-Friends, Sky, Summer, Outdoor, Trees, Clouds, Day, Sunny, Portrait, Single Person, Neutral Illumination, No Blur

Canvas, No Blur, Neutral Illumination, Small Group, No Visual Season, No Visual Place, No Visual Time

Landscape, Outdoor, Water, Trees, Sky, Day, Overexposed, No Blur, No Persons, No Visual Season

Plants, Outdoor, Partly Blurred, Macro, Animals, No Visual Time, Neutral Illumination, No Persons, Summer, Aesthetic Impression

Fig. 11.4: Example images from the visual concept detection task in 2009.

11.2.4 2009: Visual Concept Detection Task

In 2009, the Visual Concept Detection Task was carried out at a larger scale (Nowak and Dunker, 2009), with 53 hierarchically organized concepts and a database of 18,000 images from the MIR Flickr 25,000 image data set (Huiskes and Lew, 2008). Examples of the images and concepts are shown in Figure 11.4. The annotation was carried out more carefully and included a validation step as well as a test of inter–annotator agreement. 5,000 images were used for training, and the remaining 13,000 for testing. Participation continued to increase, with 19 groups submitting 73 runs. The EER and AUC evaluation measures were again used, but a new ontology–based measure (OS) (Nowak and Lukashevich, 2009) was also introduced.

11.3 Approaches to Object Recognition

Over the four years, 29 research groups participated in total. Of these, nine research groups participated in the task several times. The participation of the groups is summarized in Table 11.2. For readability, all participating groups are listed together with the group acronyms and the citations of their approaches as follows:

- **apexlab** (Nowak and Dunker, 2009): Shanghai Jiaotong University, Shanghai, China;
- **AVEIR** (Glotin et al, 2009): joint consortium of the four groups: Telecom ParisTech, LSIS, MRIM–LIG and UPMC;
- **budapest / sztaki** (Deselaers et al, 2008; Daróczy et al, 2008, 2009): Data Mining and Web search Research Group, Informatics Laboratory, Computer and Automation Research Institute, Hungarian Academy of Sciences, Hungary;
- **CEA LIST** (Deselaers and Hanbury, 2008; Nowak and Dunker, 2009): Lab of Applied Research on Software–Intensive Technologies of the CEA, France;
- **CINDI** (Clough et al, 2007): Concordia University in Montreal, Canada;

- **DEU** (Clough et al, 2007): Department of Computer Engineering of the Dokuz Eylul University in Tinaztepe, Turkey;
- **FIRST** (Binder and Kawanabe, 2009): Fraunhofer FIRST, Berlin, Germany;
- **HJFA** (Jiang et al, 2008): Microsoft Key Laboratory of Multimedia Computing and Communication of the University of Science and Technology, China;
- **HUTCIS** (Deselaers et al, 2008): Adaptive Informatics Research Centre / Laboratory of Computer and Information Science, Helsinki University of Technology, Finland;
- **I2R** (Deselaers and Hanbury, 2008; Ngiam and Goh, 2009): IPAL French–Singaporean Joint Lab of the Institute for Infocomm Research, Singapore.
- **IAM** (Hare and Lewis, 2009): Intelligence, Agents, Multimedia Group of the University of Southampton, UK;
- **INAOE TIA** (Deselaers et al, 2008; Deselaers and Hanbury, 2008; Escalante et al, 2009): TIA Research Group, Computer Science Department, National Institute of Astrophysics, Optics and Electronics, Tonantzintla, Mexico;
- **ISIS** (van de Sande et al, 2009): Intelligent Systems Lab of the University of Amsterdam, The Netherlands;
- **Kameyama** (Sarin and Kameyama, 2009): Graduate School of Global Information and Telecommunication Studies, Waseda University, Japan;
- **LEAR** (Douze et al, 2009): LEAR team of INRIA, Montbonnot, France;
- **LSIS** (Zhao and Glotin, 2008; Dumont et al, 2009): Laboratory of Information Science and Systems, France;
- **Makere** (Deselaers and Hanbury, 2008): Faculty of Computing and Information Technology, Makerere University, Uganda;
- **MedGIFT** (Clough et al, 2007): University and Hospitals of Geneva, Switzerland;
- **MMIS** (Llorente et al, 2008, 2009): Knowledge Media Institute, Open University, Milton Keynes, UK;
- **MRIM-LIG** (Pham et al, 2009): Multimedia Information Modelling and Retrieval group at the Laboratoire Informatique de Grenoble, Grenoble University, France;
- **MSRA** (Deselaers et al, 2008): Microsoft Research Asia;
- **NTU** (Deselaers et al, 2008): School of Computer Engineering, Nanyang Technological University, Singapore;
- **PRIP** (Deselaers et al, 2008): Institute of Computer–Aided Automation, Vienna University of Technology, Vienna, Austria; Intelligent Systems Lab Amsterdam, University of Amsterdam, The Netherlands;
- **RWTH** (Clough et al, 2007; Deselaers et al, 2008; Deselaers and Hanbury, 2008): Human Language Technology and Pattern Recognition Group from the RWTH Aachen University, Germany;
- **Telecom ParisTech** (Ferecatu and Sahbi, 2009): Institut TELECOM, TELECOM ParisTech, Paris, France;
- **UAIC** (Iftene et al, 2009): Faculty of Computer Science of Alexandru Ioan Cuza University, Romania;

Table 11.2: Participation in the object retrieval task over the years. The rows denote in which year the single groups participated and the number illustrates the number of run configurations that were submitted. Please note that in 2009 the maximum number of runs was restricted to five.

Group	CINDI	DEU	RWTH	MedGIFT	budapest / sztaki	HUTCIS	INAOE TIA	MSRA	NTU	PRIP	CEA LIST	HJFA	I2R	LSIS	MMIS	Makere	UPMC	XRCE	apexlab	AVEIR	FIRST	IAM	LEAR	ISIS	Kameyama	MRIM-LIG	Telecom ParisTech	UAIC	Wroclaw Uni
2006	4	2	2	3																									
2007			1		2	13	4	3	1	2																			
2008			1		13		7				3	1	8	7	4	1	6	2											
2009					5		5				4		2	5	5		5	1	3	4	4	3	5	5	5	4	2	1	5

- **UPMC** (Tollari et al, 2008; Fakeri-Tabrizi et al, 2009): University Pierre et Marie Curie in Paris, France;
- **Wroclaw Uni** (Nowak and Dunker, 2009): Wroclaw University of Technology, Poland;
- **XRCE** (Ah-Pine et al, 2008, 2009): Textual and Visual Pattern Analysis group from the Xerox Research Center Europe, France;

In the following, we outline commonly used techniques to solve the object retrieval and detection tasks. To this end, all 162 submissions from 29 research groups and 17 countries are analyzed. The submitted approaches are categorized regarding descriptors, codebook generation, classification methods, and post–processing.

11.3.1 Descriptors

A large variety of visual descriptors was used throughout the four cycles of this ImageCLEF task. Most groups apply combinations of descriptors. One broad distinction is whether a descriptor describes an image as a whole (global features) or only a region of the image (local features). Among the local features, different sizes of the described regions are considered: some descriptors only consider small square image regions while others consider large portions of the image. Furthermore, the positions from which local features are extracted vary widely. Local descriptors can, for example, be extracted from sparse interest points or from a dense grid. Frequently, a set of local features is further represented as a histogram over a visual codebook (e.g. ISIS, RWTH, IAM, LSIS, see Section 11.3.2).

Some groups also extract local features from image regions that were obtained by an unsupervised image segmentation engine. Here, often the entire image is covered with regions in a jigsaw–like way [budapest, INAOE TIA].

Among the global features many groups used color histograms [DEU, CINDI, HUTCIS, Makere, CEA–LIST, etc.] amongst other features or texture features such as edge histograms, Tamura histograms [MMIS] or Gabor features [LSIS, NTU]. Also the Gabor–based GIST features [kameyamalab, INRIA-LEAR, apexlab] and profile entropy features [LSIS] were applied.

Bag–of–words representations of SIFT, color–SIFT or image patches were common among the local features [ISIS, FIRST, MSRA, PRIP, IAM, HUTCIS, HJ-FA]. These features were often extracted from Harris-Laplace interest points or using a dense grid. Region-based local features also allow for using shape [Makere, budapest, INAOE TIA] and spatial layout [CEA-LIST].

Several approaches extract global and local features and analyze the combination of both feature types [LEAR, INAOE TIA, I2R, budapest, CEA LIST, kameyama, AVEIR, MRIM, HUTCIS].

Some groups tried to make use of additional information such as EXIF tags [UAIC] and concept names [Telecom Paristech, AVEIR]. Others obtained higher level features, e.g. with the application of a face detector [UAIC].

11.3.2 Feature Post–processing and Codebook Generation

While global image descriptors directly describe an entire image, local features are often summarized in a bag–of–visual–words descriptor. Many variations of bag–of–visual–words approaches were proposed. The most common approach is to cluster a set of representative local descriptors using k-means into 500–2,000 cluster prototypes. Then each image is represented by a histogram counting how many of its local descriptors belong into which of the clusters.

Such approaches were adopted by many groups over the four years. MSRA and RWTH followed this approach, while ISIS additionally investigated different settings for codebook generation. IAM uses a hierarchical k–means for clustering. LSIS's approach applies a Euclidean distance on multigrid features for visual word assignment after the k-means clustering and budapest replaces the k-means clustering step with a Gaussian Mixture Model.

11.3.3 Classifier

Given the image descriptors, a classifier is applied to predict the class(es) of the test images. The parameters of the classifier are trained on the training data and tuned using the validation data.

Classifiers are often grouped into generative, discriminative, or model–free approaches. Generative probabilistic models estimate the distribution of observations for each class and use this to predict which class is most likely for a certain observation. Discriminative approaches directly model the posterior probability for the classes. Another option is to combine or blend both approaches.

In the ImageCLEF object retrieval tasks, a large variety of discriminative and generative classifiers have been used. By far the most prominent approach was the classification with Support Vector Machines (SVMs) with different kernels, multiple-kernels, and multi-class extensions [CINDI, HUTCIS, MSRA, CEA-LIST, LSIS, I2R, INRIA-LEAR, ISIS, MRIM-LIG, UPMC and FIRST]. Other discriminative approaches include log–linear models [RWTH] and logistic regression [budapest, XRCE], fuzzy decision forest [UPMC] or random forests [INAOE TIA].

The most popular model–free approach was the nearest neighbor classifier. It has often been used as baseline for more sophisticated approaches using different distance functions [CINDI, DEU, INAOE TIA, CEA-LIST, HJ-FA, Makere, PRIP and Kameyama Lab] or weighted neighbors [INRIA-LEAR].

Furthermore, a variety of language models have been applied. MSRA uses a visual topic model and a trigram language model and IAM investigated a cross–language latent indexing method with a cosine distance decision function. Non–parametric density estimation functions [MMIS], Markov Random Fields [INAOE TIA] and Self Organizing Maps [HUTCIS] are further adopted methods.

Some groups used a number of classifiers and applied a fusion step of the results after classification, e.g. [HUTCIS, AVEIR].

11.3.4 Post–Processing

After the classification step, some groups further refined the results. This step was mainly applied in 2008 and 2009, as in these years a taxonomy and an ontology were offered as additional knowledge bases. Some participants incorporated this knowledge to improve their classifiers, partly also directly in the classification step. A popular approach was the co–occurrence and correlation analysis of concept context in the training data [MMIS, INAOE TIA, UPMC, budapest]. One group applied semantic similarities that were determined by word correlations in Google, WordNet and Wikipedia [MMIS]. Furthermore, thresholds were adapted in case of mutually exclusive concepts [I2R, XRCE].

11.4 Results

In this section, the results of the individual years are summarized. The task and the databases changed over the years, as outlined in Section 11.2. Therefore, the

Table 11.3: Results from the object annotation task in 2006 sorted by error rate.

Rank	Group ID	Descriptor	Classifier	Error Rate [%]
1	RWTH	dense BoW	log-linear	77.3
2	RWTH	sparse BoW	log-linear	80.2
3	cindi	global edge, color	SVM	83.2
4	cindi	global edge, color	SVM	85.0
5	cindi	global edge, color	SVM	85.2
6	cindi	global edge, color	KNN	87.1
7	DEU	global edge	generative Gauss	88.2
-	medGIFT	collection frequencies	GIFT-NN	90.5
-	medGIFT	collection frequencies	GIFT-NN	91.7
8	DEU	global colorlayout	generative Gauss	93.2

results of the different years cannot be compared directly to each other. However, we compare results across different years where possible.

11.4.1 2006: Object Annotation Task

Table 11.3 shows the results for three participating groups of the object annotation task in 2006. The results of MedGIFT are not ranked, because they submitted their runs after the deadline. The runs were evaluated using the error rate. Error rates are very high and range from 77.3% to 93.2%. Further analysis revealed that many of the test images could not be classified correctly by any method. Summarizing, the discriminative classification methods outperformed the others by a small amount.

11.4.2 2007: Object Retrieval Task

The submissions of the Object Retrieval Task in 2007 were evaluated according to average precision (AP) per class and ranked by the MAP over all classes. Table 11.4 presents the results. HUTCIS obtained the best result with a MAP of 2.9%. Considering the class–wise results, the best overall results were obtained for the `car` class with an AP of 11.1%. Also, the classes `person` and `bicycle` could be detected well with an AP of 8.6% and 4.1%, respectively. The worst results were achieved for the classes `dog` and `sheep`, which could be detected with an AP of just 0.1%. Except the classes `sheep` and `cat`, all best results per class were obtained by one of the SVM configurations of HUTCIS.

The low performance of all methods shows that the task is very difficult and that the varying number of relevant images per topic further complicates it.

Table 11.4: Results from the ImageCLEF 2007 object retrieval task with complete relevance information obtained by manual categorization for the whole database. All values have been multiplied by 100 to make the table more readable. The results for each class are presented in the corresponding columns. The MAP over all classes is in the last column. The highest AP per class is shown in bold. Please note that the results of the budapest group are not fully comparable as they assigned just a single class per photo instead of multiple classes and used a different, more strongly labelled training set.

Group ID	Descriptor	Classifier	Bicycle	Bus	Car	Mbike	Cat	Cow	Dog	Horse	Sheep	Person	MAP
HUTCIS	BoW (global and local)	SVM	**4.1**	**1.2**	10.6	0.4	0.0	0.6	0.1	**3.8**	0.0	8.3	2.9
HUTCIS	SIFT, color	SVM	2.6	1.0	**11.1**	1.0	0.0	1.0	0.1	3.2	0.0	8.1	2.8
HUTCIS	SIFT	SVM	2.4	1.1	10.3	**1.8**	0.0	**1.1**	0.1	3.0	0.0	8.1	2.8
HUTCIS	BoW (global and local)	SVM	3.0	1.1	4.2	0.6	0.0	0.7	**0.1**	2.5	0.0	**8.6**	2.1
HUTCIS	BoW (global and local)	SVM	1.6	0.9	0.5	0.3	0.0	0.6	0.1	1.5	0.0	8.3	1.4
HUTCIS	SIFT, color	SVM	1.4	1.0	0.7	0.3	0.0	0.5	0.1	1.1	0.0	8.4	1.4
HUTCIS	SIFT, color	SVM	2.0	0.8	0.4	0.3	0.0	0.8	0.1	1.1	0.0	8.2	1.3
HUTCIS	BoW (global and local)	SOM	0.9	0.7	4.5	0.6	0.0	0.3	0.1	0.7	0.0	5.6	1.3
MSRA	SIFT	pLSA + SVM	0.9	0.5	3.6	0.6	0.7	0.1	0.1	0.4	0.0	6.0	1.3
HUTCIS	SIFT, color	SVM	1.3	0.8	0.5	0.2	0.0	0.5	0.1	0.8	0.0	8.4	1.3
HUTCIS	BoW (global and local)	SOM	0.8	0.6	4.2	0.5	0.0	0.3	0.1	0.4	0.0	5.4	1.2
HUTCIS	SIFT	SVM	1.1	0.7	0.4	1.4	0.0	0.3	0.0	1.0	0.0	7.2	1.2
HUTCIS	SIFT	SVM	1.1	0.8	0.3	0.3	0.0	0.4	0.1	0.9	0.0	6.9	1.1
HUTCIS	SIFT	SVM	0.3	0.9	0.3	0.3	0.0	0.3	0.0	1.1	0.0	6.6	1.0
RWTH	dense BoW	log-linear	0.4	0.2	1.4	0.2	0.0	0.1	0.0	0.2	0.0	5.5	0.8
budapest	BoW	segment NN	0.1	0.1	0.8	0.0	0.0	0.2	0.0	0.2	0.0	4.1	0.5
NTU	global color, texture, shape	SVM	1.2	0.7	2.4	0.0	0.0	0.0	0.1	0.1	0.0	0.8	0.5
budapest	BoW	segment NN	0.4	0.0	0.4	0.1	0.0	0.0	0.1	0.1	0.0	3.9	0.5
MSRA	patch-based texture	tri-gram language model	0.4	0.3	0.7	0.1	0.1	0.0	0.0	0.3	0.0	2.5	0.4
MSRA	patch-based texture	tri-gram language model	0.3	0.2	0.5	0.0	0.0	0.1	0.0	0.2	**0.1**	2.5	0.4
INAOE-TIA	BoW	naïve Bayes + AdaBoost	0.1	0.0	0.1	0.0	0.0	0.2	0.0	0.2	0.0	3.2	0.4
INAOE-TIA	BoW	KNN + MRF	0.5	0.1	0.6	0.0	0.0	0.2	0.0	0.0	0.0	2.2	0.4
INAOE-TIA	BoW	KNN + MRF	0.5	0.1	0.6	0.0	0.0	0.2	0.0	0.0	0.0	2.2	0.4
PRIP	SIFT	EMD + NN	0.1	0.0	0.3	0.1	**1.4**	0.1	0.0	0.0	0.0	1.5	0.4
INAOE-TIA	BoW	KNN	0.3	0.0	0.5	0.1	0.0	0.0	0.0	0.0	0.0	2.2	0.3
PRIP	SIFT	EMD + NN	0.1	0.0	0.1	0.5	0.1	0.0	0.1	0.1	0.0	0.8	0.2

11.4.3 2008: Visual Concept Detection Task

The results of the 2008 Visual Concept Detection Task are presented in Table 11.5 and Table 11.6. Runs were ranked according to their EER and AUC scores. Table 11.5 shows the performance for the best run of each group in terms of EER and AUC and the descriptors and classifiers applied. The best scores of each group range from 16.7% EER to 49.3% EER. In terms of AUC, the best run achieved 90.7% AUC, while the values fall to 20% AUC.

Table 11.5: Summary of the results of the visual concept detection task in Image-CLEF 2008. The table shows the results for the best run per group.

GroupID	Descriptor	Classifier	rank	EER	AUC
XRCE	local color and texture	Fisher-Kernel SVM + logistic regression	1	16.7	90.7
RWTH	BoW	log linear model	3	20.5	86.2
UPMC	local color	fuzzy decision forests	4	24.6	82.7
LSIS	profile entropy features + others	SVM	5	25.9	80.5
MMIS	color, Tamura texture	non-parametric density estimation	13	28.4	77.9
CEA-LIST	color, spatial layout	NN, SVM	17	29.0	73.4
IPAL-I2R	variety of descriptors	—	19	29.7	76.4
budapest	global and local	logistic regression	20	31.1	74.9
TIA	global and local	SVM, random forest	24	32.1	55.6
HJ-FA	color, SIFT	KNN	47	45.1	20.0
Makere	luminance, color, texture, shape	NN	51	49.3	30.8

Table 11.6 presents the results per concept. For each concept, the best and the worst EER and AUC are shown, along with the average EER and AUC over all runs submitted. The best results were obtained for all concepts by the XRCE group, with budapest doing equally well on the `night` concept. The best AUC per concept is at least 80% for the concept `road/pathway` and rises up to 97.4% for the concepts `indoor` and `night`. The rather poor results for the concept `road/pathway` can be explained by the high variability in the appearance of this concept. The concept with the highest average score, in other words, the concept that was detected best in most runs is `sky`. Again, the concept with the worst average score is `road/pathway`.

Summarizing, discriminative approaches with local features achieved the best results. Further, the results demonstrate that the concept detection task could be solved reasonably well.

11.4.4 2009: Visual Concept Detection Task

The evaluation of the concept detection task in 2009 focused on two evaluation paradigms, the evaluation per concept and the evaluation per photo. The evaluation per concept was conducted with the EER and AUC as in the previous year. For the evaluation per photo, a new evaluation measure, the Ontology Score (OS), was introduced (Nowak et al, 2010).

The results are given in Table 11.7. The group with the best concept–based results, ISIS, achieves an EER of 23% and an AUC of 84% on average for their best run. The next three groups in the ranking closely follow these results with an EER of about 25% and an AUC of 82% and 81%. The performance of the groups at the end of the list goes up to 53% EER and falls to 7% AUC.

The evaluation per photo reveals scores in the range of 39% to 81% for the best run per group. The best results in terms of OS were achieved by the XRCE group with 81% annotation score over all photos. It can be seen from the table that the ranking of the groups is different than for the EER/AUC measures.

In Table 11.8, the results for each concept are illustrated in terms of EER and AUC over all runs submitted. All concepts could be detected at least with 44% EER and 58% AUC, but on average with an EER of 23% and an AUC of 84%. The majority of the concepts were classified best by the ISIS group. It is obvious that the aesthetic concepts (`Aesthetic_Impression`, `Overall_Quality` and `Fancy`) are classified worst (EER greater than 38% and AUC less than 66%.). This is not surprising due to the subjective nature of these concepts which also made the ground truthing difficult. The best classified concepts are `Clouds` (AUC: 96%), `Sunset-Sunrise` (AUC: 95%), `Sky` (AUC: 95%) and `Landscape-Nature` (AUC: 94%).

Summarizing, the groups that used local features such as SIFT achieved better results than the groups relying solely on global features. Most groups that investigated the concept hierarchy and analyzed, for example, the correlations between the concepts, could achieve better results in the OS compared to the EER. Again, the discriminative methods outperformed the generative and model–free ones.

Table 11.6: Overview of the results per concept of the visual concept detection task 2008.

	best			average		worst	
# concept	EER	AUC	group	EER	AUC	EER	AUC
00 indoor	8.9	97.4	XRCE	28.0	67.6	46.8	2.0
01 outdoor	9.2	96.6	XRCE	30.6	70.5	54.6	13.3
02 person	17.8	89.7	XRCE	35.9	62.2	53.0	0.4
03 day	21.0	85.7	XRCE	35.4	64.9	52.5	9.7
04 night	8.7	97.4	XRCE/budapest	27.6	72.5	73.3	0.0
05 water	23.8	84.6	XRCE	38.1	57.8	53.0	3.2
06 road/pathway	28.8	80.0	XRCE	42.6	50.7	56.8	0.0
07 vegetation	17.6	89.9	XRCE	33.9	67.4	49.7	30.7
08 tree	18.9	88.3	XRCE	36.1	62.8	59.5	1.0
09 mountains	15.3	93.8	XRCE	33.1	61.2	55.8	0.0
10 beach	21.7	86.8	XRCE	35.8	57.6	51.4	0.0
11 buildings	17.0	89.7	XRCE	37.4	60.8	64.0	0.5
12 sky	10.4	95.7	XRCE	24.0	78.6	50.8	37.3
13 sunny	9.2	96.4	XRCE	30.2	66.5	55.4	0.0
14 partly cloudy	15.4	92.1	XRCE	37.5	58.9	55.5	0.0
15 overcast	14.1	93.7	XRCE	32.1	67.6	61.5	0.0
16 animal	20.7	85.7	XRCE	38.2	54.2	58.4	0.0

Table 11.7: Summary of the results for the concept detection task in 2009. The table shows the EER and AUC performance for the best run per group ranked by EER for the concept–based evaluation and the performance with the OS measure for the best run per group for the photo–based evaluation. Note, that the best run for the EER measure is not necessarily the same run as for the OS measure.

Group ID	Descriptor	Classifier	Rank	EER	AUC	Rank	OS
ISIS	color SIFT	SVM	1	0.23	0.84	14	0.77
LEAR	BoW (global and local)	SVM / NN	5	0.25	0.82	12	0.77
I2R	global and local	SVM	7	0.25	0.81	2	0.81
FIRST	SIFT, color	multiple kernel SVM	8	0.25	0.82	4	0.80
XRCE	BoW	sparse logistic regression	14	0.27	0.80	1	0.81
budapest	various global and local features	logistic regression	17	0.29	0.77	35	0.68
MMIS	color, Tamura, Gabor	non-parametric density estimation	21	0.31	0.74	42	0.58
IAM	SIFT	cosine distance of visual terms	23	0.33	0.72	61	0.41
LSIS	various features	SVM(LDA) / Visual Dictionary	24	0.33	0.72	49	0.51
UPMC	HSV histogram	SVM	33	0.37	0.67	58	0.44
MRIM	RGB histogram, SIFT, Gabor	SVM	34	0.38	0.64	28	0.72
AVEIR	various global and local features, text	SVM / Visual Dictionary / canonical correlation	41	0.44	0.55	50	0.50
Wroclaw Uni	various features	Multivariate Gaussian Model + NN	43	0.45	0.22	11	0.78
Kameyama	global and local	KNN	47	0.45	0.16	7	0.80
UAIC	face detection, exif	NN + default values	54	0.48	0.11	32	0.69
apexlab	various features	KNN	56	0.48	0.07	17	0.76
INAOE TIA	various global features	KNN	57	0.49	0.10	20	0.74
Random	-	-	-	0.50	0.50	-	0.38
CEA LIST	global and local	Multiclass boosting	68	0.50	0.47	29	0.71
TELECOM	global, text features	Canonical Correlation Analysis + thresholds	72	0.53	0.46	65	0.39

11.4.5 Evolution of Concept Detection Performance

Comparisons of performance across years can best be made from 2008 to 2009. Although the database changed between these evaluation cycles, the methodology of the tasks was similar. Comparing the results from 2008 to 2009, the average AUC over all concepts for the best run drops from 90% to 84%, while increasing the number of concepts with a factor of about three. The most comparable concepts `indoor` and `outdoor` dropped by 13% and 7%, respectively, which can be explained with the third concept `NoVisualPlace` in the group in 2009. Other concepts could be annotated with a similar quality, e.g. `mountains` and `sky` -1%, `day` and `trees` +/-0%. The concept `person` was substituted by four concepts `single person`,

Table 11.8: Overview of the best results per concept over all submitted runs in 2009 in terms of the EER and AUC and the name of the group which achieved these results.

No.	Concept	AUC	EER	Group ID	No.	Concept	AUC	EER	Group ID
0	Partylife	0.83	0.24	ISIS	27	Day	0.85	0.24	ISIS
1	Family_Friends	0.83	0.24	ISIS	28	Night	0.91	0.17	LEAR
2	Beach_Holidays	0.91	0.16	ISIS	29	No_Visual_Time	0.84	0.25	ISIS
3	Building_Sights	0.88	0.20	ISIS	30	Sunny	0.77	0.30	LEAR - ISIS
4	Snow	0.87	0.21	LEAR	31	Sunset_Sunrise	0.95	0.11	ISIS
5	Citylife	0.83	0.25	ISIS	32	Canvas	0.82	0.25	XRCE
6	Landscape_Nature	0.94	0.13	ISIS	33	Still_Life	0.82	0.25	ISIS
7	Sports	0.72	0.34	FIRST	34	Macro	0.81	0.26	ISIS
8	Desert	0.89	0.18	ISIS	35	Portrait	0.87	0.21	XRCE - ISIS
9	Spring	0.83	0.25	FIRST	36	Overexposed	0.80	0.25	UPMC
10	Summer	0.81	0.26	ISIS	37	Underexposed	0.88	0.18	CVIUI2R
11	Autumn	0.87	0.21	ISIS	38	Neutral_Illumination	0.80	0.26	LEAR
12	Winter	0.85	0.23	ISIS	39	Motion_Blur	0.75	0.32	ISIS
13	No_Visual_Season	0.81	0.26	ISIS	40	Out_of_focus	0.81	0.25	LEAR
14	Indoor	0.84	0.25	ISIS	41	Partly_Blurred	0.86	0.22	LEAR
15	Outdoor	0.90	0.19	ISIS	42	No_Blur	0.85	0.23	LEAR
16	No_Visual_Place	0.79	0.29	ISIS	43	Single_Person	0.79	0.28	ISIS - LEAR
17	Plants	0.88	0.21	ISIS	44	Small_Group	0.80	0.28	ISIS
18	Flowers	0.87	0.20	ISIS - FIRST	45	Big_Group	0.88	0.21	ISIS
19	Trees	0.90	0.18	ISIS	46	No_Persons	0.86	0.22	ISIS
20	Sky	0.95	0.12	ISIS	47	Animals	0.83	0.25	ISIS
21	Clouds	0.96	0.10	ISIS	48	Food	0.90	0.19	ISIS
22	Water	0.90	0.18	ISIS	49	Vehicle	0.83	0.24	ISIS
23	Lake	0.91	0.16	ISIS	50	Aesthetic_Impression	0.66	0.38	ISIS
24	River	0.90	0.17	ISIS	51	Overall_Quality	0.66	0.38	ISIS
25	Sea	0.94	0.13	ISIS	52	Fancy	0.58	0.44	ISIS
26	Mountains	0.93	0.14	ISIS					

`small group`, `big group` and `no person` and dropped on average by 7%. Concepts that achieved better scores in 2009 are `beach` +4%, `clouds` +4% and `water` +5%. In case of `clouds`, the 2009 task was easier, because the concepts `overcast` and `partly cloudy` were combined in one concept.

11.4.6 Discussion

In 2006, the bag–of–visual–words approach by RWTH with a log–linear classifier performed best. In 2007, HUTCIS obtained the best result by combining various descriptors (color, edge, SIFT, combinations) and SVM classifiers. In 2008, XRCE achieved the best result using local color and texture features and a combination of Fisher–kernel SVMs and logistic models. In 2009, ISIS obtained the best result using a large variety of local descriptors extracted from different interest points and grids represented in a bag–of–words–descriptor and χ^2-SVM classifiers. For the photo-based evaluation, the XRCE group achieved the best results in 2009 with a system similar to their 2008 approach.

Over all years, the best results were obtained using discriminative classifiers. The classifier itself varied throughout the years and also the features differed. The knowledge provided in form of a taxonomy and an ontology in 2008 and 2009, respectively, was not further considered by most groups. Only in the post–processing step of the XRCE run in 2009, was the probability of the presence of a particular concept adapted by analyzing its likely relationships.

11.5 Combinations with the Photo Retrieval Task

In 2008, two automatic runs provided by participants of the visual concept detection task were made available to the participants of the photo retrieval task. These contained annotations for the database of 20,000 photos used in the photo retrieval task with the VCDT concepts. Two groups that participated in the photo retrieval task of ImageCLEF made use of these annotations. UPMC applied VCDT annotations provided by their own algorithm. They used the detected visual concepts to re-rank the first 50 results returned by text retrieval approaches. The concepts to use for the re–ranking were chosen by two approaches: (i) the concept word appears in the query text and (ii) the concept word appears in the list of synonyms (obtained by WordNet) of the words in the query text. The first approach improved the results of all the queries for which it was applicable, while the second resulted in worse results for some topics. Both approaches achieved a better overall performance than using text alone: the F–measure for the best text only run (using TF–IDF) is 0.273, while the F–measure for the run re–ranked using the first approach is 0.289.

The NII group (Inoue and Grover, 2008) made use of both provided VCDT concept annotation sets. They also used the concepts to re–rank results returned by a text retrieval approach. The best results were obtained by a re–ranking based on a hierarchical clustering which uses distances between vectors to encode the VCDT concepts. This re–ranking decreased the P20 metric while increasing the CR20 metric, resulting in an increase of the F–measure from 0.224 for text only to 0.230 after the re–ranking.

INAOE TIA used one of the provided VCDT concept annotation sets as one part of a group of visual retrieval algorithms whose results were integrated in a late fusion process. It is therefore not possible to determine the effect of only the VCDT concepts on the results.

11.6 Conclusion

This chapter presents an overview of the object and concept recognition tasks of ImageCLEF in the four years from 2006 to 2009. The tasks varied strongly over the years reflecting the objective to start with a flat classification task and going towards a full image annotation that can be used for content–based access to photo repositories. Over the years, 29 groups participated in total and submitted over 163 runs, processing a total of 35,100 test images.

For the future, we will continue to pose challenging tasks for object and concept annotation. In 2010, the task will consider Flickr User Tags, so that the participants can decide whether they solve the concept detection task purely visually, purely based on social data or if they prefer to follow multi–modal approaches. The aim is to analyze if the multi–modal annotation approaches can outperform text only or visual only approaches and which approach is best suited to which type of concepts. Furthermore, the systems are trained and evaluated on 93 concepts, containing also more subjective annotations such as `boring` or `cute` and event concepts such as `birthday` or `work`.

Acknowledgements We would like to thank the organizers of CLEF and ImageCLEF for the support of the object and concept recognition tasks. This work has been partly supported by grant No. 01MQ07017 of the German research program THESEUS funded by the Ministry of Economics and the European Commission Network of Excellence MUSCLE (FP6–507752).

References

Ah-Pine J, Cifarelli C, Clinchant S, Csurka G, Renders J (2008) XRCE's Participation to ImageCLEF 2008. In: Working Notes of CLEF 2008, Aarhus, Denmark

Ah-Pine J, Clinchant S, Csurka G, Liu Y (2009) XRCE's Participation in ImageCLEF 2009. In: Working Notes of CLEF 2009, Corfu, Greece

Binder A, Kawanabe M (2009) Enhancing Recognition of Visual Concepts with Primitive Color Histograms via Non–sparse Multiple Kernel Learning. In: Peters C, Tsikrika T, Müller H, Kalpathy-Cramer J, Jones J, Gonzalo J, Caputo B (eds) Multilingual Information Access Evaluation Vol. II Multimedia Experiments: Proceedings of the 10th Workshop of the Cross–Language Evaluation Forum (CLEF 2009), Revised Selected Papers. Lecture Notes in Computer Science (LNCS). Springer, Corfu, Greece

Braschler M, Peters C (2003) CLEF methodology and metrics. In: Peters C, Braschler M, Gonzalo J, Kluck M (eds) Evaluation of Cross–Language Information Retrieval Systems, Evaluation of Cross–Language Information Retrieval Systems. Lecture Notes in Computer Science (LNCS), vol 2406. Springer, Darmstadt, Germany, pp 394–404

Clough PD, Müller H, Sanderson M (2005) The CLEF 2004 cross–language image retrieval track. In: Peters C, Clough P, Gonzalo J, Jones G, Kluck M, Magnini B (eds) Multilingual Information Access for Text, Speech and Images Fifth Workshop of the Cross–Language Evaluation Forum, CLEF 2004. Lecture Notes in Computer Science (LNCS), vol 3491. Springer, Bath, UK, pp 597–613

Clough PD, Grubinger M, Deselaers T, Hanbury A, Müller H (2007) Overview of the ImageCLEF 2006 photographic retrieval and object annotation tasks. In: Peters C, Clough P, Gey F, Karlgren J, Magnini B, Oard D, de Rijke M, Stempfhuber M (eds) Evaluation of Multilingual and Multimodal Information Retrieval 7th Workshop of the Cross–Language Evaluation Forum, CLEF 2006. Lecture Notes in Computer Science (LNCS), vol 4730. Springer, Alicante, Spain, pp 579–594

Daróczy B, Fekete Z, Brendel M, Rácz S, Benczúr A, Siklósi D, Pereszlényi A (2008) SZTAKI@ ImageCLEF 2008: visual feature analysis in segmented images. In: Peters C, Deselaers T, Ferro N, Gonzalo J, Jones G, Kurimo M, Mandl T, Peñas A, Petras V (eds) Evaluating Systems for Multilingual and MultiModal Information Access 9th Workshop of the Cross–Language

Evaluation Forum. Lecture Notes in Computer Science (LNCS), vol 5706. Springer, Aarhus, Denmark, pp 644–651

Daróczy B, Petrás I, Benczúr A, Fekete Z, Nemeskey D, Siklósi D, Weiner Z (2009) Interest Point and Segmentation-Based Photo Annotation. In: Peters C, Tsikrika T, Müller H, Kalpathy-Cramer J, Jones J, Gonzalo J, Caputo B (eds) Multilingual Information Access Evaluation Vol. II Multimedia Experiments: Proceedings of the 10th Workshop of the Cross–Language Evaluation Forum (CLEF 2009), Revised Selected Papers. Lecture Notes in Computer Science (LNCS). Springer, Corfu, Greece

Deselaers T, Hanbury A (2008) The visual concept detection task in ImageCLEF 2008. In: Peters C, Deselaers T, Ferro N, Gonzalo J, Jones G, Kurimo M, Mandl T, Peñas A, Petras V (eds) Evaluating Systems for Multilingual and MultiModal Information Access 9th Workshop of the Cross–Language Evaluation Forum. Lecture Notes in Computer Science (LNCS), vol 5706. Springer, Aarhus, Denmark, pp 531–538

Deselaers T, Hanbury A, Viitaniemi V, Benczúr A, Brendel M, Daróczy B, Escalante Balderas H, Gevers T, Hernández Gracidas C, Hoi S, Laaksonen J, Li M, Marín Castro H, Ney H, Rui X, Sebe N, Stöttinger J, Wu L (2008) Overview of the ImageCLEF 2007 Object Retrieval Task. In: Peters C, Jijkoun V, Mandl T, Müller H, Oard D, Peñas A, Petras V, Santos D (eds) Advances in Multilingual and MultiModal Information Retrieval 8th Workshop of the Cross–Language Evaluation Forum, CLEF 2007. Lecture Notes in Computer Science (LNCS), vol 5152. Springer, Budapest, Hungary, pp 445–471

Douze M, Guillaumin M, Mensink T, Schmid C, Verbeek J (2009) INRIA–LEARs participation to ImageCLEF 2009. In: Working Notes of CLEF 2009, Corfu, Greece

Dumont E, Zhao ZQ, Glotin H, Paris S (2009) A new TFIDF Bag of Visual Words for Concept Detection. In: Peters C, Tsikrika T, Müller H, Kalpathy-Cramer J, Jones J, Gonzalo J, Caputo B (eds) Multilingual Information Access Evaluation Vol. II Multimedia Experiments: Proceedings of the 10th Workshop of the Cross–Language Evaluation Forum (CLEF 2009), Revised Selected Papers. Lecture Notes in Computer Science (LNCS). Springer, Corfu, Greece

Escalante H, Gonzalez J, Hernandez C, Lopez A, Montex M, Morales E, Ruiz E, Sucar L, Villasenor L (2009) TIA–INAOE's Participation at ImageCLEF 2009. In: Working Notes of CLEF 2009, Corfu, Greece

Everingham M, Zisserman A, Williams C, Van Gool L, Allan M, et al (2006) The 2005 PASCAL Visual Object Classes Challenge. In: Machine Learning Challenges. Evaluating Predictive Uncertainty, Visual Object Classification, and Recognising Textual Entailment (PASCAL Workshop 2005). Lecture Notes in Artificial Intelligence (LNAI). Springer, Southampton, UK, pp 117–176

Everingham M, Gool LV, Williams CKI, Winn J, Zisserman A (2010) The pascal visual object classes (voc) challenge. International Journal of Computer Vision 88:303–538

Fakeri-Tabrizi A, Tollari S, Usunier N, Gallinari P (2009) Improving Image Annotation in Imbalanced Classification Problems with Ranking SVM. In: Peters C, Tsikrika T, Müller H, Kalpathy-Cramer J, Jones J, Gonzalo J, Caputo B (eds) Multilingual Information Access Evaluation Vol. II Multimedia Experiments: Proceedings of the 10th Workshop of the Cross–Language Evaluation Forum (CLEF 2009), Revised Selected Papers. Lecture Notes in Computer Science (LNCS). Springer, Corfu, Greece

Ferecatu M, Sahbi H (2009) TELECOM ParisTech at ImageCLEF 2009: Large Scale Visual Concept Detection and Annotation Task. In: Working Notes of CLEF 2009, Corfu, Greece

Glotin H, Fakeri-Tabrizi A, Mulhem P, Ferecatu M, Zhao Z, Tollari S, Quenot G, Sahbi H, Dumont E, Gallinari P (2009) Comparison of Various AVEIR Visual Concept Detectors with an Index of Carefulness. In: Working Notes of CLEF 2009, Corfu, Greece

Grubinger M, Clough P, Müller H, Deselaers T (2006) The IAPR TC–12 benchmark — a new evaluation resource for visual information systems. In: Proceedings of the International Workshop OntoImage'2006, pp 13–23

Hare J, Lewis P (2009) IAM@ImageCLEFPhotoAnnotation 2009: Naive application of a linear–algebraic semantic space. In: Working Notes of CLEF 2009, Corfu, Greece

Huiskes MJ, Lew MS (2008) The MIR Flickr Retrieval Evaluation. In: MIR 2008: Proceedings of the 2008 ACM International Conference on Multimedia Information Retrieval, ACM press

Iftene A, Vamanu L, Croitoru C (2009) UAIC at ImageCLEF 2009 Photo Annotation Task. In: Peters C, Tsikrika T, Müller H, Kalpathy-Cramer J, Jones J, Gonzalo J, Caputo B (eds) Multilingual Information Access Evaluation Vol. II Multimedia Experiments: Proceedings of the 10th Workshop of the Cross–Language Evaluation Forum (CLEF 2009), Revised Selected Papers. Lecture Notes in Computer Science (LNCS). Springer, Corfu, Greece

Inoue M, Grover P (2008) Query Types and Visual Concept–Based Post–retrieval Clustering. In: Peters C, Deselaers T, Ferro N, Gonzalo J, Jones G, Kurimo M, Mandl T, Peñas A, Petras V (eds) Evaluating Systems for Multilingual and MultiModal Information Access 9th Workshop of the Cross–Language Evaluation Forum. Lecture Notes in Computer Science (LNCS), vol 5706. Springer, Aarhus, Denmark, pp 661–668

Jiang J, Rui X, Yu N (2008) Feature Annotation for Visual Concept Detection in ImageCLEF 2008. In: Working Notes of CLEF 2008, Aarhus, Denmark

Llorente A, Overell S, Liu H, Hu R, Rae A, Zhu J, Song D, Rüger S (2008) Exploiting Term Co–occurrence for Enhancing Automated Image Annotation. In: Peters C, Deselaers T, Ferro N, Gonzalo J, Jones G, Kurimo M, Mandl T, Peñas A, Petras V (eds) Evaluating Systems for Multilingual and MultiModal Information Access 9th Workshop of the Cross–Language Evaluation Forum. Lecture Notes in Computer Science (LNCS), vol 5706. Springer, Aarhus, Denmark, pp 632–639

Llorente A, Motta E, Rüger S (2009) Exploring the Semantics Behind a Collection to Improve Automated Image Annotation. In: Peters C, Tsikrika T, Müller H, Kalpathy-Cramer J, Jones J, Gonzalo J, Caputo B (eds) Multilingual Information Access Evaluation Vol. II Multimedia Experiments: Proceedings of the 10th Workshop of the Cross–Language Evaluation Forum (CLEF 2009), Revised Selected Papers. Lecture Notes in Computer Science (LNCS). Springer, Corfu, Greece

Moellic PA, Fluhr C (2006) ImageEVAL 2006 official campaign. Technical report, ImagEVAL

Ngiam J, Goh H (2009) Learning Global and Regional Features for Photo Annotation. In: Peters C, Tsikrika T, Müller H, Kalpathy-Cramer J, Jones J, Gonzalo J, Caputo B (eds) Multilingual Information Access Evaluation Vol. II Multimedia Experiments: Proceedings of the 10th Workshop of the Cross–Language Evaluation Forum (CLEF 2009), Revised Selected Papers. Lecture Notes in Computer Science (LNCS). Springer, Corfu, Greece

Nowak S, Dunker P (2009) Overview of the CLEF 2009 Large–Scale Visual Concept Detection and Annotation Task. In: Peters C, Tsikrika T, Müller H, Kalpathy-Cramer J, Jones J, Gonzalo J, Caputo B (eds) Multilingual Information Access Evaluation Vol. II Multimedia Experiments: Proceedings of the 10th Workshop of the Cross–Language Evaluation Forum (CLEF 2009), Revised Selected Papers. Lecture Notes in Computer Science (LNCS). Springer, Corfu, Greece

Nowak S, Lukashevich H (2009) Multilabel Classification Evaluation using Ontology Information. In: The 1st Workshop on Inductive Reasoning and Machine Learning on the Semantic Web — IRMLeS 2009, co–located with the 6th Annual European Semantic Web Conference (ESWC), Heraklion, Greece

Nowak S, Lukashevich H, Dunker P, Rüger S (2010) Performance measures for multilabel evaluation: a case study in the area of image classification. In: Proceedings of the international conference on Multimedia information retrieval, ACM press, pp 35–44

Pham T, Maisonnasse L, Mulhem P, Chevallet JP, Quénot G, Al Batal R (2009) MRIM–LIG at ImageCLEF 2009: Robot Vision, Image annotation and retrieval tasks. In: Peters C, Tsikrika T, Müller H, Kalpathy-Cramer J, Jones J, Gonzalo J, Caputo B (eds) Multilingual Information Access Evaluation Vol. II Multimedia Experiments: Proceedings of the 10th Workshop of the Cross–Language Evaluation Forum (CLEF 2009), Revised Selected Papers. Lecture Notes in Computer Science (LNCS). Springer, Corfu, Greece

van de Sande K, Gevers T, Smeulders A (2009) The University of Amsterdam's Concept Detection System at ImageCLEF 2009. In: Peters C, Tsikrika T, Müller H, Kalpathy-Cramer J, Jones J, Gonzalo J, Caputo B (eds) Multilingual Information Access Evaluation Vol. II Multimedia Experiments: Proceedings of the 10th Workshop of the Cross–Language Evaluation Forum

(CLEF 2009), Revised Selected Papers. Lecture Notes in Computer Science (LNCS). Springer, Corfu, Greece

Sarin S, Kameyama W (2009) Joint Contribution of Global and Local Features for Image Annotation. In: Working Notes of CLEF 2009, Corfu, Greece

Tollari S, Detyniecki M, Fakeri-Tabrizi A, Marsala C, Amini M, Gallinari P (2008) Using visual concepts and fast visual diversity to improve image retrieval. In: Peters C, Deselaers T, Ferro N, Gonzalo J, Jones G, Kurimo M, Mandl T, Peñas A, Petras V (eds) Evaluating Systems for Multilingual and MultiModal Information Access 9th Workshop of the Cross–Language Evaluation Forum. Lecture Notes in Computer Science (LNCS), vol 5706. Springer, Aarhus, Denmark, pp 577–584

Zhao Z, Glotin H (2008) Enhancing Visual Concept Detection by a Novel Matrix Modular Scheme on SVM. In: Peters C, Deselaers T, Ferro N, Gonzalo J, Jones G, Kurimo M, Mandl T, Peñas A, Petras V (eds) Evaluating Systems for Multilingual and MultiModal Information Access 9th Workshop of the Cross–Language Evaluation Forum. Lecture Notes in Computer Science (LNCS), vol 5706. Springer, Aarhus, Denmark

Chapter 12
The Medical Image Classification Task

Tatiana Tommasi and Thomas Deselaers

Abstract We describe the medical image classification task in ImageCLEF 2005–2009. It evolved from a classification task with 57 classes on a total of 10,000 images into a hierarchical classification task with a very large number of potential classes. Here, we describe how the database and the objectives changed over the years and how state–of–the–art approaches from machine learning and computer vision were shown to outperform the nearest neighbor-based classification schemes working on full–image descriptors that were very successful in 2005. In particular the use of discriminative classification methods such as support vector machines and the use of local image descriptors were empirically shown to be important building blocks for medical image classification.

12.1 Introduction

Thanks to the rapid development of modern medical devices and the use of digital systems, more and more medical images are being generated. This has lead to an increase in the demand for automatic methods to index, compare, analyze and annotate them. In large hospitals, several terabytes of new data need to be managed every year. Typically, the databases are accessible only by alphanumeric description and textual meta information through the standard Picture Archiving and Communication System (PACS). This also holds for digital systems compliant with the Digital Imaging and Communications in Medicine (DICOM) protocol (Lehmann et al, 2005). The DICOM header contains tags to decode the body part examined, the pa-

Tatiana Tommasi
Idiap Research Institute, Centre du Parc, Rue Marconi 19, PO Box 592, 1920 Martigny, Switzerland, e-mail: `ttommasi@idiap.ch`

Thomas Deselaers
Computer Vision Laboratory, ETH Zurich, Sternwartstrasse 7, 8092 Zurich, Switzerland, e-mail: `deselaers@vision.ee.ethz.ch`

H. Müller et al. (eds.), *ImageCLEF*, The Information Retrieval Series 32,
DOI 10.1007/978-3-642-15181-1_12,

tient position and the acquisition modality. Some of these are automatically set by the digital system according to the imaging protocol used to capture the pixel data. Others are introduced manually by the physicians or radiologists during the routine documentation. This procedure cannot always be considered as reliable, since frequently some entries are either missing, false, or do not describe the anatomic region precisely (Güld et al, 2002). This issue, along with the fact that images may contain semantic information not conveyable by a textual description, has led to growing interest in image data mining and Content–Based Image Retrieval (CBIR). Using information directly extracted from images to categorize them may improve the quality of image annotation in particular, and more generally the quality of patient care.

Until 2005, automatic categorization of medical images was often restricted to a small number of classes. For instance, several algorithms have been proposed for orientation detection of chest radiographs, where lateral and frontal orientation are differentiated by means of digital image processing (Pietka and Huang, 1992; Boone et al, 1992). For this two class experiment, the error rates are below 1% (Lehmann et al, 2003a). Pinhas and Greenspan (2003) report error rates below 1% for automatic categorization of 851 medical images into eight classes. In Keysers et al (2003) six classes are defined according to the body part examined. For their test set of 1,617 images an error rate of 8% is reported. However, such low numbers of classes are not suitable for applications in evidence–based medicine or case–based reasoning. Here the image category must be determined in much more detail.

The ImageCLEF medical image annotation challenge was born in this scenario, proposing a task reflecting real–life constraints of content–based image classification in medical applications. The organizers released a large and heterogeneous x–ray image corpus and invited all the participants to compare their algorithms on it, encouraging advances in the field.

12.2 History of ImageCLEF Medical Annotation

The medical image annotation task was added to the ImageCLEF campaign in 2005 alongside the existing medical retrieval task, and further evolved in its five editions until 2009. A description of the aims and expectations for this task, together with the database used and the error evaluation scheme adopted, is given in the following sections.

12.2.1 The Aim of the Challenge

The aim of automatic image annotation is to describe the image content based on its features, both formally and in a generalized way using methods from pattern recognition and structural analysis. This description can then be used in order to compare a new image to a known data set containing a group of pre–defined classes and thus to assign the correct label to the image.

In the medical area, automatic image classification can help in inserting conventional radiographs into an existing electronic archive without interaction and therefore costly editing of diagnostic findings. Other applications include searching for images in an image database or limiting the number of query results, e.g. after a textual image search. It may even be useful for multi–lingual annotation and DICOM header corrections, or as one component of a diagnosis support system. Without any specific application in mind, the aim of the medical annotation task in ImageCLEF was to evaluate state–of–the–art techniques for automatic annotation of medical images based on their visual properties and to promote these techniques. To provide a fair benchmark, a database of fully classified radiographs was made available to the task participants and could be used to train the classification systems. The challenge consisted of annotating a set of unlabelled images released at a later stage to prevent training on the testing data.

Starting from 2005, the annotation challenge has evolved from a simple classification task with 57 classes to a task with almost 200 classes passing through an intermediate step of about 120 classes. From the very start however, it was clear that the number of classes could not be scaled indefinitely. The number of potential categories that could be recognized in medical applications is far too high to assemble sufficient training data for creating suitable classifiers (Deselaers and Deserno, 2009). One solution to address this issue is a hierarchical class structure because it supports the creation of a set of classifiers for subproblems. Therefore, from the very beginning image annotation was based on the hierarchical Image Retrieval in Medical Applications (IRMA) code (see Section 12.2.2).

In 2005 and 2006 the classes were defined by grouping similar codes into single classes and the task was to predict the group to which a test image belongs. In 2007 the objective of the task was refined to predict the complete IRMA code. The hierarchical structure was then used to describe the image content, with the evaluation scheme allowing a finer granularity of classification accuracy. In 2008, high class imbalance was added to promote the function of prior knowledge encoded into the hierarchy. The images in the test set were mainly from classes which had only a few examples in the training data, making annotation significantly harder.

In 2009, for the fifth medical image annotation challenge edition, the task was organized as a survey of the previous year's experience. The idea was to compare the scalability of different image classification techniques with growing numbers of classes, hierarchical class structures and sparsly populated classes.

12.2.2 The Database

The database for the medical image annotation task was provided by the IRMA group from the RWTH University Hospital of Aachen, Germany. It consists of medical radiographs collected randomly from daily routine work at the Department of

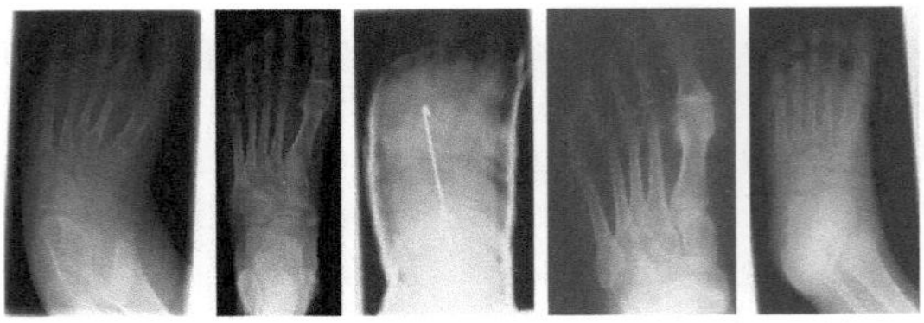

(a) (b) (c) (d) (e)

Fig. 12.1: Images from the IRMA database used for the ImageCLEF challenge (Deselaers et al, 2008). Note the high visual variability among the images. They all belong to the same class annotated as: acquisition modality 'overview image'; body orientation 'AP unspecified'; body part 'foot'; biological system 'musculosceletal'.

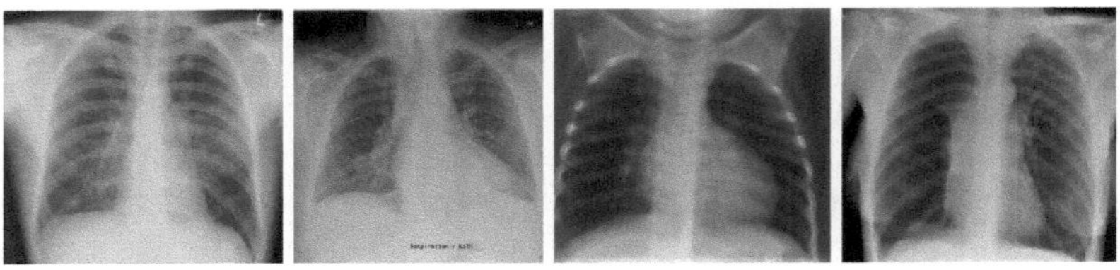

(a) (b) (c) (d)

Fig. 12.2: Images from the IRMA database used for the ImageCLEF challenge (Deselaers et al, 2008). Note the high visual similarity between the images. Each of them belongs to a different class. They all have as acquisition modality 'high beam energy', as body region 'chest unspecified', as biological system 'unspecified', but they differ for the body orientation: (*a*) 'PA unspecified', (*b*) 'PA expiration' (*c*) 'AP inspiration', (*d*) 'AP supine'.

Diagnostic Radiology. Most of the images are secondary digitalized images from plain radiography, but the database also includes images from other modalities, such as CT and ultrasound imaging. The data set contains a great variability: images of different body parts of patients from different ages, different genders, varying viewing angles, and with or without pathologies. Moreover the quality of radiographs varies considerably and there is a great within–category variability together with a strong visual similarity between many images belonging to different classes (see Figures 12.1 and 12.2). All images were provided as PNG files, scaled to fit into a 512 x 512 pixel bounding box (keeping aspect ratio) using 256 gray values.

In order to establish a ground truth, the images were manually classified by expert physicians using the IRMA code (Lehmann et al, 2003b). This method overcomes the problems of ambiguous and undetailed existing schemes considering 'is a' and 'part of' as the only possible relations between code and sub-code elements. Four aspects of the image acquisition are considered resulting in four axes:

- the technical code (T) describes the image modality;
- the directional code (D) models body orientations;
- the anatomical code (A) refers to the body region examined;

Table 12.1: Examples from the IRMA code, anatomy axis.

code	textual description
000	not further specified
...	
400	upper extremity (arm)
410	upper extremity (arm); hand
411	upper extremity (arm); hand; finger
412	upper extremity (arm); hand; middle hand
413	upper extremity (arm); hand; carpal bones
420	upper extremity (arm); radio carpal join
430	upper extremity (arm); forearm
431	upper extremity (arm); forearm; distal forearm
432	upper extremity (arm); forearm; proximal forearm
440	upper extremity (arm); elbow
...	

- the biological code (B) describes the biological system examined.

Each of them is associated with a tag with three to four characters in $\{0, \ldots, 9, a, \ldots, z\}$, where '0' denotes 'unspecified' to determine the end of a path along an axis. In this hierarchy, the more the code position differ from '0', the more detailed is the description. Thus the complete IRMA code is a string of 13 characters TTTT-DDD-AAA-BBB, a structure which can be easily extended by introducing characters in a certain code position if new image modalities are introduced. A small excerpt from the anatomy axis of the IRMA code is given in Table 12.1. Exemplar images from the database together with textual labels and their complete code are given in Figure 12.3.

In 2005, a database of 10,000 images was established. To ease the task participation, images were grouped according to their IRMA annotation at a coarse level of detail forming 57 classes. 9,000 randomly chosen images were selected as training data and given to registered participants prior to the evaluation. A remaining set of 1,000 images was published later as test data without category information. Performance was computed on the 1,000 test images and systems compared according to their ability to correctly annotate these images. In all the subsequent edition of the ImageCLEF challenge, the database was built on top of the previous year. In 2006, the 2005 set of 10,000 images was used for training and a new group of 1,000 images was collected for testing. The number of classes was more than doubled: based on the IRMA code 116 categories were defined. In 2007, the same procedure was adopted: a new set of 1,000 test images was added and the 11,000 images from 2006 were used as training data. The number of classes remained fixed at 116 but this time the task was not to predict the exact class, but to predict the code and a hierarchy–aware evaluation criterion was defined. In 2008 the data released to participants consisted of 12,076 training images (11,000 training images of 2007 + 1,000 testing images of 2007 + 76 new images) and a new test set of 1,000 samples all annotated with a total of 196 unique codes.

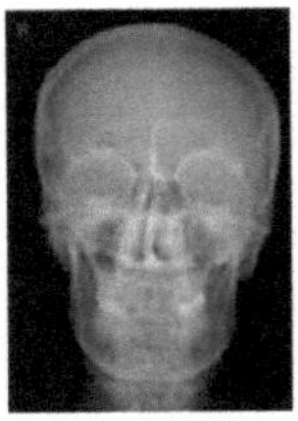

1121-120-200-700
T: plain radiography, analog, overview image
D: coronal, anteroposterior, unspecified
A: cranium, unspecified
B: musculosceletal system, unspecified

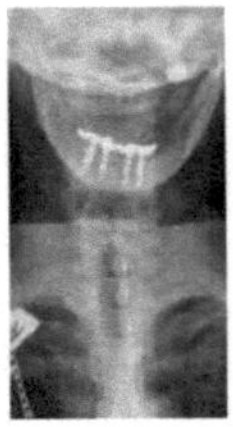

1121-120-310
T: plain radiography, analog, overview image
D: coronal, anteroposterior, unspecified
A: spine, cervical spine
B: musculosceletal system, unspecified

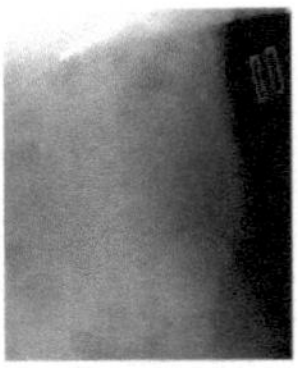

1121-127-700-500
T: plain radiography, analog, overview image
D: coronal, anteroposterior, supine
A: abdomen, unspecified
B: uropoietic system, unspecified

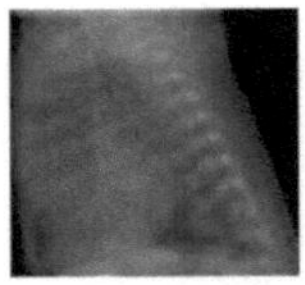

1123-211-500-000
T: plain radiography, analog, high beam energy
D: sagittal, lateral, right–left, inspiration
A: chest, unspecified
B: unspecified, unspecified

Fig. 12.3: Examples of images and corresponding labels of the IRMA database.

In all the databases used the classes were unevenly distributed reflecting the radiological routine acquisition. However in the first three editions of the challenge, each class contained at least ten images. In 2005, the largest class had 28.6% (2,860 images) share of the complete data set, the second one made up 9.6% (959 images) of the collection and there were several classes that formed only between 0.1% and 0.2% (10 to 20 images) of the complete set (Deselaers et al, 2007). In 2006 the two most populated classes had respectively 19.3% and 9.2% share of the data set, while six classes had only 1% or less (Müller et al, 2006).

Imbalance was worsened in 2008: of the total of 196 codes present in the training stage, only 187 appeared in the test set. The most frequent class in the training data consisted of more than 2,300 images but the test data had only one example from this class. The distribution of the test data was nearly uniform while for the training data the distribution was peaked on some classes (Deselaers and Deserno, 2009).

Finally in 2009 a database of 12,677 fully classified radiographs was made available as a training set (Tommasi et al, 2009). Images were provided with labels according to the classification schemes of the annotation tasks from 2005–2008:

- 57 classes as in 2005 (12,631 images) + a 'clutter' class C (46 images);
- 116 classes as in 2006 (12,334 images) + a 'clutter' class C (343 images);
- 116 IRMA codes as in 2007 (12,334 images) + a 'clutter' class C (343 images);

- 193 IRMA codes as in 2008 (12,677 images).

The 'clutter' class for a specific setting contained all the images not identifiable in that year but annotated with a higher level of code detail in the subsequent years. The test data consisted of 1,733 images. Not all the training classes have examples in this set:

2005 labels 55 classes (of 57) with 1,639 images + class C with 94 images;
2006 labels 109 classes (of 116) with 1,353 images + class C with 380 images;
2007 labels 109 IRMA codes (of 116) with 1,353 images + class C with 380 images;
2008 labels 169 IRMA codes (of 193) with 1,733 images.

Participating groups were asked to label images according to each of these schemes in order to understand how the hierarchy changes the task and how sparsely populated classes impact performance.

In 2009, the smallest class in the training data contained six images for the 2005–2007 set–ups, and only one image in the 2008 set–up. A total of 20% of the test images belong to sparsely populated training classes. Examples of the different labels are given in Figure 12.4.

12.2.3 Error Evaluation

To evaluate the performance of the runs submitted by the participants to the medical image annotation task, an error evaluation score was defined, which changed in the different editions of the ImageCLEF campaign according to the given image annotations.

In 2005 and 2006 the error was evaluated just on the capability of the algorithm to make the correct decision. Runs were ranked according to their error rates. For 2007 and 2008, the error was evaluated considering the hierarchical IRMA code.

Let an image be coded by the technical, directional, anatomical and biological independent axes. They can be analyzed separately, summing the error over the individual axes:

- let $l_1^I = l_1, l_2, \ldots, l_i, \ldots, l_I$ be the *correct* code (for one axis) of an image;
- let $\hat{l}_1^I = \hat{l}_1, \hat{l}_2, \ldots, \hat{l}_i, \ldots, \hat{l}_I$ be the *classified* code (for one axis) of an image;

where l_i is specified precisely for every position, and in $\hat{l}_i$ is allowed to say *'don't know'*, which is encoded by '*'. Note that I (the depth of the tree to which the classification is specified) may be different for different images.

Given an incorrect classification at position $\hat{l}_i$ all succeeding decisions are considered to be wrong and, given a not–specified position, all succeeding decisions are considered to be not specified. Furthermore, no error is counted if the correct code is unspecified and the predicted code is a wildcard. In that case, all remaining positions are regarded as not specified.

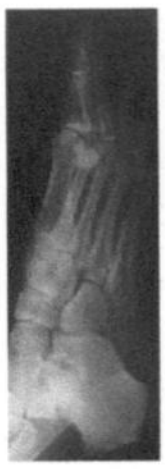

2005: 22 (11-4-91-7)
2006: 54
2007: 1121-4a0-914-700
2008: 1121-4a0-914-700

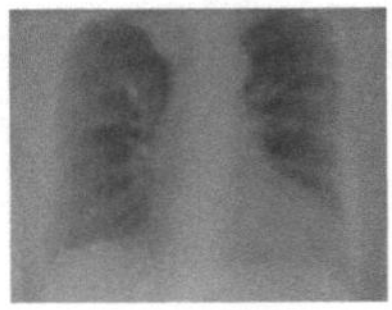

2005: 1 (11-1-50-0)
2006: 1
2007: 1123-127-500-000
2008: 1123-127-500-000

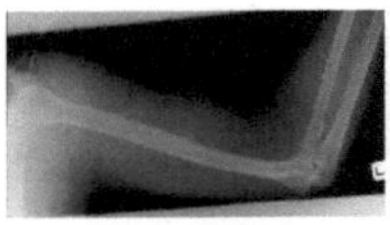

2005 in 2009: 50 (11-2-45-7)
2006 in 2009: C
2007 in 2009: CCCC-CCC-CCC-CCC
2008 in 2009: 1121-230-451-700

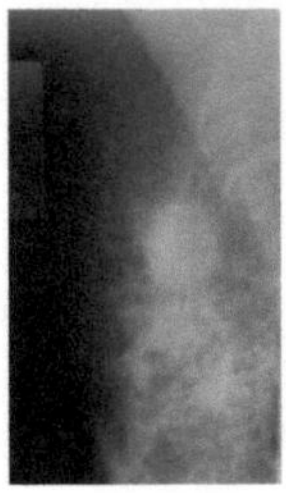

2005 in 2009: C
2006 in 2009: C
2007 in 2009: CCCC-CCC-CCC-CCC
2008 in 2009: 1127-310-600-625

Fig. 12.4: Examples of all the label settings in the different editions of the medical image annotation task in ImageCLEF.

Wrong decisions that are easy (fewer possible choices at that node) are penalized over wrong decisions that are difficult (many possible choices at that node). A decision at position l_i is correct by chance with a probability of $\frac{1}{b_i}$ if b_i is the number of possible labels for position i. This assumes equal priors for each class at each position. Furthermore, wrong decisions at an early stage in the code (higher up in the hierarchy) are penalized more than wrong decisions at a later stage in the code (lower down on the hierarchy): i.e. l_i is more important than l_{i+1}. Putting together:

$$\sum_{i=1}^{I} \underbrace{\frac{1}{b_i}}_{(a)} \underbrace{\frac{1}{i}}_{(b)} \underbrace{\delta(l_i, \hat{l}_i)}_{(c)} \tag{12.1}$$

with

$$\delta(l_i, \hat{l}_i) = \begin{cases} 0 \text{ if } l_j = \hat{l}_j \ \forall j \leq i \\ 0.5 \text{ if } l_j = * \ \exists j \leq i \\ 1 \text{ if } l_j \neq \hat{l}_j \ \exists j \leq i \end{cases} \tag{12.2}$$

where the parts of the equation:

Table 12.2: Error score evaluation for 2007 and 2008 settings. We are considering just the anatomical axis, the correct label is 463.

classified	error count
463	0.000000
46*	0.025531
461	0.051061
4*1	0.069297
4**	0.069297
47*	0.138594
473	0.138594
477	0.138594
***	0.125000
731	0.250000

(a) account for difficulty of the decision at position i (branching factor);
(b) account for the level in the hierarchy (position in the string);
(c) correct/not specified/wrong, respectively.

In addition, for every axis, the maximal possible error is calculated and the errors are normalized such that a completely wrong decision (i.e. all positions for that axis wrong) gets an error count of 0.25 and a completely correctly predicted axis has an error of 0. Thus, an image where all positions in all axes are wrong has an error count of 1, and an image where all positions in all axes are correct has an error count of 0 (see Table 12.2).

In 2009, the class 'clutter' C was introduced. Even if in the test set there were images belonging to this class, their annotation did not influence the error score for the challenge. Moreover, in 2009 the possibility to use wildcards was given even in the 2005 and 2006 settings.

12.3 Approaches to Medical Image Annotation

The ImageCLEF medical image annotation task attracted strong participation from research groups around the world since its first edition. Some of the groups have a background in data mining and retrieval systems while others specialize in object recognition and detection. The sections that follow analyze the methods according to their image representations, classification methods, the use of the hierarchy, and treatment of the unbalanced class distribution.

12.3.1 Image Representation

How to represent the image content is the first problem to face when defining an automatic annotation system. There are different strategies to extract features from images depending on which is considered the most relevant information to capture. As the x–ray images do not contain any color information, edge, shape and global texture features play an important role in this task and were used by several groups (Bo et al, 2005; Deselaers et al, 2005; Liu et al, 2006; Müller et al, 2005). Various methods used the pixel values directly and accounted for possible deformation of the images (Image Distortion Model, IDM) (Güld et al, 2005; Deselaers et al, 2005). Approaches coming from the object recognition field mostly followed the currently widely adopted assumption that an object in images consists of parts that can be modelled independently. Thus these methods considered local features extracted around interest points and used a wide variety of bag-of-features approaches (Marée et al, 2005; Liu et al, 2006; Tommasi et al, 2007; Avni et al, 2008). Generally the ordering of the visual words is not taken into account and only the frequency of the individual visual word is used to form the feature vectors. However, some groups added the spatial information to patches extracted from images (Deselaers et al, 2006; Avni et al, 2008) after observing that radiographs of a certain body part are typically taken in the same spatial arrangement. Another widely adopted strategy consists of combining different local and global descriptors into a unique feature representation (Bo et al, 2005; Liu et al, 2006; Tommasi et al, 2008a).

12.3.2 Classification Methods

Choosing the classification technique means selecting the rules that form the basis of the annotation process. Many different classification strategies were applied and while in the earlier years nearest neighbor-based approaches were most common and most successful (e.g. (Deselaers et al, 2005; Güld et al, 2005)), in 2006 and later, discriminative approaches such as log–linear models (Deselaers et al, 2006), and decision trees (Setia et al, 2008), as well as Support Vector Machines (Setia et al, 2008; Tommasi et al, 2008a; Avni et al, 2008) became more and more common and outperformed the nearest neighbor–based approaches. In many cases known Content–Based Image Retrieval (CBIR) systems are considered: both GIFT (Müller et al, 2005) and FIRE (Deselaers et al, 2005) are used in the same way. The training images are used as the image database and the test images are used to query it. For each query, the training images are ranked according to their similarity and the nearest neighbor decision rule is applied, i.e. the class of the most similar training image is chosen for every test image. Analogous to feature combination, classifier combination has also been a popular way to improve performance (Rahman et al, 2006; Tommasi et al, 2008b; Avni et al, 2008).

12.3.3 Hierarchy

From 2007, when the entire IRMA code was used for labelling, most of the proposed methods tackled the hierarchy considering four different classifiers, one for each axis. The obtained labels were then associated to give the final annotation (Gass et al, 2007; Setia et al, 2008). Other strategies consisted in defining a single classifier able to manage the knowledge encoded in the class hierarchy. Examples are the introduction of weighted distances in k–nearest neighbor classifiers (Springmann and Schuldt, 2007) or weighted splitting rules in decision trees reflecting the hierarchical error score (Setia et al, 2008). Some groups also proposed the combination of axis wise and flat annotation (the ULG group (Deselaers et al, 2008)) or to integrate the output of different classifiers considering majority voting for the characters in each position of the code (Güld and Deserno, 2007). Given the possibility to use wildcards, classifier combination was used to set a '*' when classifiers disagree (Gass et al, 2007; Güld and Deserno, 2007).

12.3.4 Unbalanced Class Distribution

One of the difficulties of the medical image annotation task was the uneven distribution of samples in the training classes. Most of the proposed strategies handled this problem by using wildcards when confidence is low. There have been only few attempts to tackle the class imbalance directly. One of the approaches focused on feature calculation: the number of patches extracted from each image to build the visual word vocabulary was set as inversely proportional to the number of images in its class (Marée et al, 2005). Another approach adapted the classifier using a k-nearest–neighbors (kNN) algorithm with a different *k* value for each class which took into account the frequency of images within the training set (Zhou et al, 2008). The presence of sparsely populated classes in the original training set was also faced by both successively dividing the data into frequency based sub–groups and training a separate SVM for each of them (Unay et al, 2009), and creating virtual examples (Tommasi et al, 2008a).

12.4 Results

In this section we focus on the methods which produced the best results over the five editions of the ImageCLEF medical image annotation task.

A total of 12 research groups participated in 2005 submitting 44 runs. The first fifteen ranked runs are summarized in Table 12.3.

The best results are obtained using the pixel values of the images directly: either by using deformation models on the complete image (scaled to a fixed size) or by using sparsely sampled image patches. Regarding the classification methods, near-

Table 12.3: Resulting error rates for the first 15 runs submitted in 2005 and 2006. (LRPM: low resolution pixel map; BOW: bag–of–words; thumb: thumbnails; Entr.: entropy; relev.: relevance evaluation on the top N retrieved images; HI: histogram intersection kernel; SPM: Spatial Pyramid Matching kernel; RBF: Radial Basis Function kernel; llc, hlc: low and high level cue combination (Tommasi et al, 2008b); oa,oo: one–vs.–all and one–vs.–one SVM multi–class extension.)

		2005		
Rank	**Group**	**Features**	**Classifier**	**ER(%)**
1	RWTH-i6	thumb. $X \times 32$ IDM	KNN k=1	12.6
2	RWTH-mi	texture JSD + thumb. $X \times 32$ IDM	KNN k=1	13.3
3	RWTH-i6	image patches, BOW	log-linear model	13.9
4	ULG	image patches	boosting	14.1
5	RWTH-mi	texture JSD + thumb. $X \times 32$ IDM	KNN k=1	14.6
6	ULG	image patches	decision trees	14.1
7	GE	texture, 8 grey lev.	GIFT + KNN k=5	20.6
8	Infocomm	texture + LRPM, llc	SVM oa + RBF	20.6
9	GE	texture, 16 grey lev.	GIFT + KNN k=5	20.9
10	Infocomm	texture + LRPM, llc	SVM oa + RBF	20.6
11	Infocomm	texture + LRPM, llc	SVM oa + RBF	20.6
12	GE	texture, 8 grey lev.	GIFT + KNN k=1	21.2
13	GE	texture, 4 grey lev.	GIFT + KNN k=10	21.3
14	MIRACLE	texture	GIFT + relev. N=20	21.4
15	GE	texture, 16 grey lev.	GIFT + KNN k=1	21.7
		2006		
Rank	**Group**	**Features**	**Classifier**	**ER(%)**
1	RWTH-i6	image patches + position, BOW	log-linear model	16.2
2	UFR	local rel. coocc. matr. 1000 p.	SVM oa + HI	16.7
3	RWTH-i6	image patches + position, BOW	SVM oa + HI	16.7
4	CISMeF	local + global texture + PCA	SVM oa + RBF	17.2
5	CISMeF	local + PCA	SVM oa + RBF	17.2
6	MSRA	global, llc	SVM oo + SPM	17.6
7	CISMeF	local + global texture + PCA	SVM oa + RBF	17.9
8	UFR	local rel. coocc. matr. 800 p.	SVM oa + HI	17.9
9	MSRA	image patches, BOW	SVM oo + SPM	18.2
10	CISMeF	local + PCA	SVM oa + RBF	20.2
11	RWTH-i6	thumb. $X \times 32$ IDM	KNN k=1	20.4
12	RWTH-mi	texture JSD + thumb. $X \times 32$ IDM	KNN k=1	21.5
13	RWTH-mi	texture JSD + thumb. $X \times 32$ IDM '05	KNN k=1	21.7
14	CINDI	local + global, hlc	SVM oo + RBF (+)	24.1
15	CINDI	local + global, hlc	SVM oo + RBF ($\times$)	24.8

est neighbor methods obtain the best results if an appropriate distance function can be defined. Most of the participating methods come from a CBIR context; however, it can be seen that those methods coming from the image classification and recognition domain field achieve good results (ranks 3, 4). The success of the deformation models by the RWTH Aachen University groups might be partly be due to their working with similar data for several years before the competition.

If we compare the winning and the second classified runs, we can see that they differ for the use of texture Tamura features. The RWTH–mi group considered this cue, comparing two images through the Jensen Shannon Divergence. It seems that adding the texture features does not help the classification, but the result may also be due to an unoptimized choice of the cue combining weights.

In 2006, 12 groups took part in the annotation task submitting 28 runs. Looking at the best 15 results (see Table 12.3) the most interesting observation is that the RWTHi6-IDM system that performed best in the previous year's task (error rate: 12.6%) obtained here an error rate of 20.4%. This decrease in performance can be explained by the larger number of classes. All better–ranked approaches use discriminative models (log–linear models or SVMs), which indicates that discriminative approaches can cope better with higher number of classes than the nearest neighbor classifier if the amount of training data is increased by only 10%.

The best–ranked approach in 2006 is a bag–of–visual words model with dense feature extraction and a dense generic visual vocabulary of 65,536 visual words incorporating the positions where features were extracted. The position information seems to be very useful and the results validate the hypothesis that as radiographs are taken under controlled conditions, the geometric layout of images showing the same body region can be assumed to be very similar. The second and third ranked approaches also incorporate spatial information into the feature vector.

Considering all the submissions in general, it can be noticed that there is an increasing trend towards the combination of multiple cues at different levels in the classification process.

In 2007, ten groups participated submitting 68 runs. Analyzing the results, it can be observed that the top performing submissions do not consider the hierarchical structure of the given task, but rather use each individual code as a whole and train a 116 class classifier (see Table 12.4). The best run using the code is on rank 6; it builds on top of the other runs from the same group using the hierarchy only in a second stage to put a wildcard where their output differs. Furthermore it can be seen that for a method, which is applied once accounting for the hierarchy/axis structure of the code and once using the straight–forward classification into 116 class approach, the one which does not know about the hierarchy outperforms the other one (runs on ranks 11 and 13, 7 and 14). Another clear observation is that methods using local image descriptors produce better results than methods using global image descriptors. In particular, the top 16 runs all use either local image features alone or local image features in combination with global descriptors. The winning run proposes very efficient local features (Scale Invariant Feature Transform) (SIFT) (Lowe, 1999) combined with global cues through a new method which performs the integration during the classification process. The method which was ranked best in 2006 was ranked 8 in 2007.

Considering the rank with respect to the applied hierarchical measure, and the ranking with respect to the error rate, it can be seen that they are quite similar. Most of the differences are clearly due to the use of the wildcard characters which can lead to an improvement for the hierarchical evaluation scheme, but will always lead to a deterioration with respect to the error rate.

Table 12.4: Resulting error rates for the first fifteen submitted runs in 2007 and 2008. The two reference run of the RWTH–mi group are also listed. (BOW: bag–of–words; HI: histogram intersection kernel; MCK: multi cue kernel (Tommasi et al, 2008b); DAS: Discriminative Accumulation Scheme (Nilsback and Caputo, 2004); llc, mlc, hlc: low, mid and high level cue combination (Tommasi et al, 2008b); prob.= probability interpretation of SVM output; vote= combination of voting in one–vs.–one multiclass SVM; AX: four different classifications, one for each axis of the IRMA code are performed and then combined; comm.: in the combination of more opinions a wildcard is used where they disagree; virt. imm.: use of virtual samples defined slightly modifying the original images; major: combination on the basis of majority voting; χ^2: chi–square kernel; RBF: Radial Basis Function kernel; Entr: entropy; oa,oo: one–vs.–all and one–vs.–one SVM multiclass extension; RF relevance feedback.)

		2007			
Rank	**Group**	**Features**	**Classifier**	**Score**	**ER(%)**
1	Idiap	local + global mlc	SVM oa + MCK χ^2	26.8	10.3
2	Idiap	local + global mlc	SVM oo + MCK χ^2	27.5	11.0
3	Idiap	local	SVM oo +χ^2	28.7	11.6
4	Idiap	local	SVM oa +χ^2	29.5	11.5
5	Idiap	local + global hlc	SVM oa +χ^2 DAS	29.9	11.1
6	RWTH-i6	comb. rank 8, 10, 11, 12	log-linear model	30.9	13.2
7	UFR	local rel. coocc. matr. 1000 p.	SVM + HI	31.4	12.1
8	RWTH-i6	patches + position, BOW	log-linear model	33.0	11.9
9	UFR	local rel. coocc. matr. 800 p.	SVM + HI	33.2	13.1
10	RWTH-i6	patches + position, BOW	log-linear model	33.2	12.3
11	RWTH-i6	patches + position, BOW	log-linear model	34.6	12.7
12	RWTH-i6	patches + position, BOW	log-linear model	34.7	12.4
13	RWTH-i6	patches + position, BOW	log-linear model	44.6	17.8
14	UFR	local rel. coocc. matr.	SVM + HI AX	45.5	17.9
15	UFR	local rel. coocc. matr.	decision tree	47.9	16.9
...					
17	RWTH-mi	texture JSD + thumb. $X \times 32$ + IDM	KNN k=5, comm.	51.3	20.0
18	RWTH-mi	texture JSD + thumb. $X \times 32$ + IDM	KNN k=5, major.	52.5	18.0
		2008			
Rank	**Group**	**Features**	**Classifier**	**Score**	
1	Idiap	local + global, llc + virt. img.	SVM oa +χ^2 comm.	74.9	
2	Idiap	local + global, llc + virt. img.	SVM oa +χ^2	83.5	
3	Idiap	local + global, llc	SVM oa +χ^2 comm.	83.8	
4	Idiap	local + global, mlc + virt. img.	SVM + MCK oa χ^2 comm.	85.9	
5	Idiap	local + global, llc	SVM oa +χ^2	93.2	
6	Idiap	local	SVM oa +χ^2	100.3	
7	TAU	patches whole img. BOW	SVM oo + RBF	105.8	
8	TAU	patches mult. res. BOW, hlc	SVM oo + RBF (prob.)	105.9	
9	TAU	patches mult. res. BOW, hlc	SVM oo + RBF (vote)	109.4	
10	TAU	patches resized img. BOW	SVM oo + RBF	117.2	
11	Idiap	local	SVM oa +χ^2	128.58	
12	RWTH-mi	texture JSD + thumb. $X \times 32$ IDM	KNN k=5 major.	182.8	
13	MIRACLE	local + global	KNN k=3	187.9	
14	MIRACLE	local + global	KNN k=2	190.4	
15	MIRACLE	local + global	KNN k=2 +RF	190.4	

In 2008, six groups participated in the image annotation task submitting 24 runs. The 15 fifteen ranked runs use discriminative models and local descriptors outperforming all the other approaches (see Table 12.4). The winning run proposes the combination of local and global features together with a technique to increase the number of images in the scarcely populated classes and evaluates the confidence of the classification decision to use wildcards opportunely. The runs on the sixth and the seventh rank positions respectively by the Idiap and TAU group, use similar features and classification methods and obtain a similar error score. This indicates that the higher performance of the first five runs is most likely due to the use of multiple cues and to the technique adopted to manage the class imbalance and to exploit the hierarchical code structure.

In 2009, seven groups took part in the challenge submitting 19 runs. The task in this last edition was to annotate a set of x–ray images using the labelling schemes from 2005, 2006, 2007 and 2008. Table 12.5 summarizes the best 15 results considering the error score sum on the four different annotation codes. The runs reaching the highest position in the rank are again by the TAU and Idiap groups as in 2008. The results indicate that their strategy suited all the data set configurations proposed in the different editions of the annotation task. In particular the runs submitted by the Idiap group are exactly the same as those in 2008, while the TAU group improved their image descriptors and optimized the kernel choice.

Finally, to evaluate the performances of the best submitted runs all over the five editions of the ImageCLEF annotation task, we can use the RWTH–mi submissions as a reference. This group participated in the competition every year proposing a baseline run (texture JSD + thumb. $X \times 32$ IDM). The ratios between the results of the first ranked run and this baseline submission are reported in Table 12.6. Remember that the error score is defined to be 1 if the code annotation of one image is completely wrong, so the ratios of two error scores can be considered as containing the same information as the ratio between two error rates. The results show an improvement over the years and give clear evidence of the advances obtained in the medical image annotation field.

12.5 Conclusion

The medical image annotation task in ImageCLEF 2005–2009 has established a standard benchmark for medical image annotation. Over the years, the task was developed from a simple classification task into a hierarchical image annotation task with a strongly imbalanced distribution of training images. By comparing the performance of the best ranked run in each year with a baseline method that was submitted in every edition, we have shown that medical image annotation has substantially advanced in the last five years.

Table 12.5: Resulting error rates for the first fifteen submitted runs in 2009. In the first part of the Table the sum score is reported for each run, in the second part there are the single error scores for each of the used label settings. (BOW: bag–of–words; llc: low level cue combination (Tommasi et al, 2008b); AX: four different classifications, one for each axis of the IRMA code are performed and then combined; comm.: in the combination of more opinions a wildcard is used where they disagree; virt. imm.: use of virtual samples defined slightly modifying the original images; major: combination on the basis of majority voting; χ^2; chi–square kernel; RBF: Radial Basis Function kernel; oa,oo: one–vs.–all and one–vs.–one SVM multiclass extension; vcad: voting based approach per axis with chopping letter by letter with descending vote.)

		2009		
Rank	**Group**	**Features**	**Classifier**	**Score**
1	TAU	patches mult. res. BOW, llc	SVM oo +χ^2	852.8
2	Idiap	local + global llc + virt. imm.	SVM oa +χ^2 comm.	899.2
3	Idiap	local + global llc	SVM oa +χ^2 comm.	899.4
4	Idiap	local + global llc	SVM oa +χ^2	1039.6
5	Idiap	local + global llc + virt. imm.	SVM oa +χ^2	1042.0
6	FEITIJS	local + global llc	bagging, rand. forest, AX	1352.6
7	VPA-Sabanci	local + block position	SVM oa + RBF, AX	1456.2
8	VPA-Sabanci	local + block position	SVM oa + RBF	1513.9
9	VPA-Sabanci	local + block position freq.	SVM oa + RBF	1554.8
10	VPA-Sabanci	local + block position freq.	SVM oa + RBF	1581.7
11	GE	texture, 8 grey lev. vcad	GIFT + KNN k=5	1633.3
12	GE	texture, 16 grey lev. vcad	GIFT + KNN k=5	1633.3
13	RWTH-mi	texture JSD + thumb. $X \times 32$ IDM	KNN k=5 major.	1994.8
14	GE	texture, 16 grey lev. + SIFT hlc	GIFT + KNN k=5	2097.6
15	VPA-Sabanci	local + block freq.	SVM oa + RBF	2744.1

Rank	**Group**	**2005**	**2006**	**2007**	**2008**
1	TAU	356	263	64.3	169.5
2	Idiap	393	260	67.2	178.9
3	Idiap	393	260	67.2	179.2
4	Idiap	447	292	75.8	224.8
5	Idiap	447	292	75.8	227.2
6	FEITIJS	549	433	128.1	242.5
7	VPA-Sabanci	578	462	155.5	261.2
8	VPA-Sabanci	578	462	201.3	272.6
9	VPA-Sabanci	587	498	169.3	300.4
10	VPA-Sabanci	587	502	172.1	320.6
11	GE	618	507	190.7	317.5
12	GE	618	507	190.7	317.5
13	RWTH-mi	790	638	207.6	359.3
14	GE	791.5	612.5	272.7	420.9
15	VPA-Sabanci	587	1170	413.1	574.0

Table 12.6: Resulting error ratios between the best run of each year and the corresponding baseline result. The error ratio for 2009 is evaluated averaging the ratios produced for each of the labelling schemes (2005, 2006, 2007, 2008).

Year	Group and Run	Error Ratio
2005	RWTH-i6, thumb. $X \times 32$ IDM & FIRE	0.947
2006	RWTH-i6, image patches + position, BOW & FIRE + max Entropy	0.747
2007	Idiap, local + global mlc & SVM oa + MCK χ^2	0.510
2008	Idiap, local + global, llc + virt. img & SVM oa $+\chi^2$ comm.	0.410
2009	TAU, patches mult. res. BOW, llc & SVM oo $+\chi^2$	0.411

Acknowledgements The authors would like to thank Thomas M. Deserno for providing the data for the medical annotation tasks from 2005–2009. We would also like to thank the organizers of CLEF and ImageCLEF for supporting the medical image annotation task over the years. T.T. was supported by the EMMA project thanks to the Hasler foundation (`http://www.haslerstiftung.ch`).

References

Avni U, Goldberger J, Greenspan H (2008) TAU MIPLAB at ImageClef 2008. In: Working Notes of CLEF 2008, Aarhus, Denmark

Bo Q, Wei X, Qi T, Chang SX (2005) Report for annotation task in ImageCLEFmed 2005. In: Working Notes of CLEF 2005, Vienna, Austria

Boone J, Seshagiri S, Steiner R (1992) Recognition of chest radiograph orientation for picture archiving and comunication system display using neural networks. Journal of Digital Imaging 5(3):190–193

Deselaers T, Deserno TM (2009) Medical image annotation in ImageCLEF 2008. In: CLEF 2008 Proceedings. Lecture Notes in Computer Science (LNCS), vol 5706. Springer, pp 523–530

Deselaers T, Weyand T, Keysers D, Macherey W, Ney H (2005) FIRE in ImageCLEF 2005: Combining content–based image retrieval with textual information retrieval. In: CLEF 2005 Proceedings. Lecture Notes in Computer Science (LNCS), vol 4022. Springer, pp 652–661

Deselaers T, Weyand T, Ney H (2006) Image retrieval and annotation using maximum entropy. In: CLEF 2006 Proceedings. Lecture Notes in Computer Science (LNCS), vol 4730. Springer, pp 725–734

Deselaers T, Müller H, Clough P, Ney H, Lehmann TM (2007) The CLEF 2005 automatic medical image annotation task. International Journal of Computer Vision 74(1):51–58

Deselaers T, Deserno TM, Müller H (2008) Automatic medical image annotation in ImageCLEF 2007: Overview, results, and discussion. Pattern Recognition Letters 29(15):1988–1995

Gass T, Weyand T, Deselaers T, Ney H (2007) FIRE in ImageCLEF 2007: Support vector machines and logistic models to fuse image descriptors for photo retrieval. In: CLEF 2007 Proceedings. Lecture Notes in Computer Science (LNCS), vol 5152. Springer, pp 492–499

Güld MO, Deserno TM (2007) Baseline results for the ImageCLEF 2007 medical automatic annotation task using global image features. In: CLEF 2007 Proceedings. Lecture Notes in Computer Science (LNCS), vol 5152. Springer, pp 637–640

Güld MO, Kohnen M, Keysers D, Schubert H, Wein BB, Bredno J, Lehmann TM (2002) Quality of DICOM header information for image categorization. In: Proceedings SPIE, vol 4685, pp 280–287

Güld MO, Christian Thies BF, Lehmann TM (2005) Combining global features for content–based retrieval of medical images. In: Working Notes of CLEF 2005, Vienna, Austria

Keysers D, Dahmen J, Ney H (2003) Statistical framework for model–based image retrieval in medical applications. Journal of Electronic Imaging 12(1):59–68

Lehmann TM, Güld O, Keysers D, Schubert H, Kohnen M, Wein BB (2003a) Determining the view position of chest radiographs. Journal of Digital Imaging 16(3):280–291

Lehmann TM, Schubert H, Keysers D, Kohnen M, Wein BB (2003b) The IRMA code for unique classification of medical images. In: Proceedings SPIE, vol 5033, pp 440–451

Lehmann TM, Güld MO, Deselaers T, Keysers D, Schubert H, Spitzer K, Ney H, Wein B (2005) Automatic categorization of medical images for content–based retrieval and data mining. Computerized Medical Imaging and Graphics 29(2):143–155

Liu J, Hu Y, Li M, Ma S, ying Ma W (2006) Medical image annotation and retrieval using visual features. In: CLEF 2006 Proceedings. Lecture Notes in Computer Science (LNCS), vol 4730. Springer, pp 678–685

Lowe DG (1999) Object recognition from local scale–invariant features. In: Proceedings of the international conference on computer vision, vol 2, p 1150

Marée R, Geurts P, Piater J, Wehenkel L (2005) Biomedical image classification with random subwindows and decision trees. In: Proceedings of the international conference on computer vision, workshop on Computer Vision for Biomedical Image Applications. Lecture Notes in Computer Science (LNCS), vol 3765. Springer, pp 220–229

Müller H, Geissbühler A, Marty J, Lovis C, Ruch P (2005) The Use of MedGIFT and EasyIR for ImageCLEF 2005. In: CLEF 2005 Proceedings. Lecture Notes in Computer Science (LNCS), vol 4022. Springer, pp 724–732

Müller H, Deselaers T, Deserno T, Kim E, Hersh W (2006) Overview of the ImageCLEFmed 2006 medical retrieval and annotation tasks. In: CLEF 2006 Proceedings. Lecture Notes in Computer Science (LNCS), vol 4730. Springer, pp 595–608

Nilsback M, Caputo B (2004) Cue integration through discriminative accumulation. In: Proceedings of the international conference on computer vision and pattern recognition, vol 2, pp 578–585

Pietka E, Huang H (1992) Orientation correction of chest images. Journal of Digital Imaging 5(3):185–189

Pinhas A, Greenspan H (2003) A continuous and probabilistic framework for medical image representation and categorization. In: Proceedings SPIE, vol 5371, pp 230–238

Rahman MM, Sood V, Desai BC, Bhattacharya P (2006) CINDI at ImageCLEF 2006: Image retrieval and annotation tasks for the general photographic and medical image collections. In: CLEF 2006 Proceedings. Lecture Notes in Computer Science (LNCS), vol 4730. Springer, pp 715–724

Setia L, Teynor A, Halawani A, Burkhardt H (2008) Grayscale medical image annotation using local relational features. Pattern Recognition Letters 29(15):2039–2045

Springmann M, Schuldt H (2007) Speeding up IDM without degradation of retrieval quality. In: Working Notes of CLEF 2007, Budapest, Hungary

Tommasi T, Orabona F, Caputo B (2007a) CLEF2007 Image Annotation Task: an SVM–based Cue Integration Approach. In: Working Notes of CLEF 2007, Budapest, Hungary

Tommasi T, Orabona F, Caputo B (2008b) CLEF2008 Image Annotation Task: an SVM Confidence–Based Approach. In: Working Notes of CLEF 2008, Aarhus, Denmark

Tommasi T, Orabona F, Caputo B (2008) Discriminative cue integration for medical image annotation. Pattern Recognition Letters 29(15):1996–2002

Tommasi T, Caputo B, Welter P, Güld MO, Deserno TM (2009) Overview of the CLEF 2009 medical image annotation track. In: Working Notes of CLEF 2009, Corfu, Greece

Unay D, Soldea O, Ozogur-Akyuz S, Cetin M, Ercil A (2009) Medical image retrieval and automatic annotation: VPA–SABANCI at ImageCLEF 2009. In: Working Notes of CLEF 2009, Corfu, Greece

Zhou X, Gobeill J, Müller H (2008) MedGIFT at ImageCLEF 2008. In: Working Notes of CLEF 2008, Aarhus, Denmark

Chapter 13
The Medical Image Retrieval Task

Henning Müller and Jayashree Kalpathy–Cramer

Abstract This chapter describes the medical image retrieval task of ImageCLEF, the image retrieval track of the CLEF. The medical task has been running for six consecutive years, beginning in 2004. Participation has increased over the years to over 45 registrations for 2010. The query topics have also evolved over the years from a starting point of using images only, via clear visual and textual information needs, and now towards case descriptions to find similar cases. The primary goal of the task is to provide challenging research questions to the scientific community to advance medical visual information retrieval on standard databases. Databases have increased significantly in size over the years to keep pace with the growing demand. The results show that textual information retrieval of images is now much further developed and produces much better results than in past years. However, visual retrieval components such as pre–classifying the images (i.e. modality detection) or improving early precision of the retrieval results can lead to an overall improvement in retrieval performance in specific domains.

13.1 Introduction

Image retrieval has been one of the most active research domains in computer vision over the past 20 years (Smeulders et al, 2000). Many approaches have been developed over that time and image retrieval has been used in a large variety of domains. Comparison of techniques has been very hard as no standard databases existed and thus techniques could only be compared with difficulty. For this reason, ImageCLEF, an image retrieval benchmark (Clough et al, 2004), was started as

Henning Müller
Business Information Systems, University of Applied Sciences Western Switzerland (HES–SO), TechnoArk 3, 3960 Sierre, Switzerland e-mail: henning.mueller@hevs.ch

Jayashree Kalpathy–Cramer
Oregon Health and Science University, Portland, OR, USA e-mail: kalpathy@ohsu.edu

H. Müller et al. (eds.), *ImageCLEF*, The Information Retrieval Series 32,
DOI 10.1007/978-3-642-15181-1_13, © Springer-Verlag Berlin Heidelberg 2010

part of the Cross Language Evaluation Forum (CLEF) to ease the comparison of the multiple techniques developed on the same databases and same tasks (Savoy, 2002).

Within ImageCLEF, a medical task was started in 2004 (Müller et al, 2007) as the medical domain has traditionally been one of the earliest application fields of image retrieval (Lowe et al, 1998; Tagare et al, 1997; Güld et al, 2004). A comprehensive overview of medical image retrieval can be found in (Müller et al, 2004a). The descriptions of the various years of the medical image retrieval task can be found in (Clough et al, 2006; Müller et al, 2009b,a, 2007, 2008; Clough et al, 2004) This chapter describes the evolution of the medical image retrieval task from 2004 to 2009. The chapter starts with the participation in the tasks and an overview of the research groups that have participated over the years (Section 13.2). Then (Section 13.3), the databases and corresponding tasks are analyzed in detail including their evolution and the reasons for this evolution. Section 13.4 describes the evolution of the techniques of the participating groups and finally, Section 13.5 details the results obtained and gives a partial analysis. The chapter finishes with conclusions and a brief outlook into the future.

13.2 Participation in the Medical Retrieval Task

When ImageCLEF started in 2003, only four research groups took part in the challenge, and all four were from the text retrieval field. One of the reasons was the databases contained mainly gray–scale images, and the query topics were geared towards semantic, text–based retrieval systems. None of the participating systems used the visual image content for retrieval.

When starting the medical task in 2004, the clear goal was to orient the task towards the visual retrieval community as no image retrieval benchmark existed at that time. The goal was also to complement the already participating text retrieval groups from the photographic retrieval tasks with visual information retrieval groups. By combining the two communities a fertile ground for mixed textual and visual retrieval could be created, and multi–modal retrieval has remained one of the main goals of ImageCLEF.

Already in the first year the participation in the medical task was important with 18 inscriptions and 11 groups submitting results. Table 13.1 shows that the number of registrations has risen strongly over the first three years and then generally remained stable, whereas the number of participants has increased more slowly but still keeps growing. For ImageCLEF 2010, so far 45 groups have registered for the medical task, which is a new record.

Starting from 2008, the number of runs per group were limited to ten runs as some groups started submitting a large numbers of runs in 2007, leading to concern about bias. This can potentially lead to problems in the pooling process, where groups with fewer runs would be disadvantaged (Zobel, 1998). A fairly large number of groups have participated over the six years including:

- Athens University of Economics, Greece (2009);

Table 13.1: Overview of registrations, participants and runs submitted to the medical image retrieval task over the years.

Year	2004	2005	2006	2007	2008	2009
Inscriptions	18	28	37	31	37	37
participants	11	13	12	13	15	17
runs submitted	43	134	100	149	130	124

- Bania Luka University, Bosnia–Hercegovina (2008);
- CWI, Netherlands (2004);
- Commissariat Energie Automique (CEA), France (2004, 2005);
- Daedalus, Spain (2004, 2005);
- Department of Medical Informatics, Aachen, Germany (2004, 2005, 2006);
- Department of Computer Science, Aachen, Germany (2004, 2005, 2006, 2007);
- Dokuz Eylul University, Izmir, Turkey (2007, 2009);
- GPLSI group, University of Alicante, Spain (2008, 2009);
- Hungarian Acadamy of Sciences, Budapest, Hungary (2008);
- I–Shou University, Taiwan (2004);
- Imperial College, UK (2004);
- Institute for Infocomm research, Singapore (2005);
- Institute for Infocomm research, Medical Imaging Lab , Singapore (2005, 2006);
- IPAL CNRS/ I2R, France/Singapore (2005, 2006, 2007);
- IRIT-Toulouse, Toulouse, France (2007, 2008);
- ISSR, Egypt (2009);
- LIRIS, INSA Lyon, France (2009);
- LITIS Lab, INSA Rouen, France (2006);
- Microsoft Research, China (2006, 2007);
- MIRACLE, Spanish University Consortium, Madrid, Spain (2007, 2008, 2009);
- MRIM-LIG, Grenoble, France (2007, 2008);
- National Chiao Tung University, Taiwan (2005);
- National Library of Medicine (NLM), National Institutes of Health NIH, Bethesda, MD, USA (2008, 2009);
- Natural Language Processing group, University Hospitals of Geneva, Switzerland (2008, 2009);
- Natural Language Processing at UNED. Madrid, Spain (2008);
- Oregon Health and Science University, USA (2004, 2005, 2006, 2007, 2008, 2009);
- State University of New York, Buffalo, USA (2004,2005, 2006, 2007);
- Tel Aviv University, Israel (2008);
- TextMess group, University of Alicante, Spain (2008);
- UIIP Minsk, Belarus, (2009);
- UNAL group, Universidad Nacional Colombia, Bogotà, Colombia (2007 2008);
- University of Applied Sciences Western Switzerland, Sierre (2009);
- University of Concordia (CINDI), Canada (2005, 2006, 2007);

- University of Fresno, USA (2009);
- University of Jaen (SINAI), Spain (2005, 2006, 2007, 2008, 2009);
- University Hospitals of Freiburg, Germany (2006);
- University Hospitals Geneva (MedGIFT), Switzerland (2004, 2005, 2006, 2007, 2008, 2009);
- University of Milwaukee, USA (2009);
- University of North Texas, USA (2009);
- University of Tilburg, Netherlands (2004);
- York University, Canada (2009).

In total, 42 different research groups from five continents have participated in ImageCLEF. OHSU and the University Hospitals of Geneva have been the only groups present in all years, but many groups have participated in several of the challenges. Main participants were PhD students or post doctoral researchers who have a chance to test their techniques and compare them with other approaches.

13.3 Development of Databases and Tasks over the Years

This section describes the databases, tasks and relevance judgments over the years. Further details on the databases can be found in Chapter 2, task developments in Chapter 3 and the relevance judgement process in Chapter 4. In general, the relevance judges are clinicians, and so domain specialists, either from OHSU or the Geneva University Hospitals.

13.3.1 2004

In 2004 the task was organized for the first time, so little experience existed. A database for image retrieval was found with a teaching file called Casimage and representative images were selected by a radiologist who had created a large portion of the database himself. Query starting points were images only, to force participants to use visual retrieval as a starting point for finding similar images. Once similar images were found, relevance feedback or automatic query expansion could use the text of the collection as well.

13.3.1.1 Database

For ImageCLEFmed 2004, the *Casimage*[1] data set was made available to participants (Müller et al, 2004b), containing almost 9,000 images of 2,000 cases (Rosset et al, 2004). Images present in Casimage include mostly radiology modalities,

[1] `http://pubimage.hcuge.ch/`

but also photographs, PowerPoint slides and illustrations. The cases are mainly in French, with around 20% being in English and 5% without annotation. Figure 13.1 shows several images from the database. The database is very varied containing mainly images from radiology but from many modalities and diseases. The images are mainly used internally for teaching, and as they have been available on the Internet, using them for image retrieval did not cause any problems as they are fully anonymized.

13.3.1.2 Topics

The topic development for 2004 was performed by a domain expert, a radiologist who knew the collection very well and created part of it himself, so all images should have at least a few examples and the variety of modalities and protocols was assumed to be present. All images chosen as queries were removed from the collection.

The goal was to use example images as queries for finding images with the same modality, the same anatomic region and, if applicable, the same view and abnormality. In Figure 13.1, 15 of the 26 topics chosen are shown, underlining the variety of images that needed to be analyzed.

13.3.1.3 Relevance Judgments

As judging all images for relevance is infeasible in such a large database, a pooling process was used, where the best N images of each system were put into a pool and then judged (Sparck-Jones and van Rijsbergen, 1975). Three persons in total were used for the relevance judgments. Images could be judged as either relevant, non–relevant or as partially relevant when relevance could not be determined clearly. Results with several sets of judgments were given, such as images judged as relevant by all judges or images that were judged relevant by at least one judge.

For the evaluation of the results, the trec_eval package was used.

13.3.2 2005–2007

13.3.2.1 Databases

From 2005, three additional data sets were added to the Casimage collection. All four databases were teaching files, thus eliminating privacy and data acquisition issues. Besides the Casimage data set, the Pathology Education Instructional Resource[2] (PEIR) database with annotation based on the Health Education Assets Li-

[2] http://peir.path.uab.edu/

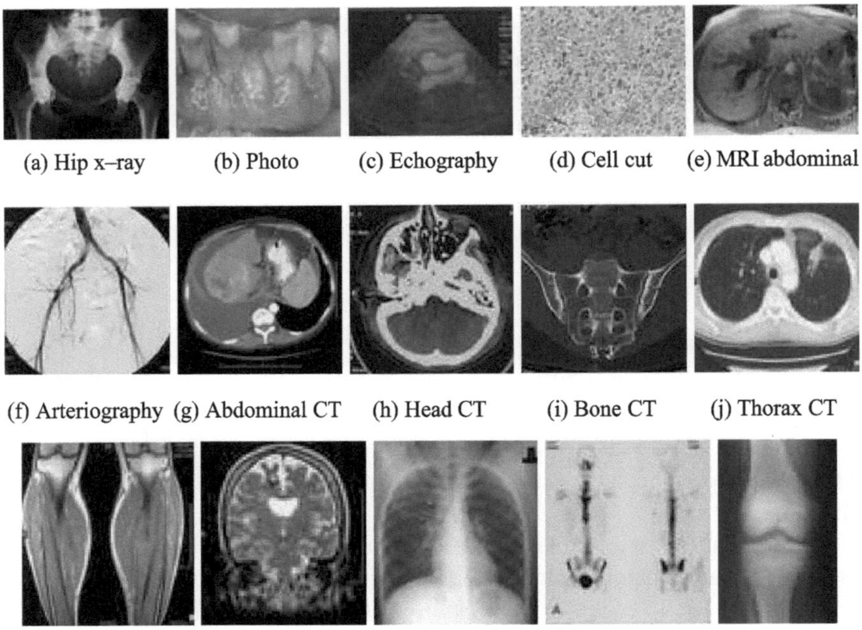

(a) Hip x–ray (b) Photo (c) Echography (d) Cell cut (e) MRI abdominal

(f) Arteriography (g) Abdominal CT (h) Head CT (i) Bone CT (j) Thorax CT

(k) MRI legs (l) Head MRI T2 Cor. (m) Thorax x–ray (n) Szintigraphie (o) X–ray

Fig. 13.1: Topics for the medical retrieval task.

brary (HEAL) project[3], mainly Pathology images (Candler et al, 2003), was used. This data set contains over 33,000 images with English annotations: these being on a per image and not per case basis as in Casimage. The nuclear medicine database of the Mallinkrodt Institute of Radiology[4] (MIR) (Wallis et al, 1995) was also made available to participants. This data set contains over 2,000 images mainly from nuclear medicine with annotations provided per case and in English. Finally, the PathoPic[5] collection (Pathology images (Glatz-Krieger et al, 2003)) was included in our data set. It contains 9,000 images with extensive annotation on a per image basis in German. A short part of the German annotation was translated into English. This data set was thus even more multi–lingual than the Casimage collection containing mainly cases in English, German, and French. Some contained case descriptions in more than one language. More details on the databases can be found in Table 13.2.

[3] `http://www.healcentral.com/`

[4] `http://gamma.wustl.edu/home.html`

[5] `http://alf3.urz.unibas.ch/pathopic/intro.htm`

Table 13.2: The databases used in ImageCLEFmed 2007.

Collection Name	Cases	Images	Annotations	Annotations by Language
Casimage	2,076	8,725	2,076	French – 1,899, English – 177
MIR	407	1,177	407	English – 407
PEIR	32,319	32,319	32,319	English – 32,319
PathoPIC	7,805	7,805	15,610	German – 7,805, English – 7,805
MyPACS	3,577	15,140	3,577	English – 3,577
Endoscopic	1,496	1,496	1,496	English – 1,496
Total	47,680	66,662	55,485	French – 1,899, English – 45,781, German – 7,805

In 2007, two more databases were added to the collection as it became apparent that collections for evaluation needed to grow to meet real–world situations. The MyPACS[6] data set contains 15,140 images of 3,577 cases, all in English. The data set contains mainly radiology images. The Clinical Outcomes Research Initiative[7] (CORI) Endoscopic image database contains 1,496 images with an English annotation per image and not per case. This database extends the spectrum of the total data sets as to date there were only a few endoscopic images in it. An overview of all six data sets can be found in Table 13.2.

This now meant we were able to use a total of more than 66,000 images, with annotations in three different languages. Through an agreement with the copyright holders, we were able to distribute these images to the participating research groups. The MyPACS database required an additional copyright agreement making the process slightly more complex than in 2004–2006.

13.3.2.2 Topics

The query topics in 2005 were based on a small survey administered for clinicians, researchers, educators, students, and librarians at Oregon Health & Science University (OHSU) (Hersh et al, 2005). A very similar survey was afterwards performed in Geneva for further topic development (**?**) Based on these surveys, topics for ImageCLEFmed were developed along the following axes:

- Anatomic region shown in the image;
- Image modality (x–ray, CT, MRI, gross pathology, ...);
- Pathology or disease shown in the image;
- Abnormal visual observation (e.g. enlarged heart).

For 2006 and 2007, topics were developed along the same axes. In 2006, the topics were based on an analysis of a search log of the medical search engine of

[6] `http://www.mypacs.net/`

[7] `http://www.cori.org/`

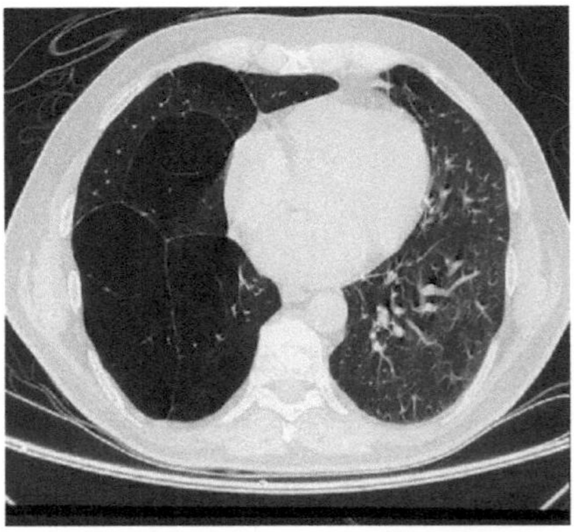

Show me chest CT images with emphysema.
Zeige mir Lungen CTs mit einem Emphysem.
Montre–moi des CTs pulmonaires avec un emphysème.

Fig. 13.2: An example of a query that is at least partly solvable visually, using the image and the text as query. However, use of annotation can augment retrieval quality. The query text is presented in three languages.

health on the net (hon[8]) (Müller et al, 2007). For 2007, an analysis of Medline queries (Müller et al, 2008) was performed to obtain candidate topics that were subsequently tested and used.

Besides using a more solid user model and input from domain experts, the queries were also classified into whether they were geared towards visual retrieval systems, textual retrieval systems or a mix of the two. The goal of this classification was to find out whether systems can adapt to these kind of queries, and where the choice of modality could be performed automatically. The goal was in general to have about the same number of queries in all categories. The textual query topics were also translated into the three languages occurring in the collection: French, English, German.

An example for a visual query of the first category can be seen in Figure 13.2. CT images of the lung are fairly homogeneous and emphysema have a very characteristic local texture.

A query topic requiring more than purely visual features is shown in Figure 13.3. As fractures can in principle appear in all bones, the anatomic region can change the image enormously, thus requiring more than a purely visual analysis.

13.3.2.3 Relevance Judgments

From 2005–2007 relevance judgments were again performed using a pooling process and taking into account the best N results of all retrieval systems to avoid bias

[8] `http://www.hon.ch/`

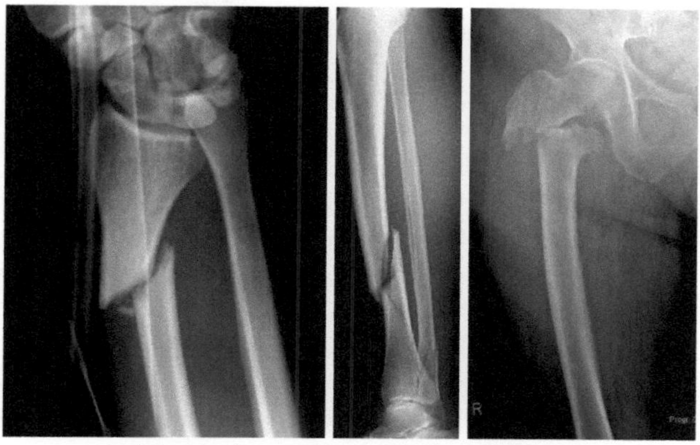

Show me all x–ray images showing fractures.
Zeige mir Röntgenbilder mit Brüchen.
Montres–moi des radiographies avec des fractures.

Fig. 13.3: A query requiring more than visual retrieval but visual features can deliver hints to good results.

of the judgments. An electronic judgement system was created to allow judges to read the image and, in this context, the text of the images. As topics were clear information needs that were deeper than modality and anatomy information, it was often not possible to judge relevance from the images alone. Again, a ternary judgement scheme was used. In 2005, a total of 25 topics was created and in 2006–2007 30 topics were created each year. All 85 topics have been combined with additional relevance judgments to create a consolidated collection (Hersh et al, 2009) as a resource for benchmarking. Baseline results on this database and the tasks are also made available to compare techniques to. The task was also made available as training data for participants.

13.3.3 2008–2009

After three years with basically the same collection a change was required as the collection was fairly well explored by 2007. Fortunately we were able to obtain rights from the Radiological Society of North America (RSNA) to use scientific journal images, one domain where image retrieval can be of particular use. This also meant a change in that the annotations were of extremely high quality, and that annotations existed both for the images in the form of captions, and for the cases in the form of the articles. As all journal articles are peer reviewed and now selected by a single clinician, the quality of cases can be expected to be higher.

13.3.3.1 Databases

The database used for the task in 2008 was made available by the RSNA. The database contains in total slightly more than 66,000 images taken from the radiological journals *Radiology* and *Radiographics*. The images are original figures used in published articles. The collection is a subset of a larger database that is available via the Goldminer[9] image search engine. For each image, the text of the figure caption was supplied as free text. However, this caption was sometimes associated with a multi–part image. In over 90% of the images the part of the caption actually referring to this sub–image was also provided. Additionally, links to HTML versions of the full–text articles were provided along with the relevant PubMed accession ID numbers. Both the full–size images as well as thumbnails were available to the participants. All texts were in English.

The image databases for 2009 and also for 2010 were basically the same collection but with newer articles from the two journals being added to the existing images and articles. It is the policy of the RSNA to make the articles of their journals available 12 months after publication.

13.3.3.2 Topics

As the database had changed significantly for 2008, the topics for this year were chosen from among the topics of the consolidated collection from 2005–2007. This also had the advantage that training data was available for participants as the relevant images for the same tasks on the previous collection were made available, albeit on a very different database.

For 2009, the topics were again created based on a survey (Radhouani et al, 2009) but this time the clinicians who filled in the questionnaires were performing tests with an actual image retrieval system that allowed visual and textual queries and browsing, making them aware of the potential and realities of retrieval systems. This can be expected to lead to more realistic tasks. On the other hand, the tasks were much more geared towards semantic retrieval than visual retrieval, although again the topics were classified into visual, mixed and semantic.

Another novelty in the topic generation were the case–based topics. These topics are expected to go one step closer to clinical routine in supplying the clinician with a case similar to the one under observation. Usually in clinical routine, the unit for retrieval is rarely the image but rather the case. These topics were developed based on the Casimage teaching files, where many classical cases exist. As topics, an abbreviated case description from Casimage was supplied in English not containing any diagnosis information but mainly the anamnesis, the images and a description of the images. The diagnosis was supplied to the relevance judges. An example for a case–based topic can be seen in Figure 13.4.

---

[9] `http://goldminer.arrs.org/`

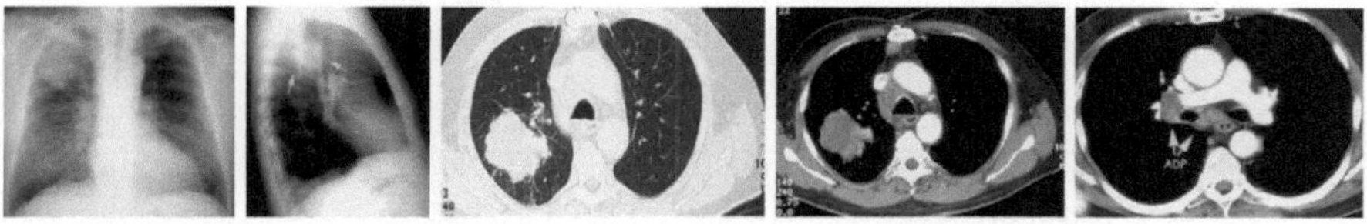

Female patient, 25 years old, with fatigue and a swallowing disorder (dysphagia worsening during a meal). The frontal chest x–ray shows opacity with clear contours in contact with the right heart border. Right hilar structures are visible through the mass. The lateral x–ray confirms the presence of a mass in the anterior mediastinum. On CT images, the mass has a relatively homogeneous tissue density.

Fig. 13.4: A case–based query topic

13.3.3.3 Relevance Judgments

In just the same way as from 2005–2007, the relevance judgments were performed by clinicians in Portland, OR at OHSU. Most of the clinicians followed a Masters degree in medical informatics and thus were familiar on the one hand with medical practice, but on the other hand with computer science methodologies. As in previous year, several topics were judged by several people allowing calculation of the kappa statistic to measure agreement amongst the judges. With the case–based topics the definition of relevance was discussed among the judges as the relevance of a case depends strongly on the knowledge of the clinician. A non–specialist can be helped with cases that are somewhat similar but not of the same diagnosis, whereas domain experts would expect the articles to be on fine differences within the same diagnosis group.

13.4 Evolution of Techniques Used by the Participants

It is extremely hard to compare the performance of the techniques over the six years of the medical retrieval task as databases and tasks have changed each year. This section only gives a broad overview of the tendencies of techniques used over the years. Table 13.3 shows the number of runs submitted for each of the visual, textual and mixed categories over the years. It becomes clear that with the increasing difficulty and a decreasing visualness of the topics, the number of visual runs has decreased. Also, the number of mixed runs has decreased over the years, with the textual runs the most frequent run type. In 2009, there was an increase in visual runs but this needs to be strengthened as visual retrieval clearly has a long way to go to obtain satisfactory results.

The manual/interactive section played an important part in the first year of ImageCLEF and also obtained the best overall results. Since then, only a few groups have invested the necessary time required in creating manual (or those involving

Table 13.3: The number of runs submitted for each of the categories over the years.

Year	2004	2005	2006	2007	2008	2009
Automatic retrieval, visual	29	31	22	28	8	16
Automatic retrieval, textual	0	15	38	40	65	59
Automatic retrieval mixed	14	88	40	81	31	30
Manual and feedback retrieval	22	6	21	3	7	13

user feedback) runs. It can be expected that such approaches could well improve results.

More on the results can be read in the six years of working notes describing the participants' approaches including a vast amount of detail.

13.4.1 Visual Retrieval

Early visual techniques used rather simple visual features such as features describing the layout of the images (quad–tree representations, downscaled versions of the images) and texture descriptions such as Tamura and Gabor filters. In 2005 this lead to good results and the best system (GIFT) obtained a good score, even achieving the best results in mixed retrieval with relevance feedback.

Since 2005, when the database grew larger and topics more complex, the features also became more complex, e.g. visual patches obtain much better results than the previous visual techniques. In an approach to use massive visual learning, a performance increase in terms of MAP from 3% to 22% was obtained in (Deselaers et al, 2007). This is partly based on the fact that similar cases from the past existed on the same database. Still, this is a remarkable results and shows that once training data is available, results can be improved in a significant way.

13.4.2 Textual Retrieval

Textual retrieval started out with fairly simple techniques of full text retrieval, often using existing system such as Lucene[10] or Terrier[11]. The first teaching files had a large number of mistakes such as spelling errors, etc., that made textual retrieval more challenging but allowed visual techniques to show their utility.

To approach multi–lingual retrieval, simple translation techniques were used by participants. Alternatively, the text of the queries and the captions could be mapped onto a medical terminology, such as MeSH (Medical Subject Headings) or the

[10] `http://lucene.apache.org/`

[11] `http://terrier.org/`

UMLS (Unified Medical Language System). As these terminologies contain the same axes as the query topics, several approaches using a mapping to ontologies obtained the best results over the years. Mapping texts onto a terminology also helped with the multi–lingual retrieval. Terminologies such as MeSH exist in all languages used for the medical task and thus the concepts in the various languages remain the same.

13.4.3 Combining Visual and Textual Retrieval

More often than not the combination of visual and textual retrieval results were limited to linear combinations and thus only a small gain can be expected. It is also clear that textual results obtain much better results than the visual techniques and thus the combinations need to be performed with care. Several groups reported mixed runs being actually worse than the textual retrieval results alone, underlining the importance of adapted fusion strategies that are detailed in Chapter 6. A very large variety of techniques have been used for the fusion, ranging from early fusion techniques in the feature space to the late fusion techniques of various results sets.

13.4.4 Case–Based Retrieval Topics

As the case–based query topics were new in 2009, the techniques used for the topics were very similar to those used for the image based topics. The texts of all images were often concatenated or the full text of the article was used. Visual retrieval results could be a simple combination of all images appearing in a case or better in an article. As there were only five case–based topics and few groups participating, it is hard to make generalizations about the task and the techniques used.

13.5 Results

When looking at the results of the techniques, it becomes clear that it is hard to compare the various techniques as the query topics and the databases have changed enormously over the years. The only reference is the use of the GIFT (GNU Image Finding Tool) system that was run in the basic configuration during all years of ImageCLEF. Thus, a comparison of the percentage gain over the GIFT system can show an evolution of the techniques. This is of coursed biased by the fact that the first competitions were largely visual whereas the query topics have become increasingly semantic over the years. Figure 13.4 show the results of the GIFT system in each year and compares with the best system for visual, textual and mixed retrieval. The comparisons are listed in absolute values and percentages.

Table 13.4: Results (in MAP) of the baseline using GIFT and improvement of the best system on each of the three categories.

Year	2004	2005	2006	2007	2008	2009
GIFT	0.31	0.094	0.0467	0.040	0.03	0.01
Visual	0.379 (22%)	0.1456 (55%)	0.0753 (61%)	**0.2328 (482%)**	0.04 (33%)	0.01 (0 %)
Textual	0 (0)	0.208 (121%)	0.2646 (466%)	0.3962 (891%)	0.29 (867%)	**0.43 (4300%)**
Mixed	0.421 (36%)	0.282 (200%)	0.3095 (562%)	0.3719 (830%)	0.29 (867%)	0.37 (3700%)

It becomes clear that only in the first year was the basic GIFT system close to the best systems in the comparison with a difference of 35% to the best overall system. From then, the difference to the best system rose each year to reach 4300%.

13.5.1 Visual Retrieval

Visual techniques clearly had the best results in the first competition, when the topics were geared towards visual retrieval. Since then, the best visual systems were better than the baseline, but only once (in 2007 with much learning) were results exceptionally improved.

Otherwise the performance increase of visual retrieval over GIFT has remained almost constant over the years. This underlines the point that visual retrieval does have better results but evolves slowly.

13.5.2 Textual Retrieval

Textual retrieval always had fairly good results but usually combinations of visual and textual retrieval have been better than purely textual retrieval from 2004–2006. In 2007 and also 2009, on the other hand, the textual techniques were better than the best mixed runs. There are several reasons for this. Textual techniques are clearly the most evolved technology of information access and the research groups have the most experience in optimizing them. On the other hand, most groups invest little time in fusion techniques. A fusion task at ICPR using the best textual and visual runs from ImageCLEF 2009 showed that the fused runs were much better than the best original submission. A review of fusion techniques of all ImageCLEF competitions can be found in Chapter 6.

13.5.3 Mixed Retrieval

The retrieval combining visual and textual features has in most years had the best overall results, but not in 2007 and 2009. This shows how little expertise there is with respect to fusion techniques for visual and textual information where the difference in performance is really important. Combining textual and visual cues for combined multimedia retrieval clearly has an important potential to improve results. Currently most techniques use simple linear combinations and in general most likely late fusion for the combination (See Chapter 6). Although the MAP of mixed runs might not always be better than the MAP of textual runs, very often the early precision is much better in such combinations. A fusion task of the best visual and textual runs of ImageCLEFmed 2009 also showed that mixed–media retrieval was better for all partners than combining only textual runs, although the quality of the text runs was much better than that of the visual runs.

13.5.4 Relevance Feedback and Manual Query Reformulation

The first ImageCLEF challenge was the only one where relevance feedback and other manual techniques had a real impact and achieved by far the best results. This was due to the fact that the queries involved no text and so text groups had to use automatic or manual query expansion techniques. Unfortunately, in the following years only very little time and effort was invested in manual techniques, although some of the techniques exhibited very good results, especially with early precision. There is still an important performance potential and using the best techniques in the challenge with a manual reformulation approach could lead to extremely good results.

13.6 Main Lessons Learned

Over the six years of ImageCLEF, techniques have shown an important development as have the query topics, becoming increasingly semantic and thus difficult, particularly for visual retrieval techniques. Visual retrieval performed well for ImageCLEF 2004, where the tasks were clearly adapted to what visual techniques can perform. In the following years, and on more semantic topics, it became clear that visual techniques alone are not capable of fulfilling the complex semantic information needs of real users that were created based on survey and/or analyzing log files of real systems. On the other hand, the image classification task showed that purely visual retrieval is very strong when classifying into a small number of classes and particularly when training data is available. Obtaining such basic information from visual data analysis can of course be of particular help if the collections used do not possess very good or incomplete annotation. Images from journal articles, on the other hand, are extremely well annotated, as are teaching files.

Text retrieval techniques have had a much better performance in ImageCLEFmed whenever there was text available describing the images. When collections are only partly annotated or when the annotations miss part of the information needs of the users, then mixed approaches combining textual and visual features can obtain best results. In most years mixed approaches actually had the best performance despite the fact that few groups actually had good textual and visual systems. Few groups actually undertook much work on the combination of features and thus the potential for results fusion was little explored. The fusion task at ICPR using ImageCLEF data showed that good fusion techniques can improve the existing results enormously. The fusion of visual and textual results always had the best results despite an enormous difference in performance between visual and textual retrieval.

Although MAP has remained the lead measure for CLEF as it is a good combination of precision and recall–based measures, the measure that most users would be interested in is early precision. Thus early precision is usually also evaluated and often shows a quite different behavior from MAP. Visual approaches are actually not as bad for early precision as they are for MAP. In particular, combinations of visual and textual retrieval were able to improve early precision and maintain a good recall and MAP at the same time. Several systems have started to optimize the results based on early precision and achieved remarkable results. One such approach is visual modality classification that can then be used to filter the text retrieval results. The utility of combining visual and textual retrieval can thus not be underestimated.

In most of the years, there was only limited learning data available as databases and tasks changed fairly frequently. The effect of massive learning was demonstrated well in (Deselaers et al, 2007). This approach was by a factor of three better than all other purely visual approaches and could rival the text retrieval approaches. Unfortunately, this approach was only tested for a single year and few other groups have worked on similar learning.

Results can be presented not only in terms of techniques used, but also in terms of the evaluation methodology ImageCLEF improved over the years.

Relevance judgments were well analyzed and an analysis of Kappa scores showed that variability is smaller than for the medical text–based task but does exist and can strongly vary between judges despite a clear description of the relevance. Relevance clearly depends on the knowledge of the clinician. The more expertise the clinician has, the more they would expect to obtain good results. Also, the less expertise a clinician has, the more related articles or cases are important.

ImageCLEF has increased in complexity over the years from mainly technical tasks in 2004 to image–based tasks in the following years including a diagnosis. In 2009 case–based retrieval tasks were started and this is clearly the right direction. Both textual and visual retrieval systems have improved over the years and the tasks need to increase in complexity to be a real challenge for participants.

13.7 Conclusions

The medical retrieval task of ImageCLEF has created over the six years of its existence a large body of data and ground truth that have been used for the evaluation of medical image retrieval systems. The availability of data have made medical retrieval accessible for a large number of research groups, many of them without a connection to a medical institution and thus without access to data. Over 40 research groups have participated in the task and many of them for several consecutive years. Over 200 publications have appeared using ImageCLEF data and several PhD theses have been completed using the data sets we have created. With help from the participants, several new challenges have been identified and the tasks have been adapted accordingly. This has at least made a start with bridging medical image retrieval and clinical practice, although clinical applications are still scarce. Image retrieval can be an important technique in medical decision–making but it needs to integrate visual and textual techniques to reach optimal results.

Acknowledgements We would like to thank the CLEF campaign for supporting the ImageCLEF initiative. We would also like to thank the RSNA and all other bodies who supplied us with images and allowed us to distribute these precious resources to our participants. This work was partially funded by the Swiss National Science Foundation (FNS) under contracts 205321–109304/1 and 200020–118638/1, the American National Science Foundation (NSF) with grant ITR–0325160, and by the University of Applied Sciences Western Switzerland (HES SO) in the context of the BeMeVIS project. Jayashree Kalpathy–Cramer was supported by the following grants from the National Library of Medicine: 2T15LM007088 and 1K99LM009889. We would also like to thank the EU projects TrebleCLEF and SemanticMining for their support.

References

Candler CS, Uijtdehaage SH, Dennis SE (2003) Introducing HEAL: The Health Education Assets Library. Academic Medicine 78(3):249–253

Clough PD, Sanderson M, Müller H (2004) The CLEF Cross Language Image Retrieval Track (ImageCLEF) 2004. In: The Challenge of Image and Video Retrieval (CIVR 2004). Lecture Notes in Computer Science (LNCS), vol 3115. Springer, pp 243–251

Clough PD, Müller H, Deselaers T, Grubinger M, Lehmann TM, Jensen J, Hersh W (2006) The CLEF 2005 cross–language image retrieval track. In: Cross Language Evaluation Forum (CLEF 2005). Lecture Notes in Computer Science (LNCS), vol 4022. Springer, pp 535–557

Deselaers T, Gass T, Weyand T, Ney H (2007) FIRE in ImageCLEF 2007. In: Working Notes of CLEF 2007, Budapest, Hungary

Glatz-Krieger K, Glatz D, Gysel M, Dittler M, Mihatsch MJ (2003) Webbasierte Lernwerkzeuge für die Pathologie — web–based learning tools for pathology. Pathologe 24:394–399

Güld MO, Keysers D, Deselaers T, Leisten M, Schubert H, Ney H, Lehmann TM (2004) Comparison of global features for categorization of medical images. In: Ratib OM, Huang HK (eds) Medical Imaging 2004: PACS and Imaging Informatics. SPIE Proceedings, vol 5371, pp 211–222

Hersh W, Müller H, Gorman P, Jensen J (2005) Task analysis for evaluating image retrieval systems in the ImageCLEF biomedical image retrieval task. In: Slice of Life conference on Multimedia in Medical Education (SOL 2005), Portland, OR, USA

Hersh W, Müller H, Kalpathy-Cramer J, Kim E, Zhou X (2009) The consolidated ImageCLEFmed medical image retrieval task test collection. Journal of Digital Imaging 22(6):648–655

Lowe HJ, Antipov I, Hersh W, Arnott Smith C (1998) Towards knowledge–based retrieval of medical images. The role of semantic indexing, image content representation and knowledge–based retrieval. In: Proceedings of the Annual Symposium of the American Society for Medical Informatics (AMIA), Nashville, TN, USA, pp 882–886

Müller H, Michoux N, Bandon D, Geissbuhler A (2004a) A review of content–based image retrieval systems in medicine—clinical benefits and future directions. International Journal of Medical Informatics 73(1):1–23

Müller H, Rosset A, Vallée JP, Terrier F, Geissbuhler A (2004b) A reference data set for the evaluation of medical image retrieval systems. Computerized Medical Imaging and Graphics 28(6):295–305

Müller H, Despont-Gros C, Hersh W, Jensen J, Lovis C, Geissbuhler A (2006) Health care professionals' image use and search behaviour. In: Proceedings of the Medical Informatics Europe Conference (MIE 2006). Studies in Health Technology and Informatics. IOS, Maastricht, The Netherlands, pp 24–32

Müller H, Boyer C, Gaudinat A, Hersh W, Geissbuhler A (2007) Analyzing web log files of the Health On the Net HONmedia search engine to define typical image search tasks for image retrieval evaluation. In: MedInfo 2007. Studies in Health Technology and Informatics, vol 12. IOS, Brisbane, Australia, pp 1319–1323

Müller H, Deselaers T, Lehmann T, Clough P, Kim E, Hersh W (2007) Overview of the ImageCLEFmed 2006 medical retrieval and medical annotation tasks. In: CLEF 2006 Proceedings. Lecture Notes in Computer Science (LNCS), vol 4730. Springer, Alicante, Spain, pp 595–608

Müller H, Deselaers T, Kim E, Kalpathy-Cramer J, Deserno TM, Clough PD, Hersh W (2008) Overview of the ImageCLEFmed 2007 medical retrieval and annotation tasks. In: CLEF 2007 Proceedings. Lecture Notes in Computer Science (LNCS), vol 5152. Springer, Budapest, Hungary, pp 473–491

Müller H, Kalpathy-Cramer J, Hersh W, Geissbuhler A (2008) Using Medline queries to generate image retrieval tasks for benchmarking. In: Medical Informatics Europe (MIE2008). Studies in Health Technology and Informatics. IOS, Gothenburg, Sweden, pp 523–528

Müller H, Kalpathy-Cramer J, Eggel I, Bedrick S, Said R, Bakke B, Kahn Jr CE, Hersh W (2009a) Overview of the CLEF 2009 medical image retrieval track. In: Working Notes of CLEF 2009, Corfu, Greece

Müller H, Kalpathy-Cramer J, Kahn Jr. CE, Hatt W, Bedrick S, Hersh W (2009b) Overview of the ImageCLEFmed 2008 medical image retrieval task. In: Peters C, Giampiccolo D, Ferro N, Petras V, Gonzalo J, Peñas A, Deselaers T, Mandl T, Jones G, Kurimo M (eds) Evaluating Systems for Multilingual and Multimodal Information Access — 9th Workshop of the Cross–Language Evaluation Forum. Lecture Notes in Computer Science (LNCS), vol 5706. Springer, Aarhus, Denmark, pp 500–510

Radhouani S, Kalpathy-Cramer J, Bedrick S, Hersh W (2009) Medical image retrieval, a user study. Tech. rep., Medical Informatics and Outcome Research, OHSU, Portland, OR, USA

Rosset A, Müller H, Martins M, Dfouni N, Vallée JP, Ratib O (2004) Casimage project — a digital teaching files authoring environment. Journal of Thoracic Imaging 19(2):1–6

Savoy J (2002) Report on CLEF–2001 experiments. In: Report on the CLEF Conference 2001 (Cross Language Evaluation Forum). Lecture Notes in Computer Science (LNCS), vol 2406. Springer, Darmstadt, Germany, pp 27–43

Smeulders AWM, Worring M, Santini S, Gupta A, Jain R (2000) Content–based image retrieval at the end of the early years. IEEE Transactions on Pattern Analysis and Machine Intelligence 22(12):1349–1380

Sparck-Jones K, van Rijsbergen C (1975) Report on the need for and provision of an ideal information retrieval test collection. British Library Research and Development Report 5266, Computer Laboratory, University of Cambridge

Tagare HD, Jaffe C, Duncan J (1997) Medical image databases: A content–based retrieval approach. Journal of the American Medical Informatics Association 4(3):184–198

Wallis JW, Miller MM, Miller TR, Vreeland TH (1995) An internet–based nuclear medicine teaching file. Journal of Nuclear Medicine 36(8):1520–1527

Zobel J (1998) How reliable are the results of large–scale information retrieval experiments? In: Croft WB, Moffat A, van Rijsbergen CJ, Wilkinson R, Zobel J (eds) Proceedings of the 21st Annual International ACM SIGIR conference on research and development in information retrieval. ACM press, Melbourne, Australia, pp 307–314

Part III
Participant reports

Selected reports of research groups participating in ImageCLEF.

Chapter 14
Expansion and Re–ranking Approaches for Multimodal Image Retrieval using Text–based Methods

Adil Alpkocak, Deniz Kilinc, and Tolga Berber

Abstract In this chapter, we present an approach to handle multi–modality in image retrieval using a Vector Space Model (VSM), which is extensively used in text retrieval. We simply extended the model with visual terms aiming to close the semantic gap by helping to map low–level features into high level textual semantic concepts. Moreover, this combination of textual and visual modality into one space also helps to query a textual database with visual content, or a visual database with textual content. Alongside this, in order to improve the performance of text retrieval we propose a novel expansion and re–ranking method, applied both to the documents and the query. When textual annotations of images are acquired automatically, they may contain too much information, and document expansion adds more noise to retrieval results. We propose a re–ranking phase to discard such noisy terms. The approaches introduced in this chapter were evaluated in two sub–tasks of ImageCLEF2009. First, we tested the multi–modality part in ImageCLEFmed and obtained the best rank in mixed retrieval, which includes textual and visual modalities. Secondly, we tested expansion and re–ranking methods in ImageCLEFWiki and the results were superior to others and obtained the best four positions in text–only retrieval. The results showed that the handling of multi–modality in text retrieval using a VSM is promising, and document expansion and re–ranking plays an important role in text–based image retrieval.

Adil Alpkocak,
Dokuz Eylul University, Department of Computer Engineering, Tinaztepe Kampusu, 35160 Buca, Izmir, Turkey e-mail: alpkocak@cs.deu.edu.tr

Deniz Kilinc,
Dokuz Eylul University, Department of Computer Engineering, Tinaztepe Kampusu, 35160 Buca, Izmir, Turkey e-mail: dkilinc@cs.deu.edu.tr

Tolga Berber
Dokuz Eylul University, Department of Computer Engineering, Tinaztepe Kampusu, 35160 Buca, Izmir, Turkey e-mail: tberber@cs.deu.edu.tr

H. Müller et al. (eds.), *ImageCLEF*, The Information Retrieval Series 32,
DOI 10.1007/978-3-642-15181-1_14,

14.1 Introduction

In recent years, there has been a tremendous increase of available multimedia data in both scientific and consumer domains as a result of rapid advances of Internet and multimedia technology. As hardware has improved and available bandwidth has grown, the size of digital image collections has reached terabytes and this amount grows constantly on a daily basis. However, the importance of this information depends on how easily we can search, retrieve and access it.

The current state–of–the–art in image retrieval has two major approaches: Content–Based Image Retrieval (CBIR) and annotation–based image retrieval. CBIR methods operate soley on images by the extraction of visual primitives such as color, texture or shape. However, there is an important shortcoming of this approach: it is not possible to extract all semantic information from images alone; known as the 'semantic gap'. Additionally, the computational cost of extracting image features such as color and shape for a large image collection is high and, furthermore, the user query has to be entered in the form of that modality with low level image features, e.g. with an example image.

Annotation–Based Image Retrieval (ABIR) uses text retrieval techniques on textual descriptions of images, which are generally performed by humans. In Web environments, much of the image content has insufficient textual metadata and it is not realistic to expect such large numbers of images to be annotated manually. A simple alternative is to use information in the form of textual metadata, accompanying an image such as its filename and HTML tags. Notably, the text surrounding images might be more descriptive and usually includes descriptions implicitly made by the page designer. All this textual data can be stored with the image itself or can be used as annotation of images associated with unstructured metadata. In fact, although content–based techniques are applied, the surrounding textual content should be considered. It is probable that surrounding text includes some form of human generated descriptions of the images, which is closer to the semantic interpretation of humans. ABIR can be an approach for image retrieval in Web resources such as Wikipedia images when surrounding text is used as annotation.

Historically, ABIR approaches were first used experimentally for image retrieval, where textual annotations were manually added to each image and the retrieval process was performed using standard database management systems (Chang and Fu, 1980; Chang and Kunil, 1981). In the early 90s, with the growth of image collections, manual annotation approaches for the images became impossible. As a result, CBIR was proposed, which is based on extracting low–level visual content such as color, texture or shape. The extracted features are then stored in a database and compared to an example query image. Many studies now exist on content–based retrieval, with different techniques used for extracting and storing features (El-Kwae and Kabuka, 2000; Ogle and Stonebraker, 1995; Wu, 1997) and on the image searching methods used (Flickner et al, 1995; Santini and Jain, 2000). With the expansion of the Web, interest in image retrieval has increased (Frankel et al, 1996; Jain et al, 1997). On the Web, the images are usually stored with an image filename, HTML tags and surrounding text. Over time, multi–modal systems have been suggested to improve image search results by using a combination of textual information with image feature information (Wu et al, 2001; Wong and Yao, 1995). In conclusion, it is hard to extract low–level features from Web images or to manually annotate them. Thus, ABIR is a reasonable retrieval approach where surrounding text is used that is an implicit description of the images and which is closer to the image semantics.

In multimedia databases, queries are semantically incomplete when they are submitted due to difficulties of describing the media. Most image queries need further tuning such as expanding queries in the ABIR approach. The text retrieval community studied query expansion extensively. However, in the literature, document expansion has not been thoroughly researched for information retrieval. From past exoerience it seems obvious that document expansion can improve the retrieval effectiveness (Singhal and Pereira, 1999; Billerbeck and Zobel, 2005). In short, the ABIR approach is promising for current state–of–the–art image retrieval; however, it requires new expansion techniques to improve its retrieval performance results.

In this chapter, we present an integrated retrieval system, which extends the well–known textual information retrieval technique with visual terms. The proposed model aims at closing the semantic gap by helping to map low–level features onto high level textual semantic concepts. Moreover, this combination of textual and visual modalities into a single model helps to query a textual database with visual content or a visual database with textual content. Consequently, images could be defined with semantic concepts instead of low–level features. Additionally, we propose a novel expansion technique for documents and queries, using WordNet (Miller et al, 1990), Word Sense Disambiguation (WSD) and similarity functions. Since document and query expansion generally result in high recall with low precision, a two–level re–ranking approach is introduced to increase precision by reordering the results sets. The first level forms a narrowing–down operation and includes re–indexing. On the second level we propose a new re–ranking approach which is Cover Coefficient (CC) based. During the initial retrieval, we combine both expanded and original documents. We evaluated the whole system on the ImageCLEF 2009 WikipediaMM task (Tsikrika and Kludas, 2009) and obtained the best four ranks in textual image retrieval.

The rest of the chapter is organized as follows: Section 14.2 introduces the details of the integrated retrieval model for multi–modality handling using the classical vector space model. Section 14.3 describes document and query expansion techniques based on WordNet. In Section 14.4, we introduce the two–level re–ranking approach which includes a narrowing–down phase and CC–based re–ranking. Section 14.5 presents the details of the results we obtained in ImageCLEF 2009. Section 14.6 concludes the chapter.

14.2 Integrated Retrieval Model

The Integrated Retrieval Model (IRM) is an extension of the classical vector space model to handle multi–modality in content-based image retrieval focusing on the integration of both textual and visual features as a single entity. IRM may also help to close the semantic gap by mapping low–level visual features to the classical VSM. In the classical VSM, a document is represented as a vector of terms. Hence, a document repository D becomes a sparse matrix with rows and columns being document and term vectors as follows:

$$D = \begin{bmatrix} t_{1,1} & t_{1,2} & \cdots & t_{1,n} \\ \vdots & \vdots & \ddots & \vdots \\ t_{m,1} & t_{m,2} & \cdots & t_{m,n} \end{bmatrix} \tag{14.1}$$

where t_{ij} is the weight of term j in document i, n and m is term and document count, respectively. The literature proposes a large number of weighting schemes (Amati and Van Rijsbergen, 2002; Beaulieu et al, 1997; Zhang et al, 2009). We used pivoted unique term weighting proposed in (Singhal et al, 1996; Chisholm and Kolda, 1999) that is a modified version of the classical cosine normalization (Salton et al, 1975) based on the term weighting aspect of modern text retrieval systems (Buckley, 1993; Manning et al, 2008). A normalization factor is added to the formula which is independent from term and document frequencies. We calculated weights of an arbitrary term, w_{ij}, using the pivoted unique normalization as follows:

$$W_{ij} = \frac{\log(dtf)+1}{sumdtf} \times \frac{U}{1+0.0118U} \times \log\left(\frac{N-nf}{nf}\right) \tag{14.2}$$

where dtf is the number of times the term appears in the document, $sumdtf$ is the sum of $\log(dtf)+1$'s for all terms in the same document, N is the total number of documents, nf is the number of documents that contain the term and U is the number of unique terms in the document. The uniqueness means that the measure of document length is based on the unique terms in the document. We used 0.0118 as pivot value.

14.2.1 Handling Multi–modality in the Vector Space Model

IRM proposes an extension to the D matrix (Eq. 14.1) by adding visual terms to represent visual content. Formally, the new document–term matrix, D' becomes as follows:

$$D' = \left[\begin{array}{cccc|cccc} t_{1,1} & t_{1,2} & \cdots & t_{1,n} & v_{1,n+1} & v_{1,n+2} & \cdots & v_{1,n+k} \\ \vdots & \vdots & \ddots & \vdots & \vdots & \vdots & \ddots & \vdots \\ t_{m,1} & t_{m,2} & \cdots & t_{m,n} & v_{m,n+1} & v_{m,n+2} & \cdots & v_{m,n+k} \end{array}\right] \tag{14.3}$$

where v_{ij} is the weight of the visual term j in document i, k is the number of visual terms. Visual and textual features are normalized independently.

In sum, IRM extends the traditional text–based VSM with visual features. Initially, we used two simple visual terms representing color information in the image. To start with, the number of pixels in a particular color or gray scale image is simply

Algorithm 14.1 *GrayScaleness*(*Image*)

Require: An Image

$count \leftarrow 0$
$channelCount \leftarrow$ Number of Channels in Image
if $channelCount = 1$ **then**
  return 1.0
else
  for $i = 1$ to *Image.height* **do**
    for $j = 1$ to *Image.width* **do**
      if $Image(i,j,0) = Image(i,j,1) \wedge Image(i,j,1) = Image(i,j,2)$ **then**
        $count \leftarrow count + 1$
      end if
    end for
  end for
  return $count / Image.totalPixelCount$
end if

counted. Then, the first visual term represents the amount of the image that is gray scale, and the second visual term is the complement of the first term. In other words, the value is the probability of color pixels in an image. Algorithm 14.1 shows the calculation of the feature.

14.3 Document and Query Expansion

Document Expansion (DE) and Query Expansion (QE) are processes to add new words to documents or queries. Expanding the queries and widening the search terms increase the recall value by retrieving more relevant documents which do not match literally with the original query. Similarly, expanding poorly defined documents by adding new terms may result in higher ranking performance. However, there is always a risk with expansion of constructing more exhaustive documents and queries than the original ones. On the other hand, both DE and QE must be used together to achieve a performance gain. The aim of expanding both documents and queries is to adapt queries to the documents and documents to queries. Thus, we used the same expansion approaches for both documents and queries.

In this chapter, we used WordNet (Miller et al, 1990) for both DE and QE phases. WordNet models the lexical knowledge of English and can also be seen as an ontology for natural language terms containing nearly 100,000 terms, divided into four taxonomic hierarchies: nouns, verbs, adjectives and adverbs. Although it is commonly argued that language semantics are mostly captured by nouns and noun term–phrases, here we considered both noun and adjective representations of terms. We also used WordNet for WSD to tune DE. We used Lesk's algorithm (Lesk, 1986) which disambiguates a target word by selecting the sense whose dictionary gloss shares the largest number of words with the glosses of neighboring words. During the DE phase, we stored the original forms of documents to compensate the similarity score computation.

Figure 14.1 illustrates the expansion of query 1 in the ImageCLEF 2009 WikipediaMM task. The query *'blue flowers'* is firstly pre–processed, the terms *'blue'* and *'flower'* are generated. Each term's sense is fetched from WordNet. In our example, *'blue'* has seven and *'flower'* has three senses. Since numerous senses exist in different domains for terms, expanding the term with all these senses results in noisy but exhaustive documents or queries. We prevent such noisy expansions by selecting the most appropriate sense with Lesk's WSD algorithm. In our example, the sense of terms is first selected. In WordNet, a sense consists of two parts: synonyms and sense definition. We used both for expansion in our work. We again pre–process the selected sense to reduce the noise level. Then, we check if the expanded terms exist in the data set. We eliminate the terms that do not exist in the dictionary. After processing as described, *'flower'* has the expanded terms; *plant, cultivated, blossom* and *bloom*. For each of the terms we calculate a similarity score between their base terms (i.e. flower). In the literature, different methods have been proposed to measure the semantic similarity between terms (Wu and Palmer, 1994; Richardson et al, 1994; Li et al, 2003; Resnik, 1999; Tversky, 1977). In this study, we use Wu and Palmer's edge counting method (Wu and Palmer, 1994). Finally, we add terms above a specific threshold to the final query or document. Threshold values for noun and adjective terms are 0.9 and 0.7, respectively. In our example, query *'blue flower'* is finally expanded as *'blue flower blueness sky bloom blossom'*.

Term phrase selection (TPS) is one of the major parts of the expansion phase. During expansion, we checked every successive word pair for existence in WordNet as a noun-phrase. If it exists, we expanded the document/query by appending the term phrase to the dictionary. For example, if a document contains *'hunting dog'*, these two successive tokens are searched in WordNet. If this phrase exists, the document is expanded with the term *'hunting dog'*. Finally the term phrase is added to the term phrase dictionary. For the WikipediaMM test collection, the numbers of new term-phrases added was 6,808.

Table 14.1 depicts the same query number 1 and its two relevant documents with IDs of 1027698 and 163477 by showing their original and expanded forms. Relevant documents are about some kind of flowers that are uploaded to Wikipedia pages. The query is *blue flowers*. Both *borage* and *lavender* are somehow related with *blue flowers* although their documents do not include these terms. In such cases, without any expansion technique, retrieval performance will not be satisfactory. The example also shows that expanding the query alone is not adequate, where only the terms of *blueness*, *sky*, *bloom* and *blossom* are added to query. However, we must also expand the documents to match. After document expansion, the terms *blue* and *flower* are added to both documents. In addition to this, *bloom* and *blossom* terms are also appended to document numbered 163477. As a result, the expansion step adds new common terms to both documents and query by using WordNet, WSD. Then, the whole VSM is rebuilt based on the new dictionary.

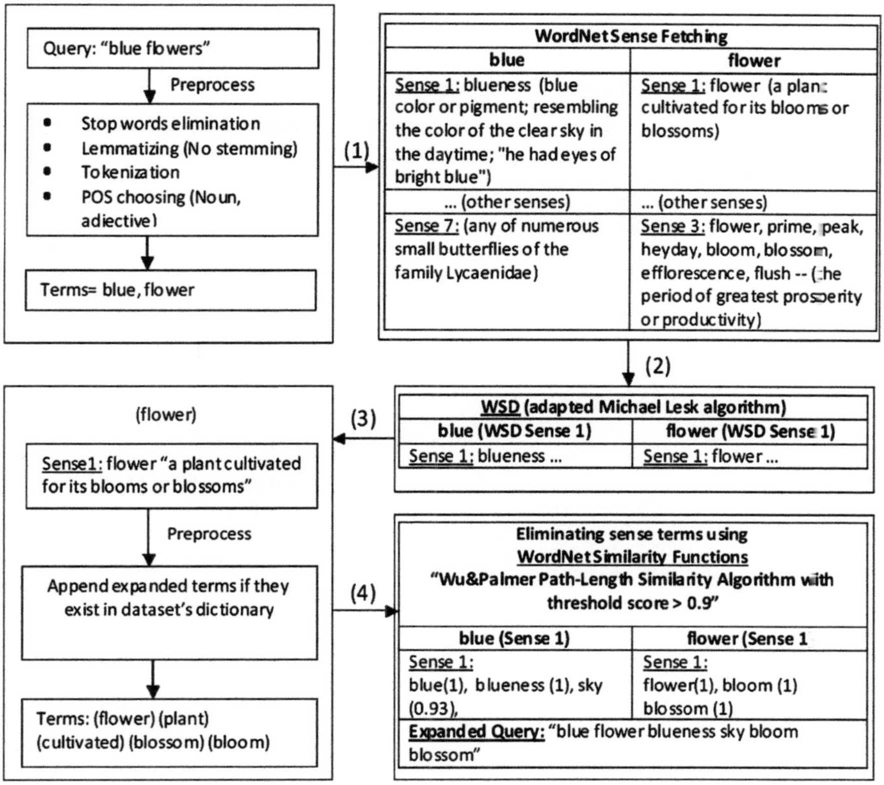

Fig. 14.1: Query expansion example.

14.4 Re–ranking

Re–ranking is a method used to reorder the initially retrieved documents to move more relevant documents to the top of the list. In the literature many re–ranking approaches have been proposed. The re–ranking approaches can be roughly classified into several groups based on the underlying methods used, such as unsupervised document clustering, semi-supervised document categorization, relevance feedback, probabilistic weighting, collaborative filtering, or combinations of different approaches. In the literature, some of the methods propose a modification in weighting scheme (Yang et al, 2006; Carbonell and Goldstein, 1998; Lingpeng et al, 2005; Callan et al, 1992). Some of them are based on clustering and use inter document similarity or user supported relevance data to re–rank documents (Balinski and Danilowicz, 2005; Allan et al, 2001; Lee et al, 2001). In re–ranking, relevant documents with low similarity scores are re-weighted and reordered to move relevant documents upward in the results set.

Table 14.1: An example document and query expansion.

Image/ Query ID	Image	Original Document / Query	Expanded Document / Query
Doc # :1027698	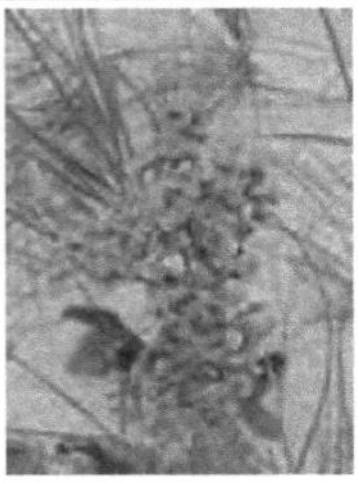	sea lavender limonium	sea lavender limonium sealavender statice various plant genus limonium temperate salt marsh spike whit mauve **flower** various old world aromatic shrub subshrub mauve **blue** cultivated division ocean body salt water enclosed land
Doc # :163477		borage flower garden made apr	borage flower garden made apr made plant cultivated **bloom blossom** tailwort hairy **blue flowered** european annual herb herbal medicine raw salad greens cooked spinach april month preceding plot ground plant cultivated
Query # :1	N/A	blue flower	blue flower blueness sky **bloom blossom**

In this chapter, we present a two–level re–ranking approach. The first level forms a narrowing-down phase of the search space, whilst the second level includes a cover coefficient based re–ranking. Before introducing our re–ranking approach, let us provide a set of preliminary definitions. We first performed an initial retrieval, called the base result, using the well–known Vector Space Model (VSM). The formula to calculate the base similarity scores is as follows:

$$r(j) = \sum_{i=1}^{n} (t_{ij} \times q_i) + \sum_{i=1}^{m} (v_{ij} \times q_{vi}) \tag{14.4}$$

where, $r(j)$ is the similarity score of jth document, n is the length of the textual vocabulary and, t_{ij} and q_i represent the weights of the jth document and query, respectively. Additionally, m is the length of the visual vocabulary, v_{ij} and q_{vi} are visual term weights for the jth document and query. The second term of the $r(j)$ formulation is for visual similarity only. When multi–modality is not required, the second term can be ignored.

Let us assume that $r'(j)$ represents the similarity score of expanded documents. To calculate overall similarity score, we use both expanded and original similarity scores by taking the averages of them with some coefficients as formulated below:

$$R_0(j) = \frac{(r(j) \times \mu) + (r'(j) \times \delta)}{2} \tag{14.5}$$

where, $R_0(j)$ show the initial similarity score of the jth document, μ and δ are coefficients to adjust results for different data sets and queries. In this study, we empirically set μ and δ values to 1 and 0.9, respectively.

14.4.1 Level 1: Narrowing-down and Re-indexing

The first level of our re–ranking approach forms a narrowing–down phase and includes re-indexing. The result sets of each query and corresponding base similarity scores, $R_0(j)$, are inputs for the re–ranking operation. In this level we first selected relevant documents using initial similarity scores, R_0. In other words, we filter out non–relevant documents based on the initial similarity scores. This operation drastically reduces both the number of documents and the number of terms. Then we construct a new VSM using these small document sets. This reduces the initial VSM data to a more manageable size so that we can perform a more complex cover coefficient–based re–ranking algorithm upon it. Following this, we calculated first level similarity scores, $R_1(j)$, as follows:

$$R_1(j) = (R_0(j) \times \alpha) + r_1(j) + \beta \tag{14.6}$$

where $r_1(j)$ is the new similarity score of the jth document in new small VSM. The value of α is the weight factor and is empirically set to 0.8. Additionally, β is set to 4 if the jth document contains the original query terms in exact order, otherwise it is set to zero.

14.4.2 Level 2: Cover Coefficient Based Re–ranking

In the second level, we present the Cover–Coefficient (CC) based re–ranking method. The content of CC was originally proposed by (Can and Ozkarahan, 1990) for text clustering. The Cover Coefficient based Clustering Methodology (C^3M) is a seed–based partitioning clustering scheme which basically consists of two different steps: (i) cluster seed selection and (ii) cluster construction. The term incidence matrix, D, is the input for C^3M, which represents documents and their terms. It is assumed that each document contains n terms and the database consists of m documents. There is a requirement to construct a C matrix, in order to employ cluster seeds for C^3M. The C matrix is a document–by–document matrix whose entries $(1 < i, j < m)$ indicate the probability of selecting any term of d_i from d_j. In other words, the C matrix indicates the relationship between documents based on a two–stage probability experiment. The experiment randomly selects terms from documents in two stages. The first stage randomly chooses a term t_k of document d_i; the second stage then chooses the selected term t_k from document d_j. For the calculation of the C matrix, c_{ij}, one must first select an arbitrary term of d_i, say, t_k, and then use this term to try to select document d_j from this term, i.e. to check if d_j contains t_k. Each row of the C matrix summarizes the results of this two-stage experiment.

Let s_{ik} indicate the event of selecting t_k from d_i at the first stage, and let s'_{jk} indicate the event of selecting d_j, from t_k at the second stage. In this experiment, the probability of the simple event, s_{ik} and s'_{jk}, that is, $P(s_{ik} \mid s'_{jk})$ can be represented as $P(s_{ik}) \times P(s'_{jk})$. To simplify the notation, s_{ik} and s'_{jk} can be used respectively, for $P(s_{ik})$ and $P(s'_{jk})$, where:

$$s_{ik} = \frac{d_{ik}}{\sum_{h=1}^{n}(d_{ih})}, \text{ and } s'_{jk} = \frac{d_{jk}}{\sum_{h=1}^{m}(d_{jh})}, \text{ where } 1 \leq i,j \leq m, 1 \leq k \leq n \quad (14.7)$$

By considering document d_i, the D matrix can be represented with respect to the two-stage probability model. Each element of the C matrix, c_{ij}, (the probability of selecting a term of d_i from d_j) can be computed by summing the probabilities of an individual path from d_i to d_j.

$$c_{ij} = \sum_{i=1}^{n} \left(s_{ik} \times s'_{jk}\right) \quad (14.8)$$

where c_{ij} shows how much i^{th} document covered by j^{th} document, for $i \neq j$, coupling of d_i with d_j.

Re–ranking starts with appending the query into new VSM as a document, and then calculating the C matrix as described above. The C matrix entries, c_{ij}, show how much the ith document is covered by j^{th} document. We considered the ith row of the C matrix, which includes how query is covered by other documents. We calculated new similarity score using both $R_1(j)$ and c_{ij} as follows:

$$R_2(j) = c_{ij} \times \frac{(\max(R^b) \times \theta)}{(100 \times \max(c_{i*}))} \quad (14.9)$$

where $\max(R^b)$ is the maximum first level similarity score for the query result set, θ is an empirical coefficient that specifies the percentage of similarity score effect, $\max(c_{i*})$ is the maximum query-by-document similarity score for the ith query. Finally, the CC based similarity score equation is as follows:

$$R(j) = R_1(j) + R_2(j) \quad (14.10)$$

where $R_1(j)$ is the first level, $R_2(j)$ is the second level and R_j is the final similarity score to be used to calculate new ranking scores.

Table 14.2: Results of experiments with the ImageCLEFmed 2009 dataset.

Run Identifier	NumRel	RelRet	MAP	P@5	P@10	P@30	P@100	Rank	Run Type
baseline	2362	1742	0.339	0.584	0.520	0.448	0.303	16	Text
IRM	2362	1754	0.368	0.632	0.544	0.483	0.324	1	Mixed

14.5 Results

The methods presented in this chapter have been tested in the WikipediaMM and ImageCLEFMed tasks. We conducted five runs with the ImageCLEFmed 2009 data set (Müller et al, 2009); however, we present only two of them. We pre–processed all 74,902 documents including combination of title and captions. First, all documents were converted into lowercase. All numbers and some punctuation characters such as dash (–) and apostrophe (') were removed. However, some of the non–letter characters such as comma (,) and slash (/) were replaced with a space. This is because the dash character conveys an important role as in x–ray and T3–MR. Then, we choose the words surrounded by spaces as index terms. For each image in the data set, we have added two visual terms as shown in the previous section. In total, the total number of indexing terms became 33,615.

After the pre–processing phase, we implemented text only retrieval on the data set. Here, we normalized text term weights as shown in (Beaulieu et al, 1997), and we simply calculated the dot product of query and document vectors as a similarity function. Then, the top 1,000 documents having the highest similarity scores were selected as result set for each query. The first row of Table 14.2, whose run identifier is baseline, shows the results we obtained from this experimentat. This was ranked in the 16th position since we used only a simple retrieval method without any enhancements.

The second experimentat focuses on the IRM, which combines two visual terms with previous experiments. The result of the IRM is shown in the second row of Table 14.2, and ranked the best position among the participants in the *mixed automatic run* track, which is multi–modal retrieval. Importantly, the result we obtained from the second experiment shows that the IRM improved the performance of baseline retrieval across all measures. Furthermore, this performance gain was obtained by using simple visual features. Figure 14.2 illustrates the precision and recall values of our experiments. IRM outperformed the classical vector space model with respect to recall at all levels of precision. Based on the results of these experimentats we can conclude that combining textual retrieval techniques with good visual features positively affects the results and improves system performance.

In the WikipediaMM task, we first performed some basic pre–processing such as deletion of punctuation, stop–word removal and lemmatization. Then, we expanded documents using WordNet and selected term phrases as described earlier in the chapter. During this phase, we take into consideration both the original and the expanded forms of the data set to calculate similarity score and converted docu-

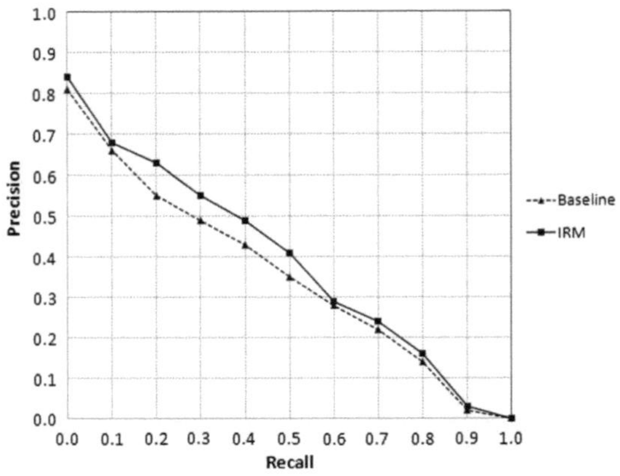

Fig. 14.2: Precision-Recall graph of baseline and IRM runs.

ment vectors before base retrieval. In addition, we also expanded queries using Term Phrase Selection (TPS) and/or WordNet for experimental purposes. Then the two–level re–ranking step begins. The first level re–ranking uses the narrowing–down approach. With the completion of the first level, the result set of each query and ranked scores are kept for the second level. The second level is based on the CC concept.

The final similarity score $R(j)$ is calculated using $R_1(j)$ from the first level and query-document similarity score, $R_2(j)$ from the second level. Finally, the two-level re–ranking process is completed and the final ranked result sets are generated.

The expansion and re–ranking approach was evaluated with the data set from ImageCLEF's WikipediaMM task, which provides a test bed for system–oriented evaluation of retrieval systems from a collection of Wikipedia Web images. The aim is to investigate and evaluate retrieval approaches in the context of a larger scale and heterogeneous collection of images that are searched for by users with diverse information needs. The data set contains 151,519 images that cover diverse topics of interest, and images are associated with unstructured and noisy textual annotations in English. The WikipediaMM 2009 sub–track data set includes 45 queries. A total of eight groups participated in the WikipediaMM Task of ImageCLEF, submitting 57 runs; 26 of them text-based retrieval and 31 also including content–based retrieval (Tsikrika and Kludas, 2009).

We participated in the WikipediaMM Task of ImageCLEF with the group name DEUCENG and conducted six runs. Before the runs, we pre–processed the data in the aforementioned ways. In addition, we also back–up the original forms of documents to calculate the similarity score as a combination of the original and expanded data sets. In all runs we used document expansion and pivoted unique normalization. The differences between the runs were based on the different expansion and re–ranking techniques. The proposed system's retrieval performance was evaluated

Table 14.3: Results of experiments on the WikipediaMM task of ImageCLEF 2009.

ID	Code	DE	QE (TPS)	QE (WN, WSD)	RR1	RR2	MAP	P@5	P@10
1	200	X					0.1861	0.3244	0.2956
2	201	X	X				0.1865	0.3422	0.2978
3	202	X	X	X			0.2358	0.4844	0.3933
4	203	X	X	X	X		0.2375	0.4933	0.4000
5	204	X	X	X	X		0.2375	0.4933	0.4000
6	205	X	X	X	X	X	0.2397	0.5156	0.4000

using MAP. We also used the P@5 and P@10 evaluation metrics. Table 14.3 shows the applied techniques for each run and their performance evaluation results.

The first run is the baseline upon which all expansion and re–ranking techniques are built. The MAP and P@5 values are 0.1861 and 0.3244, respectively. The second run includes QE with TPS only. The MAP and top precision values of the second run are slightly better than the baseline. In the third run, we expanded the queries using both TPS and WordNet with the same document expansion approaches. The third run has MAP and P@5 values of 0.2358 and 0.4844, respectively. The experimental results show that the run performs considerably better since our proposed novel expansion techniques for documents and queries are the same. The next three runs show that our re–ranking approach provides an increase in precision. In the fourth run, we conducted first–level re–ranking known as the narrowing–down approach, including reindexing. The experimental results are slightly better again, especially with the impact of the parameters tested. The MAP and P5 values are 0.2375 and 0.4933, respectively. The difference between the forth and fifth runs is that the documents in the results set above a threshold rank are used for the first level re–ranking process, but the experimental results are same. The final run (called 205) includes an additional CC based second level re–ranking approach over the result set of the fifth run. As can be seen from the MAP values in Table 14.3, the best results are obtained from the sixth run. The MAP and P@5 values are 0.2397 and 0.5156, respectively.

14.6 Conclusions

In this chapter, we have presented two new approaches for image retrieval. Firstly, a new content–based image retrieval system combining both visual and textual features of a document in the same model. We evaluate it in the ImageCLEFmed task of ImageCLEF 2009. Our method ranked first among the participants in mixed automatic runs. Results of our experiments show that the proposed multi–modality method performs better than other automatic mixed retrieval approaches, even when simple visual features are used.

Secondly, we presented a novel expansion and re–ranking approach for ABIR, and tested it in the WikipediaMM task of ImageCLEF. Here, we used new expansion techniques using WordNet, WSD and similarity functions, for both documents and queries. We also introduced a two–level re–ranking method to increase precision, based on narrowing–down and cover coefficient phases. Experiments show that our suggestion for an annotation–based image retrieval system is promising. It received the four best positions based on MAP and precision measures among all participants in the WikipediaMM task of ImageCLEF 2009.

Acknowledgements This work is partly supported by Turkish National Science Foundation (TUBITAK) under project number 107E217.

References

Allan J, Leuski A, Swan R, Byrd D (2001) Evaluating combinations of ranked lists and visualizations of inter–document similarity. Information Processing & Management 37(3):435–458

Amati G, Van Rijsbergen CJ (2002) Probabilistic models of information retrieval based on measuring the divergence from randomness. ACM Transactions on Information Systems 20(4):357–389

Balinski J, Danilowicz C (2005) Re–ranking method based on inter–document distances. Information Processing & Management 41(4):759–775

Beaulieu MM, Gatford M, Xiangji H, Robertson SE, Walker S, Williams P (1997) Okapi at TREC–5. In: Proceedings of the Fifth Text REtrieval Conference (TREC–5), National Institute of Standards and Technology, 500238, pp 143–165

Billerbeck B, Zobel J (2005) Document expansion versus query expansion for ad–hoc retrieval. In: Proceedings of the Tenth Australasian Document Computing Symposium, pp 34–41

Buckley C (1993) The importance of proper weighting methods. In: Proceedings of the workshop on Human Language Technology, Association for Computational Linguistics, pp 349–352

Callan J, Croft WB, Harding SM (1992) The INQUERY retrieval system. In: In Proceedings of the Third International Conference on Database and Expert Systems Applications, Springer, pp 78–83

Can F, Ozkarahan EA (1990) Concepts and effectiveness of the cover–coefficient–based clustering methodology for text databases. ACM Transactions of Database Systems 15(4):483–517

Carbonell J, Goldstein J (1998) The use of mmr, diversity–based reranking for reordering documents and producing summaries. In: Proceedings of the 21st annual international ACM SIGIR conference on research and development in information retrieval, ACM press, pp 335–336

Chang N, Fu K (1980) Query–by–pictorial–example. IEEE Transactions on Software Engineering 6:519–524

Chang SK, Kunil TL (1981) Pictorial data–base systems. Computer 14(11):13–21

Chisholm E, Kolda TG (1999) New term weighting formulas for the vector space method in information retrieval. Tech. rep., Oak Ridge National Laboratory

El-Kwae EA, Kabuka MR (2000) Efficient content–based indexing of large image databases. ACM Transactions on Information Systems 18(2):171–210

Flickner M, Sawhney H, Niblack W, Ashley J, Huang Q, Dom B, Gorkani M, Hafner J, Lee D, Petkovic D, Steele D, Yanker P (1995) Query by image and video content: The QBIC system. Computer 28(9):23–32

Frankel C, Swain MJ, Athitsos V (1996) Webseer: An image search engine for the world wide web. Tech. rep., University of Chicago, Chicago, IL, USA

Jain R, Lew MS, Lempinen K, Huijsmans N (1997) Webcrawling using sketches

Lee KS, Park YC, Choi KS (2001) Re–ranking model based on document clusters. Information Processing & Management 37(1):1–14

Lesk M (1986) Automatic sense disambiguation using machine readable dictionaries: how to tell a pine cone from an ice cream cone. In: Proceedings of the 5th annual international conference on Systems documentation, ACM press, pp 24–26

Li Y, Bandar ZA, McLean D (2003) An approach for measuring semantic similarity between words using multiple information sources. IEEE Transactions On Knowledge and Data Engineering

Lingpeng Y, Donghong J, Guodong Z, Yu N (2005) Improving retrieval effectiveness by using key terms in top retrieved documents. Advances in Information Retrieval 169–184

Manning CD, Raghavan P, Schtze H (2008) Introduction to information retrieval. Cambridge University Press, New York, NY, USA

Miller GA, Beckwith R, Fellbaum C, Gross D, Miller KJ (1990) Introduction to wordnet: An online lexical database. International Journal of Lexicography 3(4):235–244

Müller H, Kalpathy-Cramer J, Eggel I, Bedrick S, Radhouani S, Bakke B, Jr. C, Hersh W (2009) Overview of the ImageCLEF 2009 medical image retrieval track. In: CLEF working notes 2009

Ogle VE, Stonebraker M (1995) Chabot: Retrieval from a relational database of images. Computer 28:40–48

Resnik P (1999) Semantic similarity in a taxonomy: An information–based measure and its application to problems of ambiguity in natural language. Journal of Artificial Intelligence Research 11:95–130

Richardson R, Smeaton AF, Murphy J (1994) Using wordnet as a knowledge base for measuring semantic similarity between words. In: In Proceedings of Irish Conference on Artificial Intelligence and Cognitive Science

Salton G, Wong A, Yang C (1975) A vector space model for information retrieval. Journal of the American Society for Information Science 18(11):613–620

Santini S, Jain R (2000) Integrated browsing and querying for image databases. IEEE MultiMedia 7:26–39

Singhal A, Pereira F (1999) Document expansion for speech retrieval. In: Proceedings of the 22nd annual international ACM SIGIR conference on research and development in information retrieval, ACM press, pp 34–41

Singhal A, Buckley C, Mitra M (1996) Pivoted document length normalization. In: Proceedings of the 19th annual international ACM SIGIR conference on research and development in information retrieval, ACM press

Tsikrika T, Kludas J (2009) Overview of the wikipediamm task at ImageCLEF 2009. In: Working notes CLEF 2009, Corfu, Greece

Tversky A (1977) Features of similarity. Psychological Review 84(4):327–352

Wong SKM, Yao YY (1995) On modeling information retrieval with probabilistic inference. ACM Transactions on Information Systems 13(1):38–68

Wu JK (1997) Content–based indexing of multimedia databases. IEEE Transactions on Knowledge and Data Engineering 9:978–989

Wu Q, Iyengar SS, Zhu M (2001) Web image retrieval using self–organizing feature map. Journal of the American Society for Information Science and Technology 52(10):868–875

Wu Z, Palmer M (1994) Verbs semantics and lexical selection. In: Proceedings of the 32nd annual meeting on Association for Computational Linguistics, Association for Computational Linguistics, pp 133–138

Yang L, Ji D, Zhou G, Nie Y, Xiao G (2006) Document re–ranking using cluster validation and label propagation. In: Proceedings of the 15th ACM international conference on Information and knowledge management, ACM press, pp 690–697

Zhang R, Chang Y, Zheng Z, Metzler D, Nie Jy (2009) Search result re–ranking by feedback control adjustment for time-sensitive query. In: Proceedings of Human Language Technologies: The 2009 Annual Conference of the North American Chapter of the Association for Computational Linguistics, Companion Volume: Short Papers, Association for Computational Linguistics, pp 165–168

Chapter 15
Revisiting Sub–topic Retrieval in the ImageCLEF 2009 Photo Retrieval Task

Teerapong Leelanupab, Guido Zuccon, and Joemon M. Jose

Abstract Ranking documents according to the Probability Ranking Principle has been theoretically shown to guarantee optimal retrieval effectiveness in tasks such as ad hoc document retrieval. This ranking strategy assumes independence among document relevance assessments. This assumption, however, often does not hold, for example in the scenarios where redundancy in retrieved documents is of major concern, as it is the case in the sub–topic retrieval task. In this chapter, we propose a new ranking strategy for sub–topic retrieval that builds upon the interdependent document relevance and topic–oriented models. With respect to the topic–oriented model, we investigate both static and dynamic clustering techniques, aiming to group topically similar documents. Evidence from clusters is then combined with information about document dependencies to form a new document ranking. We compare and contrast the proposed method against state–of–the–art approaches, such as Maximal Marginal Relevance, Portfolio Theory for Information Retrieval, and standard cluster–based diversification strategies. The empirical investigation is performed on the ImageCLEF 2009 Photo Retrieval collection, where images are assessed with respect to sub–topics of a more general query topic. The experimental results show that our approaches outperform the state–of–the–art strategies with respect to a number of diversity measures.

Teerapong Leelanupab
University of Glasgow, Glasgow, G12 8RZ, United Kingdom e-mail: kimm@dcs.gla.ac.uk

Guido Zuccon
University of Glasgow, Glasgow, G12 8RZ, United Kingdom e-mail: guido@dcs.gla.ac.uk

Joemon M. Jose
University of Glasgow, Glasgow, G12 8RZ, United Kingdom e-mail: jj@dcs.gla.ac.uk

H. Müller et al. (eds.), *ImageCLEF*, The Information Retrieval Series 32,
DOI 10.1007/978-3-642-15181-1_15,

15.1 Introduction

Information Retrieval (IR) deals with finding documents relevant to a user's information need, usually expressed in the form of a query (van Rijsbergen, 1979). Documents are usually ranked and presented to the users according to the Probability Ranking Principle (PRP), that is, in decreasing order of the document's probability of relevance (Robertson, 1977). This ranking strategy is well accepted in IR, and can be justified using utility theory (Gordon and Lenk, 1999a). However, in particular scenarios, ranking documents according to the PRP does not provide an optimal ranking for the user's information need (Gordon and Lenk, 1999b). For example, this happens when redundant documents are of major concern, or when a broad view about the query topic is needed, thus aiming to retrieve all its possible sub–topics. In this situation, the PRP does not provide a satisfying ranking because it does not account for interdependent document relevance due to the assumption of independence between assessments of document relevance.

A number of recent approaches attempt to overcome PRP's limitations (Carbonell and Goldstein, 1998; Wang and Zhu, 2009; Zuccon and Azzopardi, 2010). The suggested approaches were tested on a novel retrieval task, called sub–topic retrieval (Zhai et al, 2003). In this task, documents are assessed with respect to the number of sub–topics. Interdependent document relevance is introduced in the evaluation measures, which reward strategies that retrieve all the relevant sub–topics at early ranks, while penalising unnecessary redundancy. This means promoting novelty and diversity within the ranking. The need for diversity within document rankings has been motivated by several empirical studies (Agichtein et al, 2006; Eisenberg and Berry, 2007). Addressing diversity issues allows retrieval systems to cope with poorly specified or ambiguous queries, maximizing the chance of retrieving relevant documents, and also to avoid excessive redundancy, providing complete coverage of sub–topics in the result list.

From the current strategies for sub–topic retrieval, two common patterns can be observed with respect to the modality used to achieve ranking diversification:

Interdependent document relevance paradigm: Relationships between documents are taken into account when ranking. Strategies maximise, at each rank position, a function that mixes relevance estimates and document relationships. This approach is followed by heuristics and strategies such as Maximal Marginal Relevance (MMR) (Carbonell and Goldstein, 1998), interpolating document relevance and diversity estimation, and Portfolio Theory (PT) (Wang and Zhu, 2009), combining relevance estimations and document correlations with respect to the previous ranked documents.

Topic–oriented paradigm: Relationships between retrieved documents are used to model sub–topics regardless of document relevance. Documents are thus characterised with respect to the sub–topics they cover using techniques such as clustering (Deselaers et al, 2009; van Leuken et al, 2009), classification (Huang et al, 1998), LDA (Blei et al, 2003), probabilistic latent semantic indexing (pLSI) (Hofmann, 1999), or relevance models (Lavrenko and Croft, 2001; Carterette and

Chandar, 2009). In our study, we are only interested in unsupervised clustering techniques although other techniques might be used. When considering cluster–based diversification methods, each cluster is assumed to represent a sub–topic of the information need, and thus diversification is achieved by retrieving documents belonging to different clusters.

Intuitively, the ranking strategies belonging to the first paradigm do not explicitly identify the sub–topic to be covered. Inter–document relationships, often based on statistical features of the documents, are used when accounting for novelty and diversity. Even though the number of relevant sub–topics is of importance when evaluating retrieval strategies, there is no explicit estimation of the number of sub–topics. In order to maximise the number of sub–topics, the retrieval systems based on this paradigm mainly rely on retrieving relevant documents that contain low redundant information, i.e. they are different enough to each other that they might refer to different sub–topics.

By contrast, clustering–based diversification methods, interleaving documents belonging to different clusters, explicitly cover all possible (or identified) sub–topics in top rankings. Nevertheless, this paradigm lacks an explicit model of relevance: how to combine relevance and the information regarding sub–topic clusters is still an open challenge. Documents that are selected after the clustering process to build the ranking can contain lots of new information, but they might be non–relevant; or even contain relevant information, but this might be redundant in the context of the document ranking. Consequently, ranking documents by either paradigm might produce unsatisfactory results.

Documents can cover several sub–topics (clusters), and these might even to some extent overlap. For example, the topic 'Victoria' can contain a set of documents regarding people (Queen Victoria or Victoria Beckham) and places (the state in Australia or the memorial in London) from a topical point of view. A document, clustered into the sub–topic 'Victoria Beckham', can contain information about her appearance in the Victoria state, Australia. Motivated by these considerations, we aim to alleviate the deficiencies of the two paradigms by combining their merits together. To the best of our knowledge, no empirical study has been performed comparing and integrating these two ranking paradigms in the context of sub–topic retrieval.

In this chapter, we propose a novel ranking strategy which explicitly models possible sub–topics and promotes diversity among documents in a ranking. Our strategy enables the development of a variety of algorithms for integrating statistical similarity, diversity structure, conveyed by clusters, and document dependencies. The paradigm relies on the cluster hypothesis (van Rijsbergen, 1979; Hearst and Pedersen, 1996), which assumes that relevant documents tend to be more similar to each other than non–relevant documents. When clustering documents, topically coherent groups of documents are formed by encoding possible aspects (i.e. sub–topics) of a more general topic. Relevance and diversity evidence is then merged in a ranking function (e.g. MMR), which selects documents from different clusters. The result is a document ranking that covers all the identified sub–topics, conveying at the same time relevant and novel information. This ranking approach provides insights for integrating two ranking paradigms for sub–topic retrieval: this can generally be applied to any method for estimating sub–topic classes (e.g. LDA, pLSI, relevance models) and for considering document dependencies (e.g. QPRP, PT, MMR). This study aims to investigate the performance gained from the ranking strategy produced by the integration of two ranking models, not to propose a new specific ranking model.

The contributions of this chapter are to:

- analyse and discuss the current state–of–the–art methods for diversity–aware information retrieval;
- investigate a new ranking strategy which is able to model sub–topics by means of clustering techniques and promote diversity among documents in a document ranking;
- conduct an empirical study comparing state–of–the–art ranking models for sub–topic retrieval (e.g. MMR, PT, static and dynamic clustering) against the integration models we introduce based on MMR and clustering;
- discuss the results of this study, and show that our proposed solutions outperform state–of–the–art strategies.

The chapter is structured as follows. In Section 15.2 we frame the sub–topic retrieval problem and present existing strategies for encoding novelty and diversity in document rankings. Next, we illustrate our approach based on clustering that considers sub–topic evidence when ranking using an MMR–inspired approach (Section 15.3). In Section 15.4, we present the methodology of our empirical investigation, which employes the imageCLEF 2009 Photo collection (Paramita et al, 2009), an image collection suited for the sub–topic retrieval task. The results obtained in the empirical investigation are illustrated and discussed in Section 15.5, while Section 15.6 concludes the paper stating the major contributions of our work in the light of the results obtained on the ImageClef 2009 Photo Retrieval Task collection together with lines of future investigation.

15.2 Background and Related Work

15.2.1 Sub–topic Retrieval

Conventional IR systems employ the PRP to rank documents with respect to the user's queries. Systems based on the PRP ignore interdependencies among documents ranked in the search results. These systems are generally appropriate when there are very few relevant documents and high–recall is required. An example of this situation is topic distillation in Web search, where a typical user wishes to find very few relevant core websites rather than every relevant Web page (Soboroff, 2004).

The assumption of independence in document relevance assessments that accompanied IR evaluation since the adoption of the Cranfield paradigm and on which the PRP is based, have recently been questioned. This generated a spate of work, not only on ranking functions that account for interdependent document relevance, such as MMR (Carbonell and Goldstein, 1998), PT (Wang and Zhu, 2009), QPRP (Zuccon and Azzopardi, 2010), but also with respect to test collections, evaluation measures, and retrieval tasks (Zhai et al, 2003; Clarke et al, 2008; Paramita et al, 2009). In fact, the relaxation of the independence assumption requires test collections to encode information about relevance dependencies between documents and measures are developed so as to account for such relationships. Research on novelty and diversity document ranking flowered from these needs, and a new retrieval task, called sub–topic document retrieval (or alternatively, diversity retrieval, novelty and diversity retrieval, facets retrieval), has been introduced. A number of collections has been realised for this task, including the one based on the Text REtrieval Conference (TREC) 6, 7, 8 interactive track (Zhai et al, 2003) and ImageCLEF 2008 and 2009 photo retrieval task collections (Sanderson, 2008; Paramita et al, 2009). In these collections, query topics induce sub–topics (also called facets, nuggets, aspects, intentions): each document contains zero or more sub–topics of a query topic, and one sub–topic can be contained in one or more documents. Note that if a document contains at least one query sub–topic, then it is relevant. No assumptions are made about the extent of the relevance of a document, i.e. grade of relevance: even if a document covers more sub–topics than another, the former yet might not be more relevant than the latter. Specifically, in all the collections produced for this retrieval task until today, relevance is treated as a binary feature: a document is either relevant or not, although it contains one or more sub–topics.

The aim of the task is to rank documents such that all the sub–topics associated with a query topic are covered as early in the ranking as possible, and sub–topics are covered with the least redundancy possible. Thus, the requirement that document rankings should cover all the sub–topics is greater than that documents should be relevant, since a document that covers a sub–topic is also relevant, but a list of relevant documents covers just one sub–topic. As a matter of fact, however, pure relevance ranking is unsuitable in this task.

This task resembles real situations. Often, in fact, there are an enormous number of potentially relevant documents containing largely similar content, resulting in partially or nearly duplicate information being presented within the document ranking. Secondly, in a large number of cases users pose a query for which the result set contains very broad topics related to multiple search facets, or has however multiple distinct interpretations. The query 'London' represents an example of a broad query. This might refer to 'London weather', 'London transport', 'London people', 'London travel', etc. The query 'Chelsea' represents an example of an ambiguous query that might be interpreted as 'Chelsea Clinton', 'Chelsea football club', or 'Chelsea area in London' etc. These are examples of situations where IR systems have to provide a document ranking that minimises redundant information and covers all the possible search facets (sub–topics).

Clarke et al (2008) identify the precise distinction between the concepts of *novelty* and *diversity* in information retrieval. Novelty is the need to avoid *redundancy* in search results, while diversity is the need to resolve queries' *ambiguity*. A popular approach for dealing with the redundancy problem is to provide diverse results in response to a search adopting an explicit ranking function, which usually requires a tuning of a user parameter. For example, MMR (Carbonell and Goldstein, 1998) and the Harmonic measure (Smyth and McClave, 2001) combine similarity and novelty in a unique ranking strategy. On the other hand, a traditional approach for coping with poorly specified or ambiguous queries relies on promoting diversity. This is motivated by the fact that the chances to retrieve relevant results can be maximised if results containing information from different query interpretations are presented within the document ranking.

15.2.2 The Probability Ranking Principle

The PRP (Robertson, 1977) is a well accepted ranking strategy that suggests presenting documents according to decreasing probability of document relevance to the user's information need; and the relevance of one document is considered independent from the rest of the documents. Formally, given a query q, if $P(x_i)$ is the relevance estimation for document x_i and $S(x_i,q)$ is the similarity function employed for producing such estimation, then the PRP suggests to present at rank $J+1$ a document d such that:

$$PRP_{J+1} \equiv \underset{x_i \in I \setminus J}{\text{argmax}}[p(x_i)] \approx \underset{x_i \in I \setminus J}{\text{argmax}}[S(x_i,q)] \tag{15.1}$$

where I is the set of results retrieved by the IR system; J is the set formed by the documents ranked until iteration J; x_i is a candidate document in $I \setminus J$, which is the set of documents that have not been ranked yet.

In the PRP, a document's relevance judgements are assumed independent and thus no relationship between documents is explicitly modelled in the ranking function. This is a known limitation of the PRP and, although it does not affect the optimality of the ranking principle for tasks such as ad hoc retrieval, it is the cause of the sub–optimality of the PRP in particular scenarios, such as sub–topic retrieval.

15.2.3 Beyond Independent Relevance

Maximal Marginal Relevance: Several techniques have been proposed for sub–topic retrieval. A simple method to address diversity between documents is that of MMR in set–based IR (Carbonell and Goldstein, 1998). Using a tuneable parameter, this ranking method balances the relevance between a candidate document and a query, e.g. the probability of relevance, and the diversity between the candidate document and all the documents ranked at previous positions. The ranking is linearly produced by maximising relevance and diversity scores at each rank. The MMR strategy is characterised by the following ranking function:

$$MMR_{J+1} \equiv \underset{x_i \in I \setminus J}{\operatorname{argmax}}[\lambda S(x_i, q) + (1-\lambda) \max_{x_j \in J} D(x_i, x_j)] \tag{15.2}$$

where x_j is a document in J, i.e. the set of documents that have been ranked already. The function $S(x_i, q)$ is a normalised similarity metric used for document retrieval, such as the cosine similarity, whereas $D(x_i, x_j)$ is a diversity metric. A value of the parameter λ greater than 0.5 assigns more importance to the similarity between document and query rather than to novelty/diversity. Conversely, when $\lambda < 0.5$, novelty/diversity is favoured over relevance.

In our work, when operationalising MMR, we modify how the diversity function impacts on the ranking: we substitute the function max with avg, which returns the average dissimilarity value between all pairs of x_i and x_j, instead of their largest value. To compute the dissimilarity between documents, we resort to estimating their similarity and then we revert this estimation. Specifically, the cosine function is used as a similarity measure between documents' term vectors obtained using the BM25 weighting schema. Since its similarity values range between -1 and 1, we can estimate the dissimilarity by the following formula:

$$\operatorname*{avg}_{x_j \in J} D(x_i, x_j) = \frac{\sum_{j=1}^{J}(-S(x_i, x_j)))}{J} \tag{15.3}$$

In Figure 15.1a we depict the document selection procedure suggested by MMR. In the figure, we simulate the possible clusters of documents that identify the sub–topics covered by those documents. Documents inserted in the ranking following the MMR strategy might come from the same group of sub–topics (i.e. x_1 and x_3), colliding with what is required in the sub–topic retrieval task.

Portfolio Theory: Wang and Zhu (2009) suggested to rank documents according to a paradigm proposed to select stocks in the financial market, PT. In the IR scenario diversification is achieved using PT by reducing the risk associated with document ranking. The intuition underlying PT is that the ideal ranking order is the one that balances the relevance of a document against the level of its risk or uncertainty (i.e. variance). Thus, when ranking documents, relevance should be maximised whilst minimising variance. The objective function that PT optimises is:

$$PT_{J+1} \equiv \underset{x_i \in I \setminus J}{\operatorname{argmax}} \left(p(x_i) - b w_{x_i} \delta^2_{x_i} - 2b \sum_{x_i \in J} w_{x_k} \delta_{x_i} \delta_{x_k} \rho_{x_i, x_k} \right) \tag{15.4}$$

where b represents the risk propensity of the user, $\delta^2_{x_i}$ is the variance associated to the probability estimation of document x_i, w_{x_i} is a weight expressing the importance of the rank position, and ρ_{x_i, x_k} is the correlation between document x_i and document x_k.

In summary, intuitively MMR and PT have a similar underlying schema for combining relevance and diversity. One common component of their ranking functions is the estimation of the probabilities of relevance. In both methods, the relevance estimation is balanced by a second component, which captures the degree of diversity between the candidate document and the ranking. In the empirical study we present in Section 15.4, we implemented both strategies and compared them to the novel paradigm we propose.

15.3 Document Clustering and Inter–Cluster Document Selection

15.3.1 Re–examining Document Clustering Techniques

It has been hypothesised that 'closely associated documents tend to be more relevant to the same requests' (van Rijsbergen, 1979). Similarly, in our work we hypothesise that clusters obtained considering the documents relevant to a user's information need have the potential to represent different sub–topics of the more general topic the user is interested in. Thus, clustering using unsupervised learning models extracts meaningful and representative information from documents, that can be used to model sub–topical diversity. The set of documents contained in each cluster is assumed to resemble what users perceive as a sub–topic. We then believe that incorporating evidence drawn from clusters of similar documents can enhance the performance of IR systems in the sub–topic retrieval task.

Although clustering can group documents containing similar contents, we do not intend to use clustering in isolation, in particular when selecting documents from such clusters. We hypothesise that clustering techniques combined with suitable criteria for document selection can improve the performances of systems in the sub–topic retrieval task. Regardless of the clustering technique used, strategies following the cluster–based paradigm can be characterised by how documents are selected from the clusters in order to output the final document ranking. In the following, two common approaches are revised.

The first approach, which is directly inspired by the cluster hypothesis, attempts to retrieve documents that are more similar to each other at higher ranks. Kurland and Lee (2004) propose a method, called the *interpolation algorithm*, to compute a retrieval score by mixing the evidence obtained from clusters together with document relevance estimations. The retrieval score of a candidate document d_i given this approach is calculated as:

$$\hat{p}(x_i,q) = \lambda p(x_i,q) + (1-\lambda) \sum_{t \varepsilon X} p(c_j,q) p(x_i,c_j) \tag{15.5}$$

where λ indicates the degree of emphasis on individual document information. In our study, we assume that $p(a,b)$ is the similarity between objects a and b[1]. Note that setting $\lambda = 0$ returns documents within the cluster with highest similarity to the query, i.e. the cluster with the highest $p(c_j,q)$. We refer to this approach as **Interp(.)**.

[1] These can be queries, documents, or clusters.

In the second approach we assume that each cluster represents a different sub–topic. Thus, to cover the whole set of sub–topics all the clusters have to be chosen at early ranks. In (van Leuken et al, 2009), three clustering methods with a dynamic weighting function are exploited to visually diversify image search results. Only representatives of visual–based clusters are presented to the users with the aim to facilitate faster browsing and retrieval. A similar work has been pursued in (Ferecatu and Sahbi, 2008), where the ranking is formed by selecting documents from clusters in a round–robin fashion, i.e. assigning an order to the cluster and selecting a document when cycling through all clusters. Cluster representatives are selected according to the order of the documents and are added to clusters [2]. The same approach may be applied to different clustering algorithms, i.e. k–means, Expectation–Maximisation (EM), Density–Based Spatial Clustering of Applications with Noise (DBSCAN).

Once sub–topical clusters are formed, several approaches can be employed to select a cluster representative. Deselaers et al (2009) propose selecting within each cluster the document that is most similar to the query, whereas Zhao and Glotin (2009) suggest choosing the document with the lowest rank within each cluster of the top retrieved results. In (Leelanupab et al, 2009; Urruty et al, 2009), the medoid[3] is assumed to be the best cluster representative. Finally, Halvey et al (2009) propose selecting the document that is most similar to other members of the selected cluster. In summary, these approaches ensure that documents are retrieved from all the clusters, and thus from all the sub–topics if these are correctly captured by the clustering process. However, document relevance and redundancy are not accounted for after clustering. Furthermore, despite the selection of documents from clusters according to their probability of relevance, documents at early ranks can contain duplicate information.

As a result, the top ranked documents might still be similar or highly correlated to each other. For example, as shown in Figure 15.1b, the distances of documents x_1, x_2, x_3, and x_4 selected using the clusters' medoids are constant and far away from the query q, in particular x_3 and x_4. Furthermore, if the closest documents to the query were to be selected, then the result will be documents that are close to each other, in particular when the query lays in the centre of the document space, which is surrounded by the clusters.

15.3.2 Clustering for Sub–topic Retrieval

As we have illustrated in the previous section, no current *cluster–based* method for sub–topic retrieval addresses novelty and relevance at the same time. Motivated by this consideration, this chapter investigates the effect of integrating intra–list dependence ranking models, i.e. ranking strategies that account for dependencies amongst ranked documents, and topic–oriented/cluster–based models, i.e. strategies that di-

[2] This is possible because the clustering algorithm in (Ferecatu and Sahbi, 2008) builds clusters iteratively by first selecting the centre of a cluster and then gathering its members.

[3] The document closest to the centroid of the cluster.

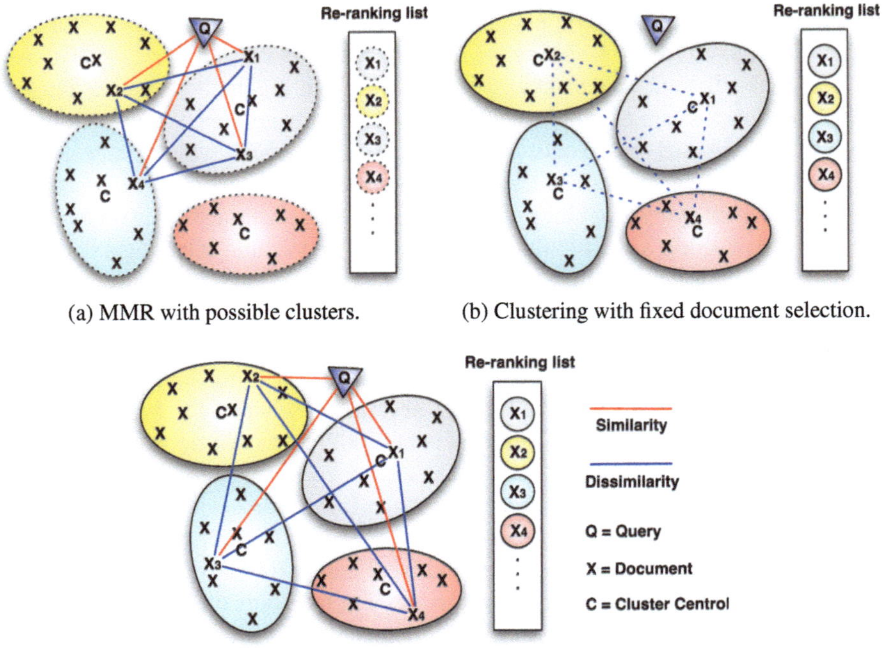

(a) MMR with possible clusters.

(b) Clustering with fixed document selection.

(c) Clustering with MMR re–ranking for document selection.

Fig. 15.1: Re–ranking methods for promoting diversity.

vide documents into sub–topic classes and consider these when ranking. Specifically, we propose two simple strategies that follow this idea, and evaluate them in the context of sub–topic retrieval. In particular, document dependencies can be exploited during the selection of representatives from sub–topic classes, obtained by employing any of the latter models. Figure 15.1c depicts the result of this approach, in which documents x_1, x_2, x_3, and x_4 are selected according to particular sub–topics, thus addressing diversity in the document ranking. We do not focus on the retrieval and relevance estimation, but we assume to have a reliable function that is able to provide an initial set of relevant documents. We suggest clustering these documents and then ranking the clusters according to the average relevance of the documents contained in each cluster. Given a query q and a cluster c_k, the average cluster relevance is defined as:

$$S_{avg}(c_k, q) = \frac{1}{I_k} \sum_{i=1}^{I} s(x_{k,i}, q) \tag{15.6}$$

where I_k is the number of documents in c_k and $X = \{x_1, ..., x_n\}$ is the initial set of relevant documents. Average cluster relevance is employed for ordering the clusters;

Algorithm 15.1 Intra-list dependency re-ranking (using MMR) on the evidence gathered from clusters.

Require: q, a user query
Require: $C = \{c_1, c_2, c_3, ..., c_k\}$, set of clusters k ranked according to average cluster relevance $S_{avg}(c_k, q)$
Require: $X_k = \{x_{k,1}, x_{k,2}, x_{k,3}, ..., x_{k,n}\}$, set of retrieved documents x within cluster c_k
Require: $j = 0$, where j is the number of documents that has been already ranked
Require: *maxDocs*, the maximum number of retrieved documents
  $J_0 = \{\}$
  while $j \leq maxDocs$ **do**
    if $j = 0$ **then**
      $J_0 = \underset{x_{k,n} \in X_k \setminus J}{\text{argmax}} [S(x_{k,n}, q)]$
    else
      $J_j = J_{j-1} \cup \underset{x_{k,n} \in X_k \setminus J}{\text{argmax}} [\lambda S(x_{k,n}, q) + (1 - \lambda) \underset{x_j \in J}{\text{avg}} D(x_{k,n}, x_j)]$
    end if
    $j = j + 1; k = k + 1$
    if $k \geq j$ **then**
      $k = 0$
    end if
  end while
  return $J_j = \{x_1, x_2, x_3, ... x_j\}$, a set of re-ranked documents to present to the user

then a round–robin approach that follows the order suggested by average cluster relevance is used in order to select individual documents within the clusters. To select a specific document within each cluster, we have to employ an intra–list dependency–based model: in our empirical study we choose to use MMR, for its simple formulation. In this step, alternative ranking functions may be employed. The complete algorithm is outlined in Algorithm 15.1: this is the same algorithm that has been implemented to produce the results reported in our empirical investigation.

15.4 Empirical Study

To empirically validate our approach and contrast it to state–of–the–art ranking strategies for sub–topic retrieval, we adopted the ImageCLEF 2009 photo retrieval collection (Paramita et al, 2009). This collection is composed of almost 500,000 images from the Belga News Agency. A text caption is associated with each image; the average length of the textual captions is 36 terms, while the total numbers of unique terms in the collection is over 260,000. Textual captions have been indexed using Terrier[4], which also served as a platform for developing the ranking strategies using Java. Before indexing the captions, we removed standard stop–words (van Rijsbergen, 1979) and applied Porter stemming. Low–level descriptors were not considered as this year's topics focus on topical, rather than visual, diversity. Moreover, the goal

[4] `http://ir.dcs.gla.ac.uk/terrier/`

```
<top>
<num> 16 </num>
<title> queen </title>
<clusterTitle> queen silvia </clusterTitle>
<clusterDesc> Relevant images will show photographs of
Queen Silvia. Relevant images may show other people if
Queen Silvia is shown in the foreground. Other images where
Queen Silvia is shown in the background are irrelevant. </clusterDesc>
<image> belga28/00059916.jpg </image>
<clusterTitle> queen rania </clusterTitle>
<clusterDesc> Relevant images will show photographs of
Queen Rania. Relevant images may show other people if Queen
Rania is shown in the foreground. Other images where Queen
Rania is shown in the background are irrelevant. </clusterDesc>
<image> belga14/01302163.jpg </image>
<clusterTitle> queen sofia </clusterTitle>
<clusterDesc> Relevant images will show photographs of Queen
Sofia. Relevant images may show other people if Queen Sofia is
shown in the foreground. Other images where Queen Sofia is shown
in the background are irrelevant. </clusterDesc>
<image> belga21/01431507.jpg </image>
<clusterTitle> queen -silvia -rania -sofia </clusterTitle>
<clusterDesc> Relevant images will show photographs of the queens
which are not included in the above clusters. </clusterDesc>
<image> belga12/01262014.jpg </image>
</top>
```


Fig. 15.2: Example of ImageCLEF 2009 database entry with sub–topic image (left) and query topic format (right).

of our study is to determine whether the integration of two diversity-aware ranking models is valid, and thus there is no major concern in ruling out visual features from the empirical investigation. Query topics have been similarly processed to the text captions. We used the set of 50 available topics, consisting of topic titles, cluster titles, cluster descriptions, and image examples. We employ only the topic titles, so as to simulate the situation where a user posts a broad or ambiguous query. Finally, we used the sub–topic judgements associated with each topic that are provided with the collection; an example of topic and image with related sub–topics is shown in Figure 15.2.

Okapi BM25 has been used to estimate document relevance given a query: its parameters have been set to standard values (Robertson et al, 1995). The same weighting schema has been used to produce document term vectors that are subsequently employed by re–ranking strategies to compute similarity/correlation. In preliminary results we have observed that this approach returns higher precision values, compared with alternative strategies, e.g. TF–IDF weighting. We experiment with several ranking lengths, i.e. 100, 200, 500, and 1,000, meaning that all the documents retrieved at ranks lower than these thresholds are discarded. In this chapter we report results for ranking up to 500 documents. Other ranking thresholds have shown similar results, and are not reported here.

Once estimates of document relevance are obtained using Okapi BM25, we produce an initial document ranking according to the probability ranking principle, i.e. we order documents with respect to decreasing probability of relevance. We denote this run with **PRP**, and it represents the baseline for every re–ranking strategy. Furthermore, the initial document ranking obtained using the PRP is used as input of the re–ranking functions. The runs obtained by implementing the maximal marginal relevance heuristic and the portfolio theory approach are denoted with **MMR** and **PT**, and they represent the state–of–the–art strategies for interdependent document relevance in the context of this investigation.

For MMR, we investigated the effect on retrieval performances of the hyper–parameter λ by varying it in the range [5] [0,1). The ranking function of MMR has been instantiated as discussed in Section 15.2.3.

When testing PT, we set b, the risk propensity parameter, as ranging from ± 1 to ± 9; we treat the variance of a document as a parameter that is constant with respect to all the documents, similarly to (Wang and Zhu, 2009). We experiment with variance values δ^2 ranging from 10^{-1} to 10^{-9}, and select the ones that achieve best performances in combination with the values of b through a grid search of the parameter space. The correlation between textual captions is computed as Pearson's correlation between the term vectors associated to the textual captions themselves.

Regarding the runs based on the topic–oriented paradigm, we adopt two different static and dynamic clustering algorithms: **k-means** and expectation maximization (**EM**), although alternative strategies may be suitable. For each query, the number of clusters required in k–means was set according to sub–topic relevance judgements for that query. In contrast, we allow the EM algorithm to determine the optimal number of clusters using cross validation, and set the minimum expected number of clusters using the sub–topic relevance judgements. After clusters are formed, documents are ranked according to techniques for selecting documents within clusters as illustrated in Section 15.3, specifically:

Interp(.): selects documents that maximise the interpolation algorithm for cluster–based retrieval;
PRP(.) : selects documents with the highest probability of relevance in the given clusters;
Mediod(.): selects the medoids of the given clusters as cluster representatives;
MMR(.) : selects documents according to maximal marginal relevance, as an example of a strategy based on an interdependent document relevance model.

Techniques that implement PRP(.) and Medoid(.) do not require any parameter tuning, whereas when instantiating Interp(.) and MMR(.), we varied their hyper–parameters in the range [0,1), and selected the value that obtained the best performance. In total, the combination of clustering algorithms and document selection criteria forms eight experimental runs that we tested in this study, i.e. Interp(k–means), PRP(k–Means), Medoid(k–means), MMR(k–means), Interp(EM), PRP(EM), Medoid(EM), and MMR(EM).

We employed three measures to assess the effectiveness of different ranking strategies in sub–topic retrieval. The first measure, called **α–NDCG**, extends the normalised discounted cumulative gain to the case of the sub–topic retrieval task; the parameter α ranges between 0 and 1 and directly accommodates novelty and

[5] We excluded the value $\lambda = 1$, since MMR's ranking function would be equivalent to that of PRP.

diversity (Clarke et al, 2008). We set $\alpha = 0.5$, as it is common practice (Clarke et al, 2009b): intuitively, this means that novelty and relevance are equally accounted for in the measure. Novelty and rank biased precision (**NRBP**) (Clarke et al, 2009a) integrate nDCG, rank–biased precision (RBP) and intention aware measures in order to model the decreasing interest of the user examining documents at late rank positions. Sub–topic recall (**S-R**) (Zhai et al, 2003) monitors sub–topic coverage in the document ranking.

15.5 Results

In Table 15.1 we report the results obtained by the instantiations of the ranking strategies considered in our empirical investigation and evaluated them using α-NDCG@10, α-NDCG@20, NRBP, and S–recall. Due to the presence of varying parameters that require tuning, we only report the best results obtained by each strategy with respect to α-NDCG@10. Percentage improvements over the PRP of each re–ranking strategy are reported in the table. The instantiations of the approaches we propose in this study, i.e. MMR(k–means) and MMR(EM), underlined in the table, provide an example of the results the integration approach, based either on static or dynamic clustering, and MMR, can achieve. Statistical significance against MMR and PT using a t–test is calculated for each of the eight runs reported in the lower part of Table 15.1, and it is indicated with *, w.r.t. MMR, and †, w.r.t. PT.

The results suggest that the integration of either static or dynamic clustering techniques with interdependent document ranking approaches can improve the performance in sub–topic retrieval. In particular, the percentage improvements of MMR(k–means) and MMR(EM) are greater than the one obtained by their peers in all the evaluation measures, except in NRBP, for which however a consistent trend can not be extracted. Furthermore, it can be observed that even applying the integration paradigm on a simple lightweight clustering algorithm such as k-means, which can be executed in runtime, can increase the retrieval performance when compared to MMR or k–means alone.

Our empirical investigation also suggests that selecting clusters in a round–robin fashion when ranking, as is done in PRP(k–means), PRP(EM), or Medoid(k–means), Medoid(EM), outperforms other policies, such as the one implemented by Interp(.). This result is consistent for all the investigated measures. Note that ranking documents according to Interp(.) may result in documents from the same cluster being ranked consecutively: this might be the case when the probabilities of cluster relevance and of the document being in the specific cluster are high. In addition, the results suggest that the runs based on the topic–oriented paradigm produce better rankings than the ones based on the interdependent document relevance paradigm (i.e. MMR and PT) in terms of α-NDCG@10, that has been used as an objective function for parameter tuning.

In summary, the results show that integrating the two paradigms for sub–topic retrieval based on interdependent document relevance and topic–oriented models can

Table 15.1: Sub–topic retrieval performances obtained in the ImageCLEF 2009 photo retrieval collection. Percentage improvements refer to the PRP baseline. Parameters are tuned with respect to α–NDCG@10, in particular: MMR ($\lambda = 0.6$), PT ($b = 9$, $\delta^2 = 10^{-3}$), Interp(k–means) ($\lambda = 0.8$), MMR(k–means) ($\lambda = 0.8$), Interp(EM) ($\lambda = 0.9$), and MMR(EM) ($\lambda = 0.7$). The best performance improvements are highlighted in bold, and statistical significance at 0.05 level, computed using a t–test, against MMR and PT are indicated by * and † respectively.

Model	*α-NDCG@10*	α-NDCG@20	NRBP	S-R@10
PRP	0.425	0.467	0.270	0.542
MMR	0.484	0.516	0.288	0.661
	+13.88%	+10.49%	+6.67%	+21.90%
PT	0.470	0.511	0.287	0.629
	+10.59%	+9.42%	+6.30%	+16.09%
Interp(K-Mean)	0.448	0.475	0.302	0.524*†
	+5.41%	+1.71%	+11.85%	-3.32%
Medoid(K-Mean)	0.463	0.490	0.291	0.591*
	+8.94%	+4.93%	+7.78%	+8.93%
PRP(K-Mean)	0.486	0.515	0.309	0.617
	+14.35%	+10.28%	+14.44%	+13.87%
MMR(K-Mean)	0.491	0.520	0.302	0.655
	+15.53%	+11.35%	+11.85%	+20.83%
Interp(EM)	0.457	0.486	0.311	0.532*
	+7.53%	+4.07%	+15.19%	-1.84%
Medoid(EM)	0.497	0.524	**0.320**†	0.646
	+16.94%	+12.21%	+18.52%	+19.11%
PRP(EM)	0.502†	0.536	0.314	0.670
	+18.12%	+14.78%	+16.30%	+23.61%
MMR(EM)	**0.508**†	**0.539**†	0.311	**0.681**†
	+19.53%	+15.42%	+15.19%	+25.59%

deliver better performance than state–of–the–art ranking strategies. In three out of four measures, our best approach based on the integration paradigm, i.e. MMR(EM), outperforms state–of–the–art approaches with statistical significance against our instantiation of PT. Despite the integration–based strategies providing less performance increments than other re–ranking approaches with respect to NRBP, the difference is very limited and might be related to the settings of the parameters internal to NRBP. Furthermore, it is difficult to quantify how this small difference in NRBP affects the user.

15.6 Conclusions

Diversity with respect to the sub–topics of the query topic is a highly desired feature for generating satisfying search results, in particular when there is a large number of documents containing similar information, or when a user enters a very broad or ambiguous query. Common diversity–based ranking approaches are devised on the basis of interdependent document relevance or topic–oriented paradigms. In this chapter, state–of–the–art strategies for diversity–aware IR are reviewed and discussed. We propose a new ranking approach, which incorporates two ranking paradigms with the aim to explicitly model sub–topics and reduce redundancy in document ranking simultaneously. An empirical investigation was conducted using the ImageCLEF 2009 photo retrieval task collection, where we contrast state–of–the–art approaches against two instantiations of the integration approach we propose. Maximal marginal relevance and portfolio theory for IR are examined as examples of the interdependent document relevance paradigm, whilst k–means and EM clustering methods are employed to explicitly model sub–topics. Various criteria for selecting documents within clusters are investigated in our study, and their performance is compared. The interpolation algorithm assumes that relevant documents tend to be more similar to each other; whereas the selection methods based on cluster representatives or PRP stem from the hypothesis that clusters can represent the sub–topics of a query. Parametric ranking functions are tuned with respect to α–NDCG@10, which is used to measure retrieval effectiveness in the sub–topic retrieval tasks. We also evaluate the strategies in terms of NRBP and S–recall.

The results of our empirical investigation suggest that ranking strategies built upon the integration of MMR and EM clustering significantly outperform all other approaches. Furthermore, we show that the integration intuition can be ideally applied to any clustering algorithm. The comparison between interdependent document relevance and topic–oriented paradigms suggests that the cluster-based retrieval strategies perform better than the former in sub–topic retrieval. With respect to our study, the interpolation algorithm in cluster–based retrieval is not suitable for results diversification, while still being suitable for the ad hoc retrieval task (Kurland and Lee, 2004). The round–robin policy for selecting clusters performs consistently better than the interpolation strategy; furthermore, selecting documents within clusters using the PRP is better than doing so by using cluster representatives such as medoids.

In summary, the integration approach effectively improves diversity performance for sub–topic retrieval. Further investigation will be directed towards the empirical validation of our integration approach on other collections for sub–topic retrieval, such as TREC 6, 7, 8 interactive and ClueWeb 2009. Furthermore, image low–level features can be employed to refine the results from text clustering since they can enhance visual diversity in addition to topical diversity.

Acknowledgements The authors are grateful to Chales Clarke for advise on the implementation of the evaluation measures used in this work. Teerapong Leelanupab would like to thank the Royal Thai Government for providing financial support. Guido Zuccon is funded by the EPSRC Renaissance project (EP/F014384/1).

References

Agichtein E, Brill E, Dumais S (2006) Improving web search ranking by incorporating user behavior information. In: Proceedings of the 29th annual international ACM SIGIR conference on Research and development in information retrieval, pp 19–26

Blei DM, Ng AY, Jordan MI (2003) Latent dirichlet allocation. Journal of Machine Learning Research 3:993–1022

Carbonell J, Goldstein J (1998) The use of MMR, diversity-based reranking for reordering documents and producing summaries. In: Proceedings of the 21st annual international ACM SIGIR conference on research and development in information retrieval, ACM press, pp 335–336

Carterette B, Chandar P (2009) Probabilistic models of ranking novel documents for faceted topic retrieval. In: Proceeding of the 18th ACM conference on information and knowledge management, pp 1287–1296

Clarke CL, Kolla M, Cormack GV, Vechtomova O, Ashkan A, Büttcher S, MacKinnon I (2008) Novelty and diversity in information retrieval evaluation. In: Proceedings of the 31st annual international ACM SIGIR conference on research and development in information retrieval, pp 659–666

Clarke CL, Kolla M, Vechtomova O (2009a) An effectiveness measure for ambiguous and underspecified queries. In: Proceedings of the 2nd International Conference on Theory of Information Retrieval, pp 188–199

Clarke CLA, Craswell N, Soboroff I (2009b) Overview of the TREC 2009 Web Track. In: Proceedings of the Text REtrieval Conference (TREC–2009)

Deselaers T, Gass T, Dreuw P, Ney H (2009) Jointly optimising relevance and diversity in image retrieval. In: Proceeding of the ACM International Conference on Image and Video Retrieval

Eisenberg M, Berry C (2007) Order effects: A study of the possible influence of presentation order on user judgments of document relevance. Journal of the American Society for Information Science and Technology 39(5):293–300

Ferecatu M, Sahbi H (2008) TELECOM ParisTech at ImageCLEFphoto 2008: Bi-modal text and image retrieval with diversity enhancement. In: Working Notes of CLEF 2008

Gordon MD, Lenk P (1999a) A utility theoretic examination of the probability ranking principle in information retrieval. Journal of the American Society for Information Science and Technology 42(10):703–714

Gordon MD, Lenk P (1999b) When is the probability ranking principle suboptimal. Journal of the American Society for Information Science and Technology 43(1):1–14

Halvey M, Punitha P, Hannah D, Villa R, Hopfgartner F, Goyal A, Jose JM (2009) Diversity, assortment, dissimilarity, variety: A study of diversity measures using low level features for video retrieval. In: Proceedings of the 31th European Conference on IR Research on Advances in Information Retrieval, pp 126–137

Hearst M, Pedersen J (1996) Reexamining the cluster hypothesis: Scatter/gather on retrieval results. In: Proceedings of the 19th annual international ACM SIGIR conference on research and development in information retrieval, pp 76–84

Hofmann T (1999) Probabilistic latent semantic indexing. In: Proceedings of the 22nd annual international ACM SIGIR conference on research and development in information retrieval, pp 50–57

Huang J, Kumar SR, Zabih R (1998) An automatic hierarchical image classification scheme. In: Proceedings of the sixth ACM international conference on Multimedia, pp 219–228

Kurland O, Lee L (2004) Corpus structure, language models, and ad hoc information retrieval. In: Proceedings of the 27th annual international ACM SIGIR conference on research and development in information retrieval, pp 194–201

Lavrenko V, Croft WB (2001) Relevance based language models. In: Proceedings of the 24th annual international ACM SIGIR conference on research and development in information retrieval, ACM press, pp 120–127

Leelanupab T, Hopfgartner F, Jose JM (2009) User centred evaluation of a recommendation based image browsing system. In: Proceedings of the 4th Indian International Conference on Artificial Intelligence, pp 558–573

van Leuken RH, Garcia L, Olivares X, van Zwol R (2009) Visual diversification of image search results. In: Proceedings of the 18th international conference on World Wide Web, pp 341–341

Paramita ML, Sanderson M, Clough PD (2009) Developing a test collection to support diversity analysis. In: Proceedings of Redundancy, Diversity, and Interdependent Document Relevance workshop held at ACM SIGIR' 09

van Rijsbergen CJ (1979) Information Retrieval, 2nd Ed. Butterworth

Robertson SE (1977) The probability ranking principle in IR. Journal of Documentation 33:294–304

Robertson SE, Walker S, Beaulieu MM, Gatford M (1995) Okapi at TREC 4. In: Proceedings of the 4th Text REtrieval Conference (TREC–4)

Sanderson M (2008) Ambiguous queries: test collections need more sense. In: Proceedings of the 31st annual international ACM SIGIR conference on Research and development in information retrieval, pp 499–506

Smyth B, McClave P (2001) Similarity vs. diversity. In: Proceedings of the 4th International Conference on Case-Based Reasoning, pp 347–361

Soboroff I (2004) On evaluating web search with very few relevant documents. In: Proceedings of the 27th annual international ACM SIGIR conference on Research and development in information retrieval, pp 530–531

Urruty T, Hopfgartner F, D. H, Elliott D, Jose JM (2009) Supporting aspect-based video browsing - analysis of a user study. In: Proceeding of the ACM International Conference on Image and Video Retrieval

Wang J, Zhu J (2009) Portfolio theory of information retrieval. In: Proceedings of the 32nd annual international ACM SIGIR conference on Diversity, and Interdependent Document Relevance workshop, pp 115–122

Zhai CX, Cohen WW, Lafferty J (2003) Beyond independent relevance: methods and evaluation metrics for subtopic retrieval. In: SIGIR '03: Proceedings of the 26th annual international ACM SIGIR conference on Research and development in informaion retrieval, pp 10–17

Zhao ZQ, Glotin H (2009) Diversifying image retrieval by affinity propagation clustering on visual manifolds. IEEE MultiMedia 99(1)

Zuccon G, Azzopardi L (2010) Using the quantum probability ranking principle to rank interdependent documents. In: Proceedings of the 32th European Conference on IR Research on Advances in Information Retrieval, pp 357–369

Chapter 16
Knowledge Integration using Textual Information for Improving ImageCLEF Collections

Manuel Carlos Díaz–Galiano, Miguel Ángel García–Cumbreras, María Teresa Martín–Valdivia, and Arturo Montejo-Ráez

Abstract In this chapter we explain our participation at ImageCLEF from 2005 to 2009. During these years we have mainly developed systems for the ad hoc and the medical retrieval tasks. Although the different proposed tasks include both visual and textual information, the diverse approaches applied by the participants also include the use of only one type of information. The SINAI group specializes in the management of textual collections. For this reason, our main goal has been to improve the general system by taking advantage of the textual information.

16.1 Introduction

The first participation of the SINAI research group at the Cross Language Evaluation Forum (CLEF) was in 2002 presenting a multi–lingual system for the ad hoc task. Since then, we have followed the developments of CLEF and have participated in different tasks (GeoCLEF, CL-SR, etc.). Our first contribution at ImageCLEF was in 2005. In consecutive years we have mainly developed systems for the ad hoc and the medical retrieval tasks. We have modified the different models in order to adapt them to the new tasks proposed in ImageCLEF (wiki, photo, etc.), the new collections (CasImage, Pathopic, IAPR TC-12, etc.) and our areas of interest (application of machine translation, filtering of information, usage of ontologies, and knowledge integration). The changes have been addressed using the results obtained by both our systems and the other techniques presented at ImageCLEF.

Although the corpora provided by the ImageCLEF organizers include both textual and visual information, we have principally managed the textual data contained

Manuel Carlos Díaz Galiano · Miguel Ángel García Cumbreras · María Teresa Martín Valdivia · Arturo Montejo Ráez
SINAI Research group, University of Jaén, Paraje Las Lagunillas, s/n, Jaén (SPAIN), e-mail: {mcdiaz,magc,maite,amontejo}@ujaen.es

H. Müller et al. (eds.), *ImageCLEF*, The Information Retrieval Series 32,
DOI 10.1007/978-3-642-15181-1_16, © Springer-Verlag Berlin Heidelberg 2010

in the different collections. In practice, our main goal is to improve the general system by taking advantage of the textual information.

In addition, we have developed separate systems for the two main retrieval tasks at ImageCLEF: ad hoc and medical retrieval, although the best and more interesting results have been achieved with medical retrieval systems.

Thus, for the ad hoc task we were mainly interested in the different translation schemes even though we have also applied several retrieval models, weighting functions and query expansion techniques. However, from 2008 the task took a different approach to evaluate the diversity of results. Each query contained textual information and some relevant clusters. From that moment, our main interest has been to develop a clustering methodology in order to improve the result obtained. We have expanded the cluster terms with WordNet[1] synonyms. We have also introduced a clustering module based on the k–means algorithm and the creation of the final topics using the information of the title and the cluster terms. Unfortunately, the application of clustering does not improve the results.

Regarding the medical retrieval, we have investigated several methods and techniques. In our first participation in 2005 we studied different fusion methods in order to merge the results obtained from the textual Information Retrieval (IR) and Content Based Information Retrieval (CBIR) systems. In 2006, we tried to filter some features in the collections by applying Information Gain (IG) techniques. We accomplished several experiments in order to determinate the set of data that introduces less noise in the corpus. However, the results were not very relevant. Thus, in 2007 our major efforts were oriented to knowledge integration. We expanded the terms in the queries using the Medical Subject Headings (MeSH[2]) ontology. The results obtained were very good using only textual information. For this reason, in 2008 we investigated the effect of using another ontology, the Unified Medical Language System (UMLS[3]) meta–thesaurus. Surprisingly the results were not as good as we thought. Our main conclusion was that it is necessary to address the expansion of terms in a controlled way. The integration of all the terms without any filter scheme can include more noise in the final model and the system performance can be affected. Finally, the last participation at ImageCLEFmed tried to investigate the effect of expanding not only the query but also the whole collection. Again the results were not successful.

The next sections describe in a more detailed way the different systems developed for the ad hoc and medical retrieval tasks during our consecutive participation

---

[1] `http://wordnet.princeton.edu/`

[2] MeSH is the U.S. National Library of Medicine's controlled vocabulary used for indexing articles for MEDLINE/PubMed. MeSH terminology provides a consistent way to retrieve information that may use different terminology for the same concepts. `http://www.ncbi.nlm.nih.gov/mesh`

[3] The UMLS integrates and distributes key terminology, classification and coding standards, and associated resources to promote creation of more effective and interoperable biomedical information systems and services, including electronic health records. `http://www.nlm.nih.gov/research/umls/`

at ImageCLEF. Finally, Section 16.5 concludes this chapter and discusses further work.

16.2 System Description

It is often said that an image is worth 1,000 words. Unfortunately, these 1,000 words may differ from one individual to another depending on their perspective and/or knowledge of the image context. Thus, even if a 1,000-word image description were available, it is not certain that the image could be retrieved by a user with a different description.

Since 2005 we have developed and improved two independent systems: a photo retrieval system and a medical retrieval system. These systems work without user interaction (fully automatic) and they are focused on the textual information of the collections. Our aim was to develop and test different methods to improve the retrieval results, working with the associated text of the images.

It is usual for IR systems to pre–process the collections and the queries. All our approaches run this step, applying stopword removal and stemming (Porter, M.F., 1980). In addition, each non–English query is translated into English with our translation module, called SINTRAM (García–Cumbreras, M.A., Urena–López, L.A., Martínez–Santiago, F. and Perea–Ortega, J.M. 2007).

16.2.1 Photo Retrieval System

For more than ten years the SINAI group has tested and developed techniques to improve mono and multi–lingual information retrieval systems. For the ad hoc task of ImageCLEF techniques included the following:

- **IR systems**. Some IR systems have been used, selecting the ones that obtained the best results in our IR experiments (mono and multi–lingual). Different parameters have been tested, such as weighting functions (TFIDF, Okapi, InQuery), Psedo-Relevance Feedback (PRF) (Salton, G. and Buckley, G., 1990) and Query Expansion (QE).
- **Translation techniques**. Our machine translation system works with different on–line machine translators and implements several heuristics to combine them.
- **Fusion techniques**. When using several systems, the results lists have to be combined into a single combined one.
- **Expansion vs. Filtering**. Some approaches have been tested in order to expand terms from the query and the document and, also, to filter them when they are not very informative.

Figure 16.1 shows a general schema of our photo retrieval system.

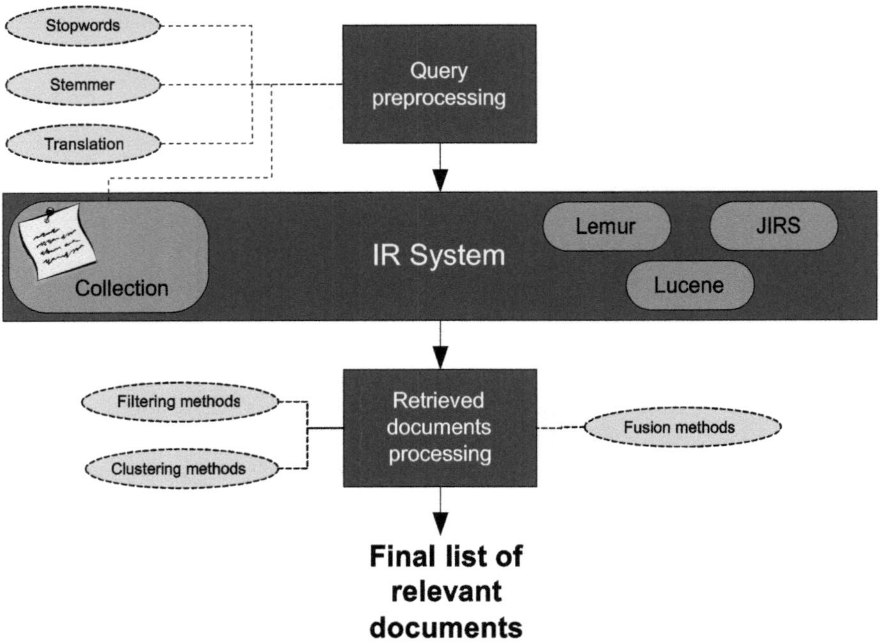

Fig. 16.1: General scheme of our photo retrieval system.

16.2.2 Medical Retrieval System

We only used textual techniques in the mixed IR system (visual and textual). For the medical retrieval task of ImageCLEF we have experimented in three aspects:

- Filtering textual information. We selected the best XML tags of the collection applying information gain (IG) metrics.
- Expanding the original query. We experimented using MeSH and UMLS ontologies.
- Combining relevant lists of textual and visual results. We applied several fusion techniques in order to merge visual and textual information.

Figure 16.2 shows a general scheme of our medical image retrieval system.

16.3 Photo Task

The main aim of the photo retrieval task is to retrieve relevant photos given a photo query. The images have associated text, normally a few words, that describe them. Our photo retrieval system only works with the associated text to retrieve relevant

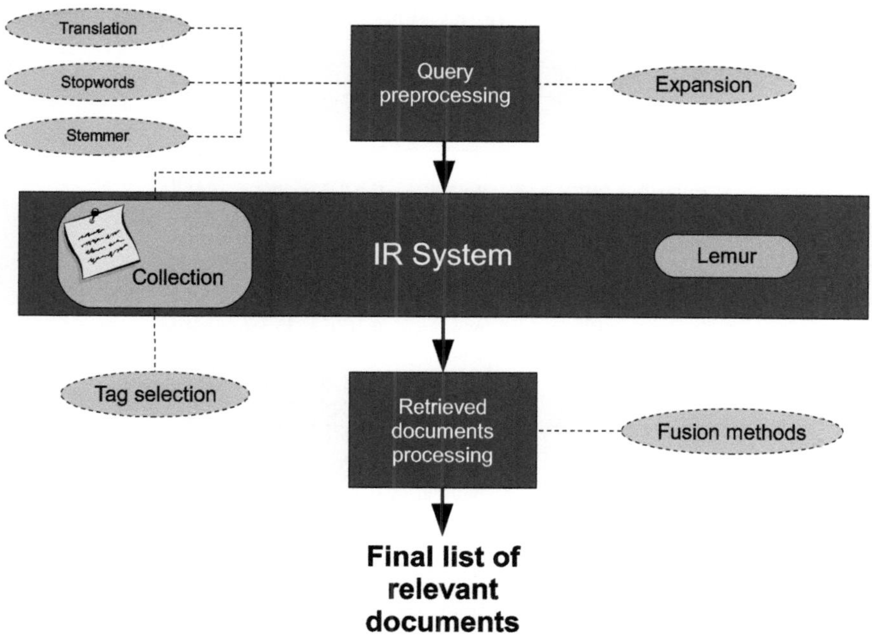

Fig. 16.2: General scheme of our medical retrieval system.

images and it only uses the text associated with each query, with or without context information.

In 2005 and 2006 the texts of the images (collection and queries) were independent phrases with a few words in the title field and a brief description in the narrative or description field. Other metadata was given for each query such as notes, dates and the location associated with the image. Some information was given in languages other than English, so it was necessary to use machine translation resources to translate them. In general, the results obtained with our system were good but it did not work well with the so–called *difficult queries*, queries with few relevant images in the collection or those with poor information.

To promote the diversity of results, with the aim of retrieving relevant images for all the queries, the query topics since 2007 included information about clusters. Each topic was clustered manually into sub–topics and the relevance judgements, to evaluate the results, included which cluster an image belonged to.

In the first developments of our system, the translation module SINTRAM was very important, because of multi–lingual queries used (English, Dutch, Italian, Spanish, French, German, Danish, Swedish, Portuguese and Russian). These first systems were composed of the following modules:

- a pre–processing module (normalization, stopword removal and stemming);

Table 16.1: Summary of results for the ad hoc task with multi–lingual queries.

Language	Initial Query	Expansion	MAP	%MONO	Rank
Dutch	title	with	0.3397	66.5%	2/15
Dutch	title	without	0.2727	53.4%	9/15
English	title + narr	with	0.3727	n/a	31/70
English	title	without	0.3207	n/a	44/70
French	title + narr	with	0.2864	56.1%	1/17
French	title + narr	without	0.2227	43.6%	12/17
German	title	with	0.3004	58.8%	4/29
German	title	without	0.2917	57.1%	6/29
Italian	title	without	0.1805	35.3%	12/19
Italian	title	with	0.1672	32.7%	13/19
Russian	title	with	0.2229	43.6%	11/15
Russian	title	without	0.2096	41.0%	12/15
Spanish	title	with	0.2416	47.3%	5/33
Spanish	title	without	0.2260	44.2%	8/33
Swedish	title	without	0.2074	40.6%	2/7
Swedish	title	with	0.2012	39.4%	3/7

- a translation module: based on the analysis of previous experiments, an automatic machine translator was defined by default for each pair of languages. For instance, Epals[4] (German and Portuguese), Prompt[5] (Spanish), Reverso[6] (French) or Systran[7] (Dutch and Italian);
- an IR module: the Lemur[8] IR system was tuned up, and PRF with the Okapi weighting function was applied.

Table 16.1 shows the best result obtained for each language with the first development. These results are presented in terms of Mean Average Precision (MAP). The first column shows the language of the queries; the second one includes the fields used (*title*, *narr*, *description*); the third one shows if there was query expansion. The %MONO column shows the loss of precision of the multi–lingual queries according to the monolingual one (English MAP). The last column shows the ranking obtained with our experiment among the rest of the participants in the ImageCLEF photo task.

The results obtained show that, in general, the IR system Lemur works well with the Okapi weighting function, and the application of query expansion improves the results. Only one Italian experiment without query expansion gets a better result. In the case of the use of only *title* or *title* + *narrative*, the results are not conclusive, but the use of only *title* seems to produce better results. Multi–lingual queries produced a loss of precision of around a 25%. Figure 16.3 shows the loss of MAP with multi–lingual queries.

[4] `http://www.epals.com/`

[5] `http://www.online-translator.com/`

[6] `http://www.reverso.net/`

[7] `http://www.systran.co.uk/`

[8] `http://www.lemurproject.org/`

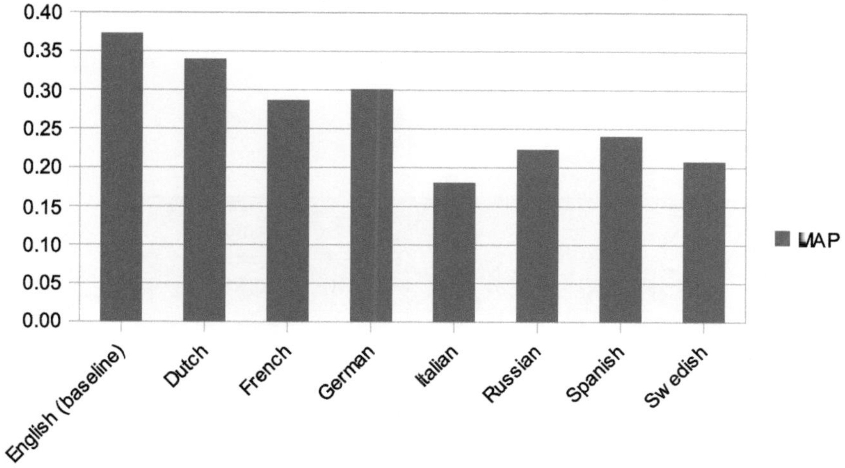

Fig. 16.3: Loss of MAP between the English queries and the multi–lingual ones.

16.3.1 Using Several IR and a Voting System

Later development of our photo retrieval system used several IR systems and a voting scheme to combine the results. Lemur and JIRS (Java Information Retrieval System) (Gómez–Soriano, J.M., Montes–y–Gómez, M., Sanchis–Arnal, E., and Rosso, P. 2005) were adapted for our system. Lemur is a toolkit that supports indexing of large-scale text databases, the construction of simple language models for documents, queries, or subcollections, and the implementation of retrieval systems based on language models as well as a variety of other retrieval models. JIRS is a passage retrieval system oriented to Question Answering (QA) tasks although it can be applied as an IR system. The complete architecture of the voting system is described in Figure 16.4.

Baseline cases with only Lemur and JIRS were run, so a final result was generated from a simple voting system with both IR systems that normalizes the scores and combines them with weights for each IR system (based on previous experiments and their evaluations). Table 16.2 shows the results obtained with the voting system (monolingual and bilingual runs).

In general, the results were poor because the set of queries was composed by only a few words. Nevertheless, our results were good in comparison with the other participants. After the analysis of these experiments, the English runs have obtained a loss of MAP of around 25%, being the worst results. Our best Spanish experiment was similar to the best one in the competition. For Portuguese we obtained the best one, and for French and Italian our runs were a bit worse: only a loss of MAP of around 8%. From these results we conclude that the Lemur IR system works better than JIRS, although the difference is not significant.

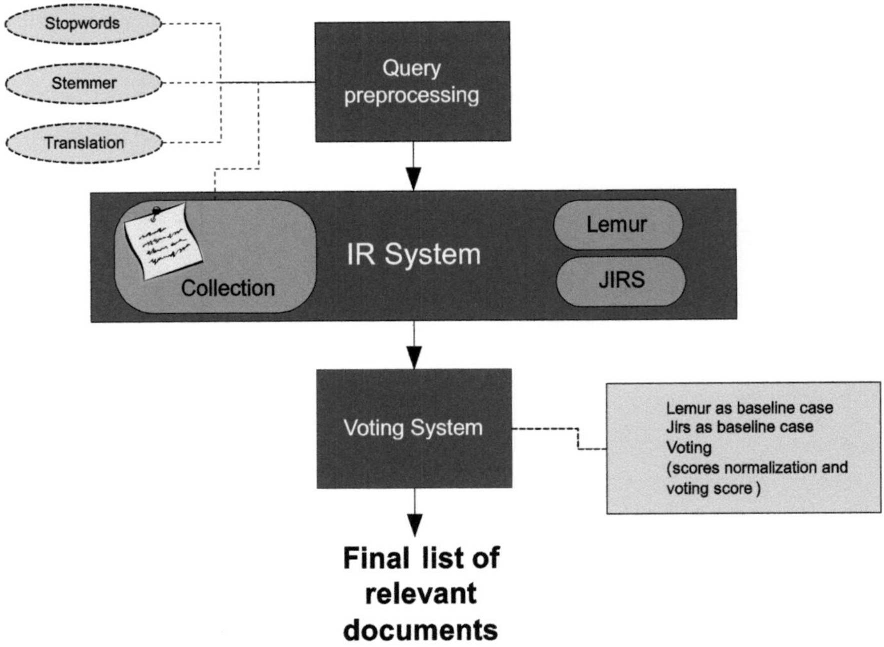

Fig. 16.4: Complete architecture of the voting system.

Fusion techniques have not improved the single ones. Lower MAP values decreased when we combined relevance lists. Other techniques must be used when the queries have few words.

16.3.2 Filtering

In the later evolution of our photo retrieval system we applied a filtering method over the results. In a first step the cluster term is expanded with its WordNet synonyms (the first sense). Then, the list of relevant documents generated by the IR system is filtered. If the relevant document contains the cluster term or a synonym, its *doc_id* (the identifier of the document) is written in another list. Finally, the new list with the filtered documents is combined with the original ones (Lemur and JIRS) in order to improve them. A simple method to do this was to double the score value of the documents in the filtered list and to add them to the original ones. The general architecture of the filtering system is shown in Figure 16.5.

The experiments carried out with the filtering system are as follows:

Table 16.2: Summary of results with the voting system (monolingual and bilingual runs).

Language	IR	MAP	Best MAP
English	Lemur	0.1591	0.2075
English	JIRS	0.1473	0.2075
English	Voting	0.0786	0.2075
Spanish	Lemur	0.1498	0.1558
Spanish	JIRS	0.1555	0.1558
Spanish	Voting	0.0559	0.1558
Portuguese	Lemur	0.1490	0.1490
Portuguese	JIRS	0.1350	0.1490
Portuguese	Voting	0.0423	0.1490
French	Lemur	0.1264	0.1362
French	JIRS	0.1195	0.1362
French	Voting	0.0323	0.1362
Italian	Lemur	0.1198	0.1341
Italian	JIRS	0.1231	0.1341
Italian	Voting	0.0492	0.1341

1. **Exp1: baseline case**. As baseline, Lemur was used as the IR system with automatic feedback and Okapi as weighting function. There was no combination of results, nor filtering method with the cluster term.
2. **Exp2: LemurJirs**. We combined the IR lists of relevant documents. Lemur with Okapi as weighting function and PRF. Before the combination of results Lemur and JIRS lists are filtered, only with the cluster term.
3. **Exp3: Lemur fb okapi**. The Lemur list of relevant documents (with Okapi and PRF) is filtered with the cluster term and its WordNet synonyms.
4. **Exp4: Lemur fb tfidf**. It is the same experiment as before, but in this case the weighting function used was TFIDF.
5. **Exp5: Lemur simple okapi**. The Lemur IR system has been run with Okapi as weighting function but without feedback. The list of relevant documents has been filtered with the cluster term and its WordNet synonyms.
6. **Exp6: Lemur simple tfidf**. The Lemur IR system has been used with TFIDF as weighting function but without feedback. The list of relevant documents has not been filtered.

The results are shown in Table 16.3. The last column represents the best F_{-} score obtained in the 2008 competition (complete automatic systems with only text).

The results show that a simple filtering method is not useful if the cluster term or related words are used to filter the IR retrieved documents. It happens because some good documents are deleted and new relevant documents are not included in the second step. In general, the results in terms of MAP or other precision values are not very different. Between the best MAP and the worse one the difference is less than 8%. Filtering methods have not improved the baseline case.

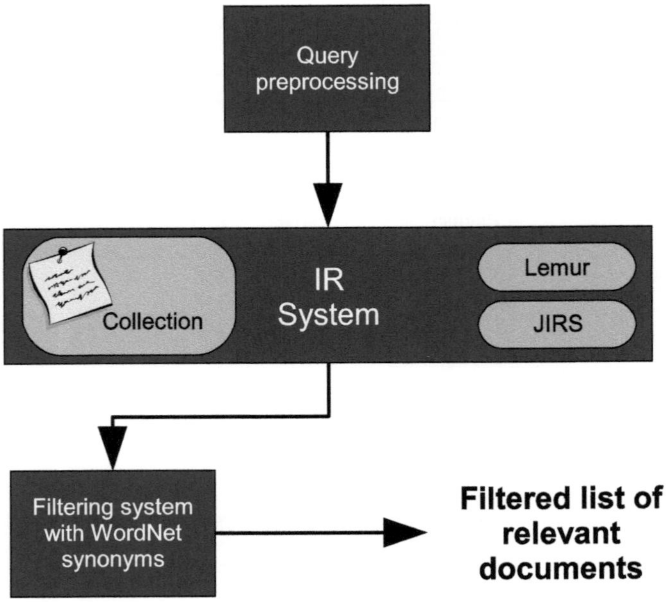

Fig. 16.5: Filtering scheme of the SINAI system.

Table 16.3: Results obtained with the filtering system.

Id	Filtering	FB	Expansion	MAP	P@5	P@10	Best F_1
Exp1	No	Yes	No	**0.2125**	**0.3744**	**0.3308**	0.2957
Exp6	No	No	No	0.2016	0.3077	0.2872	0.2957
Exp2	Yes	Yes	No	0.2063	0.3385	0.2949	0.2957
Exp3	Yes	Yes	No	**0.2089**	**0.3538**	0.3128	0.2957
Exp4	Yes	Yes	No	0.2043	0.2872	0.2949	0.2957
Exp5	Yes	No	No	0.1972	0.3385	**0.3179**	0.2957

After an analysis of the performance of filtering we can infer some reasons for this:

- Some relevant documents that appear in the first retrieval phase have been deleted because they do not contain the cluster term, so the cluster term is not useful in a filtering process.
- Other documents retrieved by the IR system that are not relevant, contain synonyms of the cluster term, so they are not deleted and the precision decreases.

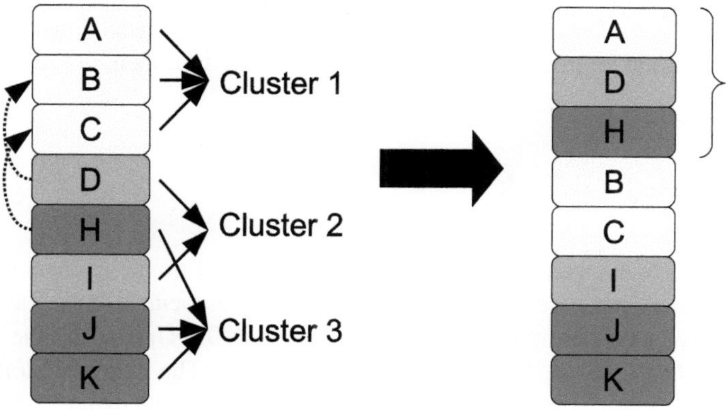

Fig. 16.6: Reordering of top results to increase variability according to clusters found.

16.3.3 Clustering

It was found that when increasing the variability of the top results in a list of documents retrieved as an answer to a query, the performance of the retrieval system increases too. Thus, in some cases it is more desirable to have less but more varied items in the results list (Chen and Karger, 2006). In order to increase variability, a clustering system has been applied. This was also used in other systems with the same aim (Ah-Pine et al, 2009). The idea behind it is rather simple: re-arrange the most relevant documents so that documents belonging to different clusters are promoted to the top of the list.

The K–means algorithm was applied on each of the lists returned by the Lemur IR system. For this, the Rapid Miner tool was used[9]. The clustering algorithm tried to group these results into four different groups, without any concern about ranking. The number of groups was established on this value as documents in the training set have this average number of clusters specified in their metadata.

Once each of the documents in the list was labeled to its computed cluster index, the list was reordered according to the described principle: we fill the list by alternating documents from different clusters. In Figure 16.6 a graphical example of this approach is given.

The list obtained with the base case was reordered according to the method described. The aim of this experiment is to increment the diversity of the retrieved results using a clustering algorithm. Results were discouraging: when no reordering

[9] Available at `http://rapid-i.com/`

of documents in the list is performed a MAP of 0.4454 was reached, whereas a MAP of 0.2233 resulted from applying our clustering based approach.

16.4 The Medical Task

The main aim of the medical task is to retrieve medical images relevant to a given query. The query has several image examples and an associated text. The collection used to search relevant images has changed since 2005. Until 2007 the collection was very heterogeneous, with several subcollections. The subcollections without XML tags were processed to mark the structure of documents using XML. We used the SINTRAM tool to translate non–English text. Each subcollection is divided up into *cases* where a case is made up of one or various images (depending on the collection), along with an associated set of textual annotations. All the collections were processed to generate one textual document per image (Díaz-Galiano et al, 2006).

In 2007 a new collection was introduced, a subset of the Goldminer[10] collection. This collection contains images from articles published in *Radiology and Radiographics* including the text of the captions and a link to the Web page of the full text article. To create the different textual collections, first we have obtained the textual information by downloading all the articles from the Web. Then, we have filtered the articles to extract different sections (title, authors, abstract, introduction, etc.). Our experiments were conducted with the LEMUR retrieval information system, applying the KL-divergence weighting scheme (Ogilvie and Callan, 2001) and PRF.

16.4.1 Metadata Selection using Information Gain

The collection used until 2007 includes a large number of XML tags. The main problem was to choose the most useful data, discarding anything that might add non-relevant information (noise) to our system. In order to automate the tag selection process we have pre–processed the collections using Information Gain (IG) (Cover and Thomas, 2006). The XML tags were selected according to the amount of information supplied. For this reason, we have used the IG measure to select the best tags in the collection, using the following formula:

$$IG(C|E) = H(C) - H(C|E) \tag{16.1}$$

[10] `http://goldminer.arrs.org/`

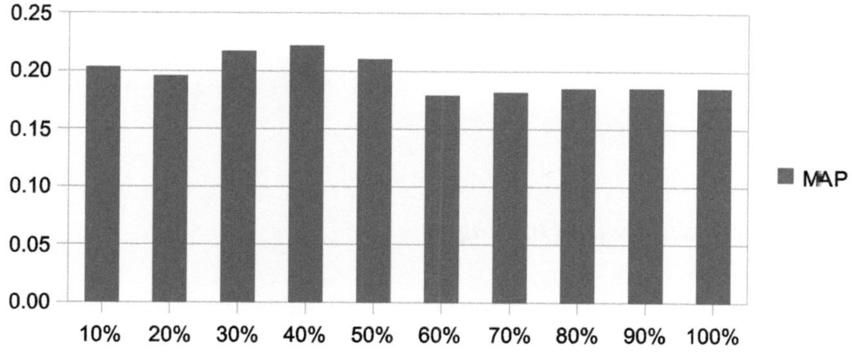

Fig. 16.7: Performance for Medical Image Retrieval in 2006.

where
C is the set of cases,
E is the value set for the E tag,
$IG(C|E)$ is the information gain for the E tag,
$H(C)$ is the entropy and of the set of cases C
$H(C|E)$ is the relative entropy of the set of cases C conditioned by the E tag

Both, $H(C)$ and $H(C|E)$ are calculated based on the frequencies of occurrence of tags according to the combination of words which they represent. The final equation for the computation of the information gain supplied by a given tag E over the set of cases C is defined as follows:

$$IG(C|E) = -\log_2 \frac{1}{|C|} + \sum_{j=1}^{|E|} \frac{|C_{e_j}|}{|C|} \log_2 \frac{1}{|C_{e_j}|} \quad (16.2)$$

where
C_{e_j} is the subset of cases in C having the tag E set to the value e_j (this value is a combination of words where order does not matter).

Since each subcollection has a different set of tags, the information gain was calculated for each subcollection individually. Then, the tags selected to compose the final collection are those showing high values of IG. We have accomplished several experiments preserving 10%, 20%...100% of tags. Figure 16.7 shows the values of MAP obtained for the Medical Image Retrieval task from 2006 using only textual information.

The results show that the collections with a low percentage of labels (between 30% and 50%) obtain the best performance, with a MAP value between 0.21 and 0.22. Therefore, this method reduces the size of the collections used and allows us to select the most significant labels within the corpus or, at least, those that provide better information. This selection system does not require external training or knowledge; it simply studies the importance of each label with regard to all the documents. Furthermore, this method is independent from the corpus as a whole since in our experiments the IG calculation has been done separately in each subcollection.

16.4.2 Expanding with Ontologies

We have experimented with two ontologies: MeSH and UMLS, performing several experiments with different expansion types. The best results have been obtained using synonyms and related terms.

To expand with the MeSH ontology we have used the *record* structure. Each record contains a representative term and a bag of synonyms and related terms. We consider that a term is a set of words (no word sequence order):

$$t = \{w_1, \cdots, w_{|t|}\} \tag{16.3}$$

where w is a word.

We have used the bag of terms to expand the queries. A bag of terms is defined as:

$$b = \{t_1, \cdots, t_{|b|}\} \tag{16.4}$$

Moreover, a term t exists in the query q ($t \in q$) if:

$$\forall w_i \in t, \exists w_j \in q / w_i = w_j \tag{16.5}$$

Therefore, if all the words of a term are in the query, we generate a new expanded query by adding all its bag of terms.

$$q \text{ is expanded with } b \text{ if } \exists t \in b / t \in q \tag{16.6}$$

In order to compare the words of a particular term to those of the query, all the words are put in lowercase and no stopword removal is applied. To reduce the number of terms that could expand the query, we have only used those that are in A, C or E categories of MeSH (A: Anatomy, C: Diseases, E: Analytical, Diagnostic and Therapeutic Techniques and Equipment): (Chevallet et al, 2006). Figure 16.8 shows an example of query expansion, with two query terms found in MeSH and their respective bags of terms.

On the other hand, to expand the queries with the UMLS meta–thesaurus, we have used the MetaMap program (Aronson, 2001) that was originally developed for information retrieval. MetaMap uses the UMLS meta–thesaurus for mapping concepts from an input text. For query expansion with MetaMap, we have mapped the terms from the query. As carried out with MeSH, in order to restrict the categories of terms that could expand the query, we have restricted the semantic types in the mapped terms (Chevallet et al, 2006) as follows:

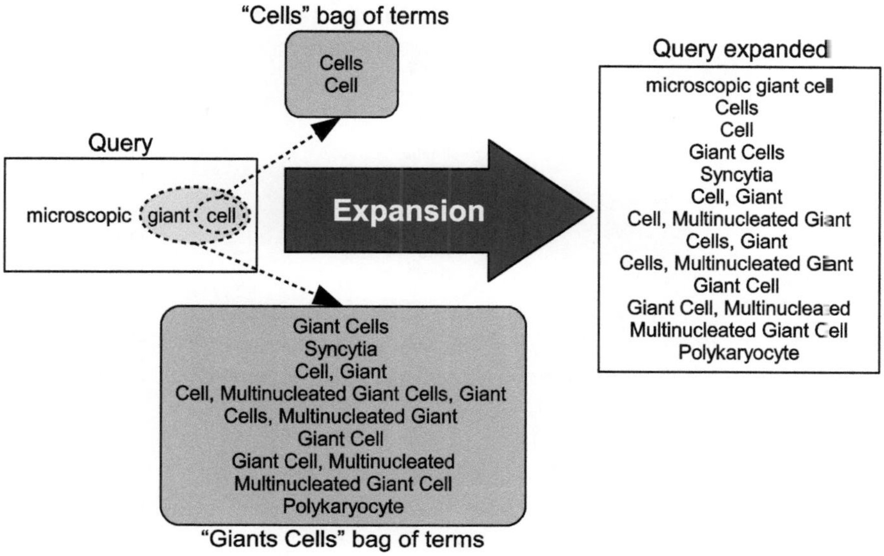

Fig. 16.8: Example of query expansion with MeSH ontology.

- bpoc: Body Part, Organ, or Organ Component;
- diap: Diagnostic Procedure;
- dsyn: Disease or Syndrome;
- neop: Neoplastic Process.

MetaMap gives two types of mapped terms: *Meta Candidates* and *Meta Mapping*. The difference between both mapped terms is that the second are the Meta Candidate with best score. For our expansion we have used the Meta Candidate terms, because these provide similar terms with differences in the words (Díaz-Galiano et al, 2006).

Prior to the inclusion of Meta Candidates terms in the queries, the words of the terms are added to a set where repeated words are deleted. All words in the set are included in the query. Figure 16.9 shows a example of query expansion using UMLS.

The organizers of the ImageCLEF medical task provided the *ImageCLEF Consolidated Test Collection* (Hersh et al, 2009). This collection combines all the collections, queries and relevance judgements used in ImageCLEFmed from 2005 to 2007. We have used this new collection to experiment with MeSH and UMLS query expansion. On the other hand, to experiment with the 2008 collection we have generated three different collections. In these collections each document contains information about each image from the original collection. The information is different for each collection. These collections are defined as follows:

- **CT**: contains *caption* of image and *title* of the article.
- **CTS**: contains *caption*, *title* and text of the *section* where the image appears.
- **CTA**: contains *caption*, *title* and text of the full *article*.

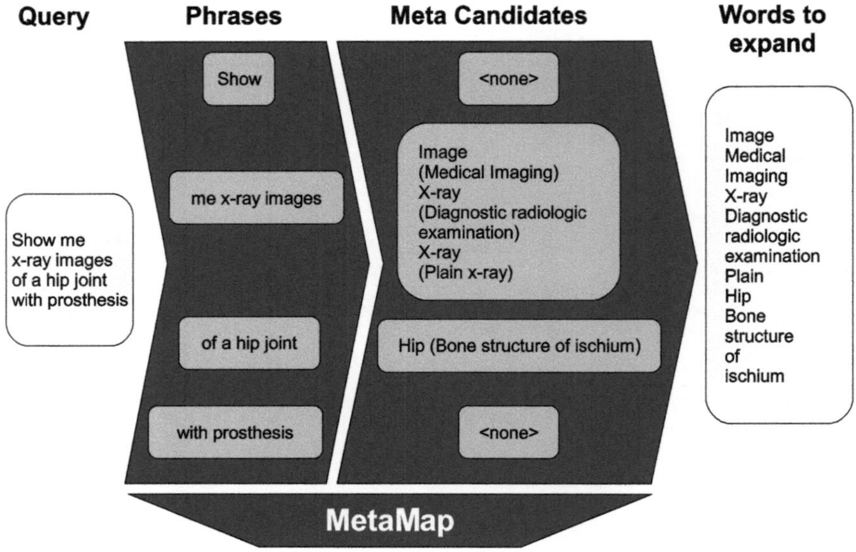

Fig. 16.9: Example of query expansion with UMLS ontology.

Table 16.4: MAP values of query expansion experiments.

Expansion	CT	CTS	CTA	Consolidated
Base	0.2480	**0.1784**	0.1982	0.2039
MeSH	**0.2792**	0.1582	**0.2057**	**0.2202**
UMLS	0.2275	0.1429	0.1781	0.1985

Table 16.4 shows the results obtained in experiments on these collections.

The MeSH expansion obtained better results than no expansion or UMLS expansion. In 2008 the University of Alicante group obtained the bests results (Navarro et al, 2008) in the textual task using a similar MeSH expansion and negative feedback. The Miracle group performed a MeSH expansion in documents and topics using the hyponyms of UMLS entities (Lana-Serrano et al, 2008) but the results obtained are worse than the baseline results. In short, the use of UMLS expansion obtained worse results than the baseline. Although the UMLS meta–thesaurus includes the MeSH ontology in the source vocabularies, MetaMap adds, in general, more terms in the queries. The MetaMap mapping was different from MeSH mapping, therefore the terms selected to expand were not the same.

One conclusion is that it is better to have less but more specific textual information. Also, including the whole section where the image appears was not a good approach. Sometimes a section contains several images, therefore the same information references different images.

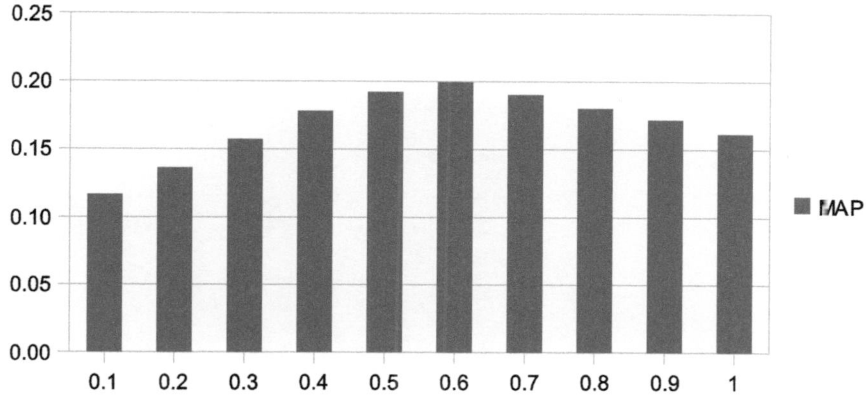

Fig. 16.10: Performance of experiments in 2005 with visual and textual fusion.

16.4.3 Fusion of Visual and Textual Lists

The fusion experiments merge the ranked lists from both systems (visual and textual) in order to obtain one final list (FL) with relevant images ranked by relevance. The merging process was accomplished giving different importance to the visual (VL) and textual lists (TL):

$$FL = TL * \alpha + VL * (1 - \alpha) \tag{16.7}$$

In order to adjust these parameters some experiments were accomplished varying α in the range [0,1] with step 0.1 (i.e.: 0.1, 0.2,...,0.9 and 1).

The next figures show the results obtained on different collections used in the ImageCLEF medical task. Figure 16.10 shows experiment results with the 2005 collection. Results with the 2007 collection are presented in Figure 16.11.

The results obtained show that the combination of heterogeneous information sources (textual and visual) improves the use of a single source. Although textual retrieval on its own overcomes visual retrieval, when used jointly the results are better than those obtained from independent retrievals.

16.5 Conclusion and Further Work

In this chapter, we have described our participation in ImageCLEF from 2005 to present. We have presented a summary of different systems developed for the photo and medical retrieval tasks.

For the photo retrieval system we have tested multiple resources and techniques: different IR systems, weighting schemes, pseudo relevance feedback, ma-

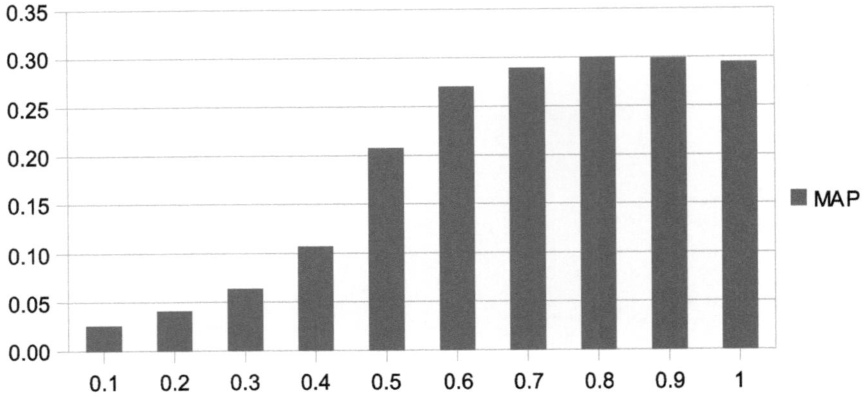

Fig. 16.11: Performance of experiment in 2007 with visual and textual fusion.

chine translators, filtering methods and clustering to increase diversity in the results. The experiments show that the translation of non–English queries introduces a loss of MAP that depends on the source language, although those multi–lingual runs achieved almost the best results in the competition. Our IR system works well, in general, with the Okapi weighting function. In addition, the application of query expansion and PRF improved the results. The applied filtering method shows that the cluster terms given in the query are not useful to filter the relevant list of images, and the applied clustering method obtained poor results in terms of MAP. However, the diversity of the relevant images was increased, so further research should be conducted on this issue.

In our future work with the photo retrieval system, we will improve the machine translation subsystem, including a new translator and heuristics to combine the results. New filtering methods are ruled out for the time being, because we are developing a new clustering module that introduces diversity in the results but taking into account the score and position of the documents in original the ranked list.

Regarding the medical task, we have applied Information Gain in order to filter tags in the collections. The best results have been obtained using around 30%-50% of the tags. In addition, it has been found that the application of fusion techniques to combine textual and visual information improves the system. Finally, several query expansion techniques have been tested using two medical resources: MeSH and UMLS. The experiments show that the expansion with less and more specific terms improves the results.

As future work we will study which resources from UMLS are more convenient for term expansion. In addition, we are interested in detecting when the query expansion is useful to improve the final results.

Acknowledgements This work has been partially supported by a grant from the Spanish Government, project TEXT-COOL 2.0 (TIN2009-13391-C04-02), a grant from the Andalusian Government, project GeOasis (P08-TIC-41999), and two grants from the University of Jaén, project RFC/PP2008/UJA-08-16-14 and project UJA2009/12/14.

References

Ah-Pine J, Bressan M, Clinchant S, Csurka G, Hoppenot Y, Renders JM (2009) Crossing textual and visual content in different application scenarios. Multimedia Tools and Applications 42(1):31–56

Aronson AR (2001) Effective mapping of biomedical text to the UMLS metathesaurus: the metamap program. In: Proceedings of the AMIA Symposium, pp 17–21

Chen H, Karger DR (2006) Less is more: probabilistic models for retrieving fewer relevant documents. In: Efthimiadis EN, Dumais ST, Hawking D, Järvelin K (eds) Proceedings of the SIGIR conference, ACM press, pp 429–436

Chevallet JP, Lim JH, Radhouani S (2006) A structured visual learning approach mixed with ontology dimensions for medical queries. In: Accessing Multilingual Information Repositories, Springer, Lecture Notes in Computer Science (LNCS), pp 642–651

Cover T, Thomas J (2006) Elements of information theory. Wiley–Interscience

Díaz-Galiano M, García-Cumbreras M, Martín-Valdivia M, Montejo-Raez A, Ureña López L (2006) SINAI at ImageCLEF 2006. In: Working Notes of CLEF 2006

García-Cumbreras MA, Urena-López LA, Martínez-Santiago F, Perea-Ortega JM (2007) BRUJA System. The University of Jaén at the Spanish task of QA@CLEF 2006. In: Lecture Notes in Computer Science (LNCS), Springer, vol 4730, pp 328–338

Gómez-Soriano JM, Montes-y-Gómez M, Sanchis-Arnal E, Rosso P (2005) A Passage Retrieval System for Multilingual Question Answering. In: 8th International Conference of Text, Speech and Dialogue 2005 (TSD 2005), Springer, Lecture Notes in Artificial Intelligence (LNCS), pp 443–450

Hersh WR, Müller H, Kalpathy-Cramer J (2009) The ImageCLEFmed medical image retrieval task test collection. Journal of Digital Imaging 22(6):648–655

Lana-Serrano S, Villena-Román J, González-Cristóbal J (2008) MIRACLE at ImageCLEFmed 2008: Evaluating Strategies for Automatic Topic Expansion. In: Working Notes of CLEF 2008

Navarro S, Llopis F, Muñoz R (2008) Different Multimodal Approaches using IR–n in Image-CLEFphoto 2008. In: Working Notes of CLEF 2008

Ogilvie P, Callan JP (2001) Experiments using the lemur toolkit. In: Proceedings of TREC

Porter MF (1980) An algorithm for suffix stripping. In: Program 14, pp 130–137

Salton G, Buckley G (1990) Improving retrieval performance by relevance feedback. Journal of American Society for Information Sciences 21:288–297

Chapter 17
Leveraging Image, Text and Cross–media Similarities for Diversity–focused Multimedia Retrieval

Julien Ah-Pine, Stephane Clinchant, Gabriela Csurka, Florent Perronnin, and Jean-Michel Renders

Abstract This chapter summarizes the different cross–modal information retrieval techniques Xerox Research Centre implemented during three years of participation in ImageCLEF Photo tasks. The main challenge remained constant: how to optimally couple visual and textual similarities, when they capture things at different semantic levels and when one of the media (the textual one) gives, most of the time, much better retrieval performance. Some core components turned out to be very effective all over the years: the visual similarity metrics based on Fisher Vector representation of images and the cross–media similarity principle based on relevance models. However, other components were introduced to solve additional issues: We tried different query– and document–enrichment methods by exploiting auxiliary resources such as Flickr or open–source thesauri, or by doing some statistical 'semantic smoothing'. We also implemented some clustering mechanisms in order to promote diversity in the top results and to provide faster access to relevant information. This chapter describes, analyses and assesses each of these components, namely: the monomodal similarity measures, the different cross–media similarities, the query and document enrichment, and finally the mechanisms to ensure diversity in what is proposed to the user. To conclude, we discuss the numerous lessons we have learnt over the years by trying to solve this very challenging task.

17.1 Introduction

Information, especially digital information, is no longer monomodal: Web pages can have text, images, animations, sound and video; audiobooks, photoblogs and videocasts are typical examples of multi–modal materials; valuable content within a photo sharing site can be found in tags and comments as much as in the actual vi-

Xerox Research Centre Europe
6 ch. de Maupertuis, 38240 Meylan, France
e-mail: FirstName.LastName@xrce.xerox.com

H. Müller et al. (eds.), *ImageCLEF*, The Information Retrieval Series 32,
DOI 10.1007/978-3-642-15181-1_17,

sual content. Nowadays, it is difficult to visit a page within a popular social network without finding a large variety of content modes surrounded by a rich structure of social information such profiles, interest groups, consumer behavior or simple conversations. This major shift in the way we access content, and the type of content we access, is largely due to the connected, easily accessible, global nature of the Internet. The democratization of the tools of production and delivery has strongly contributed to this phenomenon, one example being low cost camera–phones combined with accessible publishing tools. Such a scenario raises a strong need for tools that enable user interaction with multi–modal information.

The scientific challenge is to understand the nature of the interaction between these modalities, and in particular between text and images. How can text be associated with an image (and reciprocally an illustrative image with a text)? How can we organize and access text and image repositories in a better way than naive late fusion techniques? The main difficulty lies in the fact that visual and textual features are expressed at different semantic levels.

Naive techniques combine the scores from both text and image retrieval systems into a single relevance score: this is the late fusion approach. Departing from the classical late fusion strategy, recent approaches have considered fusion at the feature level (early fusion), estimating correspondences or joint distributions between components across the image and text modes from training data.

One of the first approaches in this family is the co–occurrence model by Mori et al (1999) where keywords are assigned to patches based on the co–occurrence of clustered image features and textual keywords in a labeled training data set. A quite similar approach proposes to find correlations between images and linked texts using Kernel Canonical Correlation Analysis (Vinokourov et al, 2003). With the development of image representation based on visual vocabularies (Sivic and Zisserman, 2003; Csurka et al, 2004), somewhat similar to textual vocabularies, new techniques appeared such as Probabilistic Semantic Analysis (Barnard et al, 2003; Monay and Gatica-Perez, 2004) or Latent Dirichlet Allocation (Blei et al, 2003). They propose to extract latent semantics from images. Machine translation models inspired Duygulu et al (2002); Iyengar et al (2005), who generalized these models to images, where the translation is done between words and image regions. Another group of work uses graph models to represent the structure of an image through a graph. For instance, Carbonetto et al (2004); Li and Wang (2003) build a Markov network to represent interactions between blobs (Carbonetto et al, 2004; Li and Wang, 2003), while Pan et al (2004) use a concept graph (Pan et al, 2004).

The use of pseudo–relevance feedback or any related query expansion mechanism has been widely used in Information Retrieval. Several works inspired by cross–lingual retrieval systems were proposed in this direction. In cross–lingual systems, a user generates his query in one language (e.g. English) and the system retrieves documents in another language (e.g. French). The analogy here is to consider the visual feature space as a language constituted of blobs or patches, simply called *visual words*.

Hence, based on query expansions models, Jeon et al (2003) proposed to extend the cross–lingual relevance models to cross–media relevance models. These models

Table 17.1: Notations.

Notation	Description
N	Number of documents in the collection
d	A document of the collection
d^T, d^V	The textual and visual part (image) of d
S	A matrix of similarities between documents
S^T, S^V	A matrix of text-based or image-based similarities
S^{VT}, S^{TV}	A matrix of cross-modal image-text or text-image similarities
q	A query
s_q	A similarity vector between the query and the documents
s_q^T, s_q^V	A text-based and image-based similarity vector
s_q^{VT}, s_q^{TV}	A cross-modal image-text and text-image similarity vector

were further generalized to continuous features by Lavrenko et al (2003) with non–parametric kernels, while Feng et al (2004) modeled the distribution of words with Bernoulli distributions.

The trans–media relevance model we describe in this chapter (see Section 17.4) can also be seen as a cross–media relevance model. The basic idea is to first use one of the media types to gather relevant multimedia information and then, in a second step, use the dual type to perform the final task (retrieval, annotation, etc). These approaches can be seen as an 'intermediate level' fusion since the media fusion takes place after a first mono–media retrieval step based on monomodal similarities (see Sections 17.2 and 17.3).

This chapter is structured in four sections: (i) visual methods, (ii) textual methods, (iii) cross-media similarities, and (iv) diversity–focused retrieval. For each of these sections, we discuss the main algorithms and show a few experimental results. Then, we draw partial conclusions on these methods before moving on to the next family of techniques. The thread of the presentation goes along with the performance of the presented technology: visual methods have generally lower performance than textual ones. Similarly, textual methods are outperformed by cross–media techniques. Finally, methods addressing diversification of the top results, to offer a better user experience, are built upon the cross–media ones. For a better understanding of different sections, we summarize our main notations in Table 17.1.

17.2 Content–Based Image Retrieval

Content–Based Image Retrieval (CBIR), also known as Query By Image Content (QBIC), consists of the application of computer vision to the image retrieval problem; that is, the problem of searching for digital images in large databases based on visual retrieval as opposed to the text– or tag–based retrieval of images. The term *content–based* means that the search analyzes the visual content of the image, where content in this context might refer to colors, shapes, textures, or any other piece of information that can be derived from the image itself. The process in-

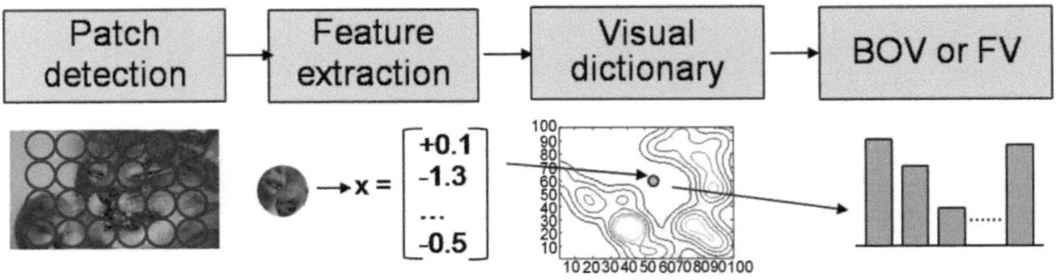

Fig. 17.1: The main steps to obtain BOV or Fisher Vector representation of images.

volves computing a feature vector for the unique characteristics of the image. While in early CBIR systems global features or rather low–level features were mainly used, recent systems tend to extract these features more locally and transform them to some higher level representations. One of the most successful approaches to transform low level image descriptors to 'higher' level descriptors is the bag–of–visual–words (BoW) representation of the images (Sivic and Zisserman, 2003, Csurka et al, 2004). When the visual vocabulary is represented by a probability density, the Fisher kernel framework proposed by Jaakkola and Haussler (1999) is applicable and the image can be represented by Fisher Vectors as proposed by Perronnin and Dance (2007).

17.2.1 Fisher Vector Representation of Images

If a probability density function (in our case a Gaussian Mixture Model or GMM) is used to model the visual vocabulary as an intermediate representation in the feature space, we can represent an image by the gradient of the log–likelihood with respect to the parameters of the model. The Fisher Vector (FV) is the concatenation of these partial derivatives and describes in which direction the parameters of the model should be modified to best fit the data (extracted image features). While this kind of representation was heavily used for image categorization, it is actually class–independent and hence suitable for image retrieval too.

The mains steps to obtain such representations are illustrated in Figure 17.1. First, local patches are either detected using interest point detectors, low level image segmentation, or simply regular sampling. Then, low–level features are computed on those patches such as color and texture histograms, Scale Invariant Feature Transform (SIFT), shape features, etc. In our experiments we sampled patches on regular grids at multiple scales and computed histograms of oriented gradients (HOG) and local color statistics (RGB means and standard deviations). The *Visual Vocabulary* can be built on a set of patches extracted from a randomly selected set of images using, for example, K–means, Mean Shift, GMMs or Random Forest. The high–

level image signature is computed by accumulating word occurrences (BoW) or by building the Fisher Vectors as described below.

In our case the visual vocabulary is a GMM with parameters $\Phi = \{\omega_m, \mu_m, \sigma_m, i = 1...M\}$ [1] trained on a set of features extracted from images to estimate their distribution in the low–level feature space:

$$p(x|\Phi) = \sum_{m=1}^{M} \omega_m p_m(x|\Phi). \tag{17.1}$$

Here, each Gaussian component $\mathcal{N}(\mu_m, \sigma_m)$ can be seen as the representation of a visual word and given a new low–level feature x_l, the probability that it was generated by the Gaussian m is:

$$\gamma_m(x_l) = \frac{\omega_m p_m(x_l|\Phi)}{\sum_{m=1}^{M} \omega_m p_m(x_l|\Phi)}. \tag{17.2}$$

In the BoW representation of the image, the low–level descriptor x_l is then transformed into a high–level M–dimensional descriptor as follows:

$$\gamma(x_l) = [\gamma_1(x_l), \gamma_2(x_l), \ldots, \gamma_M(x_l)] \tag{17.3}$$

To get a global signature (BOV) for an image or more generally the visual part of a document represented by a set of extracted low level image features $d^V = \{x_l, l = 1 \ldots L\}$, we simply average $\gamma(x_l)$ over l.

The Fisher Vector is an alternative to this BoW image representation based on the Fisher kernel framework proposed by Jaakkola and Haussler (1999). The main idea is to consider the gradient vector of the log–likelihood according to the parameters of Φ. Assuming that the x_l's were generated independently by Φ, we can write this log–likelihood as follows:

$$\log p(d^V|\Phi) = \frac{1}{L} \sum_{l=1}^{L} \nabla_\Phi \log p(x_l|\Phi). \tag{17.4}$$

We consider the gradients of $\log p(x_l|\Phi)$ with respect to the mean and standard deviation parameters (the gradient with respect to the weight parameters brings little additional information) and as suggested by Perronnin and Dance (2007), we further normalize them by the Fisher Information matrix (having a whitening effect on different dimensions):

$$F_\Phi = E_{d^V}\left[(\nabla_\Phi \log p(d^V|\Phi))\,(\nabla_\Phi \log p(d^V|\Phi))^T\right].$$

In the case of diagonal co–variance matrices and an approximation of the Fisher Information matrix, we obtain the following closed form formulas (see details in (Perronnin and Dance, 2007)):

[1] We consider diagonal covariance matrices and we denote by σ_m^2 the variance vector.

$$f_{\mu_m^r}(x_l) = \frac{\sigma_m^r}{\sqrt{\omega_m^r}} \frac{\partial \log p(x_l|\Phi)}{\partial \mu_m^r} = \gamma_m(x_l) \frac{x_l^r - \mu_m^r}{\sigma_m^r \sqrt{\omega_m^r}}, \tag{17.5}$$

$$f_{\sigma_m^r}(x_l) = \frac{\sigma_m^r}{\sqrt{2\omega_m^r}} \frac{\partial \log p(x_l|\Phi)}{\partial \sigma_m^r} = \gamma_m(x_l) \frac{(x_l^r - \mu_m^r)^2 - (\sigma_m^r)^2}{(\sigma_m^r)^2 \sqrt{2\omega_m^r}}. \tag{17.6}$$

where the superscript $r, r = 1 \dots R$ denotes the r–th dimension of a vector and R is the dimensionality of the feature space. The Fisher Vector $f_\Phi(x_l)$ of the observation x_l is the concatenation of all these partial derivatives leading to a $2 * M * R$ dimensional vector. Finally, to obtain the image representation $f_\Phi(d^V)$, we take the average over the Fisher Vectors from all the extracted patches x_l, $l = 1..L$.

We define the visual similarity between two visual documents d_1^V and d_2^V by using the L1–distance between the L1–normalized Fisher Vectors:

$$s^V(d_1^V, d_2^V) = -\|\tilde{f}_\Phi(d_1^V) - \tilde{f}_\Phi(d_2^V)\|_1 \tag{17.7}$$

where $\tilde{f}_\Phi(d_i^V)$ is $f_\Phi(d_i^V)$ after normalized it to L1–norm equal to 1.

17.2.2 Image Retrieval at ImageCLEF Photo

We used the Fisher vector–based image retrieval described above in our ImageCLEF photo retrieval experiments. Actually, as we used two types of low level features, we built two independent visual vocabularies, one for color features (local RGB statistics) and one for texture (orientation histograms). Therefore, before computing the similarity between two images using Equation 17.7 we first concatenated the two Fisher Vectors (texture and color one).

One specificity of the ImageCLEF photo retrieval challenge, compared to the classical query image based retrieval, is that for each topic there is not one, but several query images q_i^V, ($i = 1,..M$, where M is generally 3). Therefore, the main question we can ask is how to combine the information from different images to get better retrieval performance. We investigated three different strategies:

- **I1** : We considered the mean of the M Fisher Vectors (this can be seen as the concatenation of the M set of patches q_i^V into single q^V one) and used this mean Fisher Vector to query the database.
- **I2** : The database images were ranked according to each image independently and the M ranked list was combined using round–robin type selection (i.e. intermixing the M lists) and eliminating the repetitions.
- **I3** : We combine the three similarity scores (with respect to each image of the query) by averaging the scores after Student normalization.

Table 17.2 compares these three strategies on the IAPR TC–12 database used in the ImageCLEF photo retrieval 2007 and 2008. It shows results on the 39 topics created for 2008. We can see that the early fusion (mean Fisher Vector) performs worse than late score level fusions. The reason might be that the different query images

Table 17.2: Performances of different strategies for image retrieval.

Run description	MAP	P@20
I1	0.119	0.255
I2	0.130	0.301
I3	**0.151**	**0.328**

contained complementary information, and searching for images that are similar to all of them was not the best option.

While our method was the best performing CBIR system in both sessions (2007 and 2008), the overall performance is quite poor. It is therefore natural to also exploit the textual modality (as most commercial image retrieval systems do), which is done in the next section.

17.3 Text Representation and Retrieval

In this section, we summarize the techniques used during our participation in the different ImageCLEF photo sessions. Overall, we used state–of–the–art information retrieval methods: the language modeling approach[2]. The following paragraph will detail this textual information retrieval approach. We also successfully explored query expansion techniques that are described in Section 17.3.2.

17.3.1 Language Models

First the text is pre–processed: this includes tokenization, lemmatization, word decompounding and standard stopword removal. Then, starting from a BoW representation (assuming independence between words), we adopt the language modeling approach to information retrieval. The core idea is to model a document d^T by using a multinomial distribution over the words denoted by the parameter vector θ_d^T. A simple Language Model (LM) can be obtained considering the frequency of words in d^T (corresponding to the maximum likelihood estimator):

$$P_{ML}(w|d^T) = \frac{\#(w,d^T)}{|d^T|} .$$

where $\#(w,d^T)$ is the number of occurrences of word w in d^T and $|d^T|$ is the length of d^T in tokens. The probabilities should be further smoothed by the corpus language model:

[2] Other models to represent the texts such as the BM25 and DFR models could also be used in principle without altering our results significantly.

$$P_{ML}(w|C) = \frac{\sum_d \#(w, d^T)}{|C|}$$

using the Jelinek–Mercer interpolation:

$$\theta_{d^T,w} = \lambda\, P_{ML}(w|d^T) + (1-\lambda)\, P_{ML}(w|C)\,. \tag{17.8}$$

Using this language model, we can define the similarity between two documents using the cross–entropy function:

$$S^T(d_1^t, d_2^t) = \sum_w P_{ML}(w|d_1^t) \log(\theta_{d_2^t,w})) \tag{17.9}$$

17.3.2 Text Enrichment at ImageCLEF Photo

In this section, we introduce the different text enrichment techniques used during the different sessions. In fact, there are several incentives to enrich text associated to images:

- The relative sparsity of the textual representation of the photos. Textual representations of photos are usually short texts. At best, they consist of a single paragraph and at worst, images have simply very few tags. Overall, textual annotations of images are shorter than standard documents used in text collections, such as Web documents or news articles.
- The gap between lexical fields of these descriptions and the queries : queries may be expressed in a more abstract way than factual descriptions of the photos.
- Textual queries are short, often shorter than what is considered *short* for classical information retrieval benchmarks. An image and a short text can be considered as the equivalent of long queries for classical text information retrieval. Thus, queries may need some expansion to exploit associated concepts or words relevant to the queries in order to get a better recall.

In the following, the different text–enrichment mechanisms used in 2007, 2008 and 2009 are described. In short, Flickr–related tags served to enrich documents in 2007. Then, we experimented on document enrichment with the Open Office thesaurus and visual concepts. Lastly, co–occurrence measures between words were used to expand textual queries in 2009.

17.3.2.1 Year 2007: Enriching Text with Flickr

Motivated by the fact that the textual content of the documents was very poor that year (text annotations were limited to the <TITLE> fields of documents), we decided to enrich the corpus thanks to the Flickr database [3], at least for texts in English.

[3] http://www.flickr.com/services/api/

Table 17.3: Corpus Terms and their related terms from Flickr.

Corpus Term	Top 5 related Terms
Jesus	christ, church, cross, religion, god
classroom	school, class, students, teacher, children
hotel	lasvegas, building, architecture, night
Riviera	france, nice, sea, beach, french
Ecuador	galapagos, quito, southamerica, germany, worldcup

The Flickr Application Programming Interface (API) provides a function to get tags related to a given tag [4]. According to the Flickr documentation, this function returns a list of tags related to the given tag, based on clustered usage analysis. It appears that queries, on the one hand, and photographic annotations on the other hand, adopt a different level of description. Queries are often more abstract and more general than annotations. As a consequence, it is easier and more relevant to enrich the annotations than the queries: related tags are often at the same level or at the upper (more general) semantic level. Table 17.3 shows some examples of enrichment terms, related to the annotation corpus. We can observe that the related terms do encode a kind of semantic similarity, often towards a more abstract direction, but also contain noise or ambiguities.

Below is an example of an enriched document where each original term has been expanded with its top 20 related terms:

DOCNO: annotations/00/116.eng
ORIGINAL TEXT: Termas de Papallacta Papallacta Ecuador
ADDED TERMS: chillan colina sur caracalla cajon piscina snow roma italy maipo thermal nieve volcan argentina mendoza water italia montana araucania santiago quito southamerica germany worldcup soccer football bird andes wm church fifa volcano iguana cotopaxi travel mountain mountains cathedral sealion market

Enriching the text corpus partially solved the term mismatch but it also introduced a lot of noise in a document. In order to limit this noise phenomenon, the expansion can be controlled by a convex combination of the original document language model and a language modeling on the new words. (for example 0.8 for the original document and 0.2 for the news terms). Hence, most of the probabilistic mass of the language model is devoted to the the original text of a document.

17.3.2.2 Year 2008: Enriching Text with Visual Concepts and the Open Office Thesaurus

In 2008, we investigated the use of external resources in order to enrich text. Another issue that we wanted to address was the use of the visual concepts provided by the organizers as extra textual words, refining the original textual representation of the photo by higher–level visual information.

[4] `http://www.flickr.com/services/api/flickr.tags.getRelated.html`

Table 17.4: Performance of different text enrichment strategies.

	Without PRF		With PRF	
Run Description	MAP	P@20	MAP	P@20
Baseline	0.215	0.259	0.239	0.293
Document Enrichment	0.231	0.268	0.260	0.308
Query Enrichment	0.218	0.264	0.257	0.282

The first variant we developed consisted in exploiting the English Open Office thesaurus[5] to enrich the textual description of the photos and/or the queries. Several strategies can be chosen. We chose the following ones:

- Document enrichment: we added all synonyms and broader terms to the terms of the original description, when they are covered by a thesaurus entry. To give more weight to the original terms, they were artificially replicated 15 times.
- Query enrichment: we added all the synonyms and narrower terms to the terms of the original description, when available. To give more weight to the original terms, they were artificially replicated five times.

Note that we simultaneously enriched both the queries and the documents, but this resulted in performance deterioration (too much noise introduced).

As Pseudo–Relevance Feedback (PRF) is another way to do query expansion, we systematically ran experiments with and without pseudo–relevance feedback for each setting (baseline, document enrichment, query enrichment). The top ten terms of the top ten documents were used to expand the initial query language model by convex linear combination (coefficient=0.6 for the feedback model). Query model updating was based on the mixture model method (Zhai and Lafferty, 2001). The performance (MAP and Precision20) is given in Table 17.4.

It clearly appears that combining document enrichment with a thesaurus and query expansion by PRF (using the thesaurus–enriched documents in the first feedback phase) gives the best results. Performing semantic query enrichment followed by PRF (using the thesaurus–enriched query in the first feedback phase) gives slightly worse results. In any case, the use of this external resource is beneficial with respect to a standard PRF query expansion.

The second variant we developed aimed at assessing the benefits of introducing automatically detected visual concepts. These concepts were generated by the two best image categorization systems in the ImageCLEF visual concept detection task (Deselaers and Hanbury, 2008), from XRCE and RWTH and provided by the organizers for the visual photo retrieval task. Note that the XRCE method used the Fisher Vector image representation as described in Section 17.2.1.

The approach to combine these visual concepts with the text was as follows: we enriched both the documents and the queries with the visual concepts (e.g. indoor, outdoor, building, sky, night, animal, etc.) automatically associated with the images and built language models with the enriched texts. Then, we applied our retrieval

[5] Available on `http://wiki.services.openoffice.org/wiki/Dictionaries/`

Table 17.5: Performance of the combinations with automatically detected visual concepts.

	Without PRF		With PRF	
Run description	MAP	P@20	MAP	P@20
Baseline	0.215	0.259	0.239	0.293
XRCE Visual Concepts	0.241	0.297	0.269	0.334
RWTH Visual Concepts	0.232	0.271	0.258	0.308

model as described above. This can be considered as a simplistic way of doing multimedia retrieval. The obtained performance (MAP and Precision20) is given in Table 17.5. We can see clearly, that using the visual concepts increases the retrieval performance. However, as shown later, this performance is far below the results we can obtain with cross–media similarity measures (MAP=0.44, P@20=0.57).

17.3.2.3 Year 2009: Enriching Text with Lexical Entailment/Term Similarity

In 2009, textual queries were very short with a typical length of one or two words. In general, single keyword queries can be ambiguous. Query expansion techniques can help in finding several meanings or different contexts of the query word. As one of the goals was to promote diversity for the photo retrieval task, query expansion methods could help in finding new clusters. In fact, if a term has several meanings or different contexts, the most similar words to this term should partially reflect the diversity of related topics associated to it. The Chi–square statistics were used to measure the similarity between two words (Manning et al, 2000), although any other term similarity measure or lexical entailment measure could be used.

Hence, for each query word q_w, we computed the Chi–square statistics of the latter with all other words (including q_w). We kept only the top ten words and divided the scores by the maximum value (given by the inner statistic of q_w with itself). Table 17.6 displays, for some query terms, the most similar terms with the renormalized Chi–square statistics. To illustrate that co–occurrence measures can handle diversity of word senses, one can look at the most similar terms of the term *euro*. The most similar terms bear the notion of lottery, currency or football event, which were all relevant and richer than the themes indicated by the topic images (currency and euro stadium).

To sum up our models representing texts, we used standard language modeling to compute what we refer to as textual similarities. Over the years, we have also tried to compensate for the relative sparsity of texts, whether documents or queries, with the help of external resources or co–occurrence techniques. These enrichment techniques all improved the performance of the monomodal textual system. However, when the image queries are also taken into account, their impact is moderate and depends heavily on the task and the collection.

Table 17.6: Query Terms and their most similar terms.

obama 1	strike 1	euro 1
barack 0.98	hunger 0.04	million 0.05
springfield 0.16	protest 0.02	billion 0.05
illinois 0.16	worker 0.01	currency 0.03
senator 0.09	caracas 0.01	2004 0.03
freezing 0.08	led 0.01	coin 0.02
formally 0.08	venezuela 0.01	devil 0.02
ames 0.07	chavez 0.001	qualify 0.02
democrat 0.06	nationwide 0.001	qualification 0.02
paperwork 0.04	retaliatory 0.001	profit 0.01

17.4 Text–Image Information Fusion

Understanding the nature of the interaction between text and images is a real scientific challenge that has been extensively studied over the last few years. The main difficulty is to overcome the semantic gap and especially the fact that visual and textual features are expressed at different semantic levels. Here we describe the cross–media similarity measure we developed and successfully applied in the context of the ImageCLEF photo evaluation (multi–modal photo retrieval).

The main idea was to go beyond naive techniques that combine the scores from a text retrieval system and from an image retrieval system into a single relevance score (a.k.a. late fusion approach). We also wanted to avoid the early fusion models, as exploiting the correlations between the different modalities is generally quite complex and has shown rather poor performance in the past due to to variations in their level of semantic meaning (words vs. low level image features), and in dimensionality.

Our method was inspired by the trans–media pseudo feedback proposed in (Chang and Chen, 2006; Maillot et al, 2006; Clinchant et al, 2007), which is an extension of pseudo–relevance feedback, where the first retrieval step is done in one modality (e.g. textual), then the media type is switched to the other modality (e.g. visual), and the new query process is done in this new modality with a query built with the top retrieved documents in the first step. These models have shown significant improvement on retrieval performance in multi–modal databases (Clinchant et al, 2007; Ah-Pine et al, 2009c).

Cross–media similarities draw their inspiration from the trans–media relevance feedback method. However, instead of extracting words (i.e. features) with a pseudo feedback method to build a new query, cross–media similarities directly combine the monomodal similarities. They can be understood as a diffusion process of similarities, or as a particular kernel combination. These cross–media similarities are described in the next section. They were at the heart of our runs submitted in 2007, 2008 and 2009 and have proven their effectiveness (Clinchant et al, 2007; Ah-Pine et al, 2008, 2009b).

17.4.1 Cross–Media Similarities

This method, introduced by Clinchant et al (2007), assumes that two similarity matrices S^T and S^V over the same set of multimedia objects denoted $d_i = (d_i^T, d_i^V); i = 1, \ldots, N$ are precomputed on the database. The former matrix S^T is related to textual based similarities whereas the latter matrix S^V is based on visual similarities and they are both NxN matrices. Typically, we use Equation 17.9 to compute S^T and Equation 17.7 to compute S^V; however, any other textual or visual similarity can be used. Both matrices were normalized such that the proximity measure distribution of each row varies between 0 and 1.

Let us denote by $\kappa(S,k)$ the thresholding function that, for all rows of S, puts to zero all values that are lower than the kth highest value and keeps all other components to their initial value.

Accordingly, we define the cross–media similarity matrices that combine two mono–media similarity matrices as follows:

$$S^{VT} = \kappa(S^V, k^V).S^T \tag{17.10}$$

$$S^{TV} = \kappa(S^T, k^T).S^V \tag{17.11}$$

where . designates the standard matrix product. Note that the number (k^T and k^V) of the top values according to the textual, respectively visual similarities can be different. This intermediate fusion method can be seen as a graph similarity mixture through a two–step diffusion process, the first step being performed in one mode and the second step being performed in the other one (Ah-Pine et al, 2008, 2009a). This method is depicted in Figure 17.2.

In the more specific case of information retrieval, we are given a multimedia query q (q^T denoting the text part and q^V the image part of q). In that case, as far as the notations are concerned, we have the following cross–media score definition:

$$s_q^{VT} = \kappa(s_q^V, k^V).S^T \tag{17.12}$$

$$s_q^{TV} = \kappa(s_q^T, k^T).S^V \tag{17.13}$$

where s_q^T is the N dimensional similarity row vector of the textual part of the query q^T with a set of multimedia objects (their textual part d_i^T) and respectively s^V is the similarity row vector of the visual part of the query q^V with the same set of multimedia objects (but their image part d_i^V).

17.4.1.1 Fusing all Similarities

The cross–media similarities that we have described in the previous subsection, attempt to better fill the semantic gap between images and texts. They allow reinforcement of the monomedia similarities. In order to capture as much as possible of these different views, the final similarity we used is a late fusion of mono–media and

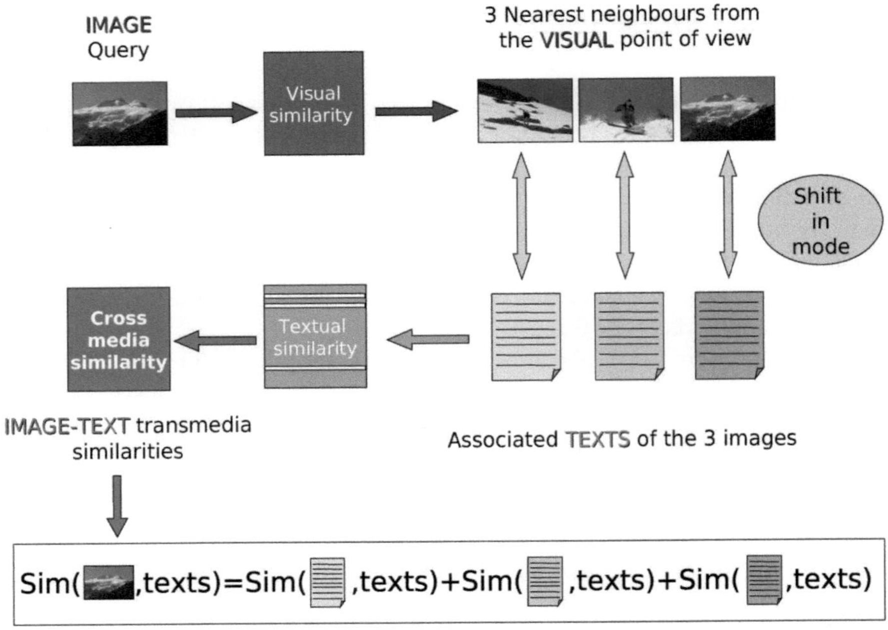

Fig. 17.2: Illustration of the trans–media pseudo feedback mechanism.

cross–media similarities. This late combination turned out to provide better results based on the scores we obtained for photo retrieval.

The final pairwise similarity matrix that evaluates the proximity between multimedia items of a set of elements is given by:

$$S = \alpha^T S^T + \alpha^V S^V + \alpha^{VT} S^{VT} + \alpha^{TV} S^{TV} \quad (17.14)$$

where $\alpha^T, \alpha^V, \alpha^{VT}, \alpha^{TV}$ are four weights that sum to 1.

Similarly, when we are given a multimedia query, the final relevance score is computed as follows:

$$s_q = \alpha^T s_q^T + \alpha^V s_q^V + \alpha^{VT} s_q^{VT} + \alpha^{TV} s_q^{TV} \quad (17.15)$$

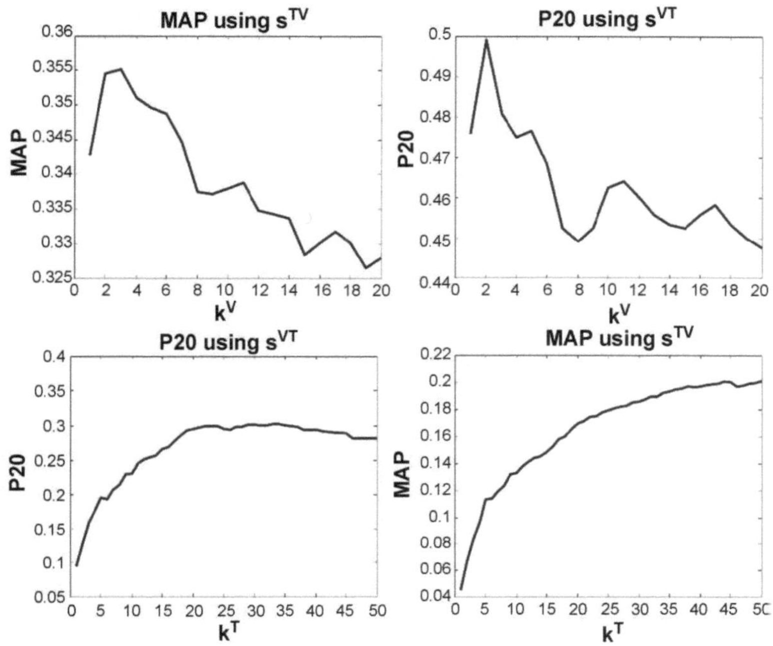

Fig. 17.3: Performance of (s^{VT}) and (s^{TV}) with variable k^V and k^T.

17.4.2 Cross–Media Retrieval at ImageCLEF Photo

17.4.2.1 The Years 2007 and 2008

The two main aspects we analyze are the effect of the number of selected documents for the trans–media feedback and the performance of cross–modal retrieval compared to monomodal retrieval. We notice that Equation 17.15 is general and, by setting the weightings at specific values, we can easily derive monomodal similarities, late fusion or cross–modal similarities (see Table 17.7). As the goal is to compare configurations of Equation 17.15 given the same S^T and S^V as inputs, we did not use the same configuration[6] as in the challenge and hence the results are not directly comparable with those reported in (Clinchant et al, 2007, 2008; Ah-Pine et al, 2008, 2009c). Nevertheless, they are about the same magnitude and of similar behavior leading to the same conclusions.

Before a comparative analysis of methods, we first analyze the effect of the number of top elements in the cross–media similarity measures given by Equations 17.12 and 17.13. Figure 17.3 shows the retrieval performance of (s^{VT}) and (s^{TV}) for vari-

[6] While the same IAPR TC–12 data was used in 2007 and 2008, in the 2007 session the image descriptions were not used. In the experiments reported they were used. On the other hand, as the 39 topics of 2008 were a subset of the 2007 topics, we perform and show average performance over all 60 topics.

Table 17.7: Comparison of the performance obtained by Equation 17.15 with different weighting parameters on the IAPR data.

Run description	α^T	α^V	α^{VT}	α^{TV}	MAP	P@20
Textual (s^T)	1	0	0	0	0.263	0.308
Visual (s^V)	0	1	0	0	0.18	0.326
Late Fusion	0.5	0.5	0	0	0.348	0.45
Cross–s^{VT}	0	0	1	0	0.354	0.499
Cross–s^{TV}	0	0	0	1	0.179	0.296
Cross–all	0.25	0.25	0.25	0.25	0.387	0.457
Cross–s^T, s^{VT}	0.5	0	0.5	0	0.411	0.522
Cross–s^T, s^V, s^{VT}	0.25	0.25	0.5	0	**0.441**	**0.573**

able k^V respectively k^T. We can see that, while using the top two or three visually similar images make a big difference, using more images decreases the performance. The main reason might be that non–relevant top images in (s^{VT}) introduce too much textual noise in the pseudo–relevance feedback. Concerning the (s^{VT}) similarity, while the performance varies more smoothly with k^T, it is globally lower than the (s^{TV}) similarity measure.

Table 17.7 shows a comparison of the ranking using equation 17.15 with different weightings. The results are averages over the 60 query topics. In the case of cross modalities, we use $k^V = 2$ and $k^T = 25$. Analyzing the table, we can see that combining images with text helps both by using a late fusion approach and by computing cross–media similarities. The only exception was when using (s^{TV}) similarities that do not seem to help the visual pseudo–relevance feedback, probably due to the noise introduced. The best results were hence obtained when we combine the cross–modal similarity s^{VT} with the late fusion $s^T + s^V$ (shown in the last row of Table 17.7). In Figure 17.4 we show the retrieval performance for different queries (limited to the first 20 topics, for readability).

17.4.2.2 The Year 2009

One of the main novelties in the 2009 photo retrieval compared to the previous years was that the new data set contained half a million images from Belga News. This was 25 times more than the IAPR TC–12 database, and hence we had to address new issues. The most important one was scalability. With such a large data set, the monomodal similarity matrices become huge (500,000 x 500,000), requiring both high computational cost and storage capacity. Even if, recently, Perronnin (2010) proposed a method to handle large scale retrieval with Fisher Vectors, we decided to adopt an alternative strategy in order to partially overcome the scalability: this consists in pre–filtering the collection by restricting it to a set of candidates that are selected uniquely from an initial textual retrieval; this is explained in a following paragraph. Before this, we present an evaluation on a sample collection, namely the

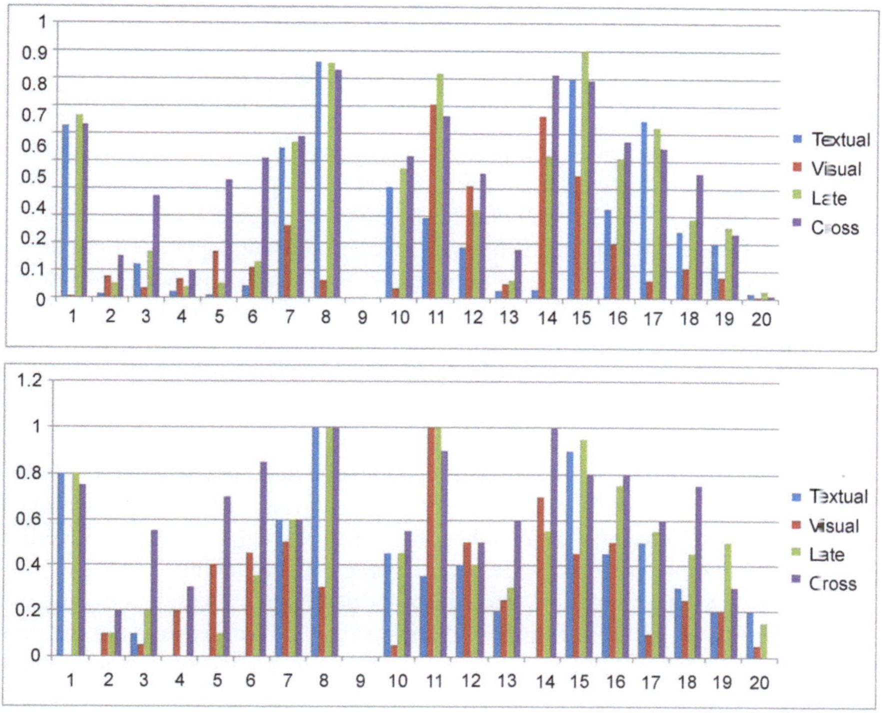

Fig. 17.4: Performance (MAP and P@20) for the first 20 topics.

Table 17.8: Comparison of the retrieval performance obtained by Equation 17.15 with different weightings on the Belga images data set.

Run description	α^T	α^V	α^{VT}	α^{TV}	MAP	P@10
Textual (s^T)	1	0	0	0	0.372	0.69
Visual (s^V)	0	1	0	0	0.012	0.146
Late Fusion	0.5	0.5	0	0	0.25	0.61
Cross-s^{VT}	0	0	1	0	0.19	0.64
Cross-s^{TV}	0	0	0	1	0.012	0.152

the 73,240 images for which relevance scores were provided by the organizers after the challenge.

As Table 17.8 shows, the performance on this subset is rather poor: neither the late fusion nor the cross–media similarities managed to extract new information from the image to improve the text–based retrieval. We have to mention that the image retrieval task from Belga News images is different from most CBIR experiments in the literature (in particular the previous ImageCLEF sessions). The main difference is that the visual similarity between images is in most cases unrelated to the semantic similarity we are seeking. Indeed a large number of query topics in 2009 were related to well known personalities. While the image representations (BoW, FV) described in Section 17.2 have shown very good performance when retrieving objects that belong to similar visual classes, scenes or (broad) locations, they are not suitable to recognize personalities in different circumstances[7]. Indeed, with this unique global image representation, two different tennis players in the field will be visually more similar to each other than the photo of the same persons in completely different circumstances (e.g. being interviewed or in a restaurant, rather than playing tennis). Hence, the visual similarity alone has real difficulties to correctly retrieve images for most topics in 2009 and explains why even the best performing systems in the challenge had poor performance (MAP= 0.014 with P10=0.15), while pure textual retrieval methods reached a MAP of 0.5 with P10 around 0.8.

As in most cases even the nearest neighbors of images were generally not semantically similar given the topic (except for near duplicates), the fusion also resulted in poor performance. However, while the original textual ranking was significantly decreased by the visual pseudo–relevance feedback due to added noise, the poor image re–ranking was significantly improved by the textual relevance feedback. Nevertheless, all combinations perform worse than the pure textual ranking.

This said, the concept of cross–modal similarities is not useless, in the sense that it is at the basis of the following pre–filtering strategy. In order to avoid the poor performances shown in Table 17.8, a natural intuition was to use the text to filter out most non-relevant images. This had the further advantage of reducing significantly the computational and storage cost: instead of computing the entire S^V similarity matrix, we only computed small sub-parts of it. In more detail, for each topic or even sub–topic, we first selected a set (a few hundred at most) of potentially relevant documents using pure text-based search. At this stage, we computed topic dependent monomodal similarities S^V_q and S^T_q on the preselected documents only. Then, we applied successfully (see (Ah-Pine et al, 2009b, 2010) and results in Section 17.5.2) our cross-similarity measures to re-rank those documents based both on visual and textual similarities, leading to a combination with better precision and diversity at the top results. Indeed, diversity seeking was a key issue in the 2008 and 2009 campaigns, and we will see in the next section that the visual, and hence cross-modal, similarities have an important role to play from this point of view as well.

17.5 Diversity–focused Multimedia Retrieval

In the 2008 and 2009 sessions, an additional sub–task to multimedia retrieval was required from the participants. It concerns diversity–focused multimedia retrieval and typically, the participants not only needed to provide relevant items to the topics but they also had to promote diversity so that the first retrieved items should be both relevant and thematically different from each other. Diversity–focused retrieval tasks

[7] In our case, the BoW or FV representations are constructed on patches extracted on the whole image, and not on specific facial locations of detected faces as in (Everingham et al, 2006).

can be encountered in different scenarios. First, we can imagine a user that has a rather general query and providing him with diverse retrieved items in the top–list is very beneficial since he can have a quick overview of the different themes related to his query. Second, we can consider a user who has a text query that is ambiguous and thus, he can give some information about the different sub–topics that he wants to retrieve using an image that illustrates each of them. In that case, the system should provide him with a top–list of items relevant to each sub–topic. ImageCLEF 2008 offered a task that belongs to the first family of scenarios, while the 2009 session belongs to the second family.

To promote diversity we basically apply a two–step approach. In the first step, we ignore the issue of diversity. In other words, we first try to find the most relevant documents using the material introduced in the previous sections. Then, in a second step, we re–rank the first relevant items by taking into account their mutual similarities in order to avoid redundancy and thus to promote diversity.

During the last two sessions, we tested various methods that are presented in Section 17.5.1 and evaluated in Section 17.5.2.

17.5.1 Re–ranking Top–Listed Documents to Promote Diversity

Among the four methods that we are introducing, the first three re-rank aiming at changing the order of the first items of a given top list so that they are not similar to each other according to a given similarity matrix.

The last method relies on the Round Robin heuristic. It implements a simple way to combine lists into a single one. This approach is used when we want to combine methods that are assumed to provide different relevant lists to the same topic or when we want to combine different lists that are relevant to several given sub–topics of a topic.

17.5.1.1 Maximal Marginal Relevance

Maximal Marginal Relevance (MMR) proposed by Carbonell and Goldstein (1998), is a re–ranking algorithm that aims to avoid redundancy amongst the first elements. It has been successfully applied in different fields such as active learning in information retrieval (Shen and Zhai, 2005; Huang et al, 2008) or in document summarization (Lin et al, 2005; Boudin et al, 2008).

We suppose that we are given a relevance vector s_q (for a given query q) as well as a similarity matrix S (for each pair of documents of the collection). The MMR framework supposes that the elements d_i should be ranked according to both s_q and S. It is a greedy algorithm: at each step (rank) r, we choose the element d_i that maximizes the following re–ranking criterion:

$$MMR_q(d_i) = \beta(r)s_q(d_i) - (1-\beta(r))\max_{j \in P_r} S(d_i,d_j) \tag{17.16}$$

where $\beta(r)$ is a mixture parameter (between 0 and 1) depending on the rank and P_r is the set of documents already selected (rank lower than r).

Traditionally, β is kept constant, but we propose a more efficient variant, where $\beta(r)$ linearly increases between $\beta(1) = \alpha$ (< 1) and $\beta(k)$=1 for some k (typically k=100), before saturating at value $\beta = 1$.

Regarding the choice of s_q, we adopted the (best) combination of mono–media and cross–media similarity measures. For S, we can take any similarity matrix (mono or cross–media) but basically we rely on the similarity matrix defined by Equation (17.14).

17.5.1.2 Clustering Based Re-ranking

We assume here that we are given an ordered top-list of documents P and a similarity matrix S between these items (both P and S could be visual, textual or cross–modal). S is normalized such that for each row, the maximal element takes the value 1 and the minimal element the value 0. We apply the Relational Analysis (RA) approach for the clustering step in order to find homogeneous themes among the set of items (Marcotorchino and Michaud, 1981; Ah-Pine et al, 2008; Ah-Pine, 2009).

The clustering function that we want to optimize with respect to X is:

$$C(S,X) = \sum_{i,j=1}^{|P|} [S(d_i,d_j) - \underbrace{\frac{1}{|\mathbb{S}^+|} \sum_{(d_i,d_j)\in\mathbb{S}^+} S(d_i,d_j)}_{\text{constant threshold}}]X(d_i,d_j) \qquad (17.17)$$

where $X(d_i,d_j) = 1$ if d_i and d_j are in the same cluster and $X(d_i,d_j) = 0$ otherwise and $\mathbb{S}^+$ is the set of pairs of documents which similarity measure is strictly positive: $\mathbb{S}^+ = \{(d_i,d_j) \in P \times P : S(d_i,d_j) > 0\}$.

From Equation 17.17, we can see that the larger the similarity between two items exceeds the mean average of strictly positive similarities, the greater the chances for them to be in the same cluster. This clustering function is based upon the central tendency deviation principle proposed by Ah-Pine (2009). In order to find a partition represented by X that maximizes the objective function, we used the clustering algorithm described in (Ah-Pine et al, 2008; Ah-Pine, 2009). This approach does not require to fix the number of clusters. This property turns out to be an advantage for finding diverse relevant themes among the documents since we do not know the number of themes for each topic.

After the clustering step, we have to define a re–ranking strategy, which takes into account the diversity provided by the clustering results. The main idea of our approach is to represent, among the first re–ranked results, elements which belong to different clusters until a stopping criterion is fulfilled. The strategy employed is described in Algorithm 17.1.

The stop criterion in Algorithm 17.1 we use is related to a parameter denoted $nbdiv \in 1,\ldots,c$, where c is the number of clusters found during the clustering pro-

Algorithm 17.1 Re–ranking strategy for a (sub-)topic

Require: A (sub-)topic q, an ordered list P according to some relevance score between q and $P_i; i = 1, \ldots, |P|$ and R the clustering results of objects in P.
Let $L1$, $L2$, $L3$ and CL be empty lists and $i = 2$.
Add P_1 as first element of the re–ranked list $L1$ and $R(P_1)$ (the cluster id of P_1) to the cluster list CL
while $i \leq |P|$ and Stopping criterion is not fulfilled **do**
  if $R(P_i) \in CL$ **then**
    Append P_i to $L2$
  else
    Append P_i to $L1$ and add $R(P_i)$ in CL
  end if
  $i = i + 1$
end while
Put if not empty the complementary list of objects from P_i to $P_{|P|}$ in $L3$.
return $L1$.append($L2$ .append($L3$))

cess. It is the maximal number of different clusters that must be represented among the first results. We assume that $nbdiv = 10$. Then, this implies that the first ten elements of the re–ranked list have to belong to ten different clusters (assuming that $c \geq 10$). Once ten different clusters are appended, the complementary list (from the 11th rank to the $|P|^{th}$ rank), is constituted of the remaining multimedia documents sorted with respect to the original list P without taking into account the cluster membership information anymore.

17.5.1.3 Density–based Re–ranking

This approach consists of identifying, among a top–list, peaks with respect to some estimated density functions. As a density measure $dens$, we used a simple one which is the sum of similarities (or distances) of the k nearest neighbors. Thus, given an object d_i, we define:

$$dens(d_i) = \sum_{d_j \in kNN_i} S(d_i, d_j) \tag{17.18}$$

where kNN_i is the set of the k nearest neighbors of d_i and S is a given similarity matrix which could be the visual–based one given by Equation 17.7 or the text–based one of Equation 17.9 or cross–media similarities as described by Equation 17.14.

Finally, we re–rank the documents according to this measure by ranking first the items that are the most dense and by discarding the near duplicates of these latter elements added to the list.

17.5.1.4 Round Robin

This method is a simple meta–heuristic approach that consists of combining multiple ranked lists into one final list. The main idea is: each ranked list takes its turn (the order of the list is chosen arbitrary) and at each turn we take the top element of the list and we append it to the final list. When a top element of a list is appended to the final list we remove it from its original list and take the next item as the new top element. The new appended documents can belong to other lists, and if it is the case we remove it from the lists so that we avoid duplicates in the final list.

The Round Robin method can be applied in the context of different scenarios. First, in the case where we have multimedia topics that are made of several sub–topics, we can consider for each of the latter a list of retrieved items and thus combine them by using the Round Robin method. Second, a more general scenario is when we have different systems that give top–lists that we want to merge. In that case too, the Round Robin approach can be used to combine the results.

17.5.2 Diversity–focused Retrieval at ImageCLEF Photo

As mentioned previously, the Round Robin method is a meta–heuristic which aims at combining lists. It is different from the other methods that we introduced previously. The three other approaches rely on the use of a similarity matrix and seek to re–rank one top–list so that topically–diverse documents are rapidly proposed to the user.

Consequently, the results that are provided by the MMR, the density–based and the clustering based re–ranking methods are comparable to each other though we did not apply all of them to both sessions. By contrast, they are not directly comparable to the Round Robin technique.

17.5.2.1 The Year 2008

In the 2008 session, we mainly applied the MMR and the clustering–based approaches to re–rank a relevant list in order to promote diversity. We recall in Table 17.9 some of the best runs we obtained. The baseline given by the third line is the run provided by Equation (17.15) with parameters $\alpha^T = \alpha^{VT} = 0.5$ and $\alpha^V = \alpha^{TV} = 0$. No re–ranking method was applied to this run. However, it provides the top–list that we aim at re–ranking in order to avoid redundancy among the first elements. Accordingly, line 1 of Table 17.9 is the run that re–ranks the baseline with respect to the clustering–based technique we described previously while line 2 used MMR. For both runs, the similarity matrix which was used to measure the thematic proximity between documents was the fused cross–media similarity given by Equation 17.14 with the same aforementioned parameters (see (Ah-Pine et al, 2008) for more details).

Table 17.9: XRCE's best runs in the 2008 session in terms of Precision at 20 (P@20) and Cluster Recall at 20 (CR@20).

Run Description	CR@20	P@20
With clust–based re–rank. (using cross–media similarities)	0.4111	0.5269
With MMR re-rank. (using cross-media similarities)	0.4015	0.5282
Without any re–ranking method (baseline)	0.3727	0.5577

We can observe in Table 17.9 that any diversity–focused method fails to increase, on average, the P@20 measure. However, any method performs better than the basic run regarding the CR@20 measure. In other words, by trying to eliminate redundancy among the first retrieved objects, unfortunately, we might push relevant objects out of the 20 first re–ranked elements and put into this final top list some irrelevant objects.

In (Ah-Pine et al, 2009c), we analyzed the behavior of the MMR and the clustering–based re–ranking methods and refer the reader to this paper for more details. Here, we underline the main observation that we made by looking at the assessment measures per query. The clustering–based strategy exhibits a consistent, stable behavior, where it systematically gives slightly lower or equal P@20 performances than the basic list, while offering CR@20 performances that are superior or equal to the baseline. The MMR method does not offer such a stability in its behavior. In fact, this method seems to take more risk in the re–ranking process than the former method, with a consequence of increased variance in the performance.

Therefore, despite comparable P@20 and CR@20 measures, the MMR technique and the clustering–based methods do not show the same behavior.

17.5.2.2 The Year 2009

In 2009, we applied the density–based, the clustering–based and the Round Robin methods.

In this session, many topics were constituted of several sub–topics which basically expressed different aspects of the main topic and gave the participants the definition of the clusters to retrieve (the topic *'brussels'* for example had sub–topics *'brussels airport'*, *'police brussels'*, *''fc brussels'* among others). Those cases represent the topics in Part 1. In this case, we treated each sub–topic as if it was independent and combined them using the Round Robin method so as to produce a single list of retrieved diverse items. The method we used to produce the top–list for each sub–topic is described in (Ah-Pine et al, 2009b). It is important to underline the fact that we first used a text–based retrieval for all sub–topics using the image captions. In other words, the results we are going to mention used a pre–filtering step which aimed at determining a preliminary set of relevant documents from a textual standpoint. After this first pass, we then used different types of similarity in order to re–rank the documents of this preliminary set by taking into account either

Table 17.10: XRCE's runs in 2009 on topics of part 1 in terms of Precision at 10 (P@10), Cluster Recall at 10 (CR@10) and F_1.

Run Description	CR@10	P@10	F_1
Text pre–filt. (captions as queries)	**83.9**	78.4	81.0
Text pre–filt. (captions as queries) + visual re-rank.	75.2	60.8	67.2
Text pre–filt. (captions as queries) + cross-modal re-rank.	83.7	**79.6**	**81.6**

text (in that case there is no re–ranking) or visual or cross–media similarities. We refer the reader to (Ah-Pine et al, 2009b) for more details.

Since the Round Robin method is the only strategy that we used to combine lists into one, we cannot outline any results analysis about this technique. However, we can comment on the results we obtained focusing on the media that performed well. Accordingly, in the case of topics in Part 1, we found that the media that gave the best results regarding the diversity assessment measure is the one based on text only. Nevertheless, the F_1 measure that combines both the precision and the diversity criteria is better for the fused cross–media similarities as given by Equation (17.14) with the parameters $\alpha^T = 5/12$, $\alpha^V = \alpha^{VT} = 1/4$ and $\alpha^{TV} = 1/12$. Text–based retrieval is by far the most important tool to achieve good performance in the multimedia task designed for 2009. Re–ranking the documents using the visual similarities after a text–based pre–filtering does not increase the results. However, combining visual and textual similarities using our cross–media techniques and re–ranking the documents with respect to the fused cross–media similarity after the text–based pre–filtering, allows us to slightly improve the P@10 and F_1 measures without hurting the CR@10. Those observations are numerically illustrated in Table 17.10.

If topics in Part 1 were already well–detailed from a diversity viewpoint since we were provided with their sub–topics, the topics of Part 2 were more challenging when seeking to promote diversity. In that case, we were only given a text query and three images. That type of multimedia topic is the kind of topic we had to deal with in 2008. For topics in Part 2, we assumed that the three image queries represented three sub–topics though it was specified that there might be more clusters to find than those three. We computed for each of them a basic top–list (in a similar way as we did for topics in Part 1, see (Ah-Pine et al, 2009b) for more details) and to each of the top–list we applied a density–based or a clustering–based re–ranking technique before fusing them with the Round Robin method. Those types of run are denoted basic runs in (Ah-Pine et al, 2009b).

Regarding the comparison between the two re–ranking techniques on the basis of the measures, we observed that density and clustering are comparable when image similarity is used, but with text similarity or fused cross-media similarity, clustering generally gives better results.

Another important observation found from the experiments is that combining different types of basic runs with the Round Robin heuristic allows us to enhance the results. This is shown in Table 17.11 where we can see that combining two basic runs can lead to more than a seven point increase in terms of the F_1 measure.

Table 17.11: Some of XRCE's best runs in 2009 on topics in part 2.

Run id	Run Description	CR@10	P@10	F_1
1	Text pre-filt. (captions as queries) + cross-modal re-rank	76.8	72.4	74.6
2	Text pre-filt. (enriched query) + Clust-based re-rank. (vis. sim.)	65.8	78.0	71.4
3	Text pre-filt. (enriched query) + Dens-based re-rank. (vis. sim.)	62.6	**83.2**	71.4
4	Round Robin of 1 and 2	82.4	78.8	80.6
5	Round Robin of 1 and 3	**82.5**	81.6	**82.0**

17.6 Conclusion

As a conclusion, we underline the main lessons learned after three participations in ImageCLEF photo:

- When dealing with multimedia text–image documents, it is beneficial to combine the text with the visual information. A simple combination strategy such as late fusion already allows us to obtain much better results than mono–media retrieval.
- The cross–media technique we designed to combine multimedia information performs better than late fusion. The good performance reached in the three last sessions of ImageCLEF shows the effectiveness and robustness of this method. We believe that it allows us to better handle the semantic gap between media.
- Text–based retrieval is fundamental as long as we have a good textual description of the images. It performed much better than visual retrieval and for 2009, we would not have been able to obtain such good results if we had not used the text as a pre–filtering before using cross–media techniques. However, visual similarities allow us to significantly gain in terms of precision and recall providing that we combine them with the text similarities in an efficient way.
- When using fused cross–media similarities as given by Equation 17.15, we consistently observed that one should give more weight to textual similarities than to visual similarities if the former performs much better than the latter, otherwise the equal weighting works well. Furthermore, generally the image–text cross–media similarities perform better than the text–image cross–media ones, and in most cases it is better not to consider the latter. We can see that the cross–media image–text is beneficial in our strategy allowing us to better bridge the gap between image and text. Finally, it is also important to mention that our cross–media similarities are dependent on a parameter k, the number of nearest neighbors considered by the pseudo–relevance feedback. Generally, considering a relatively low k (typically < 5), we avoid the risk of introducing too much noise in the cross–media similarity. This is particularly true for (s^{VT}) while the effect of this number seems to be smoother in the case of (s^{TV}).
- For text–based retrieval, we used standard language models to compute textual similarities. In order to overcome the issue with the sparsity of textual data and particularly in the context of ImageCLEF collections, we tried to enrich both the queries and the documents by using external resources or co–occurrence techniques. We showed that enriching texts data is always beneficial.

- Regarding the diversity seeking retrieval sub–task, we applied a two–step scheme, which first focuses on retrieving relevant documents and then re–ranks the top–list to avoid redundancy among the first items. This approach allowed to promote diversity without hurting the relevance of the re–ranked top–list. We used different approaches for re–ranking and each of them gave interesting results while presenting different behaviors. The clustering–based methods showed a stable behavior enabling us to consistently improve the diversity assessment while decreasing the precision measures slightly. The MMR method takes more risk, and on average allows an increase of the diversity while presenting a less stable behavior since we observed more variability of the measures at the level of queries. The density–based approach also provided good results, particularly when it is applied on the visual similarities.
- In the last session, we investigated the combination of several methods, which resulted from different techniques either at the level of the features we used for mono–media similarities or at the level of the similarities used to locally re–rank a top–list to favor diversity. It appeared that simple combination methods such as the Round Robin technique generally allow us to improve both precision and diversity. Therefore, using different text representations, enrichment techniques or similarities to re–rank objects, and combining the top–lists using the Round Robin method, is beneficial.
- While in 2007 and 2008, the collection was of around 20K multimedia documents, in 2009, it was constituted of more than 500K items. In the last session, we thus had to deal with more scalability issues than before. Indeed, it is not easy to compute the whole visual similarity matrix and an on–line method had to be designed. As mentioned previously, we first applied a text–based pre–filtering step. This strategy turned out to be a winning one since not only were we able to address the scalability issue of computing visual similarities by pre–selecting a relevant set of documents given a topic, but this text–based pre–filtering was also an efficient way to obtain a very good baseline retrieval that we were able to improve further in a second step.

Acknowledgements This work was partially supported by the European Union Projects PinView (FP7–216529) and Sync3 (FP7–231854), and the French projects Omnia (ANR–06–CIS6–01) and Fragrances (ANR–08–CORD–008).

References

Ah-Pine J (2009) Cluster analysis based on the central tendency deviation principle. In: Proceedings of the International Conference on Advanced Data Mining and Applications, pp 5–18

Ah-Pine J, Cifarelli C, Clinchant S, Csurka G, Renders J (2008) XRCE's participation to ImageCLEF 2008. In: Working Notes of CLEF 2008, Aarhus, Denmark

Ah-Pine J, Bressan M, Clinchant S, Csurka G, Hoppenot Y, Renders J (2009) Crossing textual and visual content in different application scenarios. Multimedia Tools and Applications 42(1):31–56

Ah-Pine J, Clinchant S, Csurka G, Liu Y (2009) XRCE's participation to ImageCLEF 2009. In: Working Notes of the 2009 CLEF Workshop, Corfu, Greece

Ah-Pine J, Csurka G, Renders JM (2009c) Evaluation of diversity–focused strategies for multimedia retrieval. In: Evaluating Systems for Multilingual and Multimodal Information Access, Springer, Lecture Notes in Computer Science (LNCS), vol 5706, pp 677–684

Ah-Pine J, Clinchant S, Csurka G Comparison of several combinations of multimodal and diversity seeking methods for multimedia retrieval. In: Multilingual Information Access Evaluation, Springer, Lecture Notes in Computer Science (LNCS)

Barnard K, Duygulu P, Forsyth D, de Freitas N, Jordan M (2003) Matching words and pictures. Journal of Machine Learning Research 3:1107–1135

Blei D, Jordan MI (2003) Modeling annotated data. In: Proceedings of the ACM SIGIR conference, ACM press, pp 127–134

Boudin F, El-Bèze M, Torres-Moreno J (2008) A scalable MMR approach to sentence scoring for multi–document update summarization. In: Proceedings of the international conference on computational linguistics, pp 21–24

Carbonell J, Goldstein J (1998) The use of MMR, diversity–based reranking for reordering documents and producing summaries. In: Proceedings of the ACM SIGIR conference, ACM press, pp 335–336

Carbonetto P, de Freitas N, Barnard K (2004) A statistical model for general contextual object recognition. In: European conference on computer vision, vol 1, pp 350–362

Chang YC, Chen HH (2006) Approaches of using a word-image ontology and an annotated image corpus as intermedia for cross–language image retrieval. In: Working notes CLEF 2006

Clinchant S, Renders J, Csurka G (2007) XRCE's participation to ImageCLEF 2007. In: Working Notes of CLEF 2007, Budapest, Hungary

Clinchant S, Renders JM, Csurka G (2008) Trans–media pseudo–relevance feedback methods in multimedia retrieval. In: Advances in Multilingual and Multimodal Information Retrieval, Springer, Lecture Notes in Computer Science (LNCS), vol 5152, pp 569–576

Csurka G, Dance C, Fan L, Willamowski J, Bray C (2004) Visual categorization with bags of keypoints. In: ECCV Workshop on Statistical Learning for Computer Vision, pp 59–74

Deselaers T, Hanbury A (2008) The Visual Concept Detection Task in ImageCLEF 2008. In: Working Notes of CLEF 2008

Duygulu P, Barnard K, de Freitas J, Forsyth D (2002) Object recognition as machine translation: Learning a lexicon for a fixed image vocabulary. In: European conference on computer vision, vol 4, pp 97–112

Everingham M, Sivic J, Zisserman A (2006) "hello! my name is... buffy" – automatic naming of characters in TV video. In: British machine vision conference, pp 889–908

Feng S, Lavrenko V, Manmatha R (2004) Multiple bernoulli relevance models for image and video annotation. In: International conference on computer vision and pattern recognition, vol 2, pp 1002–1009

Huang T, Dagli C, Rajaram S, Chang E, Mandel M, Poliner G, Ellis D (2008) Active learning for interactive multimedia retrieval. Proceedings of the IEEE 96(4):648–667

Iyengar G, Duygulu P, Feng S, Ircing P, Khudanpur S, Klakow D, Krause M, Manmatha R, Nock H, Petkova D, Pytlik B, Virga P (2005) Joint visual–text modeling for automatic retrieval of multimedia documents. In: Proceedings of ACM Multimedia, ACM press, pp 21–30

Jaakkola T, Haussler D (1999) Exploiting generative models in discriminative classifiers. In: Advances in Neural Information Processing Systems, MIT Press, pp 487–493

Jeon J, Lavrenko V, Manmatha R (2003) Automatic image annotation and retrieval using cross–media relevance models. In: Proceedings of the ACM SIGIR conference, ACM press, pp 119–126

Lavrenko V, Manmatha R, Jeon J (2003) A model for learning the semantics of pictures. In: Annual conference on neural information processing systems, pp 553–560

Li J, Wang JZ (2003) Automatic linguistic indexing of pictures by a statistical modeling approach. IEEE Transactions on Pattern Analysis and Machine Intelligence 25:1075–1088

Lin Z, Chua T, Kan M, Lee W, Qiu L, Ye S (2005) NUS at DUC 2007: Using evolutionary models of text. In: Document Understanding Conference

Maillot N, Chevallet JP, Valea V, Lim JH (2006) IPAL Inter–Media Pseudo–Relevance Feedback Approach to ImageCLEF 2006 photo retrieval. In: CLEF 2006 Working Notes

Manning CD, Schütze H, Lee L (2000) Review: Foundations of statistical natural language processing

Marcotorchino J, Michaud P (1981) Heuristic approach of the similarity aggregation problem. Methods of operation research 43:395–404

Monay F, Gatica-Perez D (2004) PLSA–based Image Auto–Annotation: Constraining the Latent Space. In: Proceedings of ACM Multimedia, ACM press, pp 348–351

Mori Y, Takahashi H, Oka R (1999) Image–to–word transformation based on dividing and vector quantizing images with words. In: First International Workshop on Multimedia Intelligent Storage and Retrieval Management

Pan J, Yang H, Faloutsos C, Duygulu P (2004) Gcap: Graph–based automatic image captioning. In: CVPR Workshop on Multimedia Data and Document Engineering at the computer Vision and Pattern recognition conference

Perronnin F (2010) Large–scale image retrieval with compressed fisher vectors. In: International Conference on computer vision and pattern recognition, IEEE

Perronnin F, Dance C (2007) Fisher kernels on visual vocabularies for image categorization. In: International conference on computer vision and pattern recognition, IEEE

Shen X, Zhai C (2005) Active feedback in ad hoc information retrieval. In: International ACM SIGIR conference, ACM press, pp 59–66

Sivic JS, Zisserman A (2003) Video google: A text retrieval approach to object matching in videos. In: International conference on computer vision, IEEE, vol 2, pp 1470–1477

Vinokourov A, Hardoon DR, Shawe-Taylor J (2003) Learning the semantics of multimedia content with application to web image retrieval and classification. In: Fourth International Symposium on Independent Component Analysis and Blind Source Separation

Zhai C, Lafferty JD (2001) Model–based feedback in the language modeling approach to information retrieval. In: Conference on Information and Knowledge management, pp 403–410

Chapter 18
University of Amsterdam at the Visual Concept Detection and Annotation Tasks

Koen E. A. van de Sande and Theo Gevers

Abstract Visual concept detection is important to access visual information on the level of objects and scene types. The current state–of–the–art in visual concept detection and annotation tasks is based on the bag–of–words model. Within the bag–of–words model, points are first sampled according to some strategy, then the area around these points are described using color descriptors. These descriptors are then vector–quantized against a codebook of prototypical descriptors, which results in a fixed–length representation of the image. Based on these representations, visual concept models are trained. In this chapter, we discuss the design choices within the bag–of–words model and their implications for concept detection accuracy.

18.1 Introduction

Robust image retrieval is highly relevant in a world that is adapting to visual communication. On–line services such as Flicks show that the sheer number of photos available on–line is too much for any human to grasp. Many people place their entire photo album on the Internet. Most commercial image search engines provide access to photos based on text or other metadata, as this is still the easiest way for a user to describe their visual information need. The indices of these search engines are based on the filename, associated text or tagging. This results in disappointing retrieval performance when the visual content is not mentioned, or properly reflected in the associated text. In addition, when the photos originate from non–English speaking countries, such as China, or Germany, querying the content becomes much harder.

To cater for robust image retrieval, the promising solutions from the literature are mostly concept–based, see, for example, the overview in (Snoek and Worring, 2009), where detectors are related to objects, such as *trees*, scenes, such as a *desert*, and people, such as *big group*. Any one of those brings an understanding of the

University of Amsterdam, Science Park 107, 1098 XG Amsterdam, e-mail: ksande@uva.nl, gevers@science.uva.nl

H. Müller et al. (eds.), *ImageCLEF*, The Information Retrieval Series 32, DOI 10.1007/978-3-642-15181-1_18, © Springer-Verlag Berlin Heidelberg 2010

current content. The concepts in such a lexicon allow users to query on the presence or absence of visual content elements, for example, a semantic entry into the data.

The large–scale visual concept detection task of ImageCLEF 2009, discussed in (Nowak and Dunker, 2009), evaluates 53 visual concept detectors. The concepts used are from the personal photo album domain: *beach holidays*, *snow*, *plants*, *indoor*, *mountains*, *still–life*, *small group of people*, *portrait*. For more information on the data set and concepts used, see Chapter 2.

The current state–of–the–art in visual concept detection and annotation tasks is based on the bag–of–words model (Van de Sande et al, 2010; Marszałek et al, 2007; Snoek et al, 2009; Wang et al, 2007). Within the bag–of–words, points are first sampled according to some strategy, then the areas around these points are described using color descriptors. These descriptors are then vector–quantized against a codebook of prototypical descriptors, which results in a fixed–length representation of the image. Based on these representations, visual concept models are trained.

Based on our previous work on concept detection (Van de Sande et al, 2010; Snoek et al, 2008; Uijlings et al, 2009), the participation of the University of Amsterdam within ImageCLEF has focused on improving the robustness of the visual features used in concept detectors.

Systems with the best performance in image retrieval (Van de Sande et al, 2010; Marszałek et al, 2007) and video retrieval (Snoek et al, 2008; Wang et al, 2007) use combinations of multiple features for concept detection. The basis for these combinations is formed by good color features and multiple point sampling strategies. In this chapter, we discuss the design choices within the bag–of–words model and their implications for concept detection accuracy. We focus especially on the effect of these choices on the large–scale visual concept detection and annotation task from ImageCLEF 2009 and ImageCLEF@ICPR 2010.

The remainder of this chapter is organized as follows. Section 18.2 defines components in our concept detection pipeline. Section 18.3 details our experiments and results. Finally, in Section 18.6, conclusions are drawn.

18.2 Concept Detection Pipeline

We perceive concept detection as a combined computer vision and machine learning problem. The first step is to represent an image using a fixed–length feature vector. Given a visual feature vector x_i, the aim is then to obtain a measure, which indicates whether a semantic concept C is present in photo i. We may choose from various visual feature extraction methods to obtain x_i, and use a supervised machine learning approach to learn the appearance relation between C and x_i. The supervised machine learning process is composed of two phases: training and testing. In the first phase, the optimal configuration of features is learned from the training data. In the second phase, the classifier assigns a probability $p(C|x_i)$ to each input feature vector for each semantic concept C.

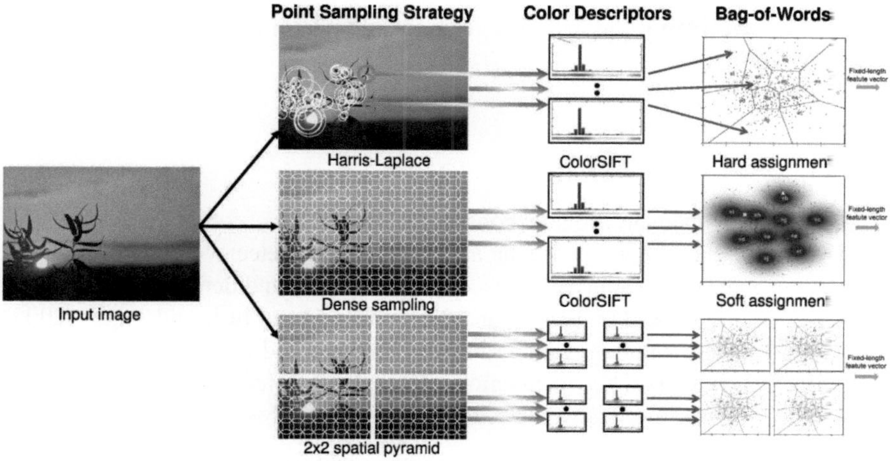

Fig. 18.1: University of Amsterdam's ImageCLEF 2009 concept detection scheme. The scheme serves as the blueprint for the organization of Section 18.2.

18.2.1 Point Sampling Strategy

The visual appearance of a concept has a strong dependency on the viewpoint under which it is recorded. Salient point methods (Tuytelaars and Mikolajczyk, 2008) introduce robustness against viewpoint changes by selecting points, which can be recovered under different perspectives. Another simpler solution is to use many points, which is achieved by dense sampling.

In the context of concept classification, two classes of concepts are identified: objects and scene types. Dense sampling has been shown to be advantageous for scene type classification, since salient points do not capture the entire appearance of an image. For object classification, salient points can be advantageous because they ignore homogenous areas in the image. If the object background is not highly textured, then most salient points will be located on the object or the object boundary.

We summarize our sampling approach in Figure 18.1: Harris–Laplace and dense point selection, and a spatial pyramid.[1]

Harris–Laplace point detector

In order to determine salient points, Harris–Laplace relies on a Harris corner detector. By applying it on multiple scales, it is possible to select the characteristic scale of a local corner using the Laplacian operator, discussed in (Tuytelaars and Mikolajczyk, 2008). Hence, for each corner the Harris–Laplace detector selects a

[1] Software to perform point sampling, color descriptor computation and the hard and soft assignment is available from `http://www.colordescriptors.com/`.

scale–invariant point if the local image structure under a Laplacian operator has a stable maximum.

Dense point detector

For concepts with many homogenous areas, like scenes, corners are often rare. Hence, for these concepts relying on a Harris–Laplace detector can be suboptimal. To counter the shortcoming of Harris–Laplace, random and dense sampling strategies have been proposed by Fei-Fei and Perona (2005) and Jurie and Triggs (2005). We employ dense sampling, which samples an image grid in a uniform fashion using a fixed pixel interval between regions. To study the effect of different parameter choices for dense sampling, we investigate three different settings:

- An interval of 6 pixels and sample at a singe scale ($\sigma = 1.2$).
- An interval of 6 pixels and sample at multiple scales ($\sigma = 1.2$ and $\sigma = 2.0$).
- An interval of 1 pixel, e.g. sample every pixel with a single scale ($\sigma = 1.2$).

Spatial pyramid

Both Harris–Laplace and dense sampling give an equal weight to all keypoints, irrespective of their spatial location in the image. To overcome this limitation, Lazebnik et al (2006) suggest repeatedly sampling fixed subregions of an image, e.g. 1 x 1, 2 x 2, 4 x 4, etc., and to aggregate the different resolutions into a so called spatial pyramid. Since every region is an image in itself, the spatial pyramid can be used in combination with both the Harris–Laplace point detector and dense point sampling, as was done in (Van de Sande et al, 2008), for example. For the ideal spatial pyramid configuration, Lazebnik et al (2006) claim 2 x 2 is sufficient, Marszałek et al (2007) suggest including 1 x 3 also. We investigate multiple divisions of the image in our experiments.

18.2.2 Color Descriptor Extraction

In the previous section, we addressed the dependency of the visual appearance of semantic concepts on the viewpoint under which they are recorded. However, the lighting conditions during photography also play an important role. Van de Sande et al (2010) analyzed the properties of color descriptors under classes of illumination changes within the diagonal model of illumination change, and specifically for data sets consisting of Flickr images. In ImageCLEF, the images used also originate from Flickr. Here, we use the four color descriptors from the recommendation table in (Van de Sande et al, 2010). The descriptors are computed around salient points obtained from the Harris–Laplace detector and dense sampling. For the color descriptors in Figure 18.1, each of those four descriptors can be inserted.

SIFT

The Scale Invariant Feature Transform (SIFT) feature proposed by Lowe (2004) describes the local shape of a region using edge orientation histograms. The gradient of an image is shift–invariant: taking the derivative cancels out offsets (see (Van de Sande et al, 2010) for details). Under light intensity changes, *i.e.* a scaling of the intensity channel, the gradient direction and the relative gradient magnitude remain the same. Because the SIFT feature is normalized, the gradient magnitude changes have no effect on the final feature. To compute SIFT features, we use the version described by Lowe (2004).

OpponentSIFT

OpponentSIFT describes all the channels in the opponent color space using SIFT features. The information in the O_3 channel is equal to the intensity information, while the other channels describe the color information in the image. The feature normalization, as effective in SIFT, cancels out any local changes in light intensity.

C–SIFT

The C-SIFT feature uses the C invariant by Geusebroek et al (2001), which can be intuitively seen as the gradient (or derivative) for the normalized opponent color space $O1/I$ and $O2/I$. The I intensity channel remains unchanged. C–SIFT is known to be scale–invariant with respect to light intensity. See (Burghouts and Geusebroek, 2009) and (Van de Sande et al, 2010) for detailed evaluation.

RGB–SIFT

For the RGB-SIFT, the SIFT feature is computed for each *RGB* channel independently. Due to the normalizations performed within SIFT, it is scale–invariant, shift–invariant, and invariant to light color changes and shift (see (Van de Sande et al, 2010) for details).

18.2.3 Bag–of–Words model

We use the well–known bag–of–words model, also known as codebook approach, see e.g. (Leung and Malik, 2001; Jurie and Triggs, 2005; Zhang et al, 2007; Van Gemert et al, 2010; Van de Sande et al, 2010). First, we assign visual descriptors to discrete codewords pre–defined in a codebook. Then, we use the frequency distribution of the codewords as a feature vector representing an image. We construct a codebook with a maximum size of 4,096 using k-means clustering. An important issue is *codeword assignment.* A comparison of codeword assignment is presented in (Van Gemert et al, 2010). Here, we only discuss two codeword assignment methods:

- **Hard assignment**. Given a codebook of codewords, the traditional codebook approach assigns each descriptor to a single best representative codeword in the codebook. Basically, an image is represented by a histogram of codeword frequencies describing the probability density over codewords.
- **Soft assignment**. The traditional codebook approach may be improved by using soft assignment through kernel codebooks. A kernel codebook uses a kernel function to smooth the hard assignment of image features to codewords. Out of the various forms of kernel–codebooks, we selected *codeword uncertainty* based on its empirical performance, shown in (Van Gemert et al, 2010).

Each of the possible sampling methods from Section 18.2.1 coupled with each visual descriptor from Section 18.2.2, and an assignment approach, results in a separate visual codebook. An example is a codebook based on dense sampling of RGB–SIFT features in combination with hard-assignment. Naturally, various configurations can be used to combine a multiple of these choices. By default, we use hard assignment in our experiments. Soft assignment is only used when explicitly stated. For simplicity, we employ equal weights in our experiments when combining different features.

18.2.4 Machine Learning

The supervised machine learning process is composed of two phases: training and testing. In the first phase, the optimal configuration of features is learned from the training data. From all machine learning approaches on offer to learn the appearance relation between C and x_i, the support vector machine by Vapnik (2000) is commonly regarded as a solid choice. We use the LIBSVM implementation by Chang and Lin (2001) with probabilistic output as described in (Lin et al, 2007). The parameter of the support vector machine we optimize is C. In order to handle imbalance in the number of positive versus negative training examples, we fix the weights of the positive and negative class by estimation from the class priors on training data. It was shown by Zhang et al (2007) that in a codebook–approach to concept detection the earth movers distance and χ^2 kernel are to be preferred. We employ the χ^2 kernel, as it is less expensive in terms of computation.

In the second machine learning phase, the classifier assigns a probability $p(C|x_i)$ to each input feature vector for each semantic concept C, i.e. the trained model is applied to the test data.

Table 18.1: Overall results of using different pyramid levels on the 3,000 images from ImageCLEF 2009 whose annotations have been made available. Evaluated over 53 concepts in the photo annotation task using MAP. The third column lists the number of concepts for which that row obtains the highest AP relative to all other rows (excluding the bottom row with the 1 x 1, 2 x 2, 1 x 3 combination). In the fourth row, the bottom row is included when determining the row with the highest AP.

Pyramid subdivisions	Average Precision	#concepts with highest AP	#concepts with highest AP
1x1	0,415	0	0
2x2	0,411	3	2
3x3	0,398	3	3
1x2	0,409	1	1
1x3	0,421	5	3
2x1	0,398	1	1
3x1	0,396	1	1
1x1, 2x2	0,427	10	6
1x1, 1x2	0,427	6	6
1x1, 1x3	0,432	14	7
1x1, 2x1	0,420	6	5
1x1, 3x1	0,421	4	4
1x1, 2x2, 1x3	0,433	not included	14

18.3 Experiments

Experiments in this section are performed using the MIR–Flickr dataset (see Section 2.4.4) using the 53 concepts annotated for ImageCLEF 2009. Concept models are trained using the ImageCLEF 2009 photo annotation task training set. Results are computed using the labels provided after the contest for 3,000 images from the test set. In subsequent years, these images became part of the training set. All results in this section will use the average precision metric, which will be the standard for the large–scale visual concept detection and annotation task from ImageCLEF 2010 onward. Higher average precision scores imply more accurate concept detection.

18.3.1 Spatial Pyramid Levels

In Table 18.1, we show results for different subdivisions of the image based on the spatial pyramid framework. These results are obtained for an intensity SIFT descriptor sampled with an interval of six pixels at two different scales and hard codebook assignment.

Inspecting the results, we see in the first seven rows that only a pyramid with horizontal bars, 1 x 3, obtains higher overall AP than just using the full image (1 x 1 subdivision). Therefore, we introduce combinations of two pyramid subdivisions, one of which is always 1 x 1. Here, we see that 1 x 1 + 2 x 2, 1 x 1 + 1 x 2 and 1 x 1 + 1 x 3 obtain the highest overall scores. Looking at per–concept results (not shown here), 1 x 1 + 2 x 2 obtains the highest AP compared to other subdivisions for ten concepts (see third column), and 1 x 1 + 1 x 3 for 14 concepts. The subdivision of 1 x 1 + 1 x 2 obtains the highest score for only six concepts. Based on these results and its popularity in the PASCAL VOC (Everingham et al, 2010), we introduce the combination of 1 x 1 + 2 x 2 + 1 x 3, which combines the two best subdivisions with two parts. If we then recount the number of concepts for which a row obtains the highest AP (see the fourth column), this new combination obtains the highest score for 14 concepts and the highest AP of all rows.

In terms of priority, the 1 x 3 subdivision (horizontal bars) is the most important, then the 1 x 1 subdivision (the whole image), and finally a 2 x 2 division. This raises the question as to why the 1 x 3 subdivision works so well on the MIR-Flickr data set. A possible explanation is the way photographers work: they attempt to center the object of interest, have a straight horizon which is approximately in the middle of the image. Therefore, the top horizontal bar of a 1 x 3 division will probably be sky, the middle horizontal bar will contain the object of interest plus the horizon, and the bottom bar will contain the ground.

Based on these results, we draw the conclusion that using a combination of spatial subdivisions of 1 x 1, 2 x 2 and 1 x 3 is a good choice for the MIR-Flickr data set. The experiments in the remainder of this section will use exactly these subdivisions.

18.3.2 Point Sampling Strategies and Color Descriptors

In Figure 18.2, results are shown for four different point sampling methods and four different color descriptors. Inspecting these results, dense sampling clearly outperforms the Harris–Laplace salient points. Sampling at two scales instead of a single scale at an interval of six pixels is better. However, when the sampling interval is set to one pixel, e.g. every pixel is described, performance at a single scale exceeds the 6-pixel multi–scale results. These observations are consistent across all descriptors. The main drawback of sampling every pixel is that 36 times more descriptors are extracted per image, which results in a significant increase in feature extraction times. A possible solution to the computational load is to use software specifically optimized for dense sampling, as is done in (Uijlings et al, 2009).

When comparing the performance of different descriptors in Figure 18.2, we observe that the RGB–SIFT descriptor yields the highest performance on the MIR–Flickr data set. The presence of rather 'artistic' photographs with large variations in lighting conditions in the data set might explain why the illumination color–invariant descriptor gets the best results. Ordering the descriptors by their performance, the OpponentSIFT descriptor is in second place, followed by SIFT and finally C–SIFT.

In conclusion, a dense sampling strategy and the RGB–SIFT descriptor together give the best results for a single feature on the MIR–Flickr data set.

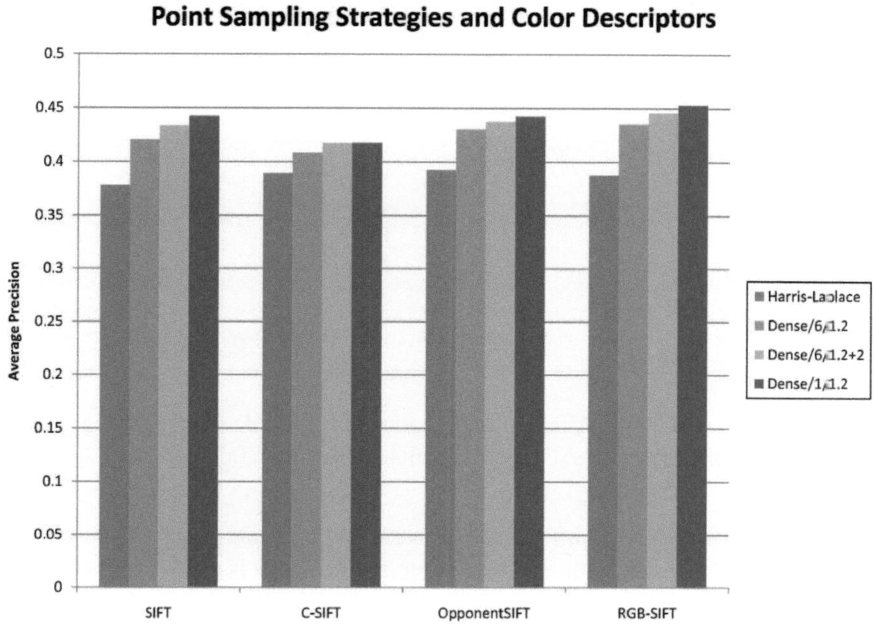

Fig. 18.2: Performance of color descriptors using either Harris–Laplace salient points or dense sampling. The dense sampling has an interval of either one or six pixels and is carried out at one (1.2) or two scales (1.2 and 2).

18.3.3 Combinations of Sampling Strategies and Descriptors

Complete concept detection systems do not use a single feature, as was done in the previous section, but use combinations of different features. The more features are added, the higher the performance becomes. In Figure 18.3 and Table 18.2, results of different system configurations are shown. The baseline is a single feature, the densely sampled SIFT descriptor with a 1 x 1, 2 x 2 and 1 x 3 spatial pyramid, i e. the best result from the spatial pyramid experiment. From the comparative experiment of point sampling strategies and descriptor, we know the RGB–SIFT descriptor is the best individual descriptor for the MIR–Flickr data set. Therefore, results for this descriptor are also listed.

The best combinations typically used in the University of Amsterdam concept detection system are as follows:

- Combination of eight features: Harris–Laplace salient points paired with each of the four ColorSIFT descriptors, and densely sampled points at multiple scales with an interval of 6 pixels, also paired with each of the four ColorSIFT descriptors.

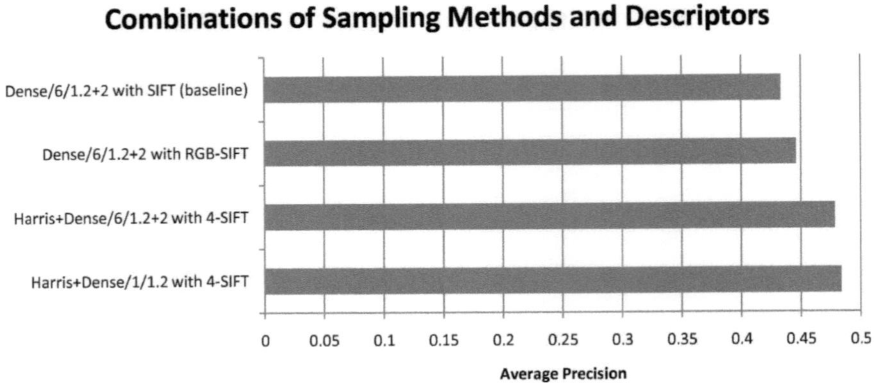

Fig. 18.3: Performance of different combinations of multiple point sampling strategies and multiple descriptors. Numeric results are given in Table 18.2.

Table 18.2: Performance of different combinations of multiple point sampling strategies and multiple descriptors. A visualization of the results is given in Figure 18.3.

Combinations of sampling strategies and descriptors	Average Precision
Dense sampling every six pixels (multi-scale) with SIFT (baseline)	0.433
Dense sampling every six pixels (multi-scale) with RGB-SIFT	0.446
Harris-Laplace and dense sampling every six pixels (multi-scale) with 4-SIFT	0.478
Harris-Laplace and dense sampling every pixel (single-scale) with 4-SIFT	**0.484**

- Combination of eight features: Harris–Laplace salient points paired with each of the four ColorSIFT descriptors, and densely sampled points at a single scale at every pixel, e.g. an interval of one pixel, also paired with each of the four ColorSIFT descriptors.

The results in Figure 18.3 and Table 18.2 show that the first combination is a relative improvement of 7% over the RGB–SIFT feature (absolute difference 0.032). The second combination is an improvement of 8% (absolute difference 0.038). These differences are significant in benchmark evaluations, and they show that using different color descriptors together is not redundant, because performance improved by combining them.

It is important to realize that about 90% of the state–of–the–art performance can be obtained by using the densely sampled RGB–SIFT feature with an interval of six pixels. The computational effort to extract this feature instead of applying the full feature combination is eight times lower when compared to the first combination, and 25 times lower compared to the second combination. For data sets orders of magnitude larger than the MIR–Flickr data set, choosing the single best feature might be more practical.

In conclusion, by combining different sampling strategies and descriptors, a performance improvement of up to 8% is possible. At the same time, when feature extraction becomes a computational bottleneck, picking a single good feature can already give up to 90% of the performance of the state–of–the–art.

18.3.4 Discussion

Based on the experiments in this section, we have found that the combination of spatial pyramid subdivisions of 1 x 1, 2 x 2 and 1 x 3 is a good choice for the MIR–Flickr data set. This confirms similar observations made on the PASCAL VOC in (Everingham et al, 2010).

In terms of point sampling strategy, a dense sampling strategy gives the best results for on the MIR–Flickr data set. This is related to the large number of scene concepts to be annotated: dense sampling has been shown to be advantageous for scene type classification, since salient points do not capture the entire appearance of an image. For dense sampling, it holds that denser sampling almost always gives higher accuracy than coarser sampling. The decision on how dense to sample should be made based on the available compute resources.

Which descriptor gives the highest performance depends on the data set used: in (Van de Sande et al, 2010) the C–SIFT descriptor gives the highest accuracy on PASCAL VOC. On the MIR–Flickr data set, the RGB–SIFT descriptor, which is invariant to illuminant color changes, gives the highest average precision.

By combining different sampling strategies and descriptors, as is done in all concept detection systems aiming for high accuracy, a performance improvement of up to 8% is possible given the features in this chapter. When limited compute resources are available, picking a single good feature, e.g. refraining from the use of combinations, can already give up to 90% of the performance of the current state–of–the–art.

18.4 ImageCLEF 2009

This section reports on the official ImageCLEF 2009 results of our concept detection system. Our focus on invariant visual features for concept detection in ImageCLEF 2009 was successful. It has resulted in the top ranking for the large–scale visual concept detection task in terms of both Equal Error Rate (EER) and Area Under the Curve (AUC).

All runs submitted to ImageCLEF 2009 use both Harris–Laplace, dense sampling with an interval of six pixels at two scales, the SVM classifier and a spatial pyramid with 1 x 1, 2 x 2 and 1 x 3 subdivisions. We do not use the EXIF metadata provided for the photos.

- **OpponentSIFT**: single color descriptor with hard assignment.
- **2–SIFT**: uses OpponentSIFT and SIFT descriptors.

Table 18.3: Overall results of the our runs evaluated over all concepts in the photo annotation task using the equal error rate (EER) and the area under the curve (AUC).

Run name	Codebook	Average EER	Average AUC
4-SIFT	Hard-assignment	**0.2345**	**0.8387**
Soft 4-SIFT	Soft-assignment	0.2355	0.8375
2-SIFT	Hard-assignment	0.2435	0.8300
OpponentSIFT	Hard-assignment	0.2530	0.8217

- **4–SIFT**: uses OpponentSIFT, C-SIFT, RGB-SIFT and SIFT descriptors. This run is equal to the first combination of the combination experiment in Section 18.3.3.
- **Soft 4–SIFT**: uses OpponentSIFT, C-SIFT, RGB-SIFT and SIFT descriptors with soft assignment. The soft assignment parameters have been taken from our PASCAL VOC 2008 system (Van de Sande et al, 2010).

In Table 18.3, the overall scores for the evaluation of concept detectors are shown. We note that the 4–SIFT run with hard assignment achieves not only the highest performance amongst our runs, but also over all other runs submitted to the large–scale visual concept detection task.

In Table 18.4, the Area Under the Curve scores have been split out per concept. We observe that the three aesthetic concepts have the lowest scores. This comes as no surprise, because these concepts are highly subjective: even human annotators only agree around 80% of the time with each other. For virtually all concepts besides the aesthetic ones, either the Soft 4–SIFT or the Hard 4–SIFT is the best run. This confirms our beliefs that these (color) descriptors are not redundant when used in combinations. Therefore, we recommend the use of these four descriptors instead of one or two. The difference in overall performance between the Soft 4–SIFT or the Hard 4–SIFT run is quite small. Because the soft codebook assignment smoothing parameter was directly taken from a different data set, we expect that the soft assignment run could be improved if the soft assignment parameter was selected with cross–validation on the training set. Together, our runs obtain the highest Area Under the Curve scores for 40 out of 53 concepts in the photo annotation task (20 for Soft 4–SIFT, 17 for 4–SIFT and three for the other runs). This analysis has shown us that our system is falling behind for concepts that correspond to conditions we have included invariance against. Our method is designed to be robust to unsharp images, so for *out–of–focus*, *partly–blurred* and *no–blur* there are better approaches possible. For the concepts *overexposed*, *underexposed*, *neutral–illumination*, *night* and *sunny*, recognizing how the scene is illuminated is very important. Because we are using invariant color descriptors, a lot of the discriminative lighting information is no longer present in the descriptors. Again, there should be better approaches possible for these concepts, such as estimating the color temperature and overall light intensity.

Table 18.4: Results per concept for our runs in the large–scale visual concept detection task using the Area Under the Curve. The highest score per concept is highlighted using a grey background. The concepts are ordered by their highest score.

Concept	4-SIFT	Soft 4-SIFT	2-SIFT	Opp.SIFT
Clouds	**0.958**	0.958	0.951	0.945
Sunset-Sunrise	0.953	**0.954**	0.947	0.946
Sky	0.945	**0.948**	0.935	0.930
Landscape-Nature	**0.944**	0.942	0.940	0.936
Sea	**0.935**	0.930	0.932	0.926
Mountains	**0.934**	0.931	0.930	0.922
Lake	0.911	0.903	**0.912**	0.900
Beach-Holidays	0.906	**0.907**	0.898	0.884
Trees	**0.903**	0.902	0.892	0.881
Water	0.901	**0.903**	0.892	0.886
Night	**0.898**	0.895	0.895	0.892
River	**0.897**	0.889	0.891	0.883
Outdoor	0.890	**0.896**	0.879	0.871
Food	**0.895**	0.895	0.881	0.877
Desert	**0.891**	0.865	0.891	0.884
Building-Sights	0.880	**0.882**	0.873	0.861
Big-Group	**0.881**	0.877	0.870	0.858
Plants	0.877	**0.881**	0.853	0.839
Flowers	0.868	**0.875**	0.846	0.836
Autumn	**0.870**	0.866	0.863	0.849
Portrait	**0.865**	0.864	0.857	0.846
Underexposed	0.858	**0.859**	0.857	0.854
No-Persons	0.850	**0.858**	0.837	0.826
Partly-Blurred	**0.852**	0.852	0.845	0.830
Winter	0.843	**0.846**	0.832	0.828
Snow	**0.846**	0.845	0.829	0.825
Day	0.841	**0.845**	0.831	0.824
No-Blur	0.843	**0.845**	0.836	0.823

Concept	4-SIFT	Soft 4-SIFT	2-SIFT	Opp.SIFT
No-Visual-Time	0.833	**0.835**	0.822	0.815
Indoor	0.830	**0.835**	0.823	0.810
Familiy-Friends	**0.834**	0.834	0.822	0.813
Partylife	**0.834**	0.834	0.831	0.819
Vehicle	**0.832**	0.832	0.832	0.822
Animals	0.818	**0.828**	0.811	0.797
Citylife	**0.826**	0.826	0.819	0.813
Still-Life	0.824	**0.825**	0.808	0.795
Spring	**0.822**	0.801	0.812	0.791
Canvas	**0.817**	0.810	0.803	0.790
Summer	**0.813**	0.813	0.791	0.782
Macro	**0.812**	0.791	0.805	0.795
No-Visual-Season	0.805	**0.806**	0.794	0.782
Small-Group	0.792	**0.795**	0.784	0.776
Single-Person	0.792	**0.795**	0.780	0.769
Out-of-focus	**0.792**	0.781	0.784	0.774
No-Visual-Place	**0.789**	0.786	0.781	0.779
Overexposed	**0.788**	0.782	0.777	0.771
Neutral-Illumination	0.778	**0.783**	0.775	0.774
Sunny	0.763	**0.765**	0.744	0.741
Motion-Blur	0.744	**0.747**	0.725	0.710
Sports	**0.695**	0.695	0.679	0.673
Aesthetic-Impression	0.658	**0.662**	0.657	0.657
Overall-Quality	0.656	0.656	0.653	**0.658**
Fancy	0.565	0.559	0.580	**0.583**
Average	0.8387	0.8375	0.8300	0.8217

18.4.1 Evaluation Per Image

For the hierarchical evaluation, overall results are shown in Table 18.5. When compared to the evaluation per concept, the Soft 4–SIFT run is now slightly better than the normal 4–SIFT run. While our method provides the best run for the per–concept evaluation, for the hierarchical evaluation measure, several other participants perform better. Discussion at the workshop has shown that exploiting the hierarchical nature of the concepts used is an interesting future direction.

18.4.2 Conclusion

The focus on invariant visual features for concept detection in ImageCLEF 2009 was successful. It resulted in the top ranking for the large–scale visual concept detection task in terms of both EER and AUC. For 40 individual concepts, the highest performance of all submissions to the task was obtained. For the hierarchical evaluation, how to exploit the hierarchical nature of the concepts is still an open question.

Table 18.5: Results using the hierarchical evaluation measures for our runs in the ImageCLEF 2009 large–scale visual concept detection task.

		Average Annotation Score	
Run name	Codebook	with agreement	without agreement
Soft 4-SIFT	Soft-assignment	**0.7647**	**0.7400**
4-SIFT	Hard-assignment	0.7623	0.7374
2-SIFT	Hard-assignment	0.7581	0.7329
OpponentSIFT	Hard-assignment	0.7491	0.7232

Table 18.6: Overall results of the our runs evaluated over all concepts in the ImageCLEF@ICPR 2010 photo annotation task using the EER and the AUC.

Run contents	Avg. EER	Avg. AUC
Harris–Laplace, dense sampling every 6 pixels (multi-scale) with 4-SIFT	0.2214	0.8538
Harris–Laplace, dense sampling every pixel (single-scale) with 4-SIFT	0.2182	0.8568
University of Surrey (enhanced machine learning)	**0.2136**	**0.8600**

18.5 ImageCLEF@ICPR 2010

The visual concept detection and annotation task has been part of a contest for the 2010 ICPR conference. The MIR–Flickr data set is used with the 53 concepts annotated for ImageCLEF 2009. However, this time there are 8,000 labelled images available for training, and the test set consists of 13,000 images.

The University of Amsterdam submitted two runs to the ICPR contest: the two good combinations which were identified in Section 18.3.3. In Table 18.6, the overall scores for the evaluation of concept detectors are shown. The first run is equal in terms of features to the best run submitted to ImageCLEF 2009. The second run, with more densely sampled SIFT, achieves higher accuracy. Compared to all other runs submitted to the task, the University of Amsterdam runs are ranked in second place. The University of Surrey achieved the highest overall accuracy. Their system uses similar visual features within a bag–of–words model but uses improved machine learning algorithms.

18.6 Conclusion

The current state–of–the–art in visual concept detection and annotation tasks is based on the bag–of–words model. Within this model, we have identified several design choices which lead to higher classification accuracy. Participation in the ImageCLEF photo annotation benchmarks was successful, and this participation was based on the following conclusions: (1) In terms of point sampling strategy, dense sampling gives the best results due to the large number of scene concepts to be annotated. (2) Increasing sampling density improves accuracy. (3) Spatial pyramid subdivisions of 1 x 1, 2 x 2 and 1 x 3 are a good choice for data sets in general. (4) The descriptor which gives the highest performance depends on the data set used; for the MIR–Flickr data set, the RGB–SIFT descriptor is recommended. (5) By combining different sampling strategies and descriptors, a performance improvement of up to 8% is possible given the features in this chapter.

Finally, when limited compute resources are available, picking a single good feature, e.g. refraining from the use of combinations, can already give up to 90% of the performance of the current state–of–the–art.

References

Burghouts GJ, Geusebroek JM (2009) Performance evaluation of local color invariants. Computer Vision and Image Understanding 113:48–62

Chang CC, Lin CJ (2001) LIBSVM: a library for support vector machines. Software available at `http://www.csie.ntu.edu.tw/~cjlin/libsvm`

Everingham M, Van Gool L, Williams C, Winn J, Zisserman A (2010) The pascal visual object classes (voc) challenge. International Journal of Computer Vision 88(2):303–338

Fei-Fei L, Perona P (2005) A bayesian hierarchical model for learning natural scene categories. In: IEEE Conference on Computer Vision and Pattern Recognition, vol 2, pp 524–531

Geusebroek JM, van den Boomgaard R, Smeulders AWM, Geerts H (2001) Color invariance. IEEE Transactions on Pattern Analysis and Machine Intelligence 23(12):1338–1350

Jurie F, Triggs B (2005) Creating efficient codebooks for visual recognition. In: IEEE International Conference on Computer Vision, pp 604–610

Lazebnik S, Schmid C, Ponce J (2006) Beyond bags of features: Spatial pyramid matching for recognizing natural scene categories. In: IEEE Conference on Computer Vision and Pattern Recognition, vol 2, pp 2169–2178

Leung TK, Malik J (2001) Representing and recognizing the visual appearance of materials using three–dimensional textons. International Journal of Computer Vision 43(1):29–44

Lin HT, Lin CJ, Weng RC (2007) A note on Platt's probabilistic outputs for support vector machines. Machine Learning 68(3):267–276

Lowe DG (2004) Distinctive image features from scale-invariant keypoints. International Journal of Computer Vision 60(2):91–110

Marszałek M, Schmid C, Harzallah H, van de Weijer J (2007) Learning object representations for visual object class recognition. Visual Recognition Challenge workshop, in conjunction with IEEE International Conference on Computer Vision

Nowak S, Dunker P (2009) Overview of the clef 2009 large scale visual concept detection and annotation task. In: Working notes CLEF 2009, Corfu, Greece

Van de Sande KEA, Gevers T, Snoek CGM (2008) A comparison of color features for visual concept classification. In: ACM International Conference on Image and Video Retrieval, ACM press, pp 141–150

Van de Sande KEA, Gevers T, Snoek CGM (2010) Evaluating color descriptors for object and scene recognition. IEEE Transactions on Pattern Analysis and Machine Intelligence 32(9)

Snoek CGM, Worring M (2009) Concept–based video retrieval. Foundations and Trends in Information Retrieval 4(2):215–322

Snoek CGM, van de Sande KEA, de Rooij O, Huurnink B, van Gemert JC, Uijlings JRR, et al (2008) The MediaMill TRECVID 2008 semantic video search engine. In: Proceedings of the TRECVID Workshop

Snoek CGM, van de Sande KEA, de Rooij O, Huurnink B, Uijlings JRR, van Liempt M, Bugalho M, Trancoso I, Yan F, Tahir MA, Mikolajczyk K, Kittler J, de Rijke M, Geusebroek JM, Gevers T, Worring M, Koelma DC, Smeulders AWM (2009) The MediaMill TRECVID 2009 semantic video search engine. In: Proceedings of the TRECVID Workshop

Tuytelaars T, Mikolajczyk K (2008) Local invariant feature detectors: A survey. Foundations and Trends in Computer Graphics and Vision 3(3):177–280

Uijlings JRR, Smeulders AWM, Scha RJH (2009) Real–time bag–of–words, approximately. In: ACM International Conference on Image and Video Retrieval. ACM press

Van Gemert JC, Veenman CJ, Smeulders AWM, Geusebroek JM (2010) Visual word ambiguity. IEEE Transactions on Pattern Analysis and Machine Intelligence 32(7):1271–1283

Vapnik VN (2000) The nature of statistical learning theory, 2nd edn. Springer

Wang D, Liu X, Luo L, Li J, Zhang B (2007) Video diver: generic video indexing with diverse features. In: ACM International Workshop on Multimedia Information Retrieval. ACM press, Augsburg, Germany, pp 61–70

Zhang J, Marszałek M, Lazebnik S, Schmid C (2007) Local features and kernels for classification of texture and object categories: A comprehensive study. International Journal of Computer Vision 73(2):213–238

Chapter 19
Intermedia Conceptual Indexing

Jean–Pierre Chevallet and Joo Hwee Lim

Abstract Indexing textual and visual documents at the same conceptual level enables the building of an effective intermedia conceptual indexing. In this chapter we present models and techniques used to achieve this conceptual indexing level. Our experiments were done on ImageCLEF medical data, using the UMLS (Unified Medical Language System) source of concepts for text, and a manually built visual ontology. We have used the UMLS meta–thesaurus as an external resource for indexing text at a conceptual level. At this level, the query can be structured using domain dimensions. The visual ontology has been built using the most frequent concepts from UMLS, and cropped images from the Web and from the corpus itself. The system is then able to index text and images using the same conceptual set.

19.1 Introduction

Medical Information Retrieval is an active research field (Boudin et al, 2010). Medical images are an integral part of medical diagnosis, research, and teaching. Medical image analysis research has focused on image registration, measurement, and visualization. Although large numbers of medical images are produced in hospitals every day, there is relatively little research in medical Content–Based Image Retrieval (CBIR) (Shyu et al, 2002a). In addition to being valuable for medical research and training, medical CBIR systems also have a role to play in clinical diagnosis (Müller et al, 2004).

A medical CBIR requires indexing of images and text with precise meaning. This implies the use of external explicit knowledge both for image and text data. For im-

Jean Pierre Chevallet
LIG, Grenoble University, e-mail: `Jean-Pierre.Chevallet@imag.fr`

Joo Hwee Lim
Institute for Infocom Research e-mail: `joohwee@i2r.a-star.edu.sg`

H. Müller et al. (eds.), *ImageCLEF*, The Information Retrieval Series 32,
DOI 10.1007/978-3-642-15181-1_19, © Springer-Verlag Berlin Heidelberg 2010

ages, this knowledge can be learned from visual examples that are manually linked to concepts. This learning is based on local features and does not rely on robust region segmentation. For text, this requires the use of an ontology to force the presence of key terms and to discard inconsistent terms. We present models and methods to extract visual features from images and to identify medical concepts from texts. We also show how concepts can be associated with image samples, and can be learned. Our ultimate goal is the definition of a unique conceptual index, i.e. that is common to image and text. We call this shared index an intermedia conceptual index.

Among the research efforts of medical CBIR, classification– or clustering–driven feature selection has received much attention. General visual cues often fail to be discriminative enough to deal with more subtle, domain–specific differences (Liu et al, 2004). Pathology–bearing regions tend to be highly localized (Dy et al, 2003). Hence, local features such as those extracted from segmented dominant image regions approximated by best fitting ellipses have been proposed (Lehmann et al, 2004). A hierarchical graph–based representation and matching scheme has been suggested to deal with multi–scale image decomposition and their spatial relationships. However, it has been recognized that pathology bearing regions cannot be segmented out automatically for many medical domains (Shyu et al, 2002b). As an alternative, a comprehensive set of perceptual categories related to pathology bearing regions and their discriminative features are manually designed and tuned to achieve superior precision rates over a brute–force feature selection approach.

Hence, it is desirable to have a medical CBIR system that represents images in terms of semantic local features that can be learned from examples rather than handcrafted with a lot of expert input and do not rely on robust region segmentation. In order to manage a large and complex set of visual entities, we propose a structured learning framework to facilitate modular design and extraction of medical visual semantics that we call VisMed terms. These VisMed terms are image regions that exhibit semantic meanings to medical practitioners and that can be learned statistically to span a new indexing space. During image indexing, VisMed terms are detected in image content, reconciled across multiple resolutions and aggregated spatially to form local semantic histograms.

The resulting compact and abstract VisMed image indexes can support both similarity–based queries and semantics–based queries efficiently. When queries are in the form of example images, both a query image and a database image can be matched based on their distributions of VisMed terms, much like the matching of feature–based histograms though the bins refer to semantic medical terms. In addition, we propose a flexible tiling (FlexiTile) matching scheme to compare the similarity between two medical images of arbitrary aspect ratios.

When a query is expressed as a text description that involves modality, anatomy, and pathology, etc., they can be translated into a visual query representation that chains the presences of VisMed terms with spatial significance via logical operators (AND, OR, NOT) and spatial quantifiers for automatic query processing based on the VisMed image indexes. This query formulation and processing scheme allows semantics–based retrieval with text queries.

By fusing the ranked lists from both the similarity– and semantics–based retrieval we can leverage the information expressed in both visual and text queries.

This chapter first presents the notion of conceptual indexing in Section 19.2 and the concept mapping on text to build an abstract index. Then in Section 19.3, we present models and techniques used to index images using visual terms of a visual ontology. Finally, in Section 19.4 we present our intermedia conceptual indexing combination framework that has produced the best results in ImageCLEF 2006.

19.2 Conceptual Indexing

In an Information Retrieval (IR) system, automatic indexing consists of describing documents in a way they can be easily retrieved. The goal of indexing in the context of IR is not to *understand* document content but rather to produce a description that enables the computation of a Relevance Status Value (RSV) as close as possible to human relevance judgment. Research activities in IR are mainly concerned about how to compute an RSV that is close to human judgment. We make the following assumption: an information retrieval system should incorporate some of the user knowledge and reasoning capabilities to propose more relevant documents. This assumption (Schank et al, 1980) has driven research on systems using domain knowledge (thesaurus, knowledge base, ontology, etc.), complex Natural Language Processing (NLP), or systems that describe document content (indexes) in a complex way (i.e. conceptual graphs (Huibers et al, 1996), terminological logics (Meghini et al, 1993)) with a match related to an uncertain logical deduction (van Rijsbergen, 1986). In our work we have endorsed this conceptual indexing paradigm, mainly for text indexing but also partially for image indexing. In this section we present our approach to conceptual indexing.

19.2.1 Concept Usage and Definition in IR

The notion of concept is difficult to define: in (Goguen, 2005), a concept is close to the notion of a category with a few mathematical constraints. We have to be pragmatic and sketch concepts close to actual needs and uses. Concepts can be defined as human understandable unique abstract notions independent from any direct material support, any language or information representation, and used to organize perception and knowledge (Chevallet et al, 2007). Concepts are abstraction units built by humans and generalized from properties of objects, facts, events, etc.

From this definition, no machine can extract concepts from any digital source but rather one can try to automatically map concepts to parts of digitalized data (image, text, etc.). In practice, concepts are identifiers with associate information to describe them, usually texts and terms, but also image patches, logical definitions or constraints. For example, CYC (Lenat, 1995) is a large set of concepts (ontology) in a machine readable format, where concepts are described by a logical expression and a set of related terms. ConceptNet (Liu and Singh, 2004) is more informal, without logical language, and it captures common sense knowledge with an emphasis on rich semantic relations. UMLS[1] describes a large set of concepts by the merging of several resources. It tries to unify terms expressing the same meaning into an identified concept. This resource is called a meta–thesaurus because it abstracts several thesauri (e.g. Medical Subject Heading or MeSH), into a comprehensive resource in the medical domain. Conceptual indexing can be defined as: using concept identifiers in an index instead of words, terms, or image features. The characteristic of such an index is to be media independent and for text to be language independent. However, there are challenges in setting up a conceptual index.

First, a resource that describes a large set of concepts is mandatory, and second is the need for an efficient tool to map concepts to documents. For text, it implies NLP tools to identify terms and replace them with concepts. Because a resource cannot include all possible variations of terms, this mapping tool must cope with linguistic variation. Concept identification is difficult because of the inherent ambiguity and flexibility of natural language. There are also many language phenomena, such as elision, that complicate the task. Selecting concepts from text means disambiguating the text, which is always a very difficult task (Ide and Veronis, 1998). Finally, a flat set of concepts can lead to a sharp decline in recall if the system is not able to establish a link from a general concept in a query (e.g. 'bone fracture'), and perhaps more precise concepts present in documents (e.g. 'fracture of the femur'). Relations in the knowledge resource are hence mandatory.

19.2.2 Concept Mapping to Text

To map concepts to text for indexing in an IR system we require the following:

- **Terminology**. A list of terms (single or multi term) from a given domain in a given language. Terms come from actual language usage in the domain. They are generally stable noun phrases (i.e. less linguistic variations than any other noun phrase) and they should have an unambiguous meaning in the restricted domain they are used in.
- **Conceptual vocabulary**. A set of concepts. A concept is a language–independent meaning (with a definition), associated with at least one term of the terminology.
- **Conceptual structure**. Each term is associated with at least one concept. Concepts are also organized into several networks. Each network links concepts using a conceptual relation.
- **Conceptual mapping algorithm**. A method that selects a set of potential concepts from a sentence using the terminology and the conceptual structure.

The terminology, conceptual vocabulary and the conceptual structure form a knowledge resource often related to a domain (such as medicine). Conceptual text index-

[1] `http://www.nlm.nih.gov/research/umls/`

ing is then the operation of transforming natural language documents into an indexing structure of concepts (sets, vectors, graphs, etc.) through a conceptual mapping algorithm using the resource.

Dealing with a precise domain may reduce some of the concepts extraction problems such as ambiguity. Term ambiguity arises when different concepts are mapped to a single term. In practice, ambiguity depends on the precision of the domain knowledge resource. If we reduce the domain we also reduce the meanings of possible terms, hence also ambiguity. For example, 'x–ray' may refer to a wave in physics but also to an image modality in radiology. Unfortunately, when we have more precise concepts (and terms) we are confronted with another form of ambiguity called structural ambiguity. At the syntactic level, structural ambiguity occurs when a phrase has more than one underlying structure. It corresponds to several ways to extract concepts. The term 'right lobe pneumonia' can be associated with a single concept but can be split into two terms associated with other concepts: 'right lobe' and 'pneumonia'.

A solution is to model concept structure equivalence. This consists of setting up a model that expresses concept compositions and relations. Some of these relations can be equivalence or subsumption. A terminological logic can be used (Sebastiani, 1994) but it is often neither simple nor possible for indexing a large set of concepts because concepts have to be expressed in the chosen formalism. A large ontology in medicine with concepts expressed in a logical format does not yet exist.

Another common difficulty for concept mapping is term variation. Terms should be stable noun phrases (Daille et al, 1996), but there are still in practice many variations in terms. It is the role of the terminology to list all term variations but in practice some variations have to be processed by the conceptual mapping algorithm.

19.2.3 Mapping Steps

In spite of these difficulties conceptual indexing can produce a high precision multi–lingual index, and can solve very precise queries. This solution is adapted to the medical domain. In the following we detail the steps for concept mapping. For IR we concentrate on noun phrases only because they support most document themes. Globally the following steps are required to extract concepts from texts:

Parts of speech tagging: morphology associates Parts Of Speech (POS) tags to every word. Without extra information such as syntax, semantics or pragmatics, some POS errors may occur. A POS tagger for IR should provide all alternatives.

Syntactic analysis: syntax is used to detect phrase boundaries. Surface syntax and a shallow parser are sufficient in IR. For noun phrases a regular expression on POS is sufficient because complexity of noun phrase syntax is low.

Term variation evaluation: variation is based on syntactic modifications, abbreviations, etc. A distance has to be computed from the original phrase because each variation could shift the original meaning.

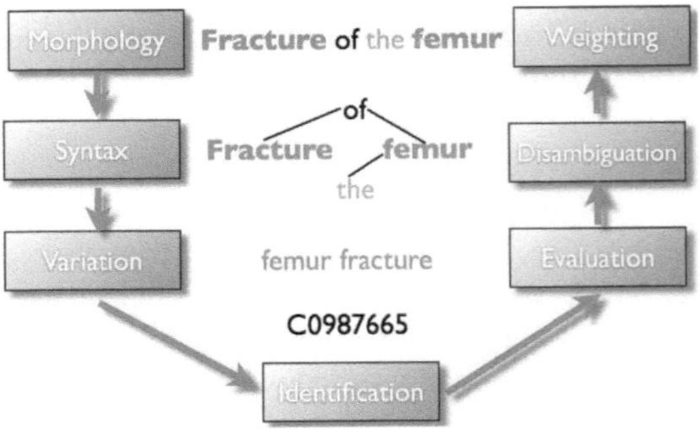

Fig. 19.1: Mapping steps of the indexing process.

Term to concept mapping: this consists of identifying possible concept candidates for each phrase variant from the terminology.

Evaluation of the mapping: evaluation expresses the probability of correct concept identification for each concept candidate.

Term disambiguation: reduces the set of candidate concepts. It can use local word or sentence information such as the concept evaluation, and global information related to the whole document such as the domain of the document.

Concept weighting: measures concept relevance, which is the IR value related to index usefulness, measuring the relevance of concept descriptors.

The mapping step, as shown in Figure 19.1, is critical because it relies on the quality of the terminology resource. A small resource will produce silence in concept detection, a larger ambiguity. The depth of a resource is also important. It refers to the size of the conceptual hierarchy and the focus of the concepts. For example, concepts in deep resources are associated with very long noun phrases such as the concept[2] C0161118 associated to the phrase:

> "Blisters with epidermal loss due to second degree burn of chest wall, excluding breast and nipple."

Deep resources will produce much more detailed and precise concepts and can lead to structural ambiguity.

We use UMLS for our domain knowledge resource because no other resource of its size currently exists. In fact, UMLS is a meta–thesaurus, i.e. a merging of existing thesauri and terminologies. Merging thesauri does not lead to an ideal conceptual structure as not all entries are terms, so not all entries can be found in actual text (like C0029537: other chest pain). Moreover, different thesaurus structures (ex: hierarchies) have to be merged into a single structure. The merging was done by exhibiting concepts linking multiple terms from multiple sources. UMLS is still a good candidate to approximate a domain knowledge resource for medical image

[2] Example from UMLS.

and text indexing. The first reason is because of its size with the 2010 version of UMLS containing more than 2.2 million concepts and 8.2 million unique concept names from over 150 source vocabularies in 17 languages. Secondly it is maintained by specialists with two updates a year. Unfortunately, UMLS is neither complete, nor consistent. In particular, the links among concepts are not equally distributed. The inter–concept relationships (such as hierarchies) are those from the source thesaurus. Hence, there is redundancy as multiple similar paths can be found between two concepts using different sources. In order to have a common categorization of this concept set, UMLS has global high level semantic categories called semantic types and semantic groups (Bodenreider and Mccray, 2003), assigned manually and independently of all thesaurus hierarchies by the meta–thesaurus editors. These structures are the only ones that are consistent for the whole data set.

To map concepts to English texts we used the MetaMap tool provided with UMLS. Here is a simplified output from a sentence:

"Show me a chest x–ray with tuberculosis."

First, the parser produces the POS (`tag`) and a simplified dependency tree with identification of the `head` and `mod`ifiers of the noun phrase.

```
phrase('Show',[verb([lexmatch([show]),
               inputmatch([Show]),tag(verb))]).
phrase(me,[pron([lexmatch([me]),inputmatch([me])
                                ,tag(pron))]).
phrase('a chest x-ray',
         [det([lexmatch([a])
              ,tag(det)])
              ,mod([lexmatch([chest]),tag(noun)]
              ,head([lexmatch([X-ray])
                    ,tag(noun)])]).
```

Only noun phrases are then mapped to concepts with an evaluation (`ev`) of each possible concept proposition. For each concept proposition, words of the associated term are mapped to the list of source terms. For example, `[2,2],[1,1]` for the third candidate means that the second term `'x'` of `'chest x-ray'` matches with `'X'` of the candidate term `'X-ray'` associated with the concept `C0034571`. The number that follows is the distance of this matching: 0 is an exact matching (exact spelling), 1 is a inflectional variant (Aronson, 2006).

```
candidates([
  ev(-923,'C0202783','Chest x-ray',
    'Radiographic procedure on chest (procedure)'
    ,[chest,x,ray]
    ,[[[1,1],[1,1],0],[2,2],[2,2],0]
      ,[3,3],[3,3],0]]),
  ev(-895,'C0856599','Breast X-ray',[breast,x,ray]
    ,[[[1,1],[1,1],4],[2,2],[2,2],0]
      ,[3,3],[3,3],0]]),
```

```
ev(-861,'C0034571','X-ray','roentgenographic'
  ,[x,ray],[[[2,2],[1,1],0],[3,3],[2,2],0]]),
ev(-861,'C0043299','X-ray'
  ,'Diagnostic radiologic examination'
  ,[x,ray],[[[2,2],[1,1],0],[3,3],[2,2],0]]),
ev(-861,'C0043309','X-ray','Roentgen Rays'
  ,[x,ray],[[[2,2],[1,1],0],[3,3],[2,2],0]]),
ev(-861,'C1306645','X-ray'
  ,'Plain film (procedure)'
  ,[x,ray],[[[2,2],[1,1],0],[3,3],[2,2],0]]),
...
```

In this example, 17 possible concepts were identified as having a possible association with the text 'a chest x–ray'. The ordering is based only on syntactic variations and it is not supposed to be a semantic distance. Hence this ordering cannot be considered a disambiguation. We are interested in precision–oriented indexing, so we decided to retain the best MetaMap proposition for indexing plus all sub partial matchings with no variation (i.e. a `0` distance value in the matching `ev` list). This is a way to take into account structural ambiguity, even if it is probably not the best representation.

19.2.4 IR Models Using Concepts

Once texts are replaced with concepts, the next step consists of producing the document index. In IR, a document index is a compact representation of a document where the RSV computation is performed. Concept mapping associates a word sequence to a set of possible concepts. We make the hypothesis that concept distributions of such a conceptual indexing approach are similar to word distributions. In reality this is incorrect. Concepts may be associated with large terms, hence statistical concept distribution may be different from single word distribution. There are also structural ambiguities, where several possible concepts are associated with a different coverage of a noun phrase. In this way we violate the usual independence assumption between items in the index that is the basis of all weighing schemes. Unfortunately, we do not have any new model to propose and we reuse existing word based IR models such as the Vector Space Model (VSM) with different weighting schemes or language models.

Using concepts as input of the indexing does not add a decisive advantage over the classical 'bag of words' approach. The real advantage for using concepts instead of words lies in the conceptual structure. Each concept is embedded in a network of semantic links and is also associated with semantic categories. These categories are used to structure document and query content. For example (Figure 19.2), we can automatically detect in the query *show me an x-ray of a fracture of a femur*, three parts: a modality (x–ray), an anatomy (the femur bone) and a pathology (fracture). This forms a conceptual structure to documents or queries. This structure can then

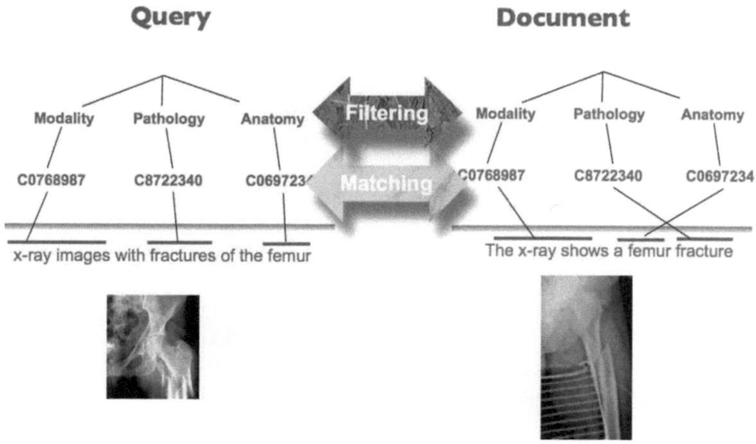

Fig. 19.2: Conceptual matching and indexing between queries and documents.

be exploited to filter the possible answers in order to constrain documents to cover the conceptual query structure.

UMLS has such a conceptual structure called semantic groups. The matching computed on the conceptual index RSV_c is completed with a conceptual structure filtering CSF:

$$RSV_c^{CSF}(Q,D) = CSF(Q,D) \times RSV_c(Q,D) \tag{19.1}$$

In our experiments, we tested two CSF functions:

Inclusion. This is a binary function that acts as a filter. Its value is 1 only when semantic groups found in Q are also in D, 0 otherwise.

Intersection. This computes the intersection size of semantic groups in Q that appear in documents D.

Both techniques improved the results as shown in the experiments in the following section.

19.2.5 Experiments using the ImageCLEF Collection

We experimented using this conceptual index approach on several versions of the ImageCLEFmed collection. For example, we show in Tables 19.1 to 19.3 the results of ImageCLEF 2006[3]. We tested various weighting schemes. Matching is computed using the inner product between the query and the document vector. Simple word or concept frequency is named FREQ. The vector space model with $tf*idf$ weighting

[3] These results are published in (Chevallet et al, 2007).

Table 19.1: Retrieval results when using word–based text indexing.

	FREQ	VSM	DFR*	BM25	BM25*
English	0.028	0.166	0.106	0.157	**0.178**
French	0.033	0.064	**0.068**	0.050	0.067
German	0.010	**0.017**	**0.017**	0.013	**0.017**
All	0.021	0.176	0.127	0.167	**0.200**

Table 19.2: Retrieval results when using conceptual text indexing.

	FREQ	VSM	DFR*	BM25	BM25*
English	0.049	0.209	0.210	0.208	**0.215**
French	0.003	**0.070**	0.047	0.043	**0.070**
German	0.007	**0.016**	0.014	0.011	**0.016**
All	0.004	0.204	0.097	0.098	**0.221**

Table 19.3: Retrieval results when using conceptual structure filtering.

	All	Inclusion		Intersection	
	MAP	MAP	Change	MAP	Change
FREQ	0.004	0.033	+725%	0.017	+325%
VSM	0.204	0.239	+17%	**0.264**	+29%
DFR	0.097	0.135	+39%	0.177	+82%
BM25	0.098	0.140	+43%	0.180	+84%
BM25*	0.221	0.247	+12%	0.250	+13%

and cosine normalization are named *VSM*. We tested two probabilistic measures: Divergence From Randomness (DFR) (Amati and Van Rijsbergen, 2002) and the BM25 probabilistic model of Robertson and Walker (1994). In DRF* and BM25*, query weights are normalized with log, idf and vector size. Tables 19.1 to 19.3 show Mean Average Precision (MAP) across all queries.

Word–based text indexing results are shown in Table 19.1. These results serve as a comparison with conceptual indexing. As one can expect, simple frequency models deliver poor results and probabilistic models are more effective.

In Table 19.2, the effect of using concepts to replace words is not stable. In fact, the benefit of using concepts instead of words is not only in the results but in the system that has a single indexing for all languages.

In Table 19.3, the first column is a copy of the conceptual indexing and results show the use of the conceptual structure filtering on documents retrieved using the conceptual indexing. Results are unambiguous, semantic filtering always provides an increase in MAP. This can be explained by the bag of words or bag of concepts models computing query and document matches using a weighted intersection. The size of this intersection has no direct influence. A document that has only two concepts matching with a query but with a high weight can be placed higher than a document with a larger matching intersection but with lower weights. The medical queries of ImageCLEF are implicitly structured (i.e. pathology, modality and anatomy). If one of the dimensions is missing in a document, this document may not be relevant at all, even if the rest of the query produces a high score.

Conceptual indexing enables the discovery of semantic structure in texts. It provides an efficient way to discard documents with high scores but which are incomplete in terms of the semantic structure and thus are less relevant. In our view, this is the key interest in using a conceptual resource (such as UMLS) for indexing compared to the classical word based indexing. Using this approach, we have obtained some of the best results among the official runs (a MAP of 0.2646) in the medical task in ImageCLEF 2006 (Lacoste et al, 2006). In this run, the $CSF(Q,D)$ function is the number of semantic dimensions common to documents and the query (size of the intersection).

19.3 Image Indexing using a Visual Ontology

Image retrieval is an active research area where a large number of techniques and models have been explored, mostly since the year 2000 (Datta et al, 2008). Images play an important role in medical diagnosis and a major evolution often follows the development of new medical imaging techniques. This is a domain where images and text co–exist in documents and can be used in queries.

We promote the use of a visual ontologies to index images using concepts. A visual ontology is an association of concepts to a set of images that represents this concept. A concept is defined in the same way as for text.

Using visual concepts for CBIR follows the idea that exogenous knowledge is mandatory to build an IR system that is able to behave in a meaningful way. Associating image examples to concepts tends to build a 'visual definition' of this concept and enables the use of machine learning algorithms to learn relations between visual features and concepts. The notion of using a visual vocabulary to represent and index the image content for more effective (i.e. semantic) query and retrieval is not new. It was proposed and applied to consumer images (Lim, 2001; Lim and Jin, 2005b).

Applications of this approach to the medical domain are fruitful because this domain is both narrow and wide at the same time. It is narrow enough to collect medical terms and concepts into a comprehensive meta–thesaurus such as UMLS. It is also wide because of the large number of terms and concepts, and the constant evolution and production of new knowledge. We promote the use of visual medical terms called VisMed that are typical composite concepts with visual appearance in medical images (e.g. x–ray–bone–fracture, CT–abdomen–liver, MRI–head–brain, photo–skin). Many concepts bear no obvious visual characteristics but a few concept combinations can be associated unambiguously to an image. Our VisMeds are defined using image region instances cropped from sample images, then modeled and built based on statistical learning. This learning consists of building a visual signature for each image sample that is fed into a supervised classifier. In our experiments, we build image signatures from color and texture features. The classifier is based on a Support Vector Machine (SVM) (Vapnik, 1998). To compute VisMed terms from training instances we use SVMs on color and texture features for an image region and denote this feature vector as z. An SVM $S_k(z)$ is a detector for VisMed term k on z. The classification vector for region z is computed via the softmax (Bishop, 1996) function as:

$$T_k(z) = \frac{\exp^{S_k(z)}}{\sum_j \exp^{S_j(z)}}. \tag{19.2}$$

i.e. $T_k(z)$ corresponds to a VisMed entry in vector T.

In the experiments, we use the YIQ color space over other color spaces (e.g. RGB, HSV, LUV) as it performed better. As texture features we adopted Gabor coefficients, which have shown to provide good results (Manjunath and Ma, 1996). A feature vector z has two parts, namely, a color feature vector z^c and a texture feature vector z^t. We compute the mean and standard deviation of each YIQ color channel and the Gabor coefficients (five scales, six orientations) respectively (Lim and Jin, 2005b). Hence, the color feature vector z^c has six dimensions and the texture feature vector z^t has 60 dimensions. Zero–mean normalization is applied to both the color and texture features. In our evaluation, we adopt Radius Basis Function (RBF) kernels with modified city–block distance between feature vectors y and z: [

$$|y - z| = \frac{1}{2}\left(\frac{|y^c - z^c|}{N_c} + \frac{|y^t - z^t|}{N_t}\right) \tag{19.3}$$

where N_c and N_t are the numbers of dimensions of the color and texture feature vectors (i.e. 6 and 60) respectively. This just–in–time feature fusion within the kernel combines the contribution of color and texture features equally. It is simpler and more effective than other feature fusion methods we have attempted.

19.3.1 Image Indexing Based on VisMed Terms

After learning, the VisMed terms are detected during image indexing from multi–scale block–based image patches without region segmentation to form semantic local histograms as described below.

Conceptually, the indexing is realized in a three–layer visual information processing architecture (Figure 19.3). The bottom layer denotes the pixel-feature maps computed for feature extraction. In our experiments, there are three color maps (i.e. YIQ channels) and 30 texture maps (i.e. Gabor coefficients of five scales and six orientations). From these maps, feature vectors z^c and z^t, compatible with those adopted for VisMed term learning (Equation 19.3), are extracted.

To detect VisMed terms with translation and scale invariance in an image to be indexed, the image is scanned with windows of different scales, similar to the strategy in view–based object detection (Sung and Poggio, 1998; Papageorgiou et al, 1998). More precisely, given an image I with resolution $M \times N$ the middle layer, a Recon-

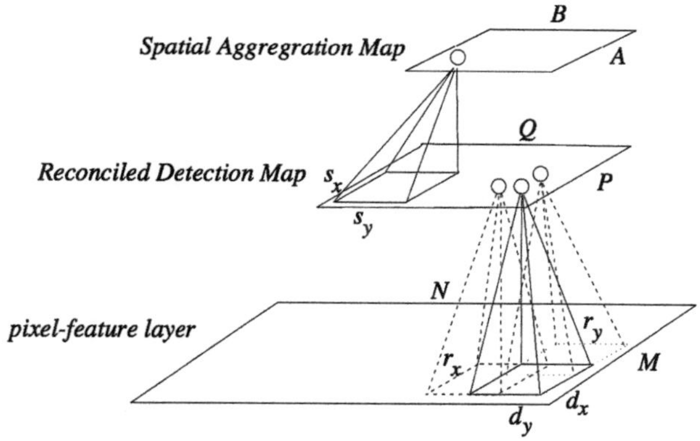

Fig. 19.3: A 3–layer architecture for image indexing.

ciled Detection Map (RDM), has a lower resolution of $P \times Q$, $P \leq M$, $Q \leq N$. Each pixel (p,q) in RDM corresponds to a two–dimensional region of size $r_x \times r_y$ in I. We further allow tessellation displacements $d_x, d_y > 0$ in X,Y directions respectively such that adjacent pixels in RDM along the X direction (along the Y direction) have receptive fields in I, which are displaced by d_x pixels along the X direction (d_y pixels along the Y direction) in I. After scanning an image, each pixel (p,q) that covers a region z in the pixel–feature layer will consolidate the classification vector $T_k(z)$ (Equation 19.2).

In our experiments, we progressively increase the window size $r_x \times r_y$ from 20×20 to 60×60 at a displacement (d_x, d_y) of $(10,10)$ pixels. For images the longer side of width and height is fixed at 360 pixels after a size normalization that preserves the aspect ratio. After the detection step we have five maps of detection of dimensions 23×35 to 19×31, which are reconciled into a common RDM as explained below.

To reconcile the detection maps across different resolutions onto a common basis we adopt the following principle: if the most confident classification of a region at resolution r is less than that of a larger region (at resolution $r+1$) that subsumes the region, then the classification output of the region is replaced by that of the larger region at resolution $r+1$. For instance, if the detection of a face is more confident than that of a building at the nose region (assuming that both face and building (but not nose) are in the visual vocabulary designed for a particular application), then the entire region covered by the face, which subsumes the nose region, should be labeled as face.

For example, a region at resolution r is covered by four larger regions at resolution $r+1$ as shown in Figure 19.4. Let $\rho = max_k max_i T_i(z_k^{r+1})$ where k refers to one of the four larger regions in the case of the example shown in Figure 19.4. Then, the principle of reconciliation says that if $max_i T_i(z^r) < \rho$, the classifica-

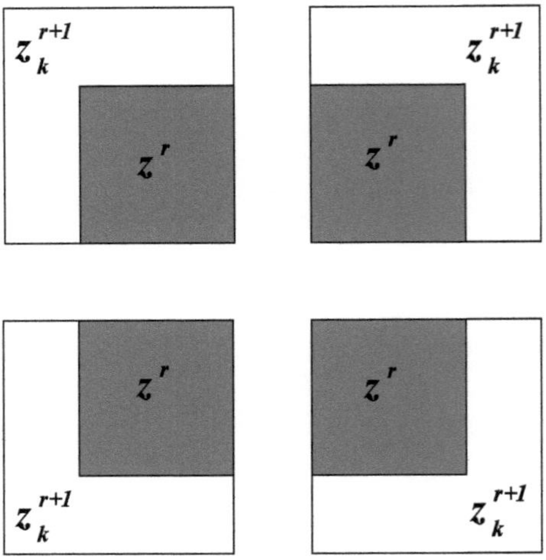

Fig. 19.4: Reconciling multi–scale VisMed detection maps.

tion vector $T_i(z^r)$ $\forall i$ is replaced by the classification vector $T_i(z_m^{r+1})$ $\forall i$ where $max_i T_i(z_m^{r+1}) = \rho$.

Using this principle, we compare detection maps of two consecutive resolutions at a time, in descending window sizes (i.e. from windows of 60×60 and 50×50 to windows of 30×30 and 20×20). After four cycles of reconciliation, the detection map that is based on the smallest scan window (20×20) has consolidated the detection decisions obtained at other resolutions for further spatial aggregation. The goal of spatial aggregation is to summarize the reconciled detection outcome in a larger spatial region. A region Z comprises n small equal regions with feature vectors $z_1, z_2, ..., z_n$ respectively. To account for the size of the detected VisMed terms in the spatial area Z, the classification vectors of the reconciled detection map are aggregated as:

$$T_k(Z) = \frac{1}{n} \sum_i T_k(z_i). \tag{19.4}$$

This is the top layer in our three–layer visual information processing architecture where a Spatial Aggregation Map (SAM) further tessellates over RDM with $A \times B$, $A \leq P$, $B \leq Q$ pixels. This form of spatial aggregation does not encode spatial relations explicitly. The design flexibility of s_x, s_y in SAM on RDM (the equivalent of r_x, r_y in RDM on I) allows us to specify the location and extent in the content to be focused and indexed. We can choose to ignore unimportant areas (e.g. margins) and emphasize certain areas with overlapping tessellation. We can even have different weights attached to the areas during similarity matching.

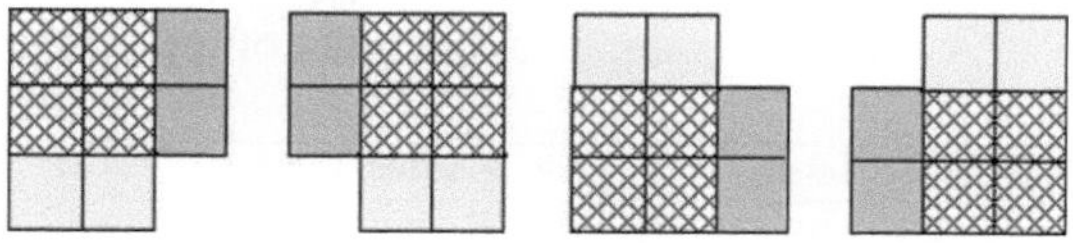

Fig. 19.5: Example to illustrate FlexiTile matching.

19.3.2 FlexiTile Matching

Given two images represented as different grid patterns, we propose a flexible tiling (FlexiTile) matching scheme to cover all possible matches. For instance, given a query image Q of a 3×1 grid and an image Z of a 3×3 grid, intuitively Q should be compared to each of the three columns in Z and the highest similarity will be treated as the final matching score. As another example, consider matching a 3×2 grid with a 2×3 grid. The four possible tiling and matching choices are shown in Figure 19.5.

The FlexiTile matching scheme is formalized as follows. A query image Q and a database image Z are represented as $M_1 \times N_1$ and $M_2 \times N_2$ grids respectively. The overlapping grid $M \times N$ where $M = \min(M_1, M_2)$ and $N = \min(N_1, N_2)$ is the maximal matching area. The similarity λ between Q and Z is the maximum matching among all possible $M \times N$ tilings,

$$\lambda(Q,Z) = \max_{m_1=1,n_1=1}^{m_1=u_1,n_1=v_1} \max_{m_2=1,n_2=1}^{m_2=u_2,n_2=v_2} \lambda(Q_{m_1,n_1}, Z_{m_2,n_2}), \tag{19.5}$$

where $u_1 = M_1 - M + 1, v_1 = N_1 - N + 1, u_2 = M_2 - M + 1, v_2 = N_2 - N + 1$ and the similarity for each tiling $\lambda(Q_{m_1,n_1}, Z_{m_2,n_2})$ is defined as the average similarity over $M \times N$ blocks as

$$\lambda(Q_{m_1,n_1}, Z_{m_2,n_2}) = \frac{\sum_i \sum_j \lambda_{ij}(Q_{m_1,n_1}, Z_{m_2,n_2})}{M \times N}, \tag{19.6}$$

and finally the similarity $\lambda_{ij}(Q_{m_1,n_1}, Z_{m_2,n_2})$ between two image blocks is computed based on the L_1 distance measure (city block distance) as,

$$\lambda_{ij}(Q_{m_1,n_1}, Z_{m_2,n_2}) = 1 - \frac{1}{2} \sum_k |T_k(Q_{p_1,q_1}) - T_k(Z_{p_2,q_2})| \tag{19.7}$$

where $p_1 = m_1 + i, q_1 = n_1 + j, p_2 = m_2 + i, q_2 = n_2 + j$ and it is equivalent to color histogram intersection except that the bins have semantic interpretation as VisMed terms.

There is a trade–off between content symmetry and spatial specificity. If we want images of similar semantics with different spatial arrangement (e.g. mirror images) to be treated as similar, we can have larger tessellated blocks in SAM (i.e. the extreme case is a global histogram). However, in applications with medical images

Table 19.4: VisMed example and number of region samples.

VisMed Terms	#	VisMed Terms	#
00-angio-aorta-artery	30	01-angio-aorta-kidney	30
02-ct-abdomen-bone	40	03-ct-abdomen-liver	20
04-ct-abdomen-vessel	30	05-ct-chest-bone	30
06-ct-chest-emphysema	30	07-ct-chest-nodule	20
08-path-alzheimer	40	09-path-kidney	50
10-path-leukemia	30	11-photo-face-eye	60
12-photo-face-mouth	30	13-photo-face-nose	30

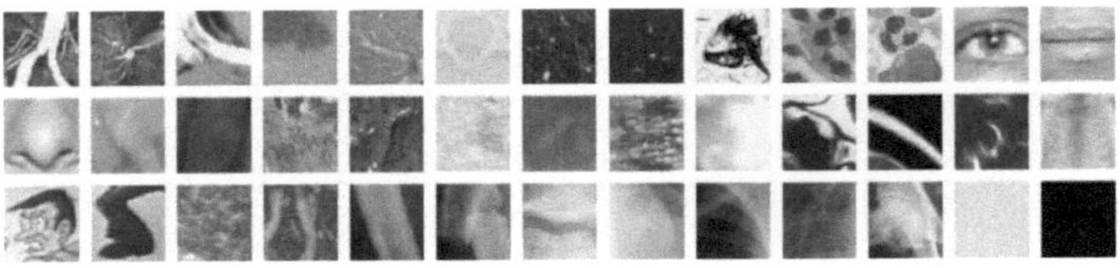

Fig. 19.6: Visual examples for VisMed terms.

where there is usually very little variance in views and spatial locations are considered differentiating across images, local histograms will provide good sensitivity to spatial specificity. Furthermore, we can attach different weights to the blocks to emphasize the focus of attention (e.g. center) if necessary.

19.3.3 Medical Image Retrieval Using VisMed Terms

We have applied the VisMed approach to the Medical Image Retrieval task in ImageCLEF. We set out VisMed terms that correspond to typical semantic regions in the medical images. We have manually designed VisMed terms relevant to the query topics. Table 19.4 lists some of the VisMed terms and Figure 19.6 illustrates visual examples.

We manually cropped image regions to train and validate VisMed terms using SVMs. As we would like to minimize the number of images selected from the test collection for VisMed term learning, we include relevant images available from the Web. For a given VisMed term, the negative samples are the union of the positive samples of all the other VisMed terms.

19.3.4 Spatial Visual Queries

A visual query language, Query By Spatial Icons (QBSI), was used to combine pattern matching and logical inference (Lim and Jin, 2005a). A QBSI query is composed as a spatial arrangement of visual semantics. A Visual Query Term (VQT) P specifies a region R where a VisMed i should appear and a query formula chains these terms up via logical operators. The truth value $\mu(P,Z)$ of a VQT P for any image Z is simply defined as:

$$\mu(P,Z) = T_i(R) \tag{19.8}$$

where $T_i(R)$ is defined in Equation 19.4.

When a query involves the presence of a VisMed term in a region larger than a single block in a grid and its semantics prefers a larger area of presence of the VisMed term to have a good match (e.g. entire kidney, skin lesion, chest x–ray images with tuberculosis), Equation 19.8 becomes:

$$\mu(P,Z) = \frac{\sum_{Z_j \in R} T_i(Z_j)}{|R|} \tag{19.9}$$

where Z_j are the blocks in a grid that cover R and $|R|$ denotes the number of such blocks. This corresponds to a spatial universal quantifier ($\forall$).

On the other hand, if a query only requires the presence of a VisMed term within a region regardless of the area of the presence (e.g. presence of a bone fracture, presence of micro nodules), then the semantics are equivalent to the spatial existential quantifier ($\exists$) and Equation 19.8 will be computed as

$$\mu(P,Z) = \max_{Z_j \in R} T_i(Z_j) \tag{19.10}$$

A QBSI query Q can be specified as a disjunctive normal form of VQT (with or without negation),

$$Q = (P_{11} \wedge P_{12} \wedge \cdots) \vee \cdots \vee (P_{c1} \wedge P_{c2} \wedge \cdots) \tag{19.11}$$

Then the query processing of query Q for any image Z is to compute the truth value $\mu(Q,Z)$ using appropriate logical operators using min/max fuzzy operations. As uncertainty values are involved in VisMed term detection and indexing, we adopt classical min/max fuzzy operations as follows:

$$\mu(\bar{P},Z) = 1 - \mu(P,Z), \tag{19.12}$$

$$\mu(P_i \wedge P_j,Z) = \min(\mu(P_i,Z),\mu(P_j,Z)), \tag{19.13}$$

$$\mu(P_i \vee P_j,Z) = \max(\mu(P_i,Z),\mu(P_j,Z)). \tag{19.14}$$

For the query processing in ImageCLEF, a query text description is manually translated into a QBSI query with the help of a visual query interface (Lim and Jin,

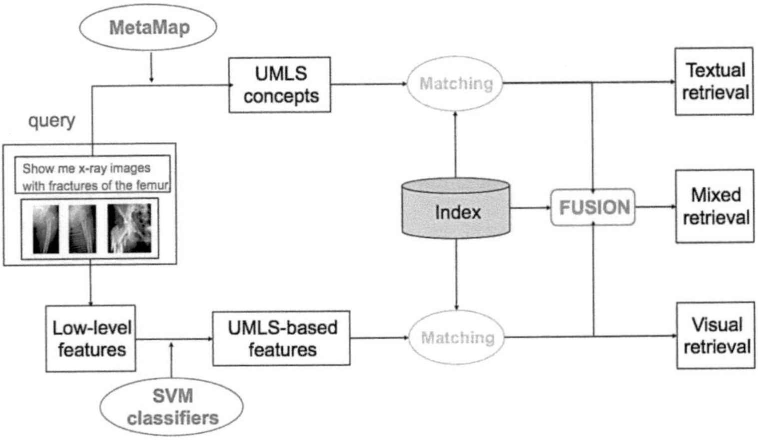

Fig. 19.7: Inter–media matching.

2005a) that outputs an XML format to state the VisMed terms, the spatial regions, the Boolean operators, and the spatial quantifiers. As an illustration, the query 'Show me x–ray images with fractures of the femur' is translated as '$\forall$ xray–bone $\in$ whole $\wedge$ $\forall$ xray–pelvis $\in$ upper $\wedge$ $\exists$ xray–bone–fracture $\in$ whole' where 'whole' and 'upper' refer to the whole image and upper part of an image respectively.

The use of VisTerms and the FlexiTile matching produce a MAP of 0.072 on the data and queries of CLEF 2005. Manual query building as proposed in Section 19.3.4 does not provide better results as the MAP is only 0.060. The two methods select different images and a linear combination can produce better results (0.092 of MAP) but with the search of the optimal weight.

The use of VisMed also requires the creation of a visual ontology. The results are then dependent on the quality of this ontology. No medical visual ontologies exist, yet, which limits the scaling up of this approach.

19.4 Multimedia and Intermedia Indexing

In order to help the creation of the visual ontology for image indexing using VisMed terms, we have built the visual ontology using a combination of UMLS concepts. This enables the building of an image and text index into the same space for indexing, and enables the building of a unique intermedia index. Hence, a single index (vector of concepts) is used to represent images and multi–lingual texts.

Because of the size difference between the index produced from images and the one produced from texts, it was more effective to compute the textual and image part of the matching separately, and to fuse the results as shown in Figure 19.7.

Because the local indexing presented previously cannot capture local image characteristics such as the modality, we have added an additional classification step. We have classified medical images using global features with a two level classification scheme. The first level corresponds to a classification for grey level versus color images. Indeed, some ambiguity can appear due to the presence of colored images, or the slightly blue or green appearance of x–ray images. This first classifier uses the first three moments in the Hue Saturation Value (HSV) color space computed on the entire image.

The second level corresponds to the classification of the modality UMLS concepts given that the image is in the grey or the color cluster. For the grey level cluster, we use grey level histogram (32 bins), texture features (mean and variance of Gabor coefficients for five scales and six orientations), and thumbnails (grey values of the 16 x 16 resized image). For the color cluster, we have adopted an HSV histogram (125 bins), Gabor texture features, and thumbnails. Zero–mean normalization (Huang et al, 1997) was applied to each feature. For each SVM classifier we adopted a RBF kernel:

$$\exp(-\gamma|x-y|^2) \tag{19.15}$$

where $\gamma = \frac{1}{2\sigma^2}$, and with a modified city–block distance:

$$|x-y| = \frac{1}{F}\sum_{f=1}^{F}\frac{|x_f - y_f|}{N_f} \tag{19.16}$$

where $x = \{x_1,...,x_F\}$ and $y = \{y_1,...,y_F\}$ are feature vectors, x_f, y_f are feature vectors of type f, N_f is the feature vector dimension, and F is the number of feature types: $F = 1$ for the grey versus color classifier, $F = 3$ for the conditional modality classifiers: color, texture, thumbnails. We use $\gamma = 1$ in all our experiments. This just–in–time feature fusion within the kernel combines the contribution of color, texture, and spatial features equally (Lim and Jin, 2006).

The probability of a modality $\textsc{mod}_i$ for an image z is given by:

$$P(\textsc{mod}_i|z) = \begin{cases} P(\textsc{mod}_i|z,C)P(C|z) & \text{if } \textsc{mod}_i \in C \\ P(\textsc{mod}_i|z,G)P(G|z) & \text{if } \textsc{mod}_i \in G \end{cases} \tag{19.17}$$

where C and G denote the color and the grey level clusters respectively, and the conditional probability $P(\textsc{mod}_i|z,V)$ is given by:

$$P(c|z,V) = \frac{\exp^{D_c(z)}}{\sum_{j\in V}\exp^{D_j(z)}} \tag{19.18}$$

where D_c is the signed distance to the SVM hyperplane that separates class c from the other classes of the cluster V.

After learning using SVM–light software (Joachims, 2002; Vapnik, 1995), each database image z is indexed according to modality given its low–level features z_f. The indexes are the probability values given by Equation (19.17). An image is represented by a semantic histogram, each bin corresponding to a modality probability.

The distance between two images is given by the Manhattan distance (i.e. city–block distance) between the two semantic histograms. This filtering alone on 2006 data, produces a MAP of 0.057. This shows that image modality matching is important in this medical collection.

VisMed terms are a combination of UMLS concepts from the modality, anatomy, and pathology semantic types. The index construction is the same as described previously. On the 2006 data, this produces an MAP of 0.048 which is lower than the detection of the modality only using global feature. A combination of these two filters using the mean of both similarity scores produces better results (MAP 0.064), but was below the top official results produced in 2006 (MAP 0.075).

19.5 Conclusions

Our experience has been that visual matching is much more difficult that text matching. The fusion of our conceptual text indexing and conceptual visual indexing, as described in this chapter, produces the best results in the year 2006 (0.309 MAP). This fusion is performed directly on a normalized RSV by the maximum. It is a linear fusion with a weight of 0.7 for image only.

Because our results has outperformed other methods in ImageCLEF 2006, we think we have proved the effectiveness of the use of a large external resource such as UMLS. In particular, a meaningful filtering based on semantic types for text and on modality for images is very effective, and is the explanation for the good results obtained.

On the image side, the difficulty of our approach lies in the availability of a visual ontology. We had to manually build a small ontology, related to the queries we needed to solve. This is a limitation, but one can expect a larger visual ontology to be available, as UMLS is available only for text .

Even if this top result is dependent on the availability of UMLS, we think our approach also proves the utility of techniques from NLP for improving Information Retrieval. In our experience, exploitation of the semantics explicitly contained is such large resource, is the only way to produce a breakthrough for future Information Retrieval systems.

Acknowledgements This work was performed during a French–Singaporean collaboration between the Institute for Infocom Research (I2R) in Singapore and the University of Grenoble in France during the establishment of the IPAL CNRS Laboratory. We would like to thank all researchers of IPAL that have participated in these experiments.

References

Amati G, Van Rijsbergen CJ (2002) Probabilistic models of information retrieval based on measuring divergence from randomness. ACM Transactions on Information Systems 20(4):357–389

Aronson AR (2006) Metamap: Mapping text to the umls metathesaurus. http://mmtx.nlm.nih.gov/docs.shtml

Bishop CM (1996) Neural networks for pattern recognition, 1st edn. Oxford University Press, USA

Bodenreider O, Mccray AT (2003) Exploring semantic groups through visual approaches. Journal of Biomedical Informatics 36(6):414–432

Boudin F, Shi L, Nie JY (2010) Improving medical information retrieval with pico element detection. In: European Conference on Information Retrieval, pp 50–61

Chevallet JP, Lim JH, Le THD (2007) Domain knowledge conceptual inter-media indexing, application to multilingual multimedia medical reports. In: ACM Sixteenth Conference on Information and Knowledge Management. ACM press, Lisboa, Portugal

Daille B, Habert B, Jacquemin C, Royauté J (1996) Empirical observation of term variations and principles for their description. Terminology 3(2):197–257

Datta R, Joshi D, Li J, Wang JZ (2008) Image retrieval: Ideas, influences, and trends of the new age. ACM Computing Surveys 40(2):1–60

Dy J, Brodley C, Kak A, Broderick L, Aisen A (2003) Unsupervised feature selection applied to content–based retrieval of lung images. IEEE Transactions on Pattern Analysis and Machine Intelligence 25(3):373–378

Goguen J (2005) What is a concept ? In: Conceptual Structures: Common Semantics for Sharing Knowledge. Lecture Notes in Computer Science (LNCS). Springer, pp 52–77

Huang T, Rui Y, Mehrotra S (1997) Content–based image retrieval with relevance feedback in mars. In: International Conference on Image Processing, pp 815–818

Huibers T, Ounis I, Chevallet JP (1996) Conceptual graph aboutness. In: The 4th International Conference on Conceptual Structures. Lecture Notes in Artificial Intelligence LNAI, vol 1115, pp 130–144

Ide N, Veronis J (1998) Introduction to the special issue on word sense disambiguation: the state of the art. Computer Linguistics 24(1):2–40

Joachims T (2002) Learning to classify text using support vector machines. Kluwer

Lacoste C, Chevallet JP, Lim JH, Wei X, Raccoceanu D, Le THD, Teodorescu R, Vuillenemot N (2006) IPAL Knowledge–based Medical Image Retrieval in ImageCLEFmed 2006. In: Working Notes of CLEF 2006

Lehmann TM, Güld M, Thies C, Fischer B, Spitzer K, Keysers D, Ney H, Kohnen M, Schubert H, Wein BB (2004) Content–based image retrieval in medical application. Methods of Information in Medicine 43(4):354–361

Lenat DB (1995) Cyc: a large-scale investment in knowledge infrastructure. Communications of the ACM 38(11):33–38

Lim J, Jin J (2005a) A structured learning framework for content–based image indexing and visual query. Multimedia Systems Journal 10(4)

Lim J, Jin J (2006) Discovering recurrent image semantics from class discrimination. EURASIP Journal of Applied Signal Processing 21:1–11

Lim JH (2001) Building visual vocabulary for image indexation and query formulation. Pattern Analysis and Applications 4(2–3):125–139

Lim JH, Jin JS (2005b) A structured learning framework for content–based image indexing and visual query. Multimedia Systems 10(4):317–331

Liu H, Singh P (2004) Conceptnet a practical commonsense reasoning tool–kit. BT Technology Journal 22(4):211–226

Liu Y, Lazar N, Rothfus W, Dellaert F, Moore A, Schneider J, Kanade T (2004) Semantic based biomedical image indexing and retrieval. In: Shapiro K, Veltkamp (eds) Trends and Advances in Content–Based Image and Video Retrieval. Springer

Manjunath BS, Ma WY (1996) Texture features for browsing and retrieval of image data. IEEE Transactions on Pattern Analysis and Machine Intelligence 18(8):837–842

Meghini C, Sebastiani F, Straccia U, Thanos C (1993) A model of information retrieval based on a terminological logic. In: Proceedings of the 16th annual international ACM SIGIR conference on Research and development in information retrieval. ACM press, pp 298–307

Müller H, Michoux N, Bandon D, Geissbühler A (2004) A review of content–based image retrieval systems in medical applications — clinical benefits and future directions. International Journal of Medical Informatics 73(1):1–23

Papageorgiou P, Oren M, Poggio T (1998) A general framework for object detection. In: Proceedings of the International conference of computer vision, pp 555–562

van Rijsbergen CJ (1986) A non–classical logic for information retrieval The Computer Journal 29(6):481–485

Robertson SE, Walker S (1994) Some simple effective approximations to the 2–poisson model for probabilistic weighted retrieval. In: Proceedings of the 17th annual international ACM SIGIR conference on research and development in information retrieval. ACM press, pp 232–241

Schank RC, Kolodner JL, DeJong G (1980) Conceptual information retrieval. In: International ACM SIGIR conference. ACM press, pp 94–116

Sebastiani F (1994) A probabilistic terminological logic for modelling information retrieval. In: Croft WB, van Rijsbergen CJ (eds) Proceedings of the 17th ACM International Conference on Research and Development in Information Retrieval. ACM press, Dublin, Ireland, pp 122–130

Shyu CR, Pavlopoulou C, Kak AC, Brodley CE, Broderick LS (2002) Using human perceptual categories for content–based retrieval from a medical image database. Computer Vision and Image Understanding 88(3):119–151

Shyu CR, Pavlopoulou C, Kak AC, Brodley CE, Broderick LS (2002) Using human perceptual categories for content–based retrieval from a medical image database. Computer Vision and Image Understanding 88(3):119–151

Sung K, Poggio T (1998) Example–based learning for view–based human face detection. IEEE Transactions on Pattern Aanalysis and Machine Intelligence 20(1):39–51

Vapnik VN (1995) The nature of statistical learning theory. Springer

Vapnik VN (1998) Statistical learning theory. Wiley–Interscience

Chapter 20
Conceptual Indexing Contribution to ImageCLEF Medical Retrieval Tasks

Loïc Maisonasse, Jean–Pierre Chevallet, and Eric Gaussier

Abstract In this chapter, we study conceptual indexing using a language modeling approach to information retrieval. In particular, we propose a conceptual representation of documents that allows the use of both concepts and labelled relations when matching documents and queries. Such semantic indexing gives effective results when large ontologies are used. We first present a model derived from the language modeling approach to information retrieval based on a conceptual representation of documents. We then introduce an extension to take into account relations between concepts. Concept and relation detection methods are, however, error–prone. We thus develop an approach to limit such errors by combining different methods. In order to illustrate various aspects of the model proposed, we conducted a series of experiments on various medical ImageCLEF collections. Our experiments in ImageCLEFmed show that the conceptual model proposed here provides good results in medical information retrieval. Experiments furthermore show that combining concept extraction methods through fusion improves the standard language model by up to 17% MAP on the medical ImageCLEF collections.

Loïc Maisonasse
TecKnowMetrix, ZAC de Champfeuillet, F–38500 VOIRON, France, e-mail: `loic.maisonnasse@laposte.net`

Jean–Pierre Chevallet
University of Grenoble, Laboratoire d'Informatique de Grenoble, F–38041 Grenoble Cedex 9, France e-mail: `Jean-Pierre.Chevallet@imag.fr`

Eric Gaussier
University of Grenoble, Laboratoire d'Informatique de Grenoble, F–38041 Grenoble Cedex 9, France e-mail: `eric.gaussier@imag.fr`

H. Müller et al. (eds.), *ImageCLEF*, The Information Retrieval Series 32, DOI 10.1007/978-3-642-15181-1_20, © Springer-Verlag Berlin Heidelberg 2010

20.1 Introduction

In focused domains, linguistic resources such as ontologies or thesauri are nowadays readily available to automatically build conceptual indexes from textual documents. Furthermore, when such resources contain relational information, advanced representations of documents can be proposed in the form of graphs of concepts. Nevertheless, semantic representations detected from texts are not always accurate and their use can sometimes degrade the performance of an information retrieval system. Methods to identify concepts in texts from an ontology usually rely on a sequence of Natural Language Processing (NLP) tools to find and normalize the correct units to be mapped to concepts. If such tools exist, their performance varies greatly from one domain to another, as well as from one language to another, so that the performance of a concept–based information retrieval system depends largely on the performance of its underlying semantic detection method. To overcome the defaults of a particular graph or concept detection method, we propose to combine the results of several, complementary detection methods.

The remainder of the chapter is organized as follows: We first briefly review semantic indexing using ontologies. We then describe the conceptual model we have used in the ImageCLEF campaigns and its extension to graphs. Then, we study various ways to merge concept detection methods in the framework of our semantic language modeling approach to information retrieval. Finally, we present a pseudo–relevance feedback extension of this model. In each of these parts, we present experimental results on various ImageCLEF collections to validate the developments proposed.

20.2 Semantic Indexing Using Ontologies

Classical information retrieval models use a bag of words to represent documents such as the vector space model (Salton et al, 1975) or the language model (Ponte and Croft, 1998). Semantic indexing can reuse such models by replacing words by concepts. In addition, one can try to use semantic structures as an index using an appropriate matching model. This idea has already been investigated in different studies such as the use of terminological logics (Meghini et al, 1993), conceptual dependencies (Berrut and Chiaramella (1989)) or frame–based indexing (Benigno et al, 1986). Unfortunately, such structures are still difficult to build from texts automatically, and they are complex to use for matching. We investigate in this chapter the use of graph of concepts, similar to the ones used in (Vintar et al, 2003) and adapt the matching procedure at the core of the language modeling approach to Information Retrieval (IR) to this semantic structure.

Conceptual indexing, i.e. the use of concepts instead of words or terms in an information retrieval system, is also a nice idea to cross the language barrier and to come up with a much more meaningful document index. A conceptual index is naturally multi–lingual because concepts can be defined as human–understandable unique abstract notions independent from any direct material support, independent from any language or information representation (Chevallet et al, 2007). Moreover, relations between concepts clarify the roles of each concept in the semantic structure. For example, in the query *blood smears that include polymorphonuclear neutrophils*[1] only documents dealing with *blood smears* related to *polymorphonuclear neutrophils* through an inclusion relation will be retrieved.

Unfortunately, even if semantic indexing is quite an old idea (Schank et al, 1980), it is still difficult to perform as it requires linguistic resources (example: ontology, terminology) and natural language processing tools to tackle difficult linguistic phenomena such as ambiguity.

We propose to use very simple concept graphs, which we have deployed in the medical domain, within the ImageCLEFmed tasks. The medical field is well suited for conceptual and graph indexing, as many resources have been developed in order to accurately index the content of medical texts. An ontology without formal definitions of concepts but with numerous links between concepts and terms is indeed a valuable structure for building a complete semantic information retrieval system. The Unified Medical Language System (UMLS[2]), also known as a meta–thesaurus, is a typical example of such a weak ontology dedicated to information retrieval. This structure has been exploited in several applications, such as (Lacoste et al, 2006) in campaigns of ImageCLEFmed CLEF[3], or in (Zhou et al, 2007) for the Text REtrieval Conference (TREC) Genomics track[4].

20.3 Conceptual Indexing

20.3.1 Language Models for Concepts

In this section, we rely on a language model defined over concepts, which we refer to as a conceptual unigram model. We assume that a query q is composed of a set $\mathscr{C}$ of concepts, each concept being independent to the others conditionally on a document model, i.e.:

$$P(C|M_d) = \prod_{c_i \in C} P(c_i|M_d)^{\#(c_i,q)} \tag{20.1}$$

where $\#(c_i,q)$ denotes the number of times the concept c_i occurs in the query q. The quantity $P(c_i|M_d)$ is directly computed through maximum likelihood using Jelinek–Mercer smoothing:

[1] Query from the medical task of ImageCLEF 2006

[2] Unified Medical Language System http://umlsinfo.nlm.nih.gov)

[3] http://www.clef-campaign.org/

[4] http://trec.nist.gov/

$$P(c_i|M_d) = (1-\lambda_u)\frac{|c_i|_d}{|*|_d} + \lambda_u\frac{|c_i|_{\mathscr{D}}}{|*|_{\mathscr{D}}}$$

where $|c_i|_d$ (respectively $|c_i|_{\mathscr{D}}$) is the frequency of concept c_i in the document d (respectively in the collection $\mathscr{D}$), and $|*|_d$ (respectively $|*|_{\mathscr{D}}$) is the size of d, i.e. the number of concepts in d (respectively in the collection).

20.3.2 Concept Detection

UMLS is a good linguistic resource candidate for medical text indexing. It is more than a terminology because it describes terms with associated concepts. The resource is large: more than 1 million concepts, 5.5 million terms in 17 languages. Unfortunately, UMLS is built from different sources (thesauri, terms lists) that are neither complete nor consistent. In fact UMLS resembles more a meta–thesaurus, i.e. a merger of existing thesauri, than a complete ontology. Nevertheless, the large set of terms and term variants in UMLS (more than 1 million concepts associated with 5.5 million terms) restricted to the medical domain, allows us to build on top of it a full scale conceptual indexing system. In UMLS, all concepts are assigned to at least one semantic type from the semantic network. This provides consistent categorization of all concepts at a relatively general level. This also enables the detection of general semantic relations between concepts present in the network. UMLS has been updated several times in the past. In the following experiments, we use the version available in 2007.

The detection of concepts in a document is a relatively well established process. It consists of four major stages:

1. Morpho–syntactic analysis (*POS tagging*) of documents with a lemmatization of inflected word forms;
2. Filtering empty words on the basis of their grammatical class;
3. Detection of words or phrases which are potential concepts;
4. Selection of relevant concepts.

For the first stage, various tools can be used depending on the language. In this work, we use MiniPar (Lin, 1998) for English, as this tool is fast and gives good results. We also use TreeTagger[5], which is available for the English, French and German languages, the three languages retained in our study. Once the documents are parsed, the second and third stages are implemented directly, on the one hand by filtering grammatical words (prepositions, determinants, pronouns, conjunctions), on the other hand by a look–up of word sequences in UMLS. This last stage will find all term variants of a concept. It should be noted that we have not used all of UMLS for the third stage: The NCI and PDQ thesauri were not taken into account

[5] `http://www.ims.uni-stuttgart.de/projekte/corplex/TreeTagger/`

because they are related to areas different from the one covered by the collection[6]. Such a restriction is also used in (Huang et al, 2003). The fourth step of the indexing process aims at eliminating errors generated by the above steps. However, this elimination is difficult to perform: Our experiments at ImageCLEFmed (Maisonnasse et al,2008) show that it is preferable to retain a larger number of concepts for information retrieval.

We also used MetaMap (Aronson, 2001), a tool associated with UMLS and widely used in conceptual indexing, and which directly yields a set of concepts from a text document. MetaMap first detects all the concepts matching a sub–phrase, possibly with variants, then proposes the best variants that cover the phrase. Tests made on this tool show that the best results are obtained by using concept matching without variants and so that is the way we use this tool in this study.

We finally obtain three concepts detection methods, which we refer to as:

- (mm) uses MetaMap, for English only;
- (mp) uses our term mapping tool with MiniPar, for the morpho–syntactic analysis of English texts;
- (tt) uses our term mapping tool with TreeTagger for the morpho–syntactic analysis of French, English and German texts.

20.3.3 Concept Evaluation Using ImageCLEFmed 2005–07

The experiment described here was carried out on the ImageCLEFmed collection. We used the collection comprising the years 2005, 2006 and 2007. This collection consists of 55,485 multi–lingual medical reports associated with medical images and 85 queries with relevance judgements (each year has respectively 25, 30 and 30 queries). Relevance judgements on the collection are made at the image level; we consider in the following that a diagnosis is relevant if at least one of its associated images is relevant. This allows us to directly evaluate our model at the textual level.

In order to estimate the parameters (smoothing coefficients) of the different models, we have divided the 85 queries into two subgroups, and have retained 43 queries (selected at random) for part 1 and 42 for part 2. We have alternated training and testing on these two parts. Two measures were retained for evaluation: the Mean Average Precision (MAP) that provides a general overview of IR results, and precision at five documents (P@5) that provides information on how the systems behaves at the top of the list of retrieved documents.

We first evaluate the results of the different concept detectors with queries and documents analyzed using only one method. The results, presented in Table 20.1, correspond to the conceptual unigram model defined by Equation 20.1. As these results are obtained with the use of a different concept detection method used in isolation, they serve as a baseline against which we will assess the performance of the

[6] This is justified by the fact that these thesauri focus on specific questions of cancer while the collection is considered more general and covers all diseases.

Table 20.1: Results for MAP and precision at five documents (P@5) with different concept detection methods. Model parameters are learned on the learning part; tests are performed on the evaluation part. The line corresponding to the baseline, *MM*, is italicized.

documents	query	λ_u	learning part 1	evaluation part 2	λ_u	learning part 2	evaluation part 1
MAP							
MM	*E.MM*	*0.1*	*0.260*	*0.246*	*0.4*	*0.251*	*0.259*
MP	E.MP	0.3	0.285	0.246	0.4	0.246	**0.284**
TT	E.TT	0.1	0.264	**0.258**	0.2	0.258	0.263
P@5							
MM	*E.MM*	*0.8*	*0.428*	*0.357*	*0.4*	*0.433*	*0.419*
MP	E.MP	0.2	0.493	0.424	0.1	0.433	**0.488**
TT	E.TT	0.1	0.451	**0.462**	0.1	0.462	0.451

different combinations we have described in Section 20.5. The results presented in Table 20.1 show that one should rely on TreeTagger for part 2 and MiniPar for part 1. The MetaMap analysis, considered as a state–of–the–art indexing method in the medical domain, provides the lower results on both MAP and on P@5. On this collection, a standard word language model (with POS filtering and Porter stemming) applied to part 2 yields a MAP of 0.244, which is slightly lower that our conceptual approach. The P@5 reaches 0.448, which is within the average of our conceptual approach.

20.4 From Concepts to Graphs

Conceptual indexing provides good results in MAP but does not perform as well in P@5. In order to improve the precision of our method we proposed using a more precise document representation that uses relations between concepts. Several researchers have looked at taking into account relations between concepts. For instance Vintar et al (2003) index documents and queries of a medical corpus on the basis of UMLS. In his work, a relationship between two concepts exists if the two concepts appear in the same sentence and are connected in the meta–thesaurus. We are in line with this work and describe our graph indexing process in the following sections.

20.4.1 A Language Model for Graphs

The probability for a query graph $G_q =< C,E >$ to be generated by the model of the document graph M_{Gd} can be written as:

$$P(G_q|M_{Gd}) = P(C|M_{Gd}) \times P(E^q|C, M_{Gd}) \tag{20.2}$$

where the probability of generating query concepts from the document model ($P(C|M_{Gd})$) is similar to the one computed in Section 20.3.1. Following a similar process for the relations leads to:

$$P(E|M_{Gd}) \propto \prod_{(c,c',l)\in\mathscr{C}^2\times\mathscr{L}} P(L(c,c') = l|C, M_{Gd})^{\#(c,c',l,q)} \tag{20.3}$$

where $L(c,c')$ is a variable with values in $\mathscr{L}$ reflecting the possible relation labels between c and c'. As before, the parameters of the model $P(L(c,c') = l|C, M_{Gd})$ are estimated by maximum likelihood with Jelinek–Mercer smoothing, giving:

$$P(L(c,c') = l|C, M_{Gd}) = (1-\lambda_e)\frac{\#(c,c',l,d)}{\#(c,c',.,d)} + \lambda_e\frac{\#(c,c',l,D)}{\#(c,c',.,D)} \tag{20.4}$$

where $\#(c,c',l,d)$ represents the number of times concepts c and c' are linked to label l in the graph of the document, and where $\#(c,c',.,d) = \sum_{l\in\mathscr{L}} \#(c,c',l,d)$. By convention, when at least one of the two concepts does not appear in the graph of document d:

$$\frac{\#(c,c',l,d)}{\#(c,c',.,d)} = 0$$

Here again, the quantities $\#(c,c',l,D)$ are similar but defined on the whole collection (i.e. as previously over the union of all the graphs from all the documents in the collection). The model we have just presented is inspired by the model defined in (Maisonnasse et al, 2008).

20.4.2 Graph Detection

The concept detection step is followed by a relation detection between concepts. Relations used are those defined in the semantic network of UMLS. We make the assumption that a relation exists between two concepts if these concepts are detected in the same sentence, and if the semantic network defines a relation between these two concepts. To detect the existence of relations, we first associate semantic categories to each concept, and then we add the semantic relations linking the two concept categories.

Table 20.2: Results for MAP and P@5 with different graph detection methods. A * indicates that the difference with the baseline is significant (Wilcoxon test, with $p = 0.05$).

Documents	Queries	λ_u	λ_e	learning part 1	evaluation part 2	λ_u	λ_e	learning part 2	evaluation part 1
MAP									
MM	E.MM	0.2	0.9	0.264	0.252*	0.8	0.8	0.256	0.255
MP	E.MP	0.5	0.6	0.294	0.253	0.5	0.6	0.253	**0.294**
TT	E.TT	0.2	0.5	0.274	**0.264**	0.1	0.7	0.264	0.271
P@5									
MM	E.MM	0.2	0.9	0.433	0.409	0.6	0.3	0.448	0.400
MP	E.MP	0.1	0.4	0.530	0.452	0.1	0.4	0.452	**0.530**
TT	E.TT	0.1	0.4	0.484	**0.495***	0.1	0.1	0.514	0.470

20.4.3 Graph Results on ImageCLEFmed 2005–07

The results presented in Table 20.2 show that using a graph representation increases the results both in MAP and P@5. In both cases, a significant improvement over the MetaMap baseline is obtained.

20.5 Mixing Concept Sources

If extracting concepts is a hard task, the impact of errors during this extraction process is not likely to be the same for documents and queries. As documents contain many sentences, it is possible that an error in one place will be compensated by a correct detection in another. The situation differs for queries, as they usually contain very few words. A single error in this case can significantly degrade the recall of the system. We showed in (Maisonnasse et al, 2008) that mixing concept detection methods on the ImageCLEF collection can improve the results of a concept–based IR system. We study here different ways to combine concept detection methods, on both queries and documents.

As we are interested in merging several concept detection methods, we do not have a single document model associated to a document d but several, each corresponding to one detection method. We denote by M_d^* the set of document models ($M_d^* = \{M_d^1, ..., M_d^p\}$). Similarly, a query will consist of a set C^* of sets of concepts, each set of concepts resulting from the application of a concept detection method ($C^* = \{C^1, ..., C^p\}$). The final retrieval value of the query (RSV) is thus given by:

$$RSV(q,d) = P(C^*|M_d^*) \tag{20.5}$$

Our problem is thus to decompose $P(C^*|M_d^*)$ according to the different concept sets and document models. As all elements in C^* and M_d^* are obtained independently of each other, we assume in the remainder that these elements are independent of each other[7].

20.5.1 Query Fusion

On the query side, we propose two ways to decompose the set C^*. In the first one, we consider that a relevant document must generate all the analyses of query q, which leads to:

$$P(C^*|M_d^*) \propto \prod_{C \in C^*} P(C|M_d^*) \tag{20.6}$$

In the second one, we consider that, to be relevant, a document has to generate at least one analysis of the query, and not all of them. This leads to:

$$P(C^*|M_d^*) \propto \sum_{C \in C^*} P(C|M_d^*) \tag{20.7}$$

Because the two previous equations propose a decomposition of C^*, we refer to them as query fusion methods. Armed with these decompositions, we can now proceed to the decomposition of M_d^*.

20.5.2 Document Model Fusion

In the language modeling approach, a language model is computed according to the document. As we use different concept extractors, a document will have different conceptual representations. There are several possibilities to merge the output of all concept extractors for a given document, which we are going to review now.

Using Bayes rule, one obtained the following rewriting of $P(C|M_d^*)$:

$$P(C|M_d^*) = \frac{P(C)}{P(M_d^*)} P(M_d^*|C)$$

In the context of IR, we are computing a sorted list of documents, through their retrieval value. The term $P(C)$, common to all documents, does not influence this ranking. Having no a priori knowledge on how each concept detection method performs, we assume that the document models in M_d^* are equiprobable, and that the probabilities $P(M_d^*)$ are the same for all the documents. We can thus write:

$$P(C|M_d^*) \propto P(M_d^*|C) \tag{20.8}$$

[7] This assumption is obviously a (useful) simplification, as all the concept detection methods we consider use the same knowledge bases.

As before, we have several ways to decompose $P(M_d^*|C)$. One can demand that each document model be associated with a given concept set C, or one can demand that at least one document model be associated to C. The former results in collecting the contribution from the different document models in a product, whereas the latter results in a sum over the contribution from the document models. Moreover, in the second approach, one can also try to rely on the best document model only, which results in taking the maximum over the contribution of the document models. The last two decompositions, involving the sum and the maximum, are in fact similar when the probability $P(M_d|C)$ is peaked on one model, and the maximum is often used as a substitute for the sum when this is difficult to compute. We consider the maximum for completeness but do not use it in the query decomposition methods as the sum can be computed efficiently in our case. The decompositions provided by these different approaches are summarized below:

$$P(M_d^*|C) \propto \begin{cases} \prod_{M_d \in M_d^*} P(M_d|C) \\ \sum_{M_d \in M_d^*} P(M_d|C) \\ \max_{M_d \in M_d^*} P(M_d|C) \end{cases} \tag{20.9}$$

Applying Bayes rule to the term $P(M_d|C)$ gives:

$$P(M_d|C) = \frac{P(M_d)}{P(C)} P(C|M_d)$$

As before, in the context of IR and with the assumptions made, the above quantity simplifies to:

$$P(M_d|C) \propto P(C|M_d)$$

Substituting this expression in Equations 20.9 and 20.8 gives:

$$P(C|M_d^*) \propto \begin{cases} \prod_{M_d \in M_d^*} P(C|M_d) \\ \sum_{M_d \in M_d^*} P(C|M_d) \\ \max_{M_d \in M_d^*} P(C|M_d) \end{cases} \tag{20.10}$$

a set of decompositions that can be combined directly with the ones given previously for the query. There is however another way to decompose $P(C^*|M_d^*)$ which we want to present now.

20.5.3 Joint Decomposition

Instead of trying to decompose the query sets first and then the document models, one can try to decompose the document models first. This can be done using Bayes formula, as:

$$P(C^*|M_d^*) = \frac{P(C^*)}{P(M_d^*)} P(M_d^*|C^*)$$

and, with the assumptions made in the context or IR:

$$P(C^*|M_d^*) \propto P(M_d^*|C^*)$$

Again, one can decompose M_d^* as a product, a sum or a maximum over the document models, leading to:

$$P(M_d^*|C^*) \propto \begin{cases} \prod_{M_d \in M_d^*} P(M_d|C^*) \\ \sum_{M_d \in M_d^*} P(M_d|C^*) \\ \max_{M_d \in M_d^*} P(M_d|C^*) \end{cases}$$

Using Bayes rule again and relying on our assumption leads, using the same development as the one used for Equation 20.10, to:

$$P(C^*|M_d^*) \propto \begin{cases} \prod_{M_d \in M_d^*} P(C^*|M_d) \\ \sum_{M_d \in M_d^*} P(C^*|M_d) \\ \max_{M_d \in M_d^*} P(C^*|M_d) \end{cases} \tag{20.11}$$

Combining the decompositions provided by Equations 20.6, 20.7, 20.10 and 20.11 finally leads to ten decompositions (as several decompositions are identical), the names of which are given in parentheses (*Fus* stands for fusion):

$$P(C^*|M_d^*) \propto \begin{cases} \prod_{C \in C^*} & \prod_{M_d \in M_d^*} & P(C|M_d) & (Fus1) \\ \prod_{C \in C^*} & \sum_{M_d \in M_d^*} & P(C|M_d)) & (Fus2) \\ \prod_{C \in C^*} & \max_{M_d \in M_d^*} & P(C|M_d)) & (Fus3) \\ \sum_{M_d \in M_d^*} & \prod_{C \in C^*} & P(C|M_d)) & (Fus4) \\ \max_{M_d \in M_d^*} & \prod_{C \in C^*} & P(C|M_d)) & (Fus5) \\ \max_{M_d \in M_d^*} & \sum_{C \in C^*} & P(C|M_d)) & (Fus6) \\ \sum_{C \in C^*} & \prod_{M_d \in M_d^*} & P(C|M_d)) & (Fus7) \\ \sum_{C \in C^*} & \sum_{M_d \in M_d^*} & P(C|M_d)) & (Fus8) \\ \sum_{C \in C^*} & \max_{M_d \in M_d^*} & P(C|M_d)) & (Fus9) \\ \prod_{M_d \in M_d^*} & \sum_{C \in C^*} & P(C|M_d)) & (Fus10) \end{cases} \tag{20.12}$$

in which the quantity $P(C|M_d)$ is computed by Equation 20.1. However, not all the merging strategies given above are interesting. In particular, we previously showed that collecting the query graphs in a sum was a poor information retrieval strategy on conceptual unigram models (Maisonnasse et al, 2009). We will thus focus only on the first five strategies (*fus1* to *fus5*) in this chapter.

In addition to the above ways to combine representations from different concept detection methods, we also directly merge the different representations of a document into a single pseudo–document d^*, leading to a single model for it noted as $M_{M_d^*}$. We call this simple merging early fusion, which corresponds to the idea that all analyses belong to the same document and thus must be used to create one single model combining the different analysis.

Table 20.3: Results for mixing concept detections on queries. The percentages in parentheses represent the difference with the baseline (conceptual unigram model with MetaMap). A * indicates that the difference with this baseline is significant (Wilcoxon test, with $p = 0.05$).

Documents	Queries	λ_u	learning part 1	evaluation part 2	λ_u	learning part 2	evaluation part 1
MAP							
MM	E.mix	0.4	0.267	**0.271** (+10.2%*)	0.3	0.271	0.265(+2.3%)
MP	E.mix	0.3	0.295	0.269 (+9.3%)	0.4	0.270	**0.293**(+13.1%)
TT	E.mix	0.2	0.263	0.267 (+8.5%)	0.3	0.268	0.261 (+0.7 %)
P@5							
MM	E.mix	0.4	0.423	0.481 (+34.7%*)	0.3	0.481	0.414 (-1.2%)
MP	E.mix	0.1	0.512	0.457 (+28.0%)	0.4	0.467	**0.484** (+15.5%)
TT	E.mix	0.2	0.470	**0.495** (+38.6%*)	0.3	0.509	0.446 (+6.4%)

20.5.4 Results on ImageCLEFmed 2005–07

We first test the merging of concept detection methods on queries only. This means that documents are analyzed with a single method, whereas queries are represented with different analyses resulting from different concept detection methods. We call the combination of three previous English concept sets (E.Mix) corresponding to (MM)(MP)(TT). As mentioned previously, the decomposition of queries based on Equation (20.7) have shown to perform poorly in information retrieval (Maisonnasse et al, 2009). We thus present in this chapter the results obtained with the decomposition based on Equation 20.6 only. Table 20.3 displays the results obtained for concepts. As one can note, combining different analyses can yield a significant improvement, both on MAP (line MM) and P@5 (lines MM and TT).

Table 20.4 shows the results obtained with a single pseudo–document resulting from the concatenation of the different document analyses. As one can note, merging analyses in one document model does not improve the results when a single analysis is used on the query side. However, when several analyses are used on the query side (line Concat–E_mix), the pseudo–document fusion strategy significantly improves the baseline. Finally, Table 20.5 presents the results with the complete merging strategies presented earlier (*fus1* to *fus5* of equation 20.12). The use of these merging strategies significantly improves both MAP and P@5 of our information retrieval system, even if a single detection method is used for the query (see for example the line E.TT in the two tables). One can also note that, due to their decomposition, *Fus2-3* are equal to *Fus4-5* when a single detection is used on the query. The best MAP results are obtained with the sum on the document models associated with the product (*Fus4*) on E_mix queries for part 2, but with the product on both queries and documents (*Fus1*) on part 1. On concepts, all the results that use fusion methods on both queries and documents provide similar results that are significantly better than the baseline, and better than all the results obtained so far.

Table 20.4: Results for mixing concept detections on documents by using one model over all analysis (concatenation). A * indicates that the difference with the baseline (concept detected with MetaMap) is significant (Wilcoxon test, with $p = 0.05$).

Documents	Queries	λ_u	learning part 1	evaluation part 2	λ_u	learning part 2	evaluation part 1
MAP							
Concat	E.MM	0.2	0.260	0.249	0.3	0.250	0.260
Concat	E.MP	0.1	0.260	0.242	0.3	0.244	0.258
Concat	E.TT	0.1	0.269	0.266	0.1	0.266	0.269 *
Concat	E.mix	0.1	0.277	**0.279***	0.4	0.279	**0.274**
P@5							
Concat	E.MM	0.1	0.409	0.419	0.5	0.424	0.400
Concat	E.MP	0.1	0.437	0.409	0.1	0.409	0.437
Concat	E.TT	0.1	0.442	0.438	0.1	0.438	0.442
Concat	E.mix	0.2	0.451	**0.486***	0.4	0.490	**0.451**

Mixing only the document models gives a significant improvement, but it requires analyzing the whole collection several times. Mixing only query analyses yields a lower improvement, but is easier to perform. On the query side, the fusion based on the sum of analyses decreases the performance of the IR system. Queries are short most of the time, so that few errors in an analysis can lead to big differences in the performance.

20.6 Adding Pseudo–Feedback

The previous section shows that using conceptual indexing within the language modeling approach to IR provides good results. Many articles have successfully proposed extending the usual language model to handle pseudo–relevance feedback. In this final section, we complete our model by proposing a pseudo–relevance feedback extension. We base this extension on the results obtained with our query combination model and we test this model on the ImageCLEF 2009 collection.

20.6.1 Pseudo–Relevance Feedback Model

We first form a new query Q_{fd} by merging the first n documents retrieved from the original query Q. For each document, we then combine its score (RSV) with the original query Q and the new query Q_{fd} to get its final score, denoted PRF. We rely on a simple linear combination:

$$PRF(Q,d) = (1-\lambda_{\text{prf}})RSV(Q,d) + (\lambda_{\text{prf}})RSV(Q_{\text{fd}},d) \qquad (20.13)$$

Table 20.5: Results combining concept detection methods in both documents and queries. Best results are in bold. A * indicates that the difference with the baseline is significant (Wilcoxon test, p=0.05).

part2

Documents	Query	*fus1*	*fus2*	*fus3*	*fus4*	*fus5*
MAP						
E_mix	E.MM	0.272*	0.261	0.260	0.261	0.260
E_mix	E.MP	0.265	0.257	0.254	0.257	0.254
E_mix	E.TT	0.283	0.280	0.277	0.280	0.277
E_mix	E_mix	0.285*	0.292*	0.289*	**0.301***	0.299*
MM MP	MM MP	0.285	0.275	0.277	0.287	0.289
P@5						
E_mix	E.MM	0.448*	0.414	0.409	0.414	0.409
E_mix	E.MP	0.452	0.443	0.433	0.443	0.433
E_mix	E.TT	0.476*	0.462	0.462	0.462	0.462
E_mix	E_mix	0.500*	0.500*	0.495*	0.524*	**0.529***
MM MP	MM MP	0.500*	0.462	0.462	0.481	0.481

part1

Documents	Query	*fus1*	*fus2*	*fus3*	*fus4*	*fus5*
MAP						
E_mix	E.MM	0.283	0.276	0.259	0.276	0.259
E_mix	E.MP	0.298	0.275	0.273	0.275	0.273
E_mix	E.TT	**0.310***	0.294	0.293	0.294	0.293
E_mix	E_mix	0.299*	0.301	0.289	0.299	0.300
MM MP	MM MP	0.295	0.298	0.299	0.313	0.312
P@5						
E_mix	E.MM	0.460	0.442	0.442	0.442	0.442
E_mix	E.MP	0.479	0.446	0.446	0.446	0.446
E_mix	E.TT	0.474	0.484	0.484	0.484	0.484
E_mix	E_mix	0.437	0.493	0.516	**0.521**	0.497
MM MP	MM MP	0.475	0.456	0.460	0.488	0.488

where the *RSV* is computed following Equation 20.6. λ_{prf} is a weight that allows one to control the importance of the original query with respect to the new one. If different collection analyses are used, we merge the results using a maximum fusion (fus 5).

20.6.2 Results

To evaluate our approach, we trained our models on the ImageCLEFmed 2008 corpus, and have run the best models obtained on the ImageCLEFmed 2009 corpus (Müller et al, 2009). The results, presented in Table 20.6, show that the number of documents to be retained in order to form the new query is important (100).

size of the	MPTT		MMMPTT		MPTTFA	MMMPTTFA
pseudo query (n)	2008	2009	2008	2009	2009	2009
20	0.279	-	0.281	-	-	-
50	0.289	-	0.290	-	-	-
100	0.292	**0.429**	0.299	0.416	0.424	0.418

Table 20.6: Results for different size of pseudo–relevance feedback with the Kullback-Leiber divergence and with different query analysis

20.7 Conclusions

This chapter explores a complete framework to handle conceptual indexing in the language model framework. We study different extensions that use the flexibility of the language model proposed to improved IR results. The results and participation in the ImageCLEFmed campaign show that the conceptual language model proposed provides good performance in medical IR. This model merging conceptual analysis improves the results and such approaches have obtained the best results at ImageCLEFmed in 2007. This merging improved with a pseudo–relevance feedback has obtained the best results in ImageCLEFmed 2009. These results show the effectiveness of conceptual indexing, and that the language model is a good framework to handle such indexing specificities.

References

Aronson A (2001) Effective mapping of biomedical text to the UMLS metathesaurus: The MetaMap program. In: Proceedings of the AMIA symposium, pp 17–21

Benigno MKD, Cross GR, deBessonet C (1986) COREL — A Conceptual Retrieval System. In: Proceedings of the 9th Annual International ACM SIGIR conference. ACM press, pp 144–148

Berrut C, Chiaramella Y (1989) Indexing medical reports in a multimedia environment: the rime experimental approach. In: Proceedings of the 12th annual international ACM SIGIR conference. ACM press, pp 187–197

Chevallet JP, Lim JH, Le THD (2007) Domain knowledge conceptual inter-media indexing, application to multilingual multimedia medical reports. In: ACM Sixteenth Conference on Information and Knowledge Management. ACM press

Huang Y, Lowe HJ, Hersh WR (2003) A pilot study of contextual UMLS indexing to improve the precision of concept-based representation in XML–structured clinical radiology reports. Journal of the American Medical Informatics Association 10(6):580–587

Lacoste C, Chevallet JP, Lim JH, Wei X, Raccoceanu D, Le THD, Teodorescu R, Vuillenemot N (2006) IPAL Knowledge–based Medical Image Retrieval in ImageCLEFmed 2006. In: Working Notes of CLEF 2006

Lin D (1998) Dependency–based evaluation of MiniPar. In: Workshop on the Evaluation of Parsing Systems, Granada, Spain, May. ACM press

Maisonnasse L, Gaussier E, Chevallet J (2008) Multiplying concept sources for graph modeling. In: Advances in Multilingual and Multimodal Information Retrieval. Lecture Notes in Computer Science (LNCS), vol 5152. Springer, pp 585–592

Maisonnasse L, Gaussier E, Chevallet JP (2009) Model fusion in conceptual language modeling. In: 31st European Conference on Information Retrieval

Meghini C, Sebastiani F, Straccia U, Thanos C (1993) A model of information retrieval based on a terminological logic. In: Proceedings of the 16th annual international ACM SIGIR conference. ACM press, pp 298–307

Müller H, Kalpathy-Cramer J, Eggel I, Bedrick S, Radhouani S, Bakke B, Kahr Jr CE, Hersh W (2009) Overview of the CLEF 2009 medical image retrieval track. In: Working Notes of the 2009 CLEF Workshop, Corfu, Greece

Ponte JM, Croft WB (1998) A language modeling approach to information retrieval. In: Proceedings of the 21st annual international ACM SIGIR conference on research and development in information retrieval. ACM press, pp 275–281

Salton G, Wong A, Yang CS (1975) A vector space model for automatic indexing. Communications of ACM 18(11):613–620

Schank RC, Kolodner JL, DeJong G (1980) Conceptual information retrieval. In: Proceedings of the ACM SIGIR conference. ACM press, Kent, UK, pp 94–116

Vintar S, Buitelaar P, Volk M (2003) Semantic relations in concept–based cross–language medical information retrieval. In: In Proceedings of the ECML/PKDD Workshop on Adaptive Text Extraction and Mining (ATEM)

Zhou W, Yu C, Smalheiser N, Torvik V, Hong J (2007) Knowledge–intensive conceptual retrieval and passage extraction of biomedical literature. In: Proceedings of the 30th annual international ACM SIGIR conference on research and development in information retrieval. ACM press, pp 655–662

Chapter 21
Improving Early Precision in the ImageCLEF Medical Retrieval Task

Steven Bedrick, Saïd Radhouani, and Jayashree Kalpathy–Cramer

Abstract Oregon Health and Science University has participated in the ImageCLEFmed medical image retrieval task since 2005. Over the years of our participation, our focus has been on exploring the needs of medical end users, and developing retrieval strategies that address those needs. Given that many users of search systems never look beyond the first few results, we have attempted to emphasize early precision in the performance of our system. This chapter describes several of the approaches we have used to achieve this goal, along with the results we have seen in doing so.

21.1 Introduction

Medical images form a vital component of a patient's health record. Effective medical image retrieval systems can play an important role aiding in diagnosis and treatment; they can also be effective educational tools for healthcare students, instructors and patients. As a result of advances in digital imaging technologies, there has been a large growth in the number of digital images stored in recent years. In addition to the Picture Archival and Communication Systems (PACS) that have become omnipresent in hospitals and clinics, there are numerous on–line collections of medical images. On–line atlases of images can be found for many medical domains including dermatology, radiology and gastroenterology. The sheer volume of medical image data provides numerous challenges and opportunities in the arena of medical image retrieval.

Steven Bedrick
Oregon Health and Science University, Portland, OR, USA e-mail: bedricks@ohsu.edu

Saïd Radhouani
Koodya sàrl, 8, rue d'Algérie, 8170 Bou Salem, Tunisia e-mail: Said.Radhouani@unige.ch

Jayashree Kalpathy–Cramer
Oregon Health and Science University, Portland, OR, USA e-mail: kalpathy@ohsu.edu

H. Müller et al. (eds.), *ImageCLEF*, The Information Retrieval Series 32,
DOI 10.1007/978-3-642-15181-1_21, © Springer-Verlag Berlin Heidelberg 2010

Medical image retrieval systems have traditionally been text–based, relying on the annotation or captions associated with the images as the input to the retrieval system. The last few decades, however, have seen significant advancements in the area of Content–Based Image Retrieval (CBIR) (Smeulders et al, 2000). CBIR systems have had some success in fairly constrained medical domains, including pathology, mammography, and certain MR and CT retrieval tasks (Müller et al, 2004). However, purely content–based image retrieval systems currently have limitations in more general medical image retrieval situations, especially when the query includes information about pathology (Müller et al, 2007; Hersh et al, 2006). Systems that use both textual and visual techniques (so–called 'mixed' systems), on the other hand, show more promising behavior (Hersh et al, 2006). One way in which they do this is by providing the user with more precise search results, particularly among the higher–ranking results ('early precision'; see Section 21.1.1 for a more detailed discussion of the meaning of this term).

During the course of this chapter, we discuss several techniques that we have developed for improving early precision in medical image retrieval systems. We will also describe the medical image retrieval system that we have built for use as a test bed for these techniques, as well as some of the ImageCLEF–related experiments we have conducted.

21.1.1 What is Early Precision?

What do we mean by 'early precision'? Recall that 'precision' refers to the proportion of a set of retrieved results that are truly 'relevant'. If a search system returns 100 results for a particular query, 75 of which are 'relevant' to that query and 25 of which are 'not relevant,' we would then say that the system had achieved a precision of 0.75. Typically, when a precision score is reported it is accompanied by an indication of the rank at which the score was computed. For example, a precision score of 0.6 at rank ten would mean that out of the first ten results, six were relevant. A similar score at rank 20 would refer to the precision of the first 20 results, and so on. For the purposes of this chapter, we will use the notation 'P@n' to refer to a result set's precision at rank n; for example, P@10 would refer to the precision calculated within the first ten results retrieved by a given system.

Early precision refers to precision scores calculated using low-ranking results, i.e. scores calculated using the first few results in a set. While there is no hard–and–fast cutoff, for the purposes of this chapter we will use the term 'early precision' to refer to precision scores that are calculated for the first 20 or fewer documents in a result set.

21.1.2 Why Improve Early Precision?

Why might we wish to focus our analysis on such a limited number of results? After all, a `trec_eval`–style run typically includes hundreds if not thousands of results. Our reason for focusing so heavily on early precision is a direct consequence of observed user behavior with modern search engines. It is well–accepted that many users of search engines never look beyond the first page of results (Hearst, 2009), and as such any changes in a system's precision beyond the first twenty or so results will be irrelevant to most users. We have therefore focused our efforts on improving our systems' performance with respects to their early precision, thereby hopefully providing our users with a better experience.

21.2 ImageCLEF

The medical image retrieval task within ImageCLEF has provided both a forum as well as a series of test collections to benchmark image retrieval techniques. The ImageCLEF campaign has been a part of the Cross Language Evaluation Forum (CLEF) since 2003 (Müller et al, 2007). CLEF itself is an offshoot from the Text REtrieval Conference (TREC). In 2004, ImageCLEFmed, a domain–specific task, was added to evaluate medical image retrieval algorithms and techniques.

The primary task of the medical track has been ad hoc retrieval of medical images given textual queries and sample images. However, the track has included more focused tasks, as well. For example, in 2005, a medical image annotation task was added to ImageCLEF. The goal of this task was to correctly classify 1,000 test images into 116 classes given a set of 10,000 training images. The classes differed primarily in anatomy[1] and view[2] of the image. It should be noted, however, that the images used for this task were primarily of a single modality (x–rays). The goal of the ImageCLEF medical image retrieval task of 2006 was to retrieve relevant images for thirty topics from a test collection of about 50,000 human–annotated images of different modalities. In more recent years, the test collection has grown to incorporate more than 70,000 images along with their textual annotations. The latest version of the test collection features images that were originally published as figures in articles from the journals of the Radiological Society of North America (RSNA), and therefore the image annotations include publication–derived metadata: figure captions, article titles, and links to originating articles' entries in MEDLINE. The topics (queries) used for the ImageCLEF medical tasks are typically derived from real–world query logs from production image retrieval systems, as well as from user studies with clinical users of medical imagery. The topics typically take the form of short snippets of text (e.g. 'CT images of a thoracic aortic dissection') and are categorized by whether the organizers feel that they lend themselves best

[1] I.e., the anatomical region on which a given image focused

[2] As in, angle of view— lateral vs. frontal, etc.

to visual or textual analysis (i.e., whether they will be most easily addressed by analyzing images' content or annotations).

Oregon Health and Science University (OHSU) has participated in the medical track of ImageCLEF since 2005. While our focus has been on the ad hoc retrieval tasks, we have also contributed runs for other medical tasks, including automated image annotation.

21.3 Our System

As part of our participation in the medical track of ImageCLEF, we have developed an experimental platform for testing different image retrieval strategies. Our group's mission is to better understand the needs and optimal implementation of systems for users in biomedical tasks, including research, education, and clinical care. Therefore, from the beginning, we designed our system to be interactive (as opposed to running in a batch–process mode). This has allowed us to use it for formal user studies as well as for ImageCLEF, and has also allowed us to experiment with more dynamic retrieval methods (described in more detail in Section 21.4.3).

21.3.1 User Interface

Our system is a Web–based search engine, somewhat similar in user interface to most general–purpose image retrieval systems (see Figure 21.1). The system and user interface are both written in the open source Ruby programming language[3] using the (also open source) Ruby on Rails framework[4]. Our system has evolved over time; early versions offered relatively few user–accessible controls, whereas the current version provides users with a variety of controls. The system is primarily designed to act in an interactive mode in which queries are entered by hand into a text box and results are viewed as thumbnails and document surrogates.

However, since we did design the system with the intent of using it for ImageCLEF, we built in several other convenience features. In addition to displaying results in a human–meaningful way as described, the system can also display its results in a `trec_eval`–compliant text format, which may be directly used as a run submission for automated evaluation. Additionally, although the primary query entry mode is the text box at the top of the main search page, queries may also be uploaded as text files. This allows users to easily submit multiple topics to the system and retrieve `trec_eval`–compliant output for each one simultaneously. Obviously, these features are next to useless for clinical end users; rather, they are simply

[3] `http://www.ruby-lang.org/`

[4] `http://www.rubyonrails.org/`

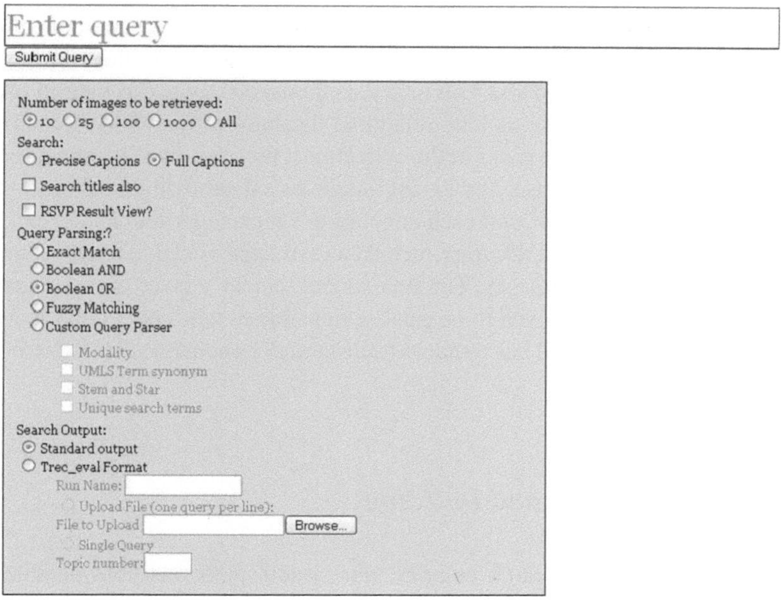

Fig. 21.1: The main query screen of our experimental image retrieval system.

convenience features for ourselves and any other researchers using our system to participate in TREC–style activities.

21.3.2 Image Database

Our system uses the open source PostgreSQL relational database management system[5] to store its index of images and their annotations. Originally, the test collection for the ImageCLEF medical retrieval task included images from a variety of sources, each of which had its own data schema and included annotations in multiple languages (English, German, and French). Generally, collections tended to use cases as their basic unit, with each case including at least one image. Some collections annotated images directly, others only annotated at the case level. This difference is especially significant for text–based retrieval, as images of different modalities or anatomies or pathologies could be linked to the same case–level annotation. In this situation, even though only one image from a case containing many images might be relevant to a query (based on the annotation), all images for the case would be retrieved in a purely text based system, reducing the precision of the search.

[5] `http://www.postgresql.org/`

Of course, the specific annotation fields also varied greatly from collection to collection. Some featured highly structured annotations, whereas others simply used a single field of unstructured text. Our final data model had to incorporate all of these different considerations. We use the relational database to maintain the mappings between the collections, the cases in the collections, the case–based annotations, the images associated with a case, any extant image–based annotations, along with the language of each annotation, and each annotation's metadata (field name, etc.). Our model also allows us to link arbitrary metadata (extracted visual features, computed class assignment, etc.) to images. The final model that we arrived at was somewhat complex, but ultimately proved to be easily extensible to new collections as well as relatively easy to index, and has remained substantially unchanged from 2006 to the present day.

21.3.3 Query Parsing and Indexing

As is the case for most search engines, our system performs considerable pre–processing of user queries. The various pre–processing steps all serve to either improve query precision or recall in one way or another. The main search screen presents a variety of search options to the user, including Boolean OR, AND, and exact match. There are also options to perform fuzzy searches, as well as to use our custom query parser. This parser forms a critical aspect of our system, and is also written in Ruby. Among other features, the custom query parser performs stopword removal using a specially–constructed list of domain–specific stopwords. These stopwords are derived from analysis of multiple years' worth of user– and ImageCLEF–derived queries. The custom query parser is highly customizable, and the user has several configuration options from which to choose.

The first such option attempts to increase query precision by restricting the modalities retrieved by the search engine. If the user selects this option, the user's query is parsed to extract the desired modality, if available. Using the modality fields described in the previous section, only those images that are of the desired modality are returned. This part of the system, along with its rationale and consequences, is described in more detail in Section 21.4.1.2.

Another option that users may choose is to perform manual or automatic query expansion using synonyms from the US National Library of Medicine's Unified Medical Language System (UMLS) meta–thesarus. Under this mode, our system identifies biomedical terms in the query and attempts to find synonyms for them in the meta–thesaurus. If the user has selected the manual query expansion mode, the system will then present them with a list of synonyms from which they may choose. Using the automatic mode, the system will simply add all retrieved synonyms to the user's query. This sort of query expansion serves to increase recall, albeit at a (sometimes significant) cost to precision.

An additional configuration option is the stem and star option, in which all the terms in the query are first stemmed, and then combined with a wildcard (*) operator. This will instruct the search sub–system to search for words containing the root of each query term. A final option allows the user to instruct the system to only use unique query terms when searching. This can be useful in combination with the UMLS query expansion option, as many of the UMLS synonyms contain significant lexical overlap.

To perform the actual textual searching, our system uses Ferret[6], a Ruby port of the popular Lucene search engine. It generates and maintains indices of both our case– and image–level annotations, takes care of some of our query parsing needs, and also handles wildcard and fuzzy searching.

21.4 Improving Precision

As discussed above, we are of the opinion that a system's early precision is strongly related to its ultimate usability. As such, we have focused our efforts on studying ways to boost our systems' early precision scores. We have found two approaches to be particularly useful:

1. *Modality filtration*, in which the system guesses the user's desired image modality from their query and only retrieves results of that modality; and
2. *Interactive result reordering*, a dynamic form of relevance feedback in which the user indicates a particularly exemplary image and the system reorders its results in response.

Both approaches take advantage of both visual and textual information, and modality filtration additionally relies on some extra query processing. Both approaches result in more relevant images within the first twenty or so search results, and have been proven to work reasonably well in many situations. However, it should be noted that both approaches do depend on a given query having first achieved a useful level of recall. The phrase 'garbage in, garbage out' definitely applies: if a query's initial results contain few or no relevant images, or if there are no images of the desired modality, neither approach will increase the user's performance.

21.4.1 Modality Filtration

Medical images are captured and generated using a wide variety of techniques, and therefore come in a wide variety of modalities[7]. Common modalities include x–rays, ultrasound scans, photographs, angiograms, CT or MR images, and so on. Often, users of medical image retrieval systems specify one or more image modalities as part of their query. For example, in the query 'CT images of a thoracic aortic

[6] `http://ferret.davebalmain.com/`

[7] An image's *modality* refers to the "physical" manner in which it was initially acquired.

dissection', the user is searching for CT images. Presumably, they are therefore uninterested in seeing pathology photos of dissected aortas, images of microscopic slides showing tissue damage resulting from aortic dissection, echocardiograms of thoraces, etc. By removing these a priori irrelevant images from the query's results, we can increase precision dramatically, thereby decreasing the amount of effort the user will need to expend in order to find a relevant result.

However, accomplishing this requires two things: first, we must know the modality of each image in our collection; and second, we must determine which modality (or modalities) the user is searching for.

21.4.1.1 Detecting Image Modality

Sometimes, entries in repositories of medical images are annotated with their modality. Since they are typically the result of clinical activity, Digital Imaging and Communications in Medicine (DICOM) images typically contain metadata describing their origin. However, DICOM images typically have lost their metadata by the time they end up in teaching repositories or other such places, either as a result of overly–zealous attempts at de–identification or simply as a result of being converted to a less rich file format such as JPEG. Furthermore, there have also been reported errors in the accuracy of DICOM headers (Güld et al, 2002), so even when a metadata–rich DICOM file is available, we may not always be able to trust its contents.

Figure captions and other such annotations often contain information about their image's modality. However, it is also common to find that the annotations or captions associated with images either do not contain such information or contain it in a misleading or unreliable way. The degree to which this may be the case varies widely from collection to collection, depending on the quality of its textual annotations. In the 2009 ImageCLEF collection (in which the images almost all have high–quality captions), we have found that we can correctly identify an image's modality using only caption text approximately 75% of the time (Kalpathy-Cramer et al, 2009) using various text–mining approaches (similar to those described in Section 21.4.1.2). Other image collections, with noisier annotations, have proved to be more challenging to work with.

In situations where we cannot rely on the textual annotations to tell us about an image's modality, we can instead use visual features of the image itself. Previously Kalpathy-Cramer and Hersh (2007) have described a modality classifier that is able to determine an image's modality using a variety of input features, including both gray– and color–level histograms, texture features, and discrete cosine transform data. On well–curated image collections, we have been able to achieve $\approx 95\%$ classification accuracy. However, that work was performed using a collection with good training data and intra–class homegeneity. In more realistic collections, in which individual document images often consist of compound figures of different modalities, images taken at different scales, and using poor training data, it is more typical for our classifier to achieve classification accuracy of 75–85%

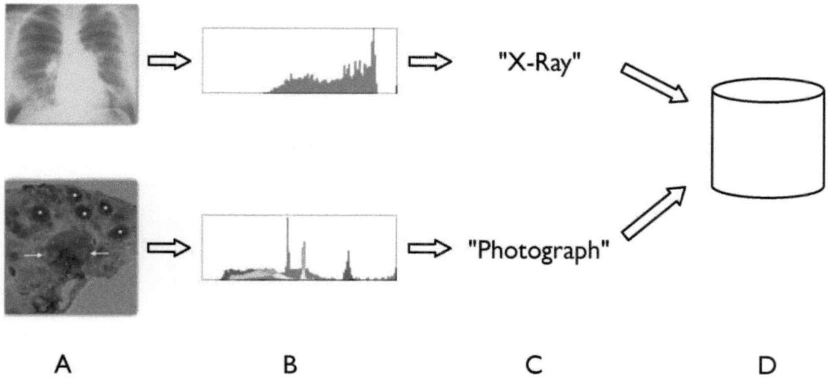

Fig. 21.2: Images in our database come from a variety of modalities (A). As part of our initial indexing, our system extracts a variety of visual features from each image (B), which it then uses to determine their modalities and assign class labels (C). The class labels are then stored in our database along with any other relevant image annotations (D).

As part of the process of loading new images into our retrieval system, we then determine each image's modality by visual, textual, or combined means, and store it alongside that image's entry in our database (see Figure 21.2). It may then be used for retrieval along with any other annotation.

Much can be accomplished using only textual or only visual information. However, by using *both* visual and textual data together, we are able to achieve much higher levels of classification accuracy. This is a pattern that we have seen in many other areas of image retrieval.

21.4.1.2 Detecting Query Modality

Once we have determined the modalities of our collections' images, we must turn to the problem of identifying and extracting the modality information contained in user queries. For this, we initially used a simple Bayesian classifier and were pleased with its performance. However, in later years, we began to experiment with simpler regular–expression–based modality detectors that would not require training data. We found their performance to be comparable to that of the Bayesian classifier. This finding is largely due to the extremely constrained textual domain of imaging modality: there are a very finite number of ways that users express their modality needs when formulating queries for image retrieval.

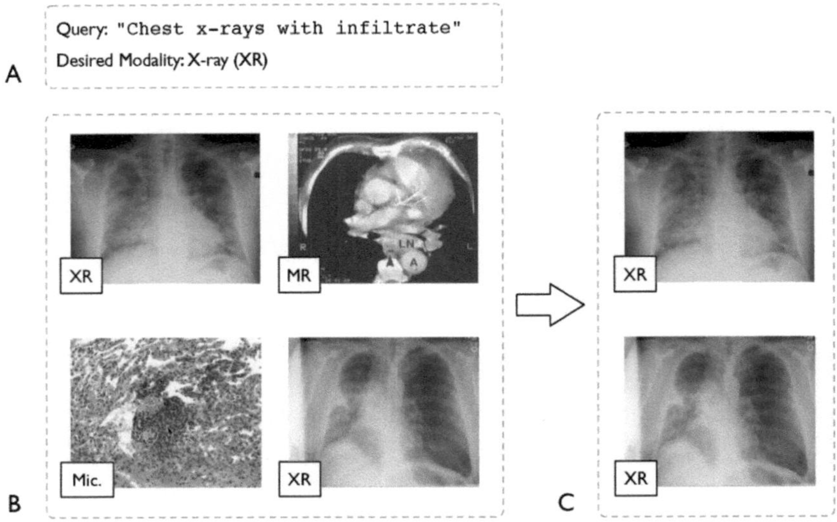

Fig. 21.3: The modality filtration process begins with the user's query, which often includes information about their desired image modality. Our system is able to extract and classify any modality–related information in the query and thereby determine what kind of image the user was looking for (A). The query is then handed off to the search system itself, which retrieves a set of candidate result images that are heterogeneous with respect to their modalities (B). Using the modality annotation obtained during pre–processing (see Figure 21.2) our system prunes any candidate results that are not of the desired modality, thereby delivering a more precise final result set to the user (C).

21.4.2 Using Modality Information for Retrieval

By using our query processing classifier, we can determine whether a given query specifies a modality and, if so, what that modality is. When combined with the fact that we know the modality of each image in our database, it becomes a simple matter to restrict our system's results to only include those of the desired modality (see Figure 21.3). This can have a substantial negative effect on recall, as we are removing a large number of results from the final set, some of which may have actually been relevant.

On the other hand, it often (in our experience) has an even more substantial positive effect on precision, as most of the images that are removed from the set are not relevant ones (assuming that our modality classifier has worked correctly). Figure 21.4 illustrates this point. This graph shows the 2007 version of our system's performance on the query 'Show me images of the oral cavity including teeth and gum tissue' both with and without modality filtration. Note the dramatic improvement in precision by using modality filtration (Kalpathy-Cramer and Hersh, 2007, 2008).

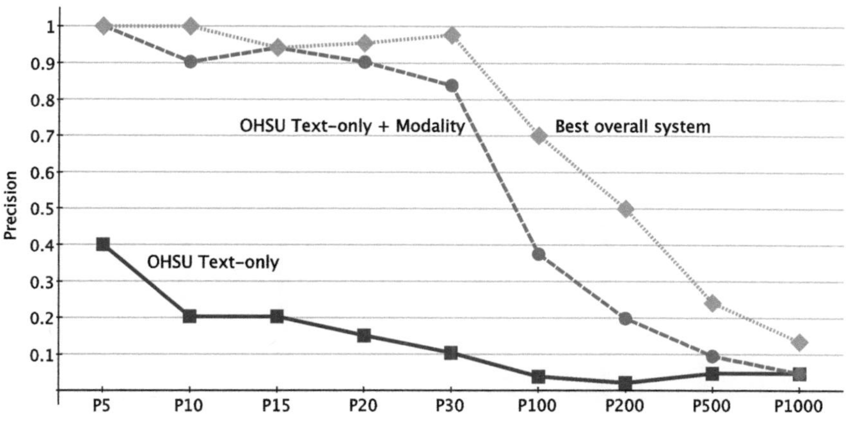

Fig. 21.4: Our system's precision performance for the query 'Show me images of the oral cavity including teeth and gum tissue' both with and without modality filtration. Note the dramatically increased early precision, as well as the marked dropoff in precision past P@30.

Of course, this effect was not present across all topics (see Figure 21.5 for an example of this phenomenon from 2009). If a topic lacked modality information, there was nothing to filter, and the results were left unimproved. Furthermore, if our classifier incorrectly determined a topic's desired modality, or if images in the collection were tagged with the wrong modality during pre–processing, result filtration actually hurt query performance, as relevant images were mistakenly removed from the set. However, in practice, we have seen major improvements in Mean Average Precision (MAP) and P@10 between baseline runs and runs using modality filtration. For example, in 2009, our baseline run achieved a P@10 of 0.380, whereas a similarly–configured run using modality filtration achieved a P@10 of 0.552. Looking only at the first five results (P@5), we saw a similar increase in performance using modality–filtration: 0.416 vs. 0.592 (Radhouni et al, 2009).

We have been able to duplicate the effect using runs from other institutions' systems. As part of a post–workshop analysis of the 2009 runs submitted by ImageCLEF medical track participants, we experimentally re–ordered each submitted run's entries such that images of the correct modality (as determined by our query and image classifiers) were ranked above images of the incorrect modality, and re–calculated precision and MAP. This resulted in a statistically significant improvement in MAP (paired t–test, $p << 0.05$). Figure 21.6 shows that virtually each run saw an increase in MAP by re–sorting using modality information, and Figure 21.7 shows the increase in early precision among all runs. Note that recall was left unaffected, since for this particular analysis we reordered results rather than removing them.

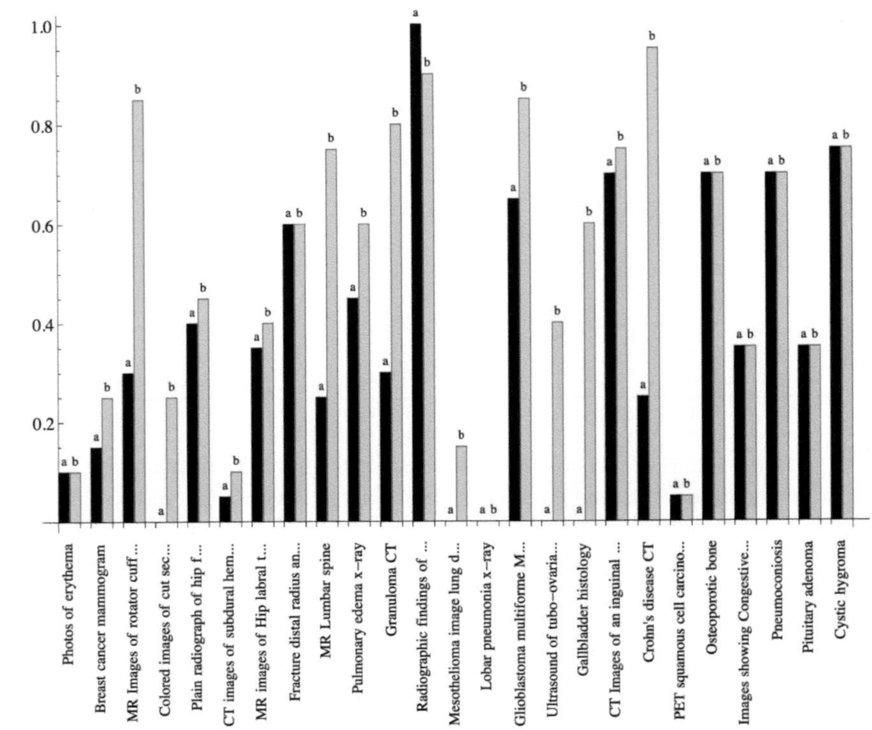

Fig. 21.5: Topic–by–topic comparison from ImageCLEF 2009 showing the change in P@20 between baseline (black bars, labeled 'a') and modality–filtered runs (gold bars, labeled 'b'). Note that topics lacking modality information (e.g. 'Osteoperotic bone') show little or no change, while topics that do include explicit modality information (e.g. 'MR lumbar spine') see major improvements.

21.4.3 Using Interactive Retrieval

Another approach we have used to improve early precision is to dynamically re–sort results according to visual similarity to an example result chosen by the user. Essentially, users of our system may select what they feel to be a visually representative image from their search's results. The system will then attempt to re–order the search results according to their degree of visual similarity with the probe image that the user selected. If the user is not satisfied with the re–ordering produced by their choice of image, they may repeat the process by selecting different probe images until they arrive at a satisfactory sorting.

To assess the visual similarity of the images within a result set, the system uses a relatively straightforward approach derived from Latent Semantic Analysis (Furnas et al, 1988). In this approach, each image in the result set is abstracted into a feature vector, which thereafter plays the same role that a document's term vector would play in classical LSA. We have experimented with sets of features derived from

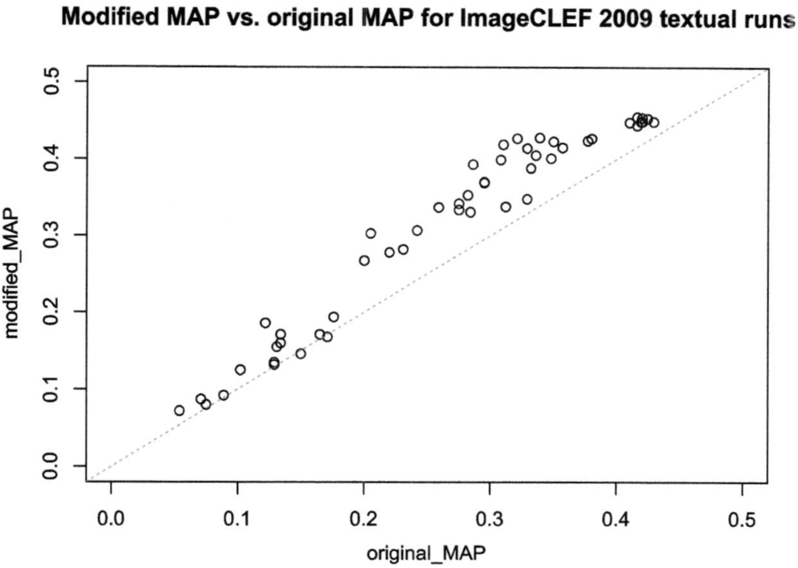

Fig. 21.6: Baseline vs. modality–reordered MAP from all textual runs submitted to the 2009 ImageCLEF medical track's ad hoc retrieval task. Reordering results based on modality resulted in a statistically significant improvement in MAP.

image color, texture, and frequency attributes; in our final system, the user is able to select which combinations of features they wish to use.

Once the feature vectors have been assembled for the images in a result set, they are combined into an $n \times m$ matrix. In this matrix, n is equal to the number of images in the result set, and m is equal to the number of features that the user has selected. Depending on the combination of features, this could be in the hundreds or low thousands. We then follow the classical LSA process, beginning by taking the Singular Value Decomposition (SVD) of our large matrix. This transforms our single matrix into three matrices that may be trivially recomposed to approximate the original matrix. The elements of one of these matrices represents the eigenvalues of the original document/term matrix; by varying the number of these elements that we use when recomposing the matrices, we may vary the fidelity of the resulting approximation.

After carrying out the SVD, we retain the first r eigenvalues of the decomposed matrix, project the probe image's m–dimensional feature vector into the new lower–dimensional space, and, finally, compute the vector distance between the probe image's new representation and that of the images in the result set. In our system, the user is able to experiment with different values for r, and may pick the one that achieves the best performance for a given set of results. The user may also quickly

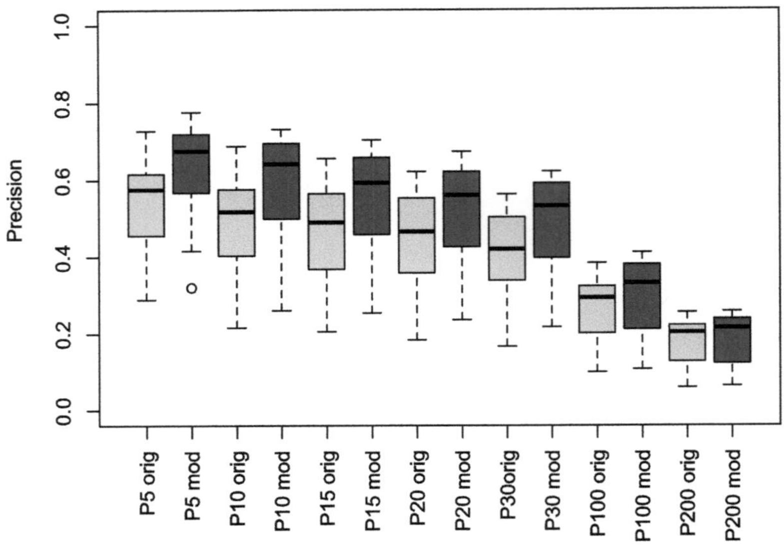

Fig. 21.7: In addition to increasing MAP, reordering submitted run results using modality information increased precision. This boxplot includes all textual runs submitted in 2009. Note the improved precision between original (orig) and modality–filtered (mod) runs.

and easily select different images to act as probe images, and can therefore evaluate many possible result sortings.

Obviously, this system's utility is variable, and depends heavily on the contents of the initial result set. In the case of a set where the desired images are simultaneously visually similar to one another and distinct from the rest of the images in the set, this visual re–sorting system works quite well. However, in the case where the desired images are visually different from one another, or where all of the results (including the non–relevant ones) are visually similar, this re–sorting system is not very useful.

For example, a result set consisting entirely of ultrasound images will not be improved very much by re–sorting. In fact, in this particular case, re–sorting the result set may hurt its precision, as any ordering imposed by our textual search engine will be lost. On the other hand, a result set in which most of the relevant images are ultrasounds and most of the non–relevant images are x–rays could benefit from being reordered based on visual similarity to a user–selected probe image.

Our present system requires the user to select a combination of features to use. This is clearly sub–optimal, and our future work could include improved feature selection methods. Similarly, the user is currently able to change the number of eigenvalues used by the algorithm. While this is a powerful tool for tuning the algorithm's performance, it is also something that we would ultimately like to automate.

In 2008, we submitted two runs using this feature. The first used it for each topic: the operator selected what he felt to be a representative image, and then submitted the resulting reordering with minimal intervention (even for topics in which it was obvious that the feature was unhelpful). Not surprisingly, this run's performance was suboptimal, and featured MAP and precision scores that were comparable to our baseline run's. Of course, in real life, the feature is meant to be used in a much more interactive and intelligent manner: presumably, a human operator would be able to decide whether or not it had improved their results, and, if not, would return to the original system–produced sorting.

We therefore submitted a second, fully interactive run, in which the operator only used the interactive re–sorting feature when he felt that it would be beneficial. The operator also took advantage of our system's interactive nature, and experimented with a variety of settings on a topic–by–topic basis. It would be meaningless to compare this run's performance to that of the more automatic runs that were submitted at the same time; however, it is worth pointing out that with the aid of the interactive re–sorting feature, our fully interactive run achieved extremely high P@10 (0.43; compared to that year's overall champion system's P@10 of 0.43).

21.5 Conclusions

The techniques we have discussed in this chapter may seem rather simplistic. While that may be the case, it is also the case that they have positive effects on our system's performance. This is a pattern that we have repeatedly observed in our work with image retrieval in general and ImageCLEF in particular: simpler and lower–tech approaches often tend to win out over more sophisticated techniques. In our case, we have had some degree of success at improving precision by focusing on a single, easily understood yet highly discriminatory feature (image modality) and then simply using it to aggressively filter our results. Furthermore, by focusing on precision, we believe that real–world end users of our system will benefit. Our next step is to attempt to carry this further: identifying other image and query features that can be easily labeled, and performing similar filtration.

We feel that it is useful to use external knowledge to interpret the semantic content of documents and queries. Indeed, most queries contain a precise description of a user need, materialized by a set of words belonging to three semantic categories: modality of the desired result images (e.g. MRI, x–ray, etc.), anatomy (e.g. leg, head, etc.), and pathology (e.g. cancer, fracture, etc.). We call these categories domain dimensions, and define them as follows: 'A dimension of a domain is a concept used to express the themes in this domain' (Radhouani, 2008). The idea behind this approach is that, in a given domain, a theme can be developed with reference to a set of dimensions of this domain. For instance, consider a physician writing a report about a medical image. They might first focus on a domain (Medicine), and then refer to specific dimensions of this domain (e.g. anatomy, pathology), during which they choose words from this dimension (e.g. femur, fracture), and finally they write their report.

In order to resolve CLEF queries, we have experimented with using these domain dimensions to interpret the queries' semantic content. To do so, we first needed to define the dimensions. For this purpose, we used external resources, such as ontologies or thesauri, to define each dimension by a hierarchy of concepts. Every concept is denoted by a set of words. Thereafter, to identify dimensions from a query, we extract query words depending on the dimension hierarchy they belong to. Once dimensions are extracted from each query, we use them to search for relevant documents. In particular, we use Boolean operators on query dimensions in order to reformulate the initial text of the query and better represent its semantic content. For instance, if we assume that a relevant document must contain all the dimensions belonging to the query, we should use the operator AND between the query's words that represent these dimensions in order to query the document collection.

This approach is still in its infancy, but in 2009 our runs utilizing domain–dimension–based result filtration saw some performance increase. Future work will explore different ways to identify domain dimensions from queries and image annotations, and also on studying how our system's behavior affects clinical end users.

Acknowledgements The authors gratefully acknowledge the support of NLM Grants 1K99LM 009889 and 2T15LM007088, as well as NSF Grant ITR–0325160.

References

Furnas GW, Deerwester S, Dumais ST, Landauer TK, Harshman RA, Streeter LA, Lochbaum KE (1988) Information retrieval using a singular value decomposition model of latent semantic structure. In: Proceedings of the 11th annual international ACM SIGIR conference on Research and development in information retrieval. ACM press, pp 465–480

Güld MO, Kohnen M, Keysers D, Schubert H, Wein BB, Bredno J, Lehmann TM (2002) Quality of DICOM header information for image categorization. In: Siegel EL, Huang HK (ed) Society of Photo–Optical Instrumentation Engineers (SPIE) Conference Series, vol 4685, pp 280–287

Hearst M (2009) Search user interfaces. Cambridge University Press, Cambridge

Hersh W, Kalpathy-Cramer J, Jensen J (2006) Medical Image Retrieval and Automated Annotation: OHSU at ImageCLEF 2006. In: Peters C, Clough P, Gey FC, Karlgren J, Magnini B, Oard DW, de Rijke M, Stempfhuber M (eds) Proceesings of the Corss–Language Evaluation Forum. Lecture Notes in Computer Science (LNCS), vol 4730. Springer, pp 660–669

Kalpathy-Cramer J, Hersh W (2007) Automatic image modality based classification and annotation to improve medical image retrieval. In: Studies in health technology and informatics, vol 129. IOS, pp 1334–1338

Kalpathy-Cramer J, Hersh W (2008) Medical Image Retrieval and Automatic Annotation: OHSU at ImageCLEF 2007. In: Peters C, Valentin J, Mandl T, Müller H, Oard D, Petras A, Petras V, Santos D (eds) Advances in Multilingual and Multimodal Information Retrieval: 8th Workshop of the Cross–Language Evaluation Forum. Lecture Notes in Computer Science (LNCS), vol 5152. Springer, pp 623–630

Kalpathy-Cramer J, Bedrick S, Lam CA, Eldredge C, Kahn Jr. CE (2009) Automated image–based classification of imaging modality. In: Proceedings of the 95th Scientific Assembly and Annual Meeting of the RSNA

Müller H, Michoux N, Bandon D, Geissbuhler A (2004) A review of content–based image retrieval systems in medical applications—clinical benefits and future directions. International Journal of Medical Informatics 73(1):1–23

Müller H, Deselaers T, Deserno T, Clough P, Kim E, Hersh W (2007) Overview of the Image-CLEFmed 2006 medical retrieval and medical annotation tasks. In: Peters C, Clough P, Gey F, Karlgren J, Magnini B, Oard D, de Rijke M, Stempfhuber M (eds) Evaluation of Multilingual and Multi–modal Information Retrieval: Seventh Workshop of the Cross–Language Evaluation Forum, CLEF 2006. Lecture Notes in Computer Science (LNCS), vol 4730. Springer, pp 595–608

Radhouani S (2008) Un modèle de recherche d'information orienté précision fondé sur les dimensions de domaine. PhD thesis, University of Geneva, Switzerland, and University of Grenoble, France

Radhouni S, Kalpathy-Cramer J, Bedrick S, Bakke B, Hersh W (2009) Multimodal medical image retrieval improving precision at ImageCLEF 2009. In: Working Notes of CLEF 2009

Smeulders A, Worring M, Santini S, Gupta A, Jain R (2000) Content–based image retrieval at the end of the early years. IEEE Transactions on Pattern Analysis and Machine Intelligence 22(12):1349–1380

Chapter 22
Lung Nodule Detection

Luca Bogoni, Jinbo Bi, Charles Florin, Anna K. Jerebko, Arun Krishnan, Sangmin Park, Vikas Raykar, and Marcos Salganicoff

Abstract The quantity of digital medical images that must be reviewed by radiologists as part of routine clinical practice has greatly increased in recent years. New acquisition devices generate images that have higher spatial resolution, both in 2–D as well as 3–D, requiring physicians to use more sophisticated visualization tools. In addition, advanced visualization systems, designed to assist the radiologist, are now part of a standard arsenal of tools which, together with workflow improvements, aid the physicians in their clinical tasks. Computer–Assisted Diagnosis (CAD) systems are one of such class of sophisticated tools to support the radiologists in tedious and time–consuming tasks such as the detection of lesions. Over the past ten years, CAD systems have evolved to reach sensitivity capabilities equivalent to or exceeding that of a radiologist, thus becoming clinically acceptable, but with limited specificity which necessitates their use as a *second reader* tool. This chapter presents one such system (LungCAD[1]) designed for the detection of nodules in the lung parenchyma. Its performance was evaluated as part of a detection challenge organized by ImageCLEF 2009.

22.1 Introduction

The introduction of computers and the subsequent invention of CT (Hounsfield, 1973) in the 1970s revolutionized medicine by introducing 3–dimensional (3–D) imaging. An x–ray source, normally set in a fixed location to generate 2–D images, was now mounted on a rotating gantry. Thus, series of x–ray projections were used to compute a 3–D image of the inside of the body from several 1–D x–ray beams acquired around a single axis of rotation. The 1–D beams from the same plane are used to reconstruct 2–D planes; the collection of 2–D planes are then stacked and

Siemens Healthcare, Imaging and Therapy Systems Division, Computer–Aided Detection Group Malvern, PA, USA, e-mail: marcos.salganicoff@siemens.com

[1] Not available commercially.

H. Müller et al. (eds.), *ImageCLEF*, The Information Retrieval Series 32,
DOI 10.1007/978-3-642-15181-1_22,

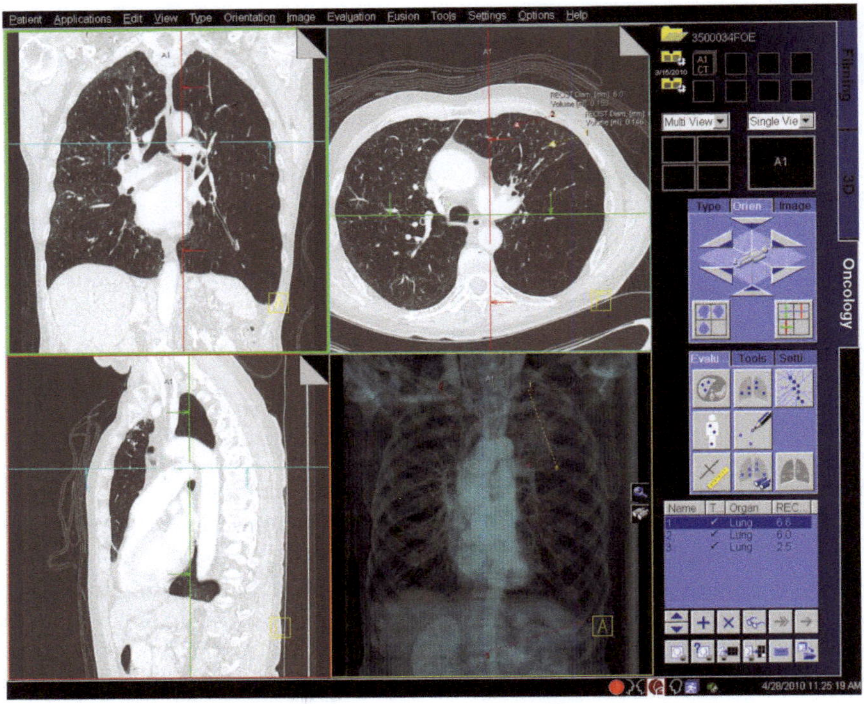

Fig. 22.1: Thoracic CT images displayed using advanced visualization tools, with CAD marks overlaid on the axial plane (top right quadrant).

presented as a 3–D volume. Radiologists were thus not only able to detect subtle variations of structures in the body, but also to locate them within a fixed frame of reference. Early CT generated images (also known as slices) were acquired and reconstructed orthogonally to the long axis of the body and then imaged on film. Modern scanners, on the other hand, allow this volume of data to be reformatted in various planes or even visualized as volumetric (3–D) representations of structures with a high degree of resolution.

In order to appreciate how these technological innovations both afford better diagnostic capabilities while introducing new challenges, we will consider how the diagnosis of lung cancer, using CT images, has benefited from the use of computer assisted detection and diagnosis technologies (CAD).

22.1.1 Lung Cancer — Clinical Motivation

Lung cancer is the most commonly diagnosed cancer worldwide, accounting for 1.2 million new cases annually (of Surveillance and Research, 2009). It is an exceptionally deadly disease: six out of ten people will die within one year of being

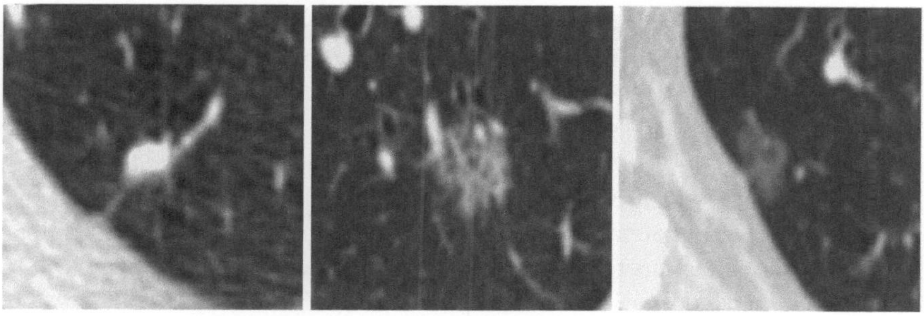

Fig. 22.2: Three different nodule morphologies, from left to right: Solid, Part–Solid and Ground Glass nodules.

diagnosed. The expected five–year survival rate for all patients is merely 15%, compared to 65% for colon, 89% for breast, and 99.9% for prostate cancer. In the United States, lung cancer is the leading cause of cancer death for both men and women and costs almost $10 billion to treat annually. However, lung cancer prognosis varies greatly depending on how early the disease is diagnosed; as with all cancers early detection provides the best prognosis. At one extreme of the prognosis spectrum are patients diagnosed with distant tumors — that have spread (metastatic) to the lung from other areas, such as colonic or pancreatic cancer, Stage IV patients — for whom the five–year survival rate is just 2%. At the other end, the prognosis of early stage lung cancer patients (Stage I) is more optimistic with a mean five–year survival rate of about 49%. Hence, if cancer is detected early, when it is relatively small in size and localized, many treatment options are then viable: surgery, radiotherapy, chemotherapy.

Today's Multi-detector CTs (MDCT) are capable of generating images which allow physicians to detect lung nodules that are 2–8 mm in size; cancers found at this early stage have an excellent prognosis. However, despite these technologies, only 24% of lung cancer cases are diagnosed at an early stage (Jemal et al, 2008) Many potentially clinically significant lesions remain, however, undetected. A contributing factor may be, actually, the increase of MDCT imaging data generated for a thoracic acquisition. Specifically, while just eight years ago, a 2–slice CT could generate 41 axial images of the thorax in a 30–second scan (single breath hold), a now state–of–the–art 64–slice dual–source CT may generate up to 480 axial slices in only 0.6 seconds for each patient, a factor of ten increase in the amount of image data to be reviewed.

Figure 22.1 illustrates typical Lung CT images for a single patient visualized in a radiology image review application. Within this application each one of hundreds of image slices must be carefully examined by a radiologist to identify if any of the marks on the image correspond to normal structures (air passages), benign tumors, lung diseases other than cancer, or early–stage lung cancer. Hence, while benefiting from the increased image resolution for diagnostic quality, radiologists face the burden and medical responsibility of having to examine an enormous number of images. Furthermore, while clinical workflow and protocols have improved, case load requirements have, for the most part, also increased. Thus, to ameliorate and mitigate the challenges, CAD tools have been developed to support in the identification and interpretation of nodules in CT scans of the lung.

Clinically, a solid nodule is defined as an area of increased opacity more than 3 mm in diameter which completely obscures normal underlying structures such as blood vessels and parenchyma of the lung. As in the case of visual identification tasks, translating this relative succinct symbolic definition into image features and data mining algorithms is a substantial challenge. Furthermore, while it is universally acknowledged that solid nodules are precursors to lung cancer, recently there has been increased interest in detecting other lesion types: Part–Solid Nodules (PSN) and Ground Glass Nodules (GGN), (see Figure 22.2). A GGN is defined as an area of a slight, homogeneous increase in density (translucent and reminiscent in appearance to a 'ground glass' surface), which does not obscure underlying bronchial and vascular markings. GGNs are known to be extremely hard to detect. Several studies (Suzuki et al, 2002) have pointed out that they are an indicator of early cancer, albeit a different variant than the one typical of solid nodules. GGN lesions present a perceptual challenge due to their subtle appearance relative to the background visual anatomical clutter of the lung, and to the fact that their appearance can also be highly variable.

22.1.2 Computer–Aided Detection of Lung Nodules

There is a growing consensus among clinical experts that the use of CAD software improves the performance of the radiologist (Suzuki et al, 2005; Summers, 2003). CAD is introduced in the workflow as a second reader[2]: the radiologist first reviews the image, then assesses any findings proposed by CAD. A CAD algorithm — typically a set of image processing algorithms followed by a classifier — may either pre–process the image or be invoked by the radiologist to generate marks that highlight structures deemed of interest. Figure 22.1 shows CAD marks superimposed on CT images. Clinical studies have demonstrated that the use of CAD software not only offers the potential to improve the detection and recognition performance of a radiologist, but also to reduce mistakes related to misinterpretation (Armato III et al, 1999; Naidich et al, 2004).

The benefit of CAD is assessed as the incremental value of CAD in normal clinical practice, such as the number of additional lesions detected using CAD. However, CAD systems must not have a negative impact on patient management. Specifically,

[2] This is the most often used and accepted paradigm of workflow integration, whereas other paradigms such as concurrent (Beyer et al, 2007) and first reader CAD (Mani et al, 2004) approaches are being explored. While the second reader paradigm approach is well established in the literature and community, these other approaches offer advantages which still need to be vetted through large studies.

a significant increase in the number of false positives may cause the radiologist to recommend unnecessary biopsies and potentially dangerous follow–ups.

The process of computer aided detection for lung nodules can also be formulated as a semantic query, by abnormality name, e.g. nodule. In this context, the goal is that of identifying Regions or Volumes Of Interest (ROI or VOI) within a current image (either 2–D or 3–D) containing a nodular abnormality. However, in the context of a differential diagnosis algorithm, as taught by medical practitioners (radiologists), detection is just one of the steps in the diagnostic chain with a highly specific search target — either the presence or absence of disease. Hence, the results of a CAD search is a list of ROIs manifested to the clinician, for instance, in the form of arrows or superimposed circles indicating the CAD proposed finding (see bottom left quadrant of Figure 22.1).

22.1.3 Ground Truth for Lesions

Fundamental to building CAD systems is the identification of what constitutes instances of disease. By selecting exemplars of true lesions, features can be selected, most often automatically, and a classification system be trained. The process of assigning labels to anatomical structures diagnosed as lung nodules relies on the availability of unequivocal evidence.

The gathering of evidence is, however, extremely difficult since the occurrence of cancer can only be ascertained by performing biopsies and obtaining a pathology report. Whereas, for example, in breast cancer virtually all suspicious lesions are routinely biopsied (providing histological ground truth), a lung biopsy is a dangerous procedure, with a 2% risk of serious complications (including death).

These limitations make obtaining definitive lung cancer ground truth infeasible, particularly for patients being evaluated for early signs of lung cancer (screening). Thus, very often CAD systems are built using ground truth based on image annotations collected from one or more expert radiologists, sometimes conflicting with one another. In clinical studies, designed to obtain regulatory approval from the US Food and Drug administration (FDA), this approach has been considered an acceptable proxy to obtaining pathological proven ground truth.

The nature of ground truth for nodules can be highly uncertain and, in this sense, differs from tasks such as normal anatomical content identification. Specifically, the queries focus on images containing specific normal anatomy, with well defined characteristics, or acquired with specific protocols having a given textual annotation. Furthermore, there are many focal abnormalities of the lung that may present a visual appearance that mimics that of a true pulmonary nodule (MacMahon et al, 2005). Their differential diagnosis may require further tests such as dimensioning, volumetry, biopsy or growth surveillance over time to ultimately determine the true pathology underlying the visual appearance within the image.

Even a single structure may be identified differently by various practitioners based on a number of factors including the patient history, training of the practitioner, pre–test probabilities, etc. These aspects have less of an impact since in other detection tasks the state of the search targets are well established — e.g. fractures and other trauma, punctuate normal anatomical landmarks, specific organs, diseases with well defined unambiguous symptoms/visual appearance.

Compounding the problem is the difficulty in obtaining the pathological gold standard for lung nodules unless they are proximal to large airways or the lung well, in which case trans–bronchial or fine needle biopsy may be possible. This problem has been addressed by CAD researchers in a number of ways (Clarke and Croft, 2004; Armato III et al, 2009; Raykar et al, In Press).

Automatic pulmonary nodule detection is also an intrinsically challenging pattern recognition task due to nodule properties and complex lung geometry. Pulmonary nodules are relatively small opacities (<30 mm), whose appearance vary greatly depending on whether they are discrete or attached to neighboring structures, whether they have well defined margins, subsolid components, calcification, etc. Furthermore, the lung tissue is interwoven with vascular and bronchial structures, a potential source of false positives for CAD. Additionally, the introduction of intravenous contrast agents such as iodine into the vasculature of the lung may modify its appearance and add further variability to its appearance. Finally, the presence of co–morbidities (emphysema, fibrosis, COPD, etc.) may overlay additional visual clutter that can make not only the human visual search task daunting, but the algorithmic one as well.

In order to improve both the accuracy and the efficiency of detecting lesions, many different approaches have been developed. A review of the current literature on lung nodule detection techniques is presented in Section 22.2. Section 22.3 introduces the CAD system evaluated in ImageCLEF 2009, and Section 22.5 presents the evaluation and the results.

22.2 Review of Existing Techniques

Pulmonary nodule detection techniques have been an area of pattern recognition research in both academic and industrial (healthcare) sectors over the past 15 years (Ko and Betke, 2001; Armato III et al, 1999; Lee et al, 2001; Farag et al, 2004; Chang et al, 2004; Paik et al, 2004). Most systems described in the literature generally consist of four steps:

1. Image pre–processing/signal conditioning;
2. Nodule candidate generation;
3. Discriminative feature computation around candidates;
4. Classification based on discriminative features.

In some applications, the candidate generation and the feature computation are merged into one single step. In this section, we review the prior approaches on the candidate generation and feature computations, and describe our approach. Furthermore, since ground truth creation is a crucial step in a successful system development, we explain a framework for handling the multiple experts annotations that contain some error.

22.2.1 Gray–Level Threshold

Gray–level thresholding based on the Hounsfield unit (HU) is one of the earliest and most basic techniques. Ko and Betke (2001) applied the several gray–level thresholds to create binary images to find nodule candidates. For each candidate, shape and location features are computed for the classification stage. Armato III et al (1999) constructs the gray–level profile for each CT slice and determine a threshold to segment the thorax. The gray–level threshold technique is effective and easy to use for CT images. However, a good specificity cannot be achieved solely through thresholding, since the HU value of pulmonary nodules is very similar to that of other structures such as blood vessels and mucous in air ways. When the process is combined with other features such as shape descriptors, the specificity can be improved.

22.2.2 Template Matching

Lee et al (2001) assumed that a Gaussian distribution can be used to approximate lung nodules and proposed the following nodule model:

$$pv_{x,y,z} = m \cdot e^{-(x^2+y^2+k \cdot z^2)/n} \tag{22.1}$$

where $pv_{x,y,z}$ is the pixel value of co–ordinate (x,y,z), and m and n are parameters representing the maximum value and variance of the distribution, respectively k regulates the scaling in z. Four reference images were generated from spherical models at various diameters (6.8 mm, 13.6 mm, 20.4 mm, and 27.2 mm). These reference images are then compared with the observed images by computing a similarity value between the model and the image defined as:

$$Similarity_{a,b} = \frac{\sum_{i=0}^{n-1}(a_i - m_a)(b_i - m_b)}{\sqrt{\sum_{i=0}^{n-1}(a_i - m_a)^2}\sqrt{\sum_{i=0}^{n-1}(b_i - m_b)^2}} \tag{22.2}$$

where

$$m_a = \frac{1}{n}\sum_{i=0}^{n-1} a_i, \quad m_b = \frac{1}{n}\sum_{i=0}^{n-1} b_i \tag{22.3}$$

The value a_i is the ith pixel in image a, the value b_i is the ith pixel in image b. a and b can be either the reference model or observed image.

Since solid nodules have gray level distributions similar to other structures such as arteries, veins, and bronchus walls, Farag et al (2004) suggested addressing the abnormality detection by geometrical template matching. Their central–symmetric Gaussian–like template is defined as

$$q(r) = q_{max}\, exp\big(-(r/\rho)^2\big) \tag{22.4}$$

where $q(r)$ is the gray level in a template point with Cartesian co–ordinates (ξ, η) with respect to the center (*i.e.*, $r^2 = \xi^2 + \eta^2$), r is the radius from the template's center, q_{max} denotes the maximum gray level for the template, and R is the template radius depending on the minimum gray level.

22.2.3 Spherical Enhancing Filters

Chang et al (2004) developed a spherical filter to enhance nodule intensities as well as a cylinder filter to suppress vessels. The cylinder filter F_{cyl} consists of cylinders aligned in different orientations, where each cylinder has a pre-defined width and length.

$$F_{cyl}(x) = \max_{\theta}\big(\min_{y \in \Omega_\theta^x} I(y)\big) \tag{22.5}$$

where Ω_θ^x is the domain of the cylinder filter centered at x with orientation θ. In order to enhance the intensities of the low contrast nodules, they also suggested a non-linear spherical filter F_{sph} with two components F_{fill} and F_{hollow}:

$$F_{sph}(x) = F_{fill}(x) - F_{hollow}(x) = \max_{y \in \Omega_{fill}^x} I(y) - \max_{y \in \Omega_{hollow}^x} I(y) \tag{22.6}$$

where Ω_{fill}^x and Ω_{hollow}^x are the domains of the filters F_{fill} and F_{hollow} centered at x, respectively. Intuitively, the response of F_{sph} is strong when a structure is isolated from other high intensity structures.

Paik et al (2004) introduced a technique called surface normal overlap to detect convex regions such as lung nodules. The inward surface normal vectors tend to intersect or nearly intersect within the tissue when the surface is convex. A 3–D array, denoted $A(x,y,z)$, counts the number of surface normals that pass through or near to each voxel. The local maxima of $A(x,y,z)$ are selected as candidate lesion locations. Since complex anatomic structures with multiple convex surface patches may generate multiple local maxima, the normal vectors are sampled on the surface with a certain distance between each vector.

22.3 Description of Siemens LungCAD System

The lung CAD algorithm evaluated in this chapter is designed as a multi–step approach with the goal of detecting parenchymal lesions with high sensitivity and specificity. The algorithm focuses on solid lesions greater than 4 mm and subsolid lesions greater than 6 mm, since most often, only lesions of at least that size are considered clinically significant (Godoy and Naidich, 2009). The algorithm has four stages: lung segmentation, candidate generation, feature computation, and classification. The first three stages: lung segmentation, candidate generation, and feature computation combine both image processing and embedded machine learning approaches, while the classification stage is exclusively a machine learning phase.

22.3.1 *Lung Segmentation*

In this stage, the lungs are identified and isolated in the thoracic CT image volume to create a region of interest for all subsequent analyses. The lungs' contour is delineated by first detecting regions in the thorax characterized by air density (initial foreground area). The collection of these regions is then processed using a series of morphological operations for filling holes in the foreground area to cover the entire lung including vascular structures and soft tissues. The result of the segmentation process is a binary foreground mask that allows the subsequent stages of the algorithm to operate only in the area of the lung.

22.3.2 *Candidate Generation*

The goal of this stage is to generate candidates with high sensitivity while keeping the number of false positives to a manageable number so as not to overburden subsequent processing stages. In the first step, the diverging gradient field response (DGFR) algorithm developed by Bogoni et al (2009) is applied to identify candidate locations. It may generate up to 300 candidates per volume. As a second step a cascading classifier reduces the number of candidates to around 80 per volume using DGFR features.

The DGFR processes the entire segmented lung area, from the previous stage, for blob–like structures. It detects isolated lesions that are consistent with the appearance of solid, subsolid and ground glass nodules. These non–solid structures are hyperattenuating regions characterized by a blob–like shape and surrounded by the background region of lower attenuation. Additionally, DGF also detects suspicious regions attached to other anatomical structures such as pleura, vessels, and airways.

In order to determine a DGFR response for a given target image, we apply Gaussian functions whose gradient fields are diverging:

$$g_\sigma(x,y,z) = \frac{1}{(2\pi\sigma^2)^{\frac{3}{2}}} e^{-\frac{x^2+y^2+z^2}{2\sigma^2}} \tag{22.7}$$

The DGFR response $\mathbf{dgfr}(f)$ is computed by the scalar product of the image and the Gaussian gradient vectors:

$$\mathbf{dgfr}(f) = \frac{\overrightarrow{\nabla} f}{\|\overrightarrow{\nabla} f\|} \cdot \overrightarrow{\nabla} g \tag{22.8}$$

where $\overrightarrow{\nabla} f$ is the gradient vector of the given image, $\overrightarrow{\nabla} g$ is the gradient vector of Gaussian function, and $\|\overrightarrow{\nabla} f\|$ is the gradient magnitude. A prototypical example of such a candidate is represented by a perfectly spherical object with hyperattenuation, and surrounded by a homogeneous lower attenuation region. If voxel attenuation is interpreted as height, then hyperattenuating regions surrounded by hypoattenuating regions appear as crests or local maxima in the cross–section. By using this interpretation, the attenuation pattern detected by the candidate generator is a roughly spherical region of locally maximal attenuation. Since the notion of local maximum is relative to the surrounding tissues, and does not rely on arbitrarily set Hounsfield Units, it allows for the detection of both solid and subsolid lesions.

In the next step, multiple classifiers are applied to the candidate lesions to successively filter off candidates that have the least correlation with typical lesions previously seen by the classifiers in a training set. At this stage of the processing, the algorithm's sensitivity is in the order of 95–100%. Subsequent stages focus on reducing the false positive rate while preserving the high sensitivity.

22.3.3 Feature Extraction

The feature extraction stage gathers the image–based discriminative features for each candidate identified in the previous stage. This information is used by the next stage (classification) to label each candidate as either a true or false positive. Discriminative features are computed using the image voxels' intensity value, from multiple adjacent image planes, neighboring a candidate. The computed features can be categorized into several groups: (a) those indicative of voxel attenuation distributions within the candidate, (b) those pertaining to the candidate's shape and curvature, and (c) those that describe the candidate's edge and margins. These features capture candidate properties that can be used to disambiguate true lesions from typical false positives. Some of the typical false positives, mimicking true nodules, are caused by pleural thickening, connective tissues between vessels, partial volume, etc.

22.4 Classification

The final classification stage differentiates candidates that are true nodules from the rest of the candidates based on the computed feature vectors. The goal of the classifier is to reduce the number of false positives without appreciable decrease in the sensitivity. The various data mining techniques successfully used for multiple CAD areas such as lung nodules, breast cancers, colon polyps, pulmonary embolism, and cardiovascular disease are described in (Rao et al, 2009). We exploit many of these jointly developed techniques in our LungCAD system. We will briefly describe some of the key techniques employed in the design of the final classifier.

22.4.1 *Multiple Instance Learning*

The candidate generation step generates a lot of potential candidates. Any candidate which is close to the radiologist mark is considered a positive example for training and the rest of the candidates are considered as negative examples. Candidates are labeled positive if they are within some pre–determined distance from a radiologist mark; some of the positively labeled candidates may actually refer to healthy structures that just happen to be near a mark, thereby introducing labeling errors in the training data. These labeling errors can potentially sabotage the learning process by confusing a classifier that is being trained with faulty labels, resulting in classifiers with poor performance. Multiple Instance Learning (MIL) is one of the effective ways to deal with this problem. Our LungCAD system utilizes the multiple instance learning–based classifier developed by Raykar et al (2008) that performs automatic feature selection and classifier design jointly.

In the MIL framework the training set consists of bags. A bag contains many instances. A bag which contains K instances is denoted by $\mathbf{x} = \{x_j\}_{j=1}^{K}$ where $x_j \in \mathbb{R}^d$ is the feature vector. All the instances in a bag share the same bag–level label denoted by $y \in \{0,1\}$. A bag is labeled positive if it contains at least one positive instance. The probability that a bag contains at least one positive instance is one minus the probability that all of them are negative. Hence the posterior probability for the positive bag can be written as

$$p(y=1|\mathbf{x}) = 1 - \prod_{j=1}^{K} \left[1 - \sigma(w^\top x_j)\right], \tag{22.9}$$

where the posterior probability for the positive class is modeled as a *logistic sigmoid* acting on the linear classifier $f_w(x_j) = w^\top x_j$. The logistic sigmoid function is defined as $\sigma(z) = 1/(1+e^{-z})$. A negative bag means that *all* examples in the bag are negative. Hence

$$p(y=0|\mathbf{x}) = \prod_{j=1}^{K} \left[1 - \sigma(w^\top x_j)\right]. \tag{22.10}$$

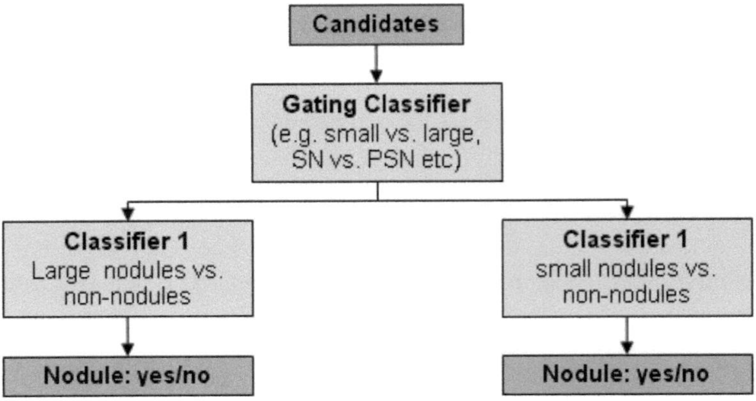

Fig. 22.3: A typical gated classification architecture.

Using this model we find the maximum likelihood estimator to learn a classification function that can predict the labels of unseen instances and/or bags. Multiple instance learning is highly relevant to our system design, since multiple candidates may point to the same abnormality, and it suffices if just one candidate is correctly classified, for the ROI will then be generated. Similarly, the CAD system can intelligently consolidate multiple candidates pointing to the same anatomical structure.

22.4.2 *Exploiting Domain Knowledge in Data–Driven Training–Gated Classifiers*

Incorporating medical domain knowledge and prior observations is critical to improving the performance of the CAD system. For example nodules have various characteristics in their shapes, sizes, and appearances. The simplest example is that lesions can be very big or small. Many of the image features are calculated by averaging over the voxels within segmented nodules. Features calculated on large lesions will hence be more accurate than those evaluated on a small one. Consequently, it may be more meaningful to construct classifiers with separate decision boundaries respectively for large and small candidates. Gating (see Figure 22.3) is a technique used to automatically learn meaningful clusters among candidates and construct classifiers, one for each cluster, to classify true candidates from false detections. This process can obviously be extended to incorporate different kinds of knowledge, for instance, to exploit differences between the properties of central versus peripheral nodules, or between vessel and pleural attachment, etc.

A novel Bayesian hierarchical mixture of experts (HME) has been developed and tested in our LungCAD system. The basic idea behind the HME is to decompose a complicated task into multiple simple and tractable sub–tasks. The HME model consists of several domain experts and a gating network that decides which experts are most trustworthy on any input pattern. In other words, by recursively partitioning the feature space into subregions, the gating network probabilistically decides which patterns fall in the domain of expertise of each expert.

In many scenarios we also know what kind of false positives our system generates. Thus, we may also have labels for the different sub–classes in the negatives. In (Dundar et al, 2008) we presented a methodology to take advantage of the sub–class information available in the negative class to achieve a more robust description of the target class. The sub–class information, which is neglected in conventional binary classifiers, provides a better insight of the data set and, when incorporated into the learning mechanism, acts as an implicit regularizer. We proposed a method to train a polyhedral classifier jointly, where each face of the polyhedron can classify each of the negative sub–classes. The linear faces of the polyhedron achieve robustness whereas multiple faces provide flexibility.

22.4.3 Ground Truth Creation: Learning from Multiple Experts

In most instances, CAD systems are built from labels assigned by multiple radiologists who identify the locations of malignant lesions.

As discussed in Section 22.1.3, when multiple experts examine medical images, a subjective, possibly noisy, version of the reference standard is introduced. In practice, there is a substantial amount of disagreement even among the best experts, and hence it is of great practical interest to determine an optimal way to train a classifier in such a setting.

Because of the intrinsic uncertainty of ground truth annotation in lung CAD, it makes sense to explicitly model the reliability of the the labelers in the creation of the ground truth by weighting their relative importance in the labelling ensemble based on their performance. In (Raykar et al, In Press), we propose a Bayesian framework for supervised learning in the presence of multiple annotators providing labels but no absolute gold standard. The proposed algorithm iteratively establishes a particular gold standard, measures the performance of the annotators given that reference standard, and then refines the ground truth based on the performance measures. Experimental results indicate that the proposed method is superior to the commonly used majority voting baseline.

When multiple experts label the lesions, the majority voting is the most common approach in the ground truth creation. It assumes that all the experts are equally good. However, if there is only one true expert among several annotators, the majority voting will be biased toward the non–experts. To resolve the issue, Raykar et al (In Press) proposed an algorithm to discover the best experts and assign a higher weight to them for the annotation consolidation.

Finally, a large number (100,000s) of candidates are produced in the candidate generation stage to uncover any suspicious regions, which results in a large amount of training data. This imposes a requirement for the scalability of the learning algorithms. Typically we have observed that linear models are more computationally tractable than sophisticated nonlinear methods. Boosting algorithms are also efficient to scale up with large data. Additionally, the learning algorithms must also deal with highly–imbalanced training sets with only a small fraction of positives, also a common problem in large scale retrieval tasks.

All these different factors have been considered and combined to produce a single solution, whose performance on the particular set of Lung Image Database Consortium (LIDC) images is presented in the following section.

22.5 ImageCLEF Challenge

22.5.1 Materials and Methods

The CAD system described in Section 22.3 was trained off–line on several hundred thoracic CT images, marked by different radiologists, from different institutions. The principal aim and benefit of this diversity is to account for inter–observer variability, technical variabilities between acquisition protocols and acquisition systems, as much as possible. The resulting CAD system has, as a consequence, lower sensitivity and specificity than it would have if tailored to data from a single institution; however, it is more robust to variabilities and hence deployable at various sites around the world. As part of the ImageCLEF Challenge, thoracic CT series of 46 patients were collected, as part of the LIDC (McNitt-Gray et al, 2007). The aim of the LIDC initiative is to provide a publicly available data set of images to benchmark CAD and other image processing algorithms. Each image was read by four thoracic radiologists who, independently, marked and measured all visible lung lesions.

The CAD system presented in this chapter was developed, trained and tested using images that were not part of the LIDC collection. The CAD system's output was compared to the marks of the four radiologists. The sensitivity and specificity statistics were computed using two different reference standards: (1) considering lesions 4 mm and above that have been reported by three out of four radiologists (majority), and (2) with lesions found by all four radiologists (consensus), see Figures 22.4 – 22.5. Some marks were placed by one or two radiologists only. These may be readers' false positives or may point to inconspicuous nodules missed by the other readers. The results presented below treat these marks systematically as false positives, see Figure 22.6. The confidence intervals were computed at 95%.

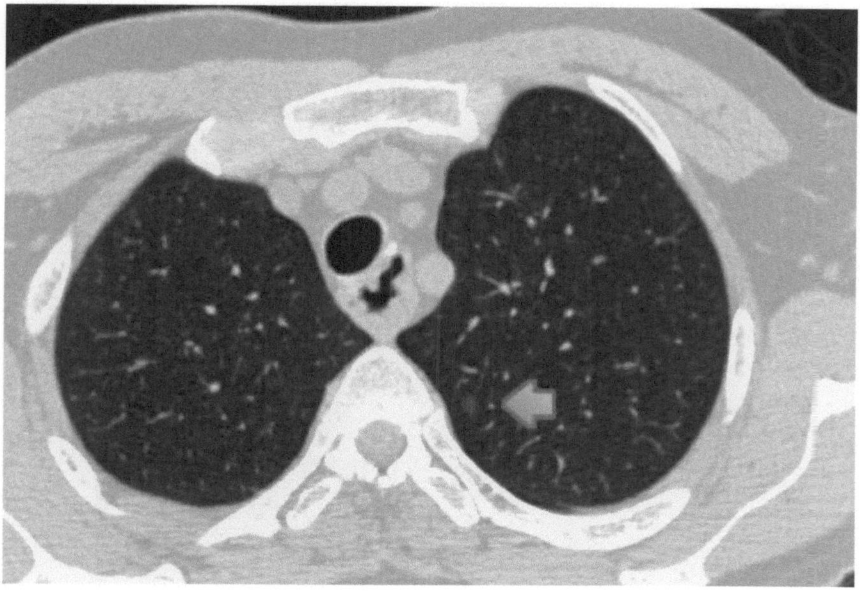

Fig. 22.4: 8 mm subsolid nodule detected by all four radiologists and CAD.

Table 22.1: CAD sensitivity per nodule attenuation

	# nodules	CAD sensitivity
Solids	29/33	87.9% ± 11.1%
Part-Solids	5/6	83.3% ± 29.8%
Total	34/39	87.2% ± 10.5%

22.5.2 Results

The CAD prototype detected 34 out of 39 (87.2% ± 10.5%) nodules marked by the majority of the readers (three or more readers) with 2.8 false positives per volume on average. Considering the readers consensus (all four readers), the CAD prototype detected 32 out of 36 nodules (88.9% ± 10.3%) with 3.0 false positives per volume on average.

When the results are analyzed by attenuation, the system detected 29 out of 33 (87.9% ± 11.1%) solid and five out of six (83.3% ± 29.8%) subsolid lesions marked by the majority (three or more readers). Based on a consensus (all four readers) between the readers, the system detected 27 out of 30 (90% ± 10.7%) solid and five out of six subsolid lesions. See Table 22.1

These results are consistent with other studies (Opfer and Wiemker, 2007), showing that CAD tends to detect lesions that are the most frequently reported as nodules.

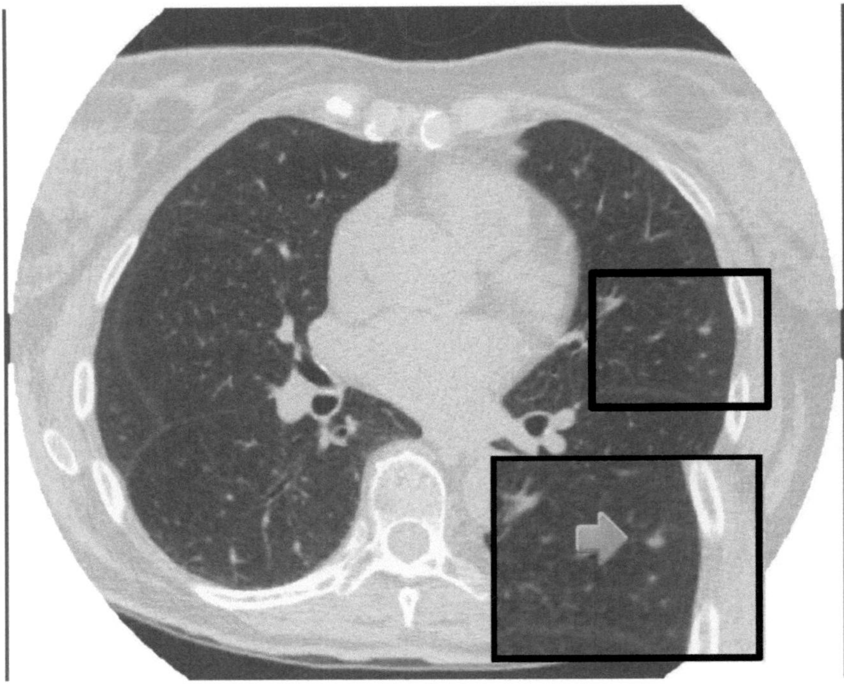

Fig. 22.5: 6 mm subsolid nodule detected by all four radiologists and missed by CAD.

22.6 Discussion and Conclusions

22.6.1 Clinical Impact

The true impact for a medical image mining system is not measured in terms of its stand–alone accuracy, rather by the benefit derived by radiologists using the software. An interesting difference between typical image retrieval systems and CAD systems is that, while in the former, it is sufficient to characterize the stand–alone performance of retrieval systems in terms of recall and precisions, in the latter, clinical impact must also be considered. Namely, similar terms–of–art such as sensitivity and specificity are used in the diagnostic realm, regulatory approval of diagnostic devices using CAD (at least in the US via the FDA) requires that the said service should be evaluated in combination with the user (the person–machine system). The net diagnostic improvement of the combined system must be demonstrated objectively via the use of multi–reader multi–case fully–crossed receiver–operating characteristic analysis with strong statistical power (McClish et al, 2002).

In a clinical validation study completed in 2004 and submitted to the FDA using an earlier version of the algorithm designed specifically for solid nodules, we an-

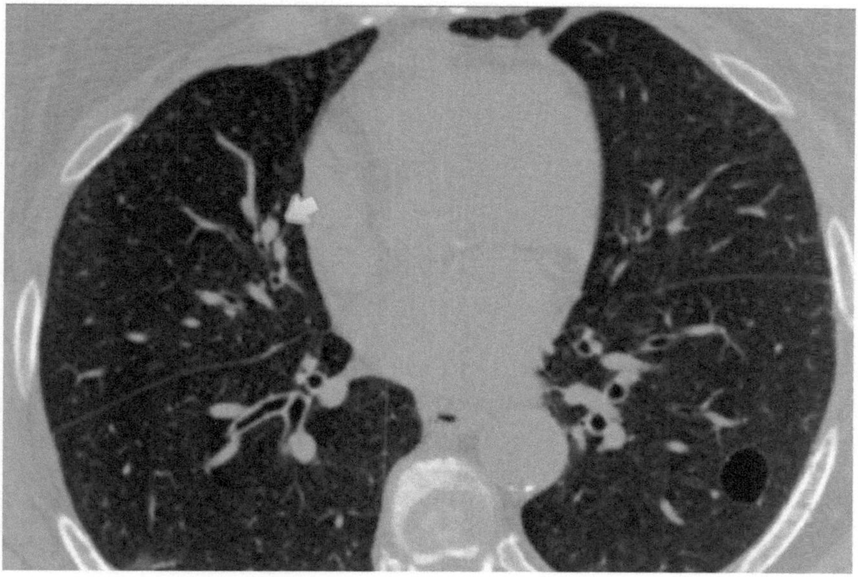

Fig. 22.6: Vessel–attached inconspicuous 8 mm nodule detected by only two of the four radiologists and CAD. Although this structure is undoubtly a lung nodule, it was overlooked by two readers and thus was counted as false positive for CAD.

alyzed a retrospective sample of 196 cases from four large research hospitals. CT scans were collected from patients referred for routine assessment of clinically or radiographically known or suspected pulmonary nodules. These cases contained a total of 1,320 nodules as confirmed by a majority of a panel of five expert radiologists. The cases were interpreted independently by 17 general radiologists, first without and then with the use of our LungCAD product. Every one of these 17 radiologists improved their detection of solid nodules > 3 mm to a statistically significant extent. The average reader improvement in Area Under Curve (AUC) using the nonparametric Receiver Operating Characteristic (ROC) technique for detecting nodules was 0.048 ($p<0.001$) with a 95% confidence interval of (0.036, 0.059). This study showed a statistically significant improvement in the area under the nonparametric ROC curve with the use of our LungCAD software for detection of lung nodules.

A subsequent clinical study (Godoy et al, 2008) was performed using a later version of the LungCAD prototype on 54 chest CT scans reviewed by two radiologists at New York University Medical Center and Seoul National University Bundang Hospital with the goal of evaluating the impact of our most recent LungCAD system in the detection of different kinds of lung nodules. The 54 cases used in the study had total of 395 nodules of which 234 were solid nodules, 29 were part–solid nodules, and 132 ground glass nodules. Two readers read the 54 cases first without CAD and then with CAD. The study showed that the CAD software resulted in a significant increase in sensitivity for each reader and for each attenuation level. The use of CAD did not increase the number of false positives for any of the readers.

The LungCAD algorithm was also deployed on PACS systems, allowing the radiologists to review the CAD marks directly from their PACS stations without interruption from their normal clinical workflow. Godoy et al (2009b) demonstrated that such a system has a statistically significant impact for all readers using it.

Newer research prototype systems have also been evaluated although they have not yet been distributed commercially. A study (Godoy et al, 2009a) presented at the recent American Roentgen Ray Society (ARRS) 2009 annual meeting concluded that the use of our research prototype significantly increased the mean reader sensitivity for all types of attenuation (solid, part-solid, ground-glass) ($p < 0.001$). Based on these and many other clinical studies, we have demonstrated that the use of CAD as a second reader improves radiologist's detection of different kinds of pulmonary nodules.

22.6.2 Future Extensions of CAD

Current CAD systems do not provide query by example to find other similar structures intra–image although this could have significant clinical value (Tao et al, 2009). In the particular context of image series taken at different time–points, a search for the same lesions across multiple studies would allow for automatic detection of growth and characterization of morphological changes.

Another extension is the processing of very large image sets, such as the ones currently archived by healthcare facilities, for the detection of images with certain anatomical structures. The challenges in this task can be basically summarized by the absence of any efficient pre-indexing of data on RIS and PACS systems. Furthermore, no query specification language/image words or vocabulary has been established so far; which means that any detection system presently needs to be built on top of a large a priori labeled set of training data.

The interplay between image processing and data mining components is crucial, and it is important to understand the impact of each component in order to jointly optimize the overall product. Indeed good image processing algorithms created the features that made subsequent data mining algorithms successful, and often a deep analysis of the fundamental ideas behind these algorithms could lead to a much better understanding of the statistical issues that would be faced by the classifier.

The organ–specific localization and segmentation queries could form the basis for rapid prototyping of new CAD algorithms. This, along with integration of evidence over multiple scans/timepoints and modalities are fertile areas for future extensions of CAD. For instance, one could easily grasp the benefits of multiple disease–specific CAD systems interacting to deliver comprehensive diagnostic information.

References

Armato III S, Giger M, Moran C, Blackburn J, Doi K, MacMahon H (1999) Computerized detection of pulmonary nodules on CT scans. RadioGraphics 19:1303–1311

Armato III S, Roberts R, Kocherginsky M, Aberle D, Kazerooni E, MacMahon H, van Beek E, Yankelevitz D, McLennan G, McNitt-Gray M, Meyer C, Reeves A, Caligiuri P, Quint L, Sundaram B, Croft B, Clarke L (2009) Assessment of radiologist performance in the detection of lung nodules: dependence on the definition of truth. Academic Radiology 16:28–38

Beyer F, Zierott L, Fallenberg EM, Juergens KU, Stoeckel J, Heindel W, Wormanns D (2007) Comparison of sensitivity and reading time for the use of computer-aided detection (cad) of pulmonary nodules at MDCT as concurrent or second reader. European Radiology 17(11):2941–2947

Bogoni L, Liang J, Periaswamy S (2009) System and method for toboggan based object segmentation using divergent gradient field response in images. Technical Report US7526115B2, Siemens Medical solutions USA, Inc., Malvern, PA

Chang S, Emoto H, Metaxas DN, Axel L (2004) Pulmonary micronodule detection from 3D Chest CT. Lecture Notes in Computer Sciences (LNCS) 3217:821–828

Clarke L, Croft B (2004) Development of public resources to support quantitative imaging methods in cancer. Academic Radiology 14:1438–1440

Dundar MM, Wolf M, Lakare S, Salganicoff M, Raykar VC (2008) Polyhedral classifier for target detection: a case study: colorectal cancer. In: Proceedings of the 25th international conference on Machine learning. ACM press, pp 288–295

Farag AA, El-Baz A, Gimel'farb GG, Falk R, Hushek SG (2004) Automatic detection and recognition of lung abnormalities in helical ct images using deformable templates. Lecture Notes in Computer Science (LNCS) 3217:856–864

Godoy MC, Naidich DP (2009) Subsolid pulmonary nodules and the spectrum of peripheral adenocarcinomas of the lung: Recommended interim guidelines for assessment and management. Radiology 606–622

Godoy MC, Kim TJ, Ko JP, Florin CH, Jerebko AK, Naidich DP (2008) Computer-aided detection of pulmonary nodules on CT: Evaluation of a new prototype for detection of ground–glass and part–solid nodules. In: Proceedings of the RSNA

Godoy MC, Bonavita J, Girvin F, O'Sullivan P, Wickstrom M, Naidich D (2009a) Role of a computer–assisted diagnosis (cad) within pacs for identifying nodules on low dose screening ct studies: A prospective evaluation. In: Proceedings of the RSNA

Godoy MC, Ko JP, Kim TJ, Naidich DP, Bogoni L, Florin CH, Vlahos I, Park S, Salganicoff M (2009b) Effect of computer–aided diagnosis on radiologists' detection performance of subsolid pulmonary nodules on ct: Initial results. In: Proceedings of the American Roentgen Ray Society annual meeting

Hounsfield GN (1973) Computerised transverse axial scanning (tomography) part 1: Description of system. British Journal of Radiology 46:1016–1022

Jemal A, Siegel R, Ward E, Hao Y, Xu J, Murray T, Thun MJ (2008) Cancer statistics, 2008. CA: A Cancer Journal for Clinicians 58(2):71–96

Ko JP, Betke M (2001) Chest ct: Automated nodule detection and assessment of change over time-preliminary experience. Radiology 218(1):267–273

Lee Y, Hara T, Fujita H, Itoh S, Ishigaki T (2001) Automated detection of pulmonary nodules in helical ct images based on an improved template-matching technique. IEEE Transactions of Medical Imaging 20(7):595–604

MacMahon H, Austin JHM, Gamsu G, Herold CJ, Jett JR, Naidich DP, Patz Jr EF, Swensen SJ (2005) Guidelines for management of small pulmonary nodules detected on ct scans: A statement from the fleischner society. Radiology 395–400

Mani A, Napel S, Paik D, Jeffrey RJ, Yee J, Olcott E, Prokesch R, Davila M, Schraedley-Desmond P, Beaulieu C (2004) Computed tomography colonography: feasibility of computer–

aided polyp detection in a "first reader" paradigm. Journal of Computer Assisted Tomography 28(3):318–326

McClish D, Zhou X, Obuchowski N (2002) Statistical methods in diagnostic medicine. Wiley Series in Probability and Statistics, vol 414. Wiley, p 464

McNitt-Gray MF, Armato SG, Meyer CR, Reeves AP, McLennan G, Pais RC, Freymann J, Brown MS, Engelmann RM, Bland PH, Laderach GE, Piker C, Guo J, Towfic Z, Qing DPY, Yankelevitz DF, Aberle DR, van Beek EJR, MacMahon H, Kazerooni EA, Croft BY, Clarke LP (2007) The lung image database consortium (lidc) data collection process for nodule detection and annotation. Academic Radiology 14(12):1464–1474

Naidich DP, Ko JP, Stoeckel J, Abinanti N, Lu S, Moses D, Moore W, Vlahos I, Novak CL (2004) Computer–aided diagnosis: impact on nodule detection among community level radiologists. a multi-reader study. In: CARS, pp 902–907

Opfer R, Wiemker R (2007) Performance analysis for computer-aided lung nodule detection on lidc data. Proceedings of the SPIE 6515:65151

Paik DS, Beaulieu CF, Rubin GD, Acar B, Jeffrey Jr RB, Yee J, Dey J, Napel S (2004) Surface normal overlap: A computer-aided detection algorithm with application to colonic polyps and lung nodules in helical ct. IEEE Transactions on Medial Imaging 23(6):661–675

Rao RB, Fung G, Krishnapuram B, Bi J, Dundar M, Raykar VC, Yu S, Krishnan S, Zhou X, Krishnan A, Salganicoff M, Bogoni L, Wolf M, Jerebko A, Stoeckel J (2009) Mining medical images. In Proceedings of the Third Workshop on Data Mining Case Studies and Practice Prize, Fifteenth Annual SIGKDD International Conference on Knowledge Discovery and Data Mining (KDD 2009)

Raykar V, Krishnapuram B, Bi J, Dundar M, Bharat R (2008) Bayesian multiple instance learning: Automatic feature selection and inductive transfer. In Proceedings of the 25th International Conference on Machine Learning 307:808–815

Raykar VC, Yu S, Zhao LH, Valadez GH, Florin C, Bogoni L, Moy L (In Press) Learning from crowds. Journal of Machine Learning Research

Summers RM (2003) Road maps for advancement of radiologic computer-aided detection in the 21st century. Radiology 229(1):11–13

of Surveillance D, Research HP (2009) Cancer facts & figures. American Cancer Society

Suzuki K, Asamura H, Kusumoto M, Kondo H, Tsuchiya R (2002) "Early" peripheral lung cancer: prognostic significance of ground glass opacity on thin-section computed tomographic scan. Annals of Thoracic Surgery 74(5):1635–1639

Suzuki K, Li F, Sone S, Doi K (2005) Computer–aided diagnostic scheme for distinction between benign and malignant nodules in thoracic low–dose CT by use of massive training artificial neural network. IEEE Transactions of Medical Imaging 24(9):1138–1150

Tao Y, Zhou XS, Bi J, Jerebko A, Wolf M, Salganicoff M, Krishnan A (2009) An adaptive, knowledge-driven medical image search engine for interactive diffuse parenchymal lung disease quantification. In: SPIE proceedings, vol 7260

Chapter 23
Medical Image Classification at Tel Aviv and Bar Ilan Universities

Uri Avni, Jacob Goldberger, and Hayit Greenspan

Abstract We present an efficient and accurate image categorization system, applied to medical image databases within the ImageCLEF medical annotation task. The methodology is based on local representation of the image content, using a bag–of–visual–words approach. We explore the effect of different parameters on system performance, and show best results using dense sampling of simple features with spatial content in multiple scales, combined with a nonlinear kernel based Support Vector Machine classifier. The system was ranked first in the ImageCLEF 2009 medical annotation challenge, with a total error score of 852.8.

23.1 Introduction

In this chapter we describe a visual Bag–of–Words (BoW) framework and its application to image classification. The visual BoW is related to the more traditional BoW model, commonly used in natural language processing and information retrieval for text documents (Blei et al, 2003; Manning et al, 2008). In this model a document is statistically modeled as an instance of a multinomial word distribution and is represented as a frequency of occurrence word histogram. The representation as a frequency vector of word occurrences does not take grammar rules or word order into account. It does, however, preserve key information about the content of the document. This representation can be used to compare documents, and to identify document topics. The BoW representation is also the basis for more complex probabilistic models, e.g. the highly successful Latent Dirichlet Allocation

Uri Avni
Tel Aviv University, Tel Aviv 69978, Israel, e-mail: uriavni@post.tau.ac.il

Jacob Goldberger
Bar Ilan University, Ramat–Gan 52900, Israel e-mail: goldbej@eng.biu.ac.il

Hayit Greenspan
Tel Aviv University, Tel Aviv 69978, Israel e-mail: hayit@eng.tau.ac.il

H. Müller et al. (eds.), *ImageCLEF*, The Information Retrieval Series 32,
DOI 10.1007/978-3-642-15181-1_23,

(LDA) approach (Blei et al, 2003). The BoW representation is successfully used in document classification, clustering and retrieval tasks and is the cornerstone of all Internet search engines.

In recent years the BoW approach has been successfully applied to general scene and object recognition tasks. Varma and Zisserman (2003) introduced the idea of using joint distribution of intensity values over compact neighborhoods for the task of texture classification. In their work, the image representation was learned from local patches in the data. They showed that using the local patches they can outperform previous work based on large filter banks. Fei-Fei and Perona (2005) learned natural scene categories using a set of localized visual *words* which were either grayscale patches or Scale Invariant Feature Transform (SIFT) descriptors (Lowe, 1999), sampled on a grid, randomly, or at interest points. A generative hierarchical model was trained to describe the resulting visual words distribution. In (Sivic and Zisserman, 2008) vector quantization of invariant local image descriptors was used to form clusters, and the clusters were referred to as the set of visual words. They then searched for objects throughout a movie sequence in a similar way as in text retrieval. Nowak et al (2006) focused on comparing the performance of various competing strategies for image representation in the BoW approaches to visual categorization. Their study compared different image sampling, codebook generation and histogram normalization methods on five popular databases. Lazebnik et al (2006) extended the BoW approach to take into account geometrical correspondence by introducing 'spatial pyramids' - a technique of partitioning the image into increasingly fine subregions, and computing histograms of local features within each subregion. They demonstrated significant performance improvement over traditional BoW in global scene classification and object recognition tasks. Zhang et al (2007) presented a large–scale evaluation of the visual words approach for texture classification and object recognition. Images were represented as distributions of features extracted from a sparse set of local keypoints. After examining different keypoint detectors and descriptors, as well as different kernels and classifiers, the findings showed state–of–the–art results on four texture classification and five object recognition databases.

23.1.1 Visual Words in Medical Archives

Approaches using patch–based, visual BoW concepts are gradually emerging in medical tasks. The MedGIFT system is an open–source based retrieval engine for medical images (Müller et al, 2005). It uses texture features in the form of local Gabor filter responses and color features as the mode color of blocks in various sizes, with a combination of global color and texture histograms. An application for mammography tissue classification and segmentation is shown in (Bosch et al, 2006). The use of BoW techniques for large scale radiograph archive categorization can be found in several systems that participated in the ImageCLEF international competition (Tommasi et al, 2009; Deselaers and Deserno, 2009; Avni et al,

2009). It is interesting to note that in the last few years, approaches based on local patch representation achieved the highest scores for categorization accuracy. In 2006, Deselaers et al (2006) displayed the best medical annotation results using the visual BoW approach, where the features were local patches of different sizes taken at every position and scaled to a common size. Patch dimensionality was reduced to between six and eight components using the Principal Components Analysis (PCA) transformation. Patch x,y co–ordinates were added as two additional components. In this work, no dictionary was used; rather the feature space was quantized uniformly in every dimension and the image was represented as a sparse histogram in the quantized space. Several classification techniques were examined, including the nearest neighbor classifier, maximum entropy classifier, and Support Vector Machine (SVM). Tommasi et al (2008) had the highest score in 2007 and 2008. In this work both global and local features were used. The global features were downscaled versions of the images (32×32). The local features were modified SIFT descriptors (128 values), sampled randomly. The set of local features was represented as a histogram over a dictionary, built using the K–means algorithm (in a 128 dimensional space) on randomly selected feature vectors from the entire database (K=500). Four image quadrants were learned and represented separately. The final representation for a given image was thus the (32×32) pixel values of the global image along with four times the (500) histogram bins. Classification was done with SVM ('one vs. one', 'one vs. all') with different integration techniques for global and local features.

23.2 The Proposed TAU–BIU Classification System Based on a Dictionary of Visual–Words

We next describe the system we have been developing, as a joint effort of the Tel–Aviv University and Bar–Ilan University research groups. To represent an image using the BoW model, the image must be treated as a document. Unlike in the text world, there is no natural concept for a word or a dictionary. We thus need to find a way to break down the image into a list of visual elements, and a way to discretize the visual element space, since the number of possible visual elements in an image is enormous. In the visual BoW model, an image representation step usually takes place in a three–step procedure involving local feature detection, feature description and codebook generation. The visual word model can thus take the form of a histogram representation of the image, based on a collection of its local features. Each bin in the histogram is a codeword index out of a finite vocabulary of visual codewords, generated in an unsupervised way from the data. Images are compared and classified based on this discrete and compact histogram representation. We next review the image representation part of the TAU–BIU system. Key components are shown in the flow–diagram in Figure 23.2.

23.2.1 Patch Extraction

Given an image, feature detection is used to extract several small local patches. Each small patch shows a localized view of the image content. These patches are considered as candidates for basic elements, or 'words'. The patch size needs to be larger than a few pixels across, in order to capture higher–level semantics such as edges or corners. At the same time, the patch size should not be too large if it is to serve as a common building block for many images. Common feature detection approaches include using a regular sampling grid, a random selection of points, or the selection of points with high information content using salient point detectors. We utilize all the information in the image, by sampling rectangular patches of size 9×9 around every pixel. This simple feature detection approach has been shown to be effective (Nowak et al, 2006).

23.2.2 Feature Space Description

Following the feature detection step, the feature representation method involves representing the patches using feature descriptors. In this step, a large random subset of images is used (ignoring their labels). We extract patches using a regular grid, and normalize each patch by subtracting its mean gray level, and dividing it by its standard deviation. This step ensures invariance to local changes in brightness, provides local contrast enhancement and augments the information within a patch. Patches that have a single intensity value are abundant in x–ray images. These patches are common in all categories, much like stopwords in text documents. These patches are ignored. We are left with a large collection of several million vectors. To reduce both the computational complexity of the algorithm and the level of noise, we apply a PCA procedure to this initial patch collection. The first few components of the PCA, which are the components with the largest eigenvalues, serve as a basis for the information description.

A popular alternative approach to raw patches is the SIFT representation (Lowe, 1999) which is advantageous in scenery images (Fei-Fei and Perona, 2005; Zhang et al, 2007), where object scales can vary. We examine this option in the experiments defining the system parameter set. In addition to patch content information represented either by PCA coefficients or SIFT descriptors, we add the patch center coordinates to the feature vector. This introduces spatial information into the image representation, without the need to explicitly model the spatial dependency between patches. Special care should be taken when combining features having different units, such as coordinates and PCA coefficients. The relative feature weights were tuned experimentally on a cross–validation set (see Section 23.3).

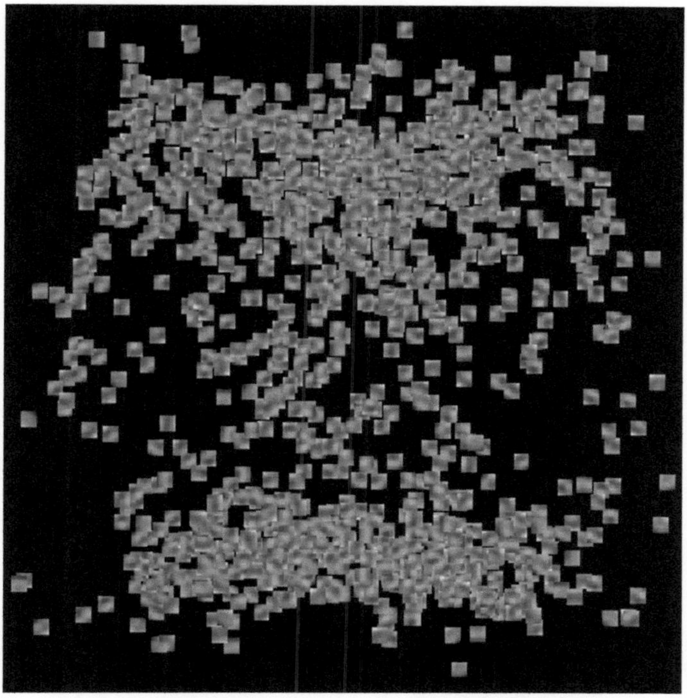

Fig. 23.1: Visual dictionary of 1,000 words. Representative image patches are shown at their respective spatial coordinates.

23.2.3 Quantization

The final step of the bag–of–words model is to convert vector represented patches into *visual words* and to generate a representative *dictionary*. A visual word can be considered as a representative of several similar patches. A frequently–used method is to perform K–means clustering over the vectors of the initial collection, and then cluster them into K groups in the feature space. The resultant cluster centers serve as a vocabulary of K visual words. A sample dictionary of 1,000 visual words generated by this process is shown in Figure 23.1. Due to the fact that we included spatial coordinates as part of the feature space, the visual words have a localization component in them, which is reflected as a spatial spread of the words in the image plane. Words are denser in areas with greater variability across images in the database. In order to accelerate the look–up process, dictionary words are stored in a kd–tree indexed by the spatial coordinates.

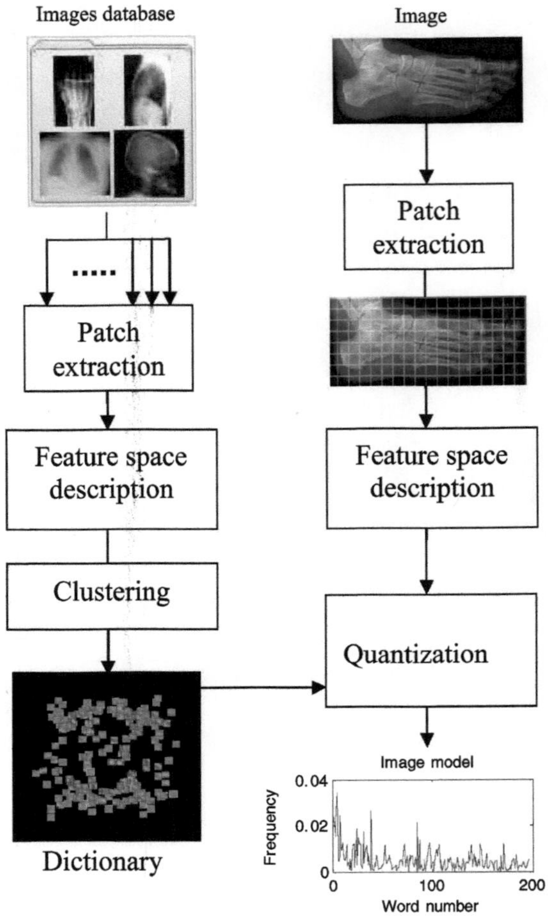

Fig. 23.2: Dictionary building and image representation flow chart.

23.2.4 From an Input Image to a Representative Histogram

A given (training or testing) image can now be represented by a unique distribution over the generated dictionary of words. In our implementation, patches are extracted from every pixel in the image. For an x–ray image measuring 512×512 pixels there are typically several hundred thousand non–empty patches. The patches are projected into the selected feature space, and translated (quantized) to indices by looking up the most similar feature–vector in the generated dictionary. Using the spatial indexation of dictionary words, the dictionary look–up process is accelerated by comparing a new patch only to dictionary words at a certain radius from it. The dictionary generation process and the transformation from an image to its representative histogram are shown in Figure 23.2 left column and right column,

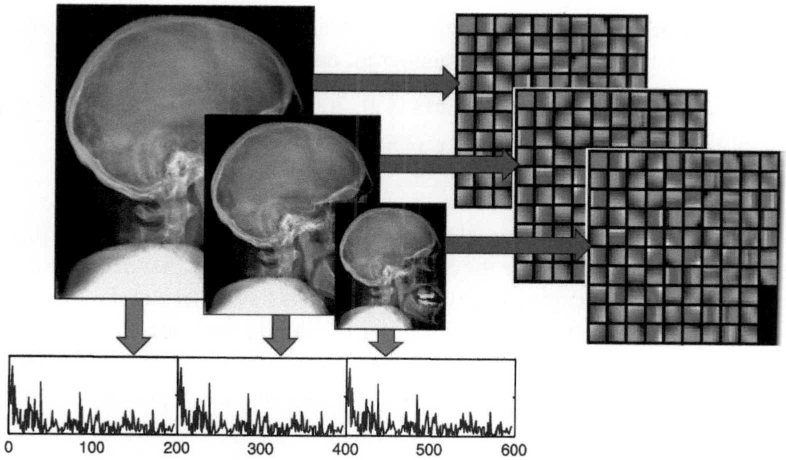

Fig. 23.3: Image representation at multiple scales.

respectively. Note that as a result of including spatial features, both the image local content and spatial layout are preserved in the discrete histogram representation.

Multi–scale image information may in some cases provide additional information that supports the required discrimination. To address this, we repeat the dictionary building process for scaled–down replications of the input image, using the same patch size. The image representation in this case is a 1-D concatenation of histograms from varying scales. This process, illustrated in Figure 23.3, provides a richer image representation. It does not imply scale invariance, as in (Deselaers et al, 2006). In our experiments we found that objects of interest in radiographs appear at roughly a similar size–range across all images, thus invariance to the scale is not a necessity.

23.2.5 Classification

We examined two classification approaches: one based on the K–nearest–neighbor classifier, using the image–to–image distance with different distance measures. The second is a non–linear multi–class SVM with different kernels. For the nearest neighbor classifier, we examined four popular choices of image–to–image distances: the Symmetric Kullback–Leibler (SKL) distance, the Jeffrey Divergence (JD), L_1 and L_2:

- $SKL(I,J) = \sum_i I_i \log \frac{I_i}{J_i} + J_i \log \frac{J_i}{I_i}$
- $JD(I,J) = \sum_i I_i \log \frac{I_i}{(I_i+J_i)} + J_i \log \frac{J_i}{(I_i+J_i)}$
- $L_p(I,J) = \sum_i |I_i - J_i|^p$

The second approach is a multi–class SVM classifier. We examined several non–linear kernels commonly used with histogram data:

- Histogram intersection kernel (Barla et al, 2003): $K(x,y) = \exp(-\sum_i min(x_i, y_i))$
- Radial Basis Function kernel: $K(x,y) = e^{-\gamma \|x-y\|^2}$
- χ^2 kernel: $K(x,y) = \exp(-\gamma \sum_i \frac{|x_i - y_i|^2}{|x_i + y_i|})$

In the histogram intersection kernel there are no free parameters. The optimization is therefore one dimensional over the SVM cost parameter, which makes it convenient for fast parameter evaluation. The two other kernels have a free trade-off parameter γ, and require careful optimization. In order to classify multiple categories, we use the one–vs.–one extension of the binary classifier, where $N(N-1)/2$ binary classifiers are trained for all pairs of categories in the data set. Whenever an unknown image is classified with a binary classifier it casts one vote for its preferred class, and the final result is the class with the most votes. Since each binary classifier runs independently, parallelization of both training and testing phases of the SVM is straightforward. It is implemented as a parallel enhancement of the LIBSVM (Chang and Lin, 2001) library.

23.3 Experiments and Results

A key component in using the BoW paradigm in a categorization task is the tuning of the system parameters. An optimization step is thus required for a given task and image archive. We focus on three components of the system: finding the optimal set of local features, finding the optimal dictionary size, and optimizing the classifier parameters. We use a large generic archive of radiographs (IRMA) to tune the system parameters. We then show comparative results of automated organ and orientation detection in the ImageCLEF 2009 competition.

The IRMA database (Lehmann et al, 2004) consists of 12,667 categorized radiographs, labeled according to the IRMA coding system (Lehmann et al, 2003), with each category described by four axes: Technical axis: image modality; Directional axis: body orientation; Anatomical axis: body region examined and Biological axis: biological system examined. The IRMA data set has served algorithm development teams throughout the years, and in the past several years has been a source for the ImageCLEF medical annotation competition. Images in the IRMA database consist of scanned x–ray images, gray scale, 512 pixels long. The x–ray images are noisy with irregular brightness and contrast, and may contain dominant visual artifacts such as artificial limbs and x–ray frame borders. Some classes have large intra–variability, as seen for example in Figure 23.4, while images from different classes may be visually similar, as seen in Figure 23.5. Note the category label which consists of the four axes defined above. These properties make the automatic classification task challenging.

We optimized the system parameters by classifying subsets of the database, using several cross–validation experiments. In the following experiments, 10,667 images

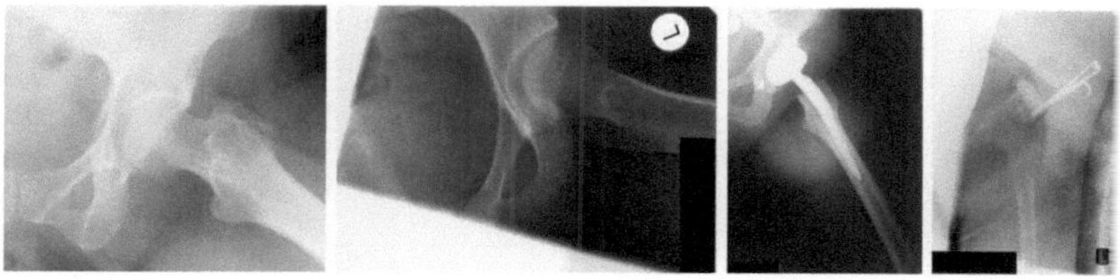

Fig. 23.4: Images from IRMA category (left to right): 'Overview image, Mediolateral, Left hip, Musculosceletal system'. Large intra–class variability can be seen.

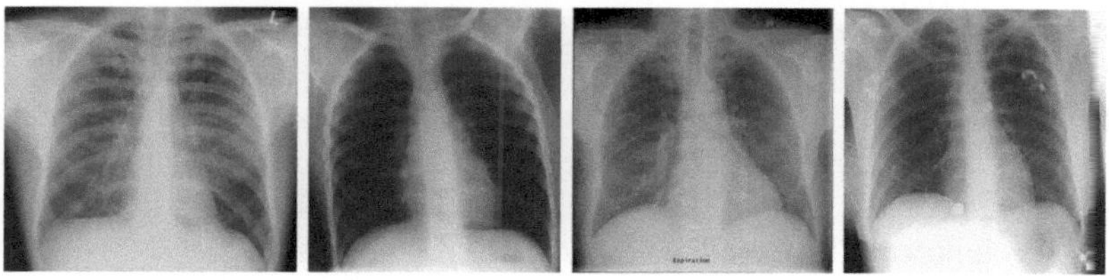

(a) (b) (c) (d)

Fig. 23.5: Visually similar frontal chest categories in the IRMA database: (a) 'High beam energy, Posteroanterior', (b) 'Child filter, Anteroposterior - inspiration', (c) 'High beam energy, Posteroanterior - expiration': (d) 'High beam energy, Anteroposterior - supine'.

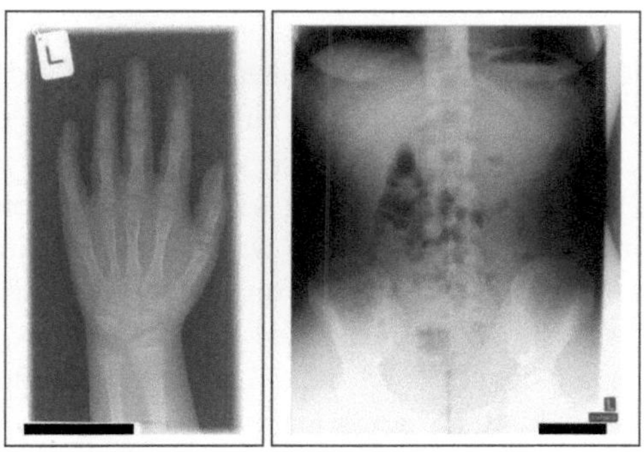

Fig. 23.6: Sample images with artifacts near the borders, such as misaligned x–ray frame, blacked out bars and various labels.

were used for training and 2,000 randomly drawn images were used for testing and verification. The optimization is performed independently in three steps: finding the optimal set of local features, finding optimal dictionary size, and optimizing classifier parameters.

Table 23.1: Comparison of different features

Features	Average %	Standard Deviation
Raw Patches	88.43	0.32
SIFT	90.80	0.41
Normalized	**91.29**	0.56

23.3.1 Sensitivity Analysis

We examined three feature extraction strategies: raw patches, raw patches with normalized variance, and SIFT descriptors. In all cases we added spatial coordinates to the feature vector. We used dense extraction of features around every pixel in the image. There are often strong artifacts near the image border that are not relevant to the image category, as seen in Figure 23.6, so a 5% margin from the image border was ignored. The feature extraction step produces about 100,000 to 200,000 features from a single image. It is our experience that x–ray images from the same category usually appear in a similar scale and orientation in a given archive. In this task the invariance of the SIFT features to scale and orientation is thus unnecessary. We used SIFT descriptors taken at a single scale, without aligning the orientation, as in (Tommasi et al, 2008). Raw patches, normalized patches and the 128 dimensional SIFT descriptors were dimensionally reduced using PCA.

Table 23.1 summarizes the classification results of the three feature sets. Normalizing patch variance improved the classification rate compared to raw patches. The gain can be attributed to the local contrast invariance achieved in this step. In this task, using normalized patches proved marginally preferable to SIFT descriptors in terms of classification accuracy. However, when using raw patches, the feature extraction step was significantly faster than with SIFT descriptors, as seen in Figure 23.7. The majority of the running time was spent in the image representation step; this step took over three seconds per image with the SIFT features, but less than half a second with the simpler variance–normalized raw patches. Time was measured on a dual quad–core Intel Xeon 2.33 GHz. In the following sections variance normalized raw patches are used as features.

Figure 23.8 depicts the effect of using four to ten components for variance–normalized raw patches. It can be seen that the number of components had a minimal effect on classification accuracy. The addition of spatial coordinates to the feature set, on the other hand, improved classification performance noticeably, as seen in Figure 23.9. We found that when using seven PCA components, the optimal range for the x, y coordinates was $[-3, 3]$. Bars show means and standard deviations from 20 cross validation experiments running on 1,000 random test images.

We next investigated the appropriate number of words in the dictionary. As Figure 23.10 shows, increasing the number of dictionary words proved useful up to 1,000 words. Adding additional words after that point increased the computational time with no evident improvement in the classification rate. Combining the above, the classification system used normalized raw patch features, with seven PCA com-

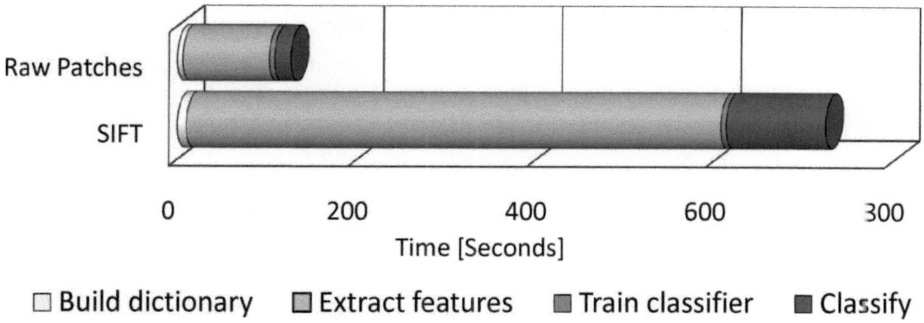

Fig. 23.7: Running time using SIFT descriptors and normalized raw patches.

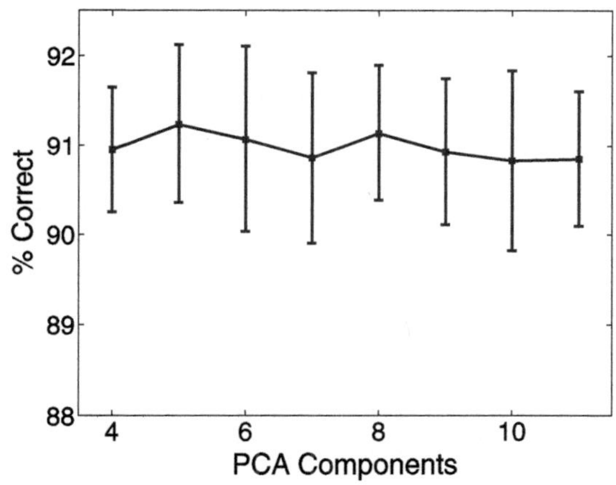

Fig. 23.8: Effect of the number of PCA components in a patch on classification accuracy

Table 23.2: Comparison of distance metrics using a k–nearest neighbor classifier (k=3).

Distance metric	Accuracy %
L_2	78.3
SKL	80.7
JD	82.0
L_1	82.6

ponents, spatial features with weight [-3,3], and 1,000 visual words. Using the SVM with a histogram intersection kernel achieved a classification accuracy of 91.29%.

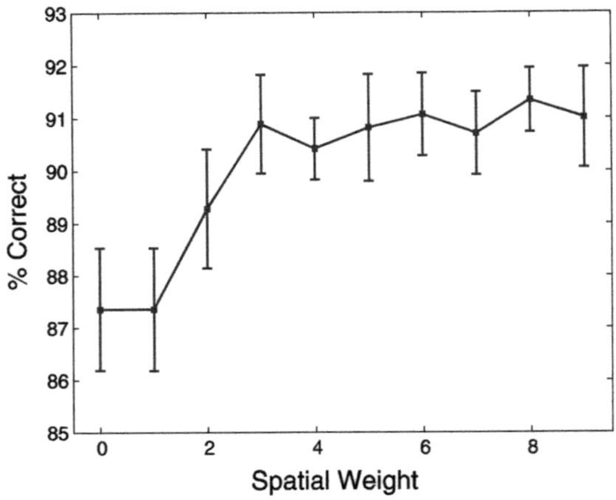

Fig. 23.9: Effect of spatial features: Weight of spatial features (x–axis); Classification accuracy (y–axis).

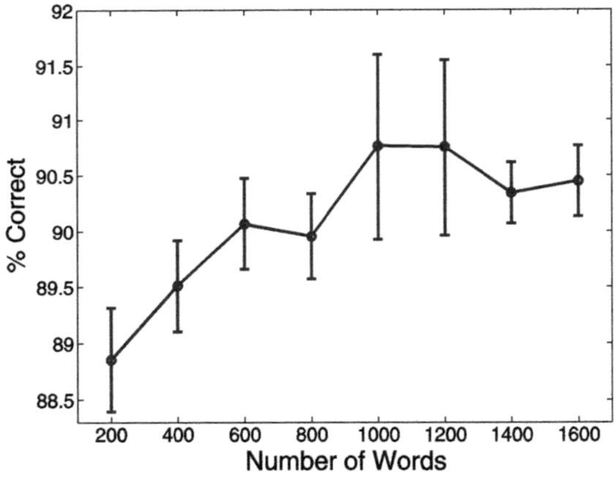

Fig. 23.10: Effect of dictionary size on classification accuracy.

23.3.2 Optimizing the Classifier

We used the k–Nearest Neighbor (KNN) classifier with several distance metrics: the Symmetric Kullback-Leibler (SKL) distance, the Jeffrey Divergence (JD), L_1 and L_2. Table 23.2 summarizes the classification success rate for these metrics. L_1 metric showed the best performance, while L_2 was the weakest by a margin of over 4%.

Note that the SVM classifier achieved over 90% accuracy in our earlier experiments, and is therefore clearly superior to the best KNN classifier. We next exam-

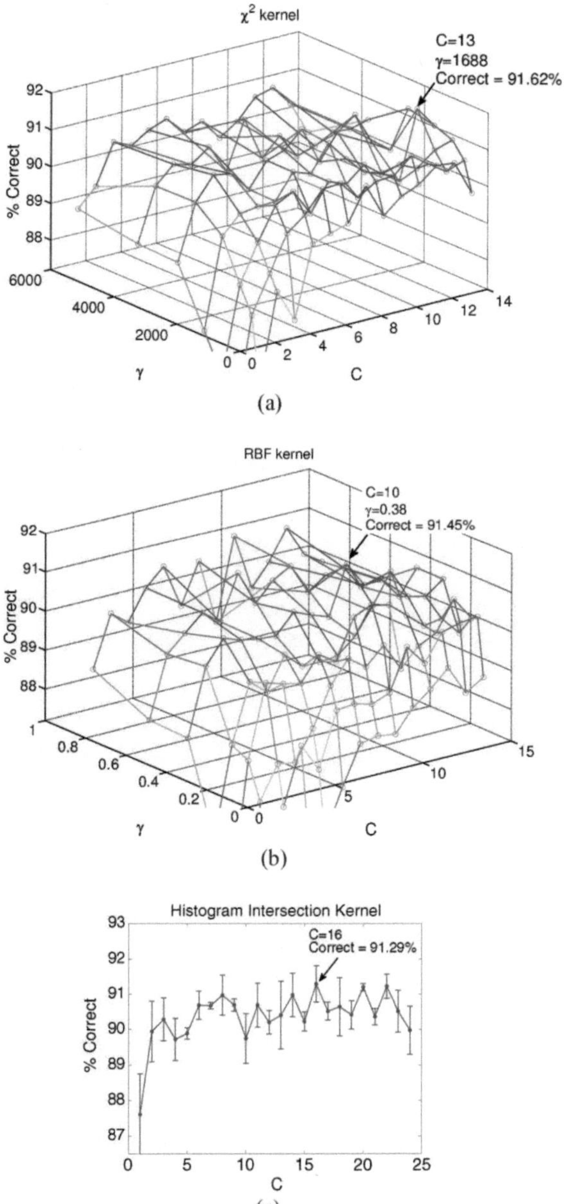

Fig. 23.11: Cross validation of SVM parameters using (a): χ^2 kernel (b): RBF kernel (c): Histogram intersection kernel.

ined two additional kernel types with the SVM classifier: the Radial Basis Function (RBF) and the χ^2 kernels. We used the optimal features and dictionary size consistently across all experiments. For these kernel types the SVM cost parameter C, and

Table 23.3: Comparison of SVM kernel types, for 1-scale and 3-scale models.

Kernel	Average % 1-scale	Average % 3-scales
Radial Basis	91.45	91.59
Histogram Intersection	91.29	91.89
χ^2	91.62	**91.95**

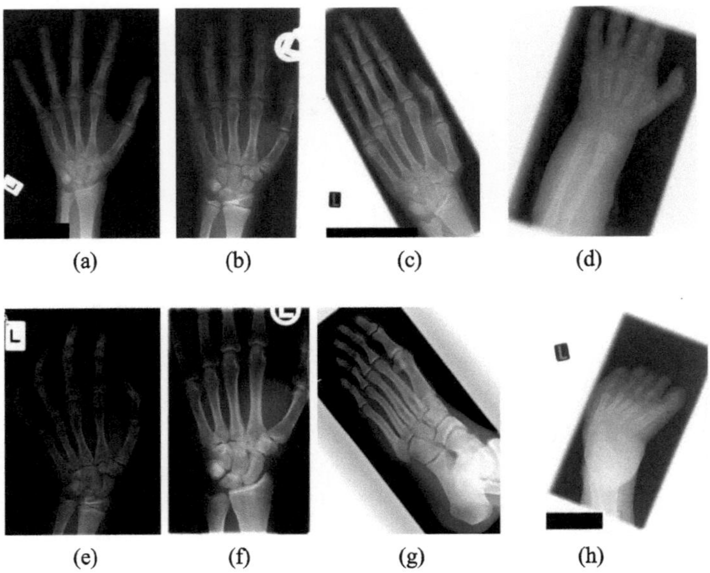

Fig. 23.12: Detecting category 'posteroanterior, left hand': (a),(b),(c),(d) Correctly classified. (e) False negative, misclassified as 'left anterior oblique, left hand'. False positives come from categories: (f) anteroposterior, left carpal joint (g) anteroposterior, left foot (h) right anterior oblique, right foot.

free kernel parameter γ, were scanned simultaneously over a grid to find the classifier's optimal working point. The histogram intersection kernel has no free kernel parameter, the optimization is one dimensional over the SVM cost parameter. These experiments are depicted in Figure 23.11. Table 23.3 summarizes the best parameters for the different kernels. The χ^2 kernel is ranked first by a small margin with 91.62% accuracy, followed by the RBF kernel with 91.45%.

In the final experiment, we took information from multiple image scales into account by repeating the dictionary creation step on scaled–down versions of the original image. The image representation was thus a concatenation of histograms built on the single scale dictionaries. We used three scales: the original image, 1/2 size and 1/8 size. Using three scales further improved the accuracy for all kernels, as seen in the right–most column of Table 23.3. The average classification accuracy with the χ^2 kernel was 91.95%.

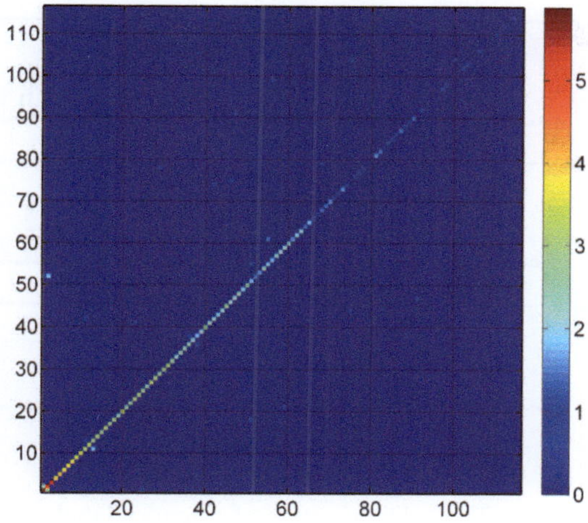

Fig. 23.13: Confusion matrix of 116 categories — logarithmic scale.

23.3.3 Classification Results

Figure 23.12 demonstrates the subtlety of the challenge by examining the classification accuracy of a single category: 'Posteroanterior, Left hand'. In this run there were 2,000 random test images, with 57 images from the examined category, out of which 56 were correctly detected by the described system as shown in Figure 23.12(a,b,c,d). Only one image, Figure 23.12(e), was falsely classified and was detected as a neighboring category – 'Left anterior oblique, Left hand' (false negative). Three images from other categories, Figure 23.12(f,g,h), were misclassified as 'Posteroanterior, Left hand' (false positives). These images have strong visual resemblance to the left hand category. A complete confusion matrix of the overall system running on 2,000 random images is displayed in Figure 23.13.

The ImageCLEF medical image annotation challenge increased in database size and labels complexity throughout the years, from 53 and 116 numerical labels in 2005 and 2006, to 116 and 193 IRMA codes in 2007 and 2008. The distribution of images across the categories is non–uniform, as seen in Figure 23.14. In 2009 the challenge was held for the last time; it used the four labeling sets of previous years, and examined classification accuracy of algorithms as the complexity of categories increases, on 1,733 previously unseen images. The error–counting scheme takes into account the hierarchical structure of the IRMA code — the penalty is greater for errors made in higher levels of the hierarchy (Deselaers et al, 2008).

In our above experiments the system parameters were tuned using only the labels from ImageCLEF 2007. Each of the 116 categories was treated as a separate label, disregarding the hierarchical nature of the IRMA code. The system was applied to

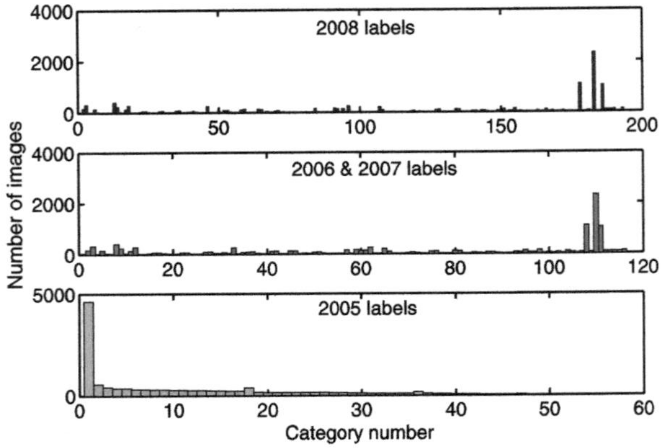

Fig. 23.14: Distribution of category labels.

Table 23.4: First and second best error scores in ImageCLEF 2009 medical annotation task. Lower is better.

Run	2005	2006	2007	2008	Sum
This work	356	263	64.3	169.5	852.8
Second best - Idiap (Tommasi et al, 2009)	393	260	67.23	178.93	899.16

the four labeling sets, and submitted to the ImageCLEF 2009 medical annotation challenge.

Table 23.4 shows the accuracy of the classification system on the four labeling sets in ImageCLEF 2009 medical annotation task and the second best result. Our system, presented in (Avni et al, 2009), was ranked first on three of the four labeling sets (2005, 2007 and 2008), and first in the overall error score.

23.4 Discussion

In this chapter we presented a visual words approach to medical image categorization. In our work we investigated the effect of different parameters on the overall classification score, and tuned the system to achieve high accuracy in classification of general x–ray images. We showed improvement of the classification score when including spatial coordinates in the feature vector and when using several dictionaries in multiple scales. In this task using dense and simple features is advantageous to using SIFT descriptors both in accuracy and computation time, with about half a second training and classification time per image. We reported state–of–the–art results in the task of organ and orientation identification in the ImageCLEF 2009 medical annotation challenge. The relatively high accuracy of this work in annotating a large medical archive with nearly 200 categories raises the motivation to explore similar approaches for pathology–level categorization, a task with possible clinical importance.

References

Avni U, Goldberger J, Greenspan H (2009) Dense simple features for fast and accurate medical x–ray annotation. In: Working notes of CLEF 2009, Corfu, Greece

Barla A, Odone F, Verri A (2003) Histogram intersection kernel for image classification. In: International conference on image processing, vol 3

Blei DM, Ng AY, Jordan MI (2003) Latent Dirichlet allocation. Journal of Machine Learning Research 3:993–1022

Bosch A, Muñoz X, Oliver A, Martí J (2006) Modeling and classifying breast tissue density in mammograms. In: Computer Vision and Pattern Recognition, pp 1552–1558

Chang CC, Lin CJ (2001) LIBSVM: a library for support vector machines. Software available at `http://www.csie.ntu.edu.tw/~cjlin/libsvm`

Deselaers T, Deserno TM (2009) Medical image annotation in ImageCLEF 2008. In: CLEF 2008 Proceedings. Lecture Notes in Computer Science (LNCS), vol 5706. Springer, pp 523–530

Deselaers T, Hegerath A, Keysers D, Ney H (2006) Sparse patch–histograms for object classification in cluttered images. In: DAGM Symposium, pp 202–211

Deselaers T, Kalpathy-Cramer J, Müller H, Deserno TM (2008) Hierarchical classification for ImageCLEF 2008 medical image annotation

Fei-Fei L, Perona P (2005) A Bayesian hierarchical model for learning natural scene categories. In: Computer Vision and Pattern Recognition, vol 2, pp 524–531

Lazebnik S, Schmid C, Ponce J (2006) Beyond bags of features: Spatial pyramid matching for recognizing natural scene categories. In: Computer vision and pattern recognition, vol 2, pp 2169–2178

Lehmann T, Güld M, Thies C, Fischer B, Spitzer K, Keysers D, H Ney MK, Schubert H, Wein B (2004) Content–based image retrieval in medical applications. Methods of Information in Medicine 43(4):354–361

Lehmann TM, Schubert H, Keysers D, Kohnen M, Wein BB (2003) The IRMA code for unique classification of medical images. In: Proceedings SPIE, pp 109–117

Lowe D (1999) Object recognition from local scale–invariant features. International conference on computer vision 2:1150–1157

Manning CD, Raghavan P, Schütze H (2008) Introduction to information retrieval. Cambridge University Press

Müller H, Lovis C, Geissbuhler A (2005) The medGIFT project on medical image retrieval. In: Gao X, Tully C, Lin C, Thom S, Müller H (eds) Medical Imaging and Telemedicine, Wuyishan, Fujian, China, pp 2–7. European Union AsiaICT Program

Nowak E, Jurie F, Triggs B (2006) Sampling strategies for bag–of–features image classification. In: European conference on computer vision. Springer, pp 490–503

Sivic J, Zisserman A (2008) Video google: a text retrieval approach to object matching in videos. In: International conference on computer vision, vol 2, pp 1470–1477

Tommasi T, Orabona F, Caputo B (2008) Discriminative cue integration for medical image annotation. Pattern Recognition Letters 29(15):1996–2002

Tommasi T, Caputo B, Welter P, Güld MO, Deserno TM (2009) Overview of the CLEF 2009 medical image annotation track. In: Working notes of CLEF 2009

Varma M, Zisserman A (2003) Texture classification: are filter banks necessary? In: Computer Vision and Pattern Recognition, vol 2, pp 691–698

Zhang J, Marszalek M, Lazebnik S, Schmid C (2007) Local features and kernels for classification of texture and object categories: A comprehensive study. International Journal of Computer Vision 73(2):213–238

Chapter 24
Idiap on Medical Image Classification

Tatiana Tommasi and Francesco Orabona

Abstract The team from the Idiap Research Institute in Martigny, Switzerland, participated in three editions of the CLEF medical image annotation task always reaching among the highest positions in the rankings. Here, we present in detailed form the successful strategies we used in the different editions of the challenge to face the inter– vs. intra–class image variability, to exploit the hierarchical labeling, and to cope with the unbalanced distribution of the classes.

24.1 Introduction

This chapter presents the algorithms and results of the Idiap participation in the ImageCLEFmed annotation task in 2007, 2008 and 2009. The goal of the challenge was to develop an automatic image annotation system able to distinguish x–ray images on the basis of the body region, the biological system examined, the body orientation and the imaging modality. The idea is to exploit content–based image analysis without making use of the textual information generally associated with medical images. A system performing this task reliably can avoid the cost of manually annotating several terabytes of image data collected annually in radiology departments and also help in image retrieval.

There are two main issues when working on large databases of medical images: intra–class variability vs. inter–class similarity and data imbalance. The first problem is due to the fact that images belonging to the same visual class might look very different, while images that belong to different visual classes might look very

Tatiana Tommasi
Idiap Research Institute, Centre du Parc, Rue Marconi 19, PO Box 592, 1920 Martigny, Switzerland, e-mail: `ttommasi@idiap.ch`

Francesco Orabona (work done while at Idiap Research Institute)
Dipartimento di Scienze dell'Informazione, Universita' degli Studi di Milano, via Comelico 39, 20135 Milano, Italy, e-mail: `francesco@orabona.com`

H. Müller et al. (eds.), *ImageCLEF*, The Information Retrieval Series 32,
DOI 10.1007/978-3-642-15181-1_24, © Springer-Verlag Berlin Heidelberg 2010

similar. Data imbalance is related to the natural statistics of the onset of diseases in the different parts of the body, thus it reflects the a priori probabilities of the routine diagnosis in a radiological clinic. To overcome both these problems, an automatic annotation system needs to use the most discriminative information from the available data; it also needs to be able to weigh properly the information coming from differently populated classes in the learning process.

For the CLEF challenge, the images were identified on the basis of the Image Retrieval in Medical Applications (IRMA) code (Lehmann et al, 2003). This is a multi–axial hierarchical scheme, which adds a further difficulty in the annotation process.

In our experience as participants of the ImageCLEFmed challenge, we tackled these problems and proposed different discriminative solutions based on Support Vector Machines (SVM) (Cristianini and Shawe-Taylor, 2000). In 2007 and 2008 our best run ranked first, while in 2009 the run reproducing the winning strategy of 2008 ranked second.

In the rest of the chapter we will focus on a number of issues as follows: Section 24.2 gives details about how we combined multiple cues to face the inter–class vs. intra–class variability; Section 24.3 introduces our confidence–based approach to exploit the hierarchical structure of the data; and Section 24.4 describes our strategy to overcome the data imbalance by creating virtual examples. Finally in Section 24.5 we describe our experimental set–up and summarize our results. Conclusions are drawn in Section 24.6.

24.2 Multiple Cues for Image Annotation

Several authors tried to address the inter-class vs. intra–class variability problem using local and global features, and more generally different types of descriptors, separately or combined together in a multiple cues approach (Müller et al, 2006; Güld et al, 2006; Florea et al, 2006). For some of these examples the performance was not very good. However, years of research on visual recognition in other domains have shown clearly that multiple cue methods outperform single–feature approaches (Matas et al, 1995; Mel, 1997; Sun, 2003). To have the maximum advantage from cue integration, each feature should represent a different aspect of the data allowing for a more informed decision. Heterogeneous and complementary visual cues, bringing different information content, were successfully used in the past (Slater and Healey, 1995; Mel, 1997; Nilsback and Caputo, 2004; Gehler and Nowozin, 2009). Regarding the integration techniques, they can all be reduced to one of these three approaches: *high–level*, *mid–level* and *low–level* integration (Sanderson and Paliwal, 2004; Polikar, 2006). Figure 24.1 illustrates schematically the basic ideas behind these methods.

Participating in the ImageCLEF challenge we proposed a discriminative approach for integration of cues by defining three strategies, one for each of the pos-

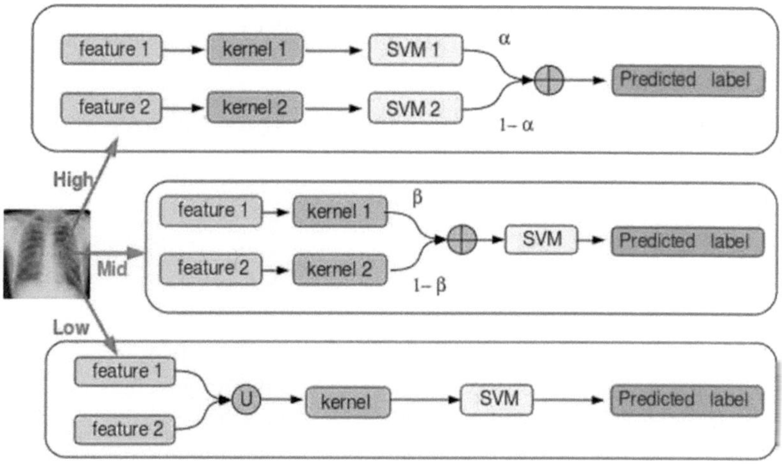

Fig. 24.1: A schematic illustration of the high–level, mid–level and low–level cue integration approaches.

sible levels of cue integration. The methods used are described in detail in the following sections.

24.2.1 High–Level Integration

High–level cue integration methods start from the output of two or more classifiers dealing with complementary information. Each of them produces an individual hypothesis about the object to be classified. All these hypotheses are then combined together to achieve a consensus decision. We applied this integration strategy using the Discriminative Accumulation Scheme (DAS) proposed first in (Nilsback and Caputo, 2004). It is based on a weak coupling method called accumulation, which does not neglect any cue contribution. Its main idea is that information from different cues can be summed together.

Suppose we are given M object classes and for each class a set of N_j training images $\{I_i^j\}_{i=1}^{N_j}$, $j = 1, \ldots M$. For each image we extract a set of P different features $T_p(I_i^j), p = 1 \ldots P$ so that for an object j we have P new training sets. For each feature we train an SVM. Kernel functions may differ from cue to cue and model parameters can be estimated via cross validation. Given a test image $\hat{I}$ and assuming $M \geq 2$, for each single–cue SVM we compute the distance from the separating hyperplane $D_j(p)$, $p = 1 \ldots P$. After collecting all the distances $\{D_j(p)\}_{p=1}^{P}$ for all the M objects and the P cues, we classify the image $\hat{I}$ using the linear combination:

$$j^* = \arg \max_{j=1 \ldots M} \left\{ \sum_{p=1}^{P} a_p D_{j(p)} \right\} . \tag{24.1}$$

The coefficients $\{a_p\}_{p=1}^{P} \in \Re^+$ are determined via cross validation during the training phase.

24.2.2 Mid–Level Integration

Combining cues at the mid–level means that the different feature descriptors are kept separated, but they are integrated in a single classifier generating the final hypothesis. To implement this approach we developed a scheme based on multi–class SVMs with a Multi Cue Kernel, K_{MC}. This new kernel combines different features ($T_p(I)$) extracted from the images (I):

$$K_{MC}(\{T_p(I_i)\}_p, \{T_p(I)\}_p) = \sum_{p=1}^{P} a_p K_p(T_p(I_i), T_p(I)), \quad \sum_{p=1}^{P} a_p = 1 . \tag{24.2}$$

The Multi Cue Kernel is a Mercer kernel, as positively weighted linear combinations of Mercer kernels are Mercer kernels themselves (Cristianini and Shawe-Taylor, 2000). In this way it is possible to perform only one classification step, identifying the best weighting factors $a_p \in \Re^+$ through cross validation while determining the optimal separating hyperplane. This means that the coefficients a_p are guaranteed to be optimal.

24.2.3 Low–Level Integration

To combine cues it is also possible to use a low–level fusion strategy, starting from the descriptors and combining them in a new representation. In this way the cue integration does not directly involve the classification step. Here we use feature concatenation: two feature vectors f_i and c_i are combined into a single feature vector $v_i = (f_i, c_i)$ that is normalized to have its sum equal to one and is then used for classification. In this way the information related to each cue is mixed without a weighting factor that allows it to control the influence of each information channel on the final recognition result. A general drawback of this method is that the dimensionality of the feature vector increases as the number of cues grows, implying longer learning and recognition times, higher memory requirements and possibly risks curse of dimensionality effects.

24.3 Exploiting the Hierarchical Structure of Data: Confidence Based Opinion Fusion

The evaluation scheme for the medical image annotation task addresses the hierarchical structure of the IRMA code considering the number of possible choices at each node and the position of each node in the hierarchy. So, wrong decisions in easy nodes were penalized more than wrong decisions in difficult nodes, and mistakes at an early stage in the code were more costly than at a later stage. Moreover, the error evaluation method allowed the classifier to decide a 'don't know' at any level of the code, independently for each of the four axes: image modality, body orientation, body region and biological system (Lehmann et al, 2003).

In 2007 and 2008 many groups participating in the ImageCLEF medical annotation task tried to exploit the hierarchical labeling of the images classifying separately on the four axes of the IRMA code. However, analyzing the results, it was observed that the top–performing runs used each individual code as a class, using a flat classification approach which did not use the hierarchy (Deselaers et al, 2008). The only way to take advantage of the hierarchy seemed to be by exploiting the use of wildcard characters. Thus models which estimate the classifier's confidence in its decisions could be useful.

Discriminative classifiers usually do not provide any out–of–the–box solutions for estimating the confidence in the decision, but in some cases they can be transformed into opinion makers on the basis of the value of the discriminative function. In the case of SVM, it can be done by considering the distances between the test samples and the classification hyperplane. This approach turns out to be very efficient due to the use of kernel functions and does not require additional processing in the training phase.

In the one–vs.–all multi–class extension of SVM, if M is the number of classes, M SVMs are trained, each separating a single class from all remaining ones. The decision is then based on the distances of the test sample, $\boldsymbol{x}$, to the M hyperplanes, $D_j(\boldsymbol{x})$, $j = 1 \dots M$. The final output is the class corresponding to the hyperplane for which the distance is largest:

$$j^* = \arg \max_{j=1 \dots M} D_j(\boldsymbol{x}) \,. \tag{24.3}$$

If now we think of the confidence as a measure of unambiguity of the decision, we can define it as the difference between the maximal and the next largest distance:

$$C(\boldsymbol{x}) = D_{j^*}(\boldsymbol{x}) - \max_{j=1 \dots M, j \neq j^*} D_j(\boldsymbol{x}) \,. \tag{24.4}$$

The value $C(\boldsymbol{x})$ can be thresholded to obtain a binary confidence measure. Hence a confident prediction is assumed if $C(\boldsymbol{x}) > \tau$, for a given threshold τ. In the cases in which the decision is not confident, we decided to compare the labels corresponding to the first two margins and to put a 'don't know' term in the points of the code in which they differ.

24.4 Facing the Class Imbalance Problem: Virtual Examples

Unbalanced data sets define a challenging problem in machine learning. Classifiers generally perform poorly on unevenly distributed data sets because they are designed to generalize from sample data and output the simplest hypothesis that best fits them. In a binary problem with negative instances which heavily outnumber the positive ones, this means classifying almost all instances as negative. On the other hand, making the classifier too specific may make it sensitive to noise and prone to overfitting. Although SVMs have shown remarkable success in many applications, their capabilities are very limited when applied to the problem of learning from multi–class databases in which some of classes are sparsely populated. There are two known approaches to solve this problem. One is to bias the classifier so that it pays more attention to samples from poorly populated classes. This can be done, for instance, by increasing the penalty associated with misclassifying the class with few data with respect to the others. The second approach is to pre–process the data by resampling methods (Akbani et al, 2004). A possible alternative to resampling consists in exploiting the known invariances of the data to generate new synthetic minority instances and rebalance the data set. We adopted this solution.

Keysers et al (2003), the creators of the IRMA corpus used for the ImageCLEF challenge, explain that small transformations of the images do not alter their class membership. Therefore to improve the classification reliability, we enriched the poorly populated classes producing virtual examples as slightly modified copies of the training images. We increased and decreased each image side (100, 50 pixels); rotated them right and left (20, 40 degrees); shifted right, left, up, down and in the four diagonal directions (50 pixels); increased and decreased the brightness (add and subtract 20 to the original gray level). Thus the number of images in the poorly populated classes (with less than ten images) was increased by a factor of 17.

24.5 Experiments

All the techniques described above were optimized on the training set released and applied on the unlabeled test set of the last three editions of the CLEF challenge. In the following subsections we summarize the specific choices made in running the experiments and the results obtained.

24.5.1 Features

To extract different and complementary information from the images, we chose two types of features that were then combined with the high–, mid– and low–level integration strategies. In 2007 we combined a local (modSIFT) and a global feature

(Raw Pixels), while in 2008 and 2009 we considered two different local cues (modSIFT and LBP).

ModSIFT. Scale Invariant Feature Transform (SIFT) (Lowe, 1999) is a well known algorithm in computer vision used to detect and describe local features in images. We decided to use it adopting a bag–of–words approach: analogous to text classification, the basic idea is to sample image patches and to match them to a set of pre–specified 'visual words'. Note that the ordering of the visual words is not important and only the frequency of appearance of each word is used to form the feature vectors. The main implementation choices are thus: (1) how to sample patches, (2) what visual patch descriptor to use, and (3) how to build the vocabulary.

Regarding point (1), we used random sampling. Due to the low contrast of the radiographs it would be difficult to use any interest point detector. Moreover it has been pointed out by different papers and systematically verified by Nowak et al (2006) that a dense random sampling is always superior to any strategy based on interest point detectors for image classification tasks.

Regarding point (2), we decided to use a modified version of the SIFT descriptor. SIFTs are designed to describe an area of an image so as to be robust to noise, illumination, scale, translation and rotation changes. Given the specific constraints of our classification task, we slightly modified the classical version of this descriptor. The SIFT rotation invariance is not relevant for the ImageCLEFmed classification task, as the various structures in the radiographs are likely to always appear with the same orientation. Moreover, the scale is not likely to change too much between images of the same class. Hence a rotation– and scale–invariant descriptor could discard useful information for the classification. Thus we extracted the points at only one octave, the one that gave us the best classification performance on a validation set, and we removed the rotation–invariance. We call the modified SIFT descriptor modSIFT.

Regarding point (3), we built the vocabulary randomly sampling 30 points of each input image and extracting a modSIFT feature at each point. The visual words are created using an unsupervised K–means clustering algorithm. Note that in this phase both training and test images could be used, because the process does not need the labels. We chose K template modSIFTs with K equal to 500 and thus defined a vocabulary with 500 words. Various sizes of vocabulary were tested ($K = 500, 1,000, 2,000$). Preliminary results on a validation set showed no significant differences in performance between these three vocabulary sizes. We chose therefore $K = 500$, the smallest, for computational reasons.

Finally, the feature vector for an image is defined by extracting a random collection of points from the images. The resulting distribution of descriptors in the feature space is then quantized in the visual words of the vocabulary and converted into a frequency histogram. To add some spatial information, we decided to divide the images into four parts, collecting the histograms separately. In this way the dimension of the input space is multiplied by four (feature vector with $500 \times 4 = 2,000$ elements) but in our tests we gained about 3% in classification performance. We extracted 1,500 modSIFTs in each sub–image: such dense sampling adds robustness to the process. Figure 24.2 shows an example of the extracted local features.

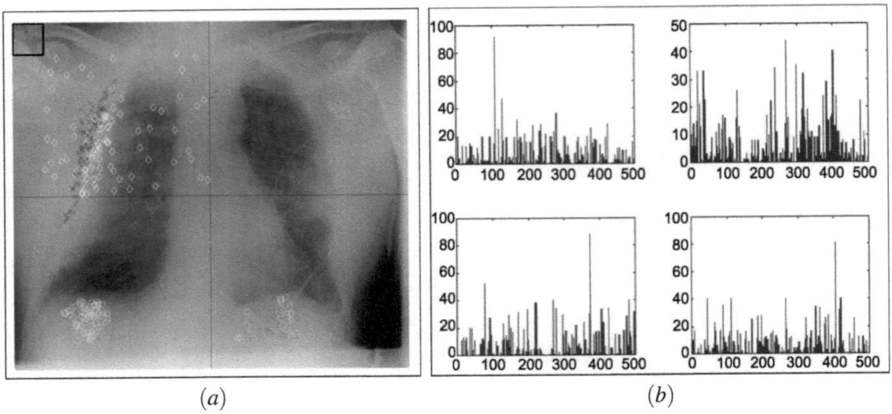

Fig. 24.2: (a) The four most present visual words in the image are drawn, each with a different color (better viewed in color). The square in the upper left corner represents the size of the patch used for computing the modSIFT descriptor. (b) Total counts of the visual words in the four sub–images.

In 2008 and 2009 we slightly modified the modSIFT feature inspired by the approach in (Lazebnik et al, 2006). We added to the original vector the histogram obtained extracting the feature from the whole image producing a final vector of 2,500 elements.

LBP. Local Binary Patterns (LBP) (Ojala et al, 2002) have been used extensively in face recognition, object classification (Ahonen et al, 2006; Zhang et al, 2007) and also in the medical area (Unay et al, 2007; Oliver et al, 2007). The basic idea of LBP is to build a binary code that describes the local texture pattern in a circular region thresholding each neighborhood on the circle by the gray value of its center. After choosing the dimension of the radius R and the number of points P to be considered on each circle, the images are scanned with the LBP operator pixel by pixel and the outputs are accumulated into a discrete histogram (Ojala et al, 2002). The operator is gray–scale invariant, moreover we used the *riu*2 rotational invariant LBP version which considers the uniform patterns with two spatial transitions ($LBP^{riu2}_{P,R}$; (Ojala et al, 2002)).

Our preliminary results on a validation set showed that the best way to use LBP on the medical image database at hand was by combining a two dimensional histogram $LBP^{riu2}_{8,8}$ with $LBP^{riu2}_{16,12}$ and concatenating it with the two dimensional histogram made by $LBP^{riu2}_{16,18}$ together with $LBP^{riu2}_{24,22}$. In this way a feature vector of 648 elements is obtained. Each image is divided into four parts, one vector is extracted from each sub–image and from the central area and then they are concatenated producing a vector of 3,240 elements (see Figure 24.3).

Raw Pixels. We used the raw pixels as simplest possible global descriptor. Preliminary results on a validation set showed that downscaling images to 32 x 32 pixels did not produce any significant difference compared to downscaling to 48 x 48 but

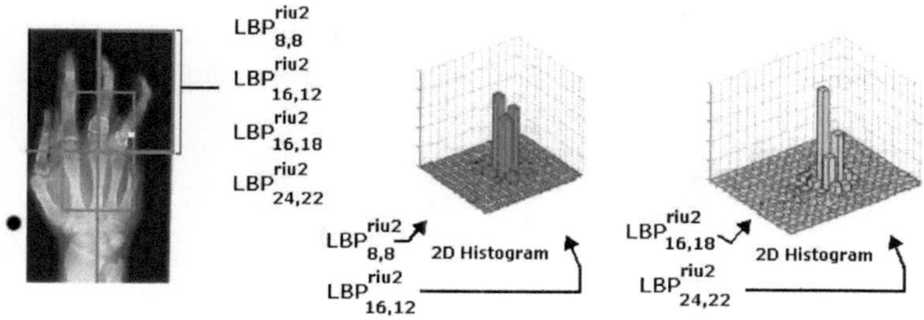

Fig. 24.3: A schematic drawing which shows how we built the texture feature vector combining the 1–dimensional histograms produced by the LBP operators in 2–dimensional histograms.

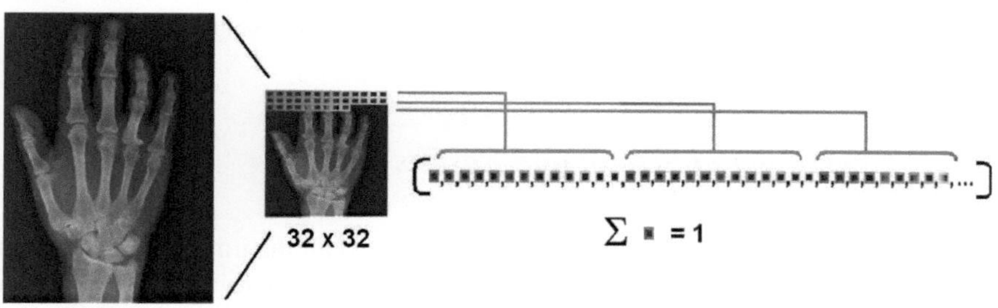

Fig. 24.4: An example showing the raw pixel representation.

the classification performance was better than that obtained on 16 x 16 images. So the images were resized to 32 x 32, regardless of the original dimension. The obtained $1,024$ pixel intensity values were then normalized to have sum equal to 1 and used as input features. Figure 24.4 shows how we built the raw pixel representation for each image.

24.5.2 Classifier

SVMs are a class of learning algorithms based on Statistical Learning Theory (Cristianini and Shawe-Taylor, 2000). Born as a linear classifier, SVM can be easily extended to nonlinear domains through the use of kernel functions. The kernels implicitly map the input space to a higher dimensional space, even with infinite dimensions. At the same time the generalization power of the classifier is kept under control by a regularization term that avoids overfitting in such high dimensional spaces (Cristianini and Shawe-Taylor, 2000).

The choice of the kernel heavily affects the performance of the SVM. We used an exponential χ^2 kernel for all the feature types and integration approaches, which is a valid kernel as proved in (Fowlkes et al, 2004):

$$K(\mathbf{x},\mathbf{y}) = \exp\left(-\gamma \sum_{i=1}^{N} \frac{(x_i - y_i)^2}{|x_i + y_i|}\right) . \qquad (24.5)$$

In our experiments we also tested the linear kernel and the Radial Basis Function (RBF) kernel, but all of them gave worse results than the χ^2. The parameter γ was tuned through cross-validation together with the SVM cost parameter C.

Even if the labels are hierarchical, we used the standard one–vs.–all and one–vs.–one multi–class approaches. We verified experimentally that with our features, the recognition rate was lower using an axis–wise classification. This could be due to the fact that each super–class has a variability so high that our features are not able to model it, while they can model the small sub–classes very well.

24.5.3 Experimental Set–up and Results

To obtain reliable results in the training phase we used all the images released from the CLEF organizers, not considering the distinction between training and validation when it was suggested. Our strategy was to create five disjoint train/test splits on which to optimize the learning parameters. The performance was evaluated on the basis of the error score, the same as used in a second stage to rank the runs submitted to the challenge.

In 2008 and 2009, to take care of the class imbalance, the released database was divided into:

- rich_set: images belonging to classes with more than ten elements. From this group we built five disjoint sets, rich_train$_i$/rich_test$_i$, where the test sets were created by randomly extracting five images for each of the classes. Note that in this way we automatically considered a normalization on the classes.
- poor_set: images belonging to classes with less than ten elements. We used the whole poor_set as a second test set.

We trained the classifier on the rich_train$_i$ set and tested both on the rich_test$_i$ and on the poor_set, for each of the five splits. In this way, although the classes with few images were not considered in the training phase, we could evaluate the performance of the classifier to assign to those images the corresponding nearest class in the hierarchy. The error score was evaluated using the program released by the ImageCLEF organizers. The score values were normalized by the number of images in the corresponding test set, producing two average error scores. They were then multiplied by 500 and summed together supposing an ideal test set of 1,000 samples constituted half by images from the rich_set and half by images from the poor_set. The average of the scores obtained on the five splits is an estimator of the expected

Table 24.1: Ranking of our runs, name, score, and gain with respect to the best run of other participants (RWTHi6-4RUN-MV3) in 2007. The Low level cues integration was used only after the challenge. 'oa' and 'oo' indicate respectively the one–vs.–all and one–vs.–one SVM multi–class extensions.

Rank	Name	Score	Gain
1	Mid_oa	26.85	4.08
	Low_oa	26.96	3.96
	Low_oo	26.99	3.93
2	Mid_oo	27.54	3.38
3	modSIFT_oo	28.73	2.20
4	modSIFT_oa	29.46	1.47
5	High	29.90	1.03
6	RWTHi6-4RUN-MV3	30.93	0
28	PIXEL_oa	68.21	−37.28
29	PIXEL_oo	72.41	−41.48

value of the score. Each parameter in our methods was found by optimizing this expected score.

To evaluate the effect of introducing virtual examples in the poor_set we extracted from it only images belonging to classes with more than one element. We called this set poor_more. From this set we created six poor_more_train$_j$/poor_more_test$_j$ splits, where the train sets were defined extracting one image from each of the classes. Each poor_more_train set was enriched with the virtual examples as described in Section 24.4. Then we combined these sets joining rich_train$_i$ and poor_more_train$_j$ to build the training set and testing separately on rich_test$_i$ and poor_more_test$_j$.

Tables 24.1, 24.2, and 24.3 summarize all the results obtained by the Idiap team runs in 2007, 2008 and 2009 with the relative gain with respect to the best result from the other participating groups. In 2009 we participated in the ImageCLEFmed challenge organization and we decided to simply reuse the best approaches proposed in 2008, submitting these as baseline runs.

24.6 Conclusions

The Idiap team participated in the CLEF medical image annotation task from 2007 to 2009 proposing discriminative approaches coming from the image classification and recognition domain. The methods used are based on a combination of different local and global features and SVM as the classifier, together with specific solutions to face the class imbalance problem and to exploit the hierarchical labeling structure of data. On the basis of the results obtained we can state that the strategies adopted are suited to solve the challenging issue of annotating a large medical image database.

Table 24.2: Ranking of our submitted runs, name, score and gain with respect to the best run of the other participants (TAU-BIOMED-svm_full) in 2008. The extension 'virtual' stands for poor class enrichment by the use of virtual examples; 'confidence' stands for the combination of the first two SVM margins for the confidence based opinion fusion. For all the runs we used the one–vs.–all SVM multi–class extension.

Rank	Name	Score	Gain
1	Low_virtual_confidence	74.92	30.83
2	Low_virtual	83.45	22.30
3	Low_confidence	83.79	21.96
4	Mid_virtual_confidence	85.91	19.84
5	Low	93.20	12.55
6	modSIFT	100.27	5.48
7	TAU-BIOMED-svm_full	105.75	0
11	LBP	128.58	−22.83

Table 24.3: Ranking of our submitted runs, name, score and gain with respect to the best run of the other participants (TAUbiomed) in 2009. The extension 'virtual' stands for poor class enrichment by the use of virtual examples; 'confidence' stands for the combination of the first two SVM margins for the confidence based opinion fusion. For all the runs we used the one–vs.–all SVM multi–class extension.

Rank	Name	Score	Gain
1	TAUbiomed	852.8	0
2	Low_virtual_confidence	899.16	−46.36
3	Low_confidence:	899.4	−46.6
4	Low	1039.63	−186.83
5	Low_virtual	1042	−189.2

Acknowledgements The authors were supported by the EMMA project thanks to the Hasler foundation (http://www.haslerstiftung.ch/).

References

Ahonen T, Hadid A, Pietikainen M (2006) Face description with local binary patterns: Application to face recognition. IEEE Transactions on Pattern Analysis and Machine Intelligence 28(12):2037–2041

Akbani R, Kwek S, Japkowicz N (2004) Applying support vector machines to imbalanced datasets. In: European Conference on Machine Learning Lecture Notes in Computer Science (LNCS), vol 3201. Springer, pp 39–50

Cristianini N, Shawe-Taylor J (2000) An introduction to support vector machines: and other kernel–based learning methods. Cambridge University Press

Deselaers T, Deserno TM, Müller H (2008) Automatic medical image annotation in ImageCLEF 2007: Overview, results, and discussion. Pattern Recognition Letters 29(15):1988–1995

Florea F, Rogozan A, Cornea V, Bensrhair A, Darmoni S (2006) MedIC/CISMeF at ImageCLEF 2006: image annotation and retrieval tasks. In: Working Notes of CLEF 2006

Fowlkes C, Belongie S, Chung F, Malik J (2004) Spectral grouping using the Nyström method. IEEE Transactions on Pattern Analysis and Machine Intelligence 26:214–225

Gehler P, Nowozin S (2009) On feature combination for multiclass object classification. In Proceedings of the IEEE International Conference on Computer Vision. IEEE Computer Society

Güld M, Thies C, Fischer B, Lehmann T (2006) Baseline results for the ImageCLEF 2006 medical automatic annotation task. In: CLEF 2006 Proceedings. Lecture Notes in Computer Science (LNCS), vol 4730. Springer, pp 686–689

Keysers D, Dahmen J, Ney H, Wein BB, Lehmann TM (2003) Statistical framework for model-based image retrieval in medical applications. Journal of Electronic Imaging 12(1):59–68

Lazebnik S, Schmid C, Ponce J (2006) Beyond bags of features: Spatial pyramid matching for recognizing natural scene categories. Proceedings of the Conference on Computer Vision and Pattern Recognition 2:2169–2178

Lehmann TM, Schubert H, Keysers D, Kohnen M, Wein BB (2003) The IRMA code for unique classification of medical images. In: Proceedings SPIE, vol 5033, pp 109–117

Lowe DG (1999) Object recognition from local scale–invariant features. In: Proceedings of the IEEE International Conference on Computer Vision, vol 2. IEEE Computer Society, p 1150

Matas J, Marik R, Kittler J (1995) On representation and matching of multi–coloured objects. Proceedings of the IEEE International Conference on Computer Vision 726

Mel BW (1997) SEEMORE: Combining color, shape, and texture histogramming in a neurally inspired approach to visual object recognition. Neural computation 9:777–804

Müller H, Gass T, Geissbuhler A (2006) Performing image classification with a frequency–based information retrieval schema for ImageCLEF 2006. In: Working Notes of CLEF 2006

Nilsback M, Caputo B (2004) Cue integration through discriminative accumulation. Proceedings of the Conference on Computer Vision and Pattern Recognition 2:578–585

Nowak E, Jurie F, Triggs B (2006) Sampling strategies for bag–of–features image classification. In: Proceedings of the European Conference of computer vision. Lecture Notes in Computer Science (LNCS). Springer, pp 490–503

Ojala T, Pietikäinen M, Mäenpää T (2002) Multiresolution gray–scale and rotation invariant texture classification with local binary patterns. IEEE Transactions on Pattern Analysis and Machine Intelligence 24(7):971–987

Oliver A, Lladó X, Freixenet J, Martí J (2007) False positive reduction in mammographic mass detection using local binary patterns. In: Medical Image Computing and Computer–Assisted Intervention — MICCAI 2007 Lecture Notes in Computer Science (LNCS), vol 4791. Springer, pp 286–293

Polikar R (2006) Ensemble based system in decision making. IEEE Circuits and Systems Magazine 6(3):21–45

Sanderson C, Paliwal KK (2004) Identity verification using speech and face information. In: Digital Signal Processing, pp 449–480

Slater D, Healey G (1995) Combining color and geometric information for the illumination invariant recognition of 3–D objects. Proceedings of the International Conference on Computer Vision 563

Sun Z (2003) Adaptation for multiple cue integration. Proceedings of the Conference on Computer Vision and Pattern Recognition 440

Unay D, Ekin A, Cetin M, Jasinschi R, Ercil A (2007) Robustness of local binary patterns in brain MR image analysis. Proceedings of the 29th Annual International Conference of the IEEE Engineering in Medicine and Biology Society 2098–2101

Zhang L, Li S, Yuan X, Xiang S (2007) Real–time object classification in video surveillance based on appearance learning. In: Proceedings of the Conference on Computer Vision and Pattern Recognition. IEEE Computer Society

Part IV
External views

Reflections on ImageCLEF and information retrieval evaluation.

Chapter 25
Press Association Images — Image Retrieval Challenges

Martin Stephens and Dhavalkumar Thakker

Abstract In order to maximise the potential benefits of large repositories of digital images available both publicly and in private collections, intelligent information retrieval systems are required. Unfortunately, most image search engines rely on free–text search that often returns non–relevant results based on the occurrence of search keywords in text accompanying the images being matched purely at a lexical, rather than a semantic, level. In this chapter we report on ongoing work at Press Association Images on building a semantically–enabled image annotation and retrieval engine that relies on methodically structured ontologies for image annotation, thus allowing for more intelligent reasoning about the image content and subsequently improving the end–user browsing experience.

25.1 Press Association Images — A Brief History

25.1.1 The Press Association

> "The Press Association[1] (Moncrieff, 2005) is the backbone and the flesh of British journalism. But it does not wear jazzy clothes. Every day since its birth in 1868 — with very few insignificant blips — the PA has poured out word and (in the last fifty years or so) pictures that, as often as not, have provided the essential material for every morning and evening newspaper in the British Isles — and, more recently, for radio and television as

Martin Stephens
Press Association Images, Pavilion House, 16 Castle Boulevard, Nottingham, NG7 1FL, UK
e-mail: martin.stephens@pressassociation.com

Dhavalkumar Thakker
Press Association Images/Nottingham Trent University, Burton Street, Nottingham, NG1 4BU, UK, e-mail: dhaval.thakker@paphotos.com

[1] http://www.pressassociation.com/

H. Müller et al. (eds.), *ImageCLEF*, The Information Retrieval Series 32,
DOI 10.1007/978-3-642-15181-1_25, © Springer-Verlag Berlin Heidelberg 2010

> well. Throughout the media industry, the PA has a reputation for speed, accuracy, fairness and flexibility."

So starts Chris Moncrieff's history of the Press Association: 'Living on a Deadline'. Moncrieff was the Press Association's Political Editor and in the foreword to the book Tony Blair, the then serving Prime Minister, describes him and the Press Association as follows:

> "For those like myself who entered Parliament in 1983, the Press Association was personified by the distinctive and seemingly ever–present figure of Chris Moncrieff, the PA's then political editor. Although many political journalists were more famous in the wider world, you quickly learnt that few were more important than Chris — and that none was more careful to ensure he did not abuse his position or the influence it gave him. What marked out Chris — and continues to mark out his successors and the PA itself — is their pride in trying to report the news quickly, accurately and fairly."

The aim of providing news, pictures and more recently video in a quick and unbiased way is at the core of what the Press Association does. It is strange to think in this era of high speed Internet and fibre optic cables that the roots of a modern news agency can be traced back to a law given royal assent by Queen Victoria on the 31st July, 1868. The bill, The Telegraph Act proposed by Disraeli's Conservative government, nationalized the private telegraph companies in the United Kingdom.

Moncrieff describes the coming of the telegraph wires as a revolution:

> "They had been put up along more than two thousand miles of railway track and were created primarily to carry messages for the train companies. But it was soon realised that they could carry news of events as well. Before their arrival, newspapers had to wait for news notes carried by sailing ships, horse riders, donkey carts and pigeons."

Prior to the act the telegraph wires were in the hands of private companies which effectively held the regional newspapers to ransom. They provided vital information such as parliamentary reports, speeches, commercial information and sport from outside the newspaper's circulation area, but with a monopoly on supply charged high prices and often provided reports that were full of errors. Any complaints from the papers were met with increased charges.

The regional newspaper owners came together as the act was being passed to form The Press Association, whose objectives were to record with accuracy and without bias all events of sufficient news interest, free from all outside pressure and to circulate the reports as quickly as possible using the telegraph wires.

The Press Association was registered as a limited company on the 6th November, 1868. It is today still owned by newspaper groups, both national and regional, and though the technology has changed its core aim remains the same. The PA is trusted to provide accurate information, be that in the form of words, video or pictures and their associated captions.

Perhaps the most famous example of the work of the PA in recent memory is the short newsflash which ran on The Press Association's wire at 04.41 on Sunday, 31 August 1997 which read: "Diana, Princess of Wales, has died, according to British sources, the Press Association has learned this morning." It is as a result of the high regard The PA is held in, that despite no official announcement being available, the world's media accepted this news and reported it. On a more lighthearted note Chris Moncrieff's position in political history was sealed in 2007 when the newly refurbished café bar at the Press Gallery in the Palace of Westminster was named Moncrieff's.

25.1.2 Images at the Press Association

For the first 77 years of its existence The Press Association's output was just words but in 1945 pictures were added as part of the regional papers' Comprehensive Service. The initial aim was to supply provincial newspapers with up to 30 pictures a week, which would either be delivered to their London offices or dispatched by train parcels to the regions. It is interesting to note that today at busy times 1945's weekly target of 30 pictures could easily be sent out within one hour.

The 26th November 1945, the day the picture service was launched, was a quiet news day. However, a picture taken just a moment's walk from The Press Association's Fleet Street offices of a local gas strike was well used by the papers, including the front page of the London Evening News. The picture, taken by Roy Illingworth — one of PA's seven staff photographers — contrasted the gloom of Fleet Street, blacked out as a result of the strike, with the blaze of light in the City of Westminster just a few yards away (see Figure 25.1).

At its launch, in addition to the photographers, darkroom technicians and messengers, the picture department staff included two very important people: a picture librarian and a salesman. It is testament to the quality of the work of the picture librarian that the picture of the gas strike can still be found in The Press Association library and shows that even in its earliest days the need for archiving and the potential value of images was recognised.

The Press Association came relatively late to photography. Many photographic press agencies had been in existence since the 19th century. Agencies such as Central News, Central Press, Barratts and Sport & General were providing picture services well before the PA started its service. Through acquisitions and representation agreements the PA's picture library, Press Association Images, has over the years become home to the collections of a number of these agencies, including the Central News collection, which takes the library's archive right back to the start of The Press Association with its earliest pictures dating from the 1860s.

These collections, counted in the millions of images and each bringing with them the idiosyncrasies of their picture librarians, are stored in the Press Association Images offices. Formats range from the earliest glass plates through to colour negatives which were shot until the advent of professional digital cameras meant the need for speed marked the end of well over 100 years of reliance on film of various types to record news events.

The Press Association's current digital archive holds about 7.5 million images, a figure that may be out of date before this sentence is finished. With its own staff photographers and with representation agreements in place with a wide range of agencies and photographers from around the world, including The Associated Press, in excess of 35,000 new images are added to the database in an average week. Figure 25.2 shows examples from the current archive.

25.2 User Search Behaviour

25.2.1 Types of Users

There are around 25,000 active users of the Press Association Images database. They access the system via a Web interface[2]. The same system is used by Press

Fig. 25.1: The first image of the PA archives.

[2] `http://www.pressassociation.com/images/`

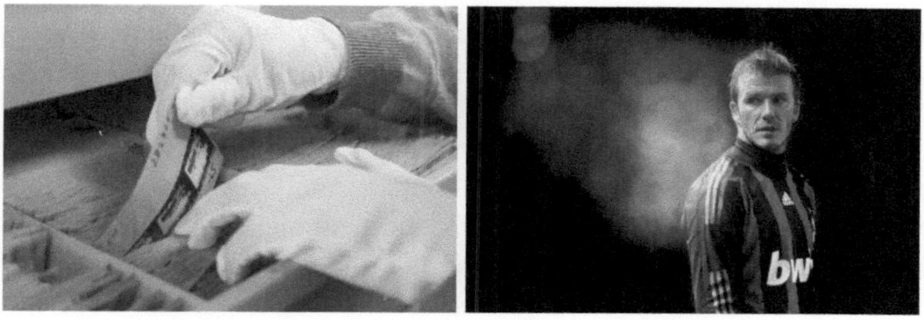

Fig. 25.2: Example images from the Press Association's historical and contemporary archives.

Association Images staff, those within the larger PA Group and the organisation's customers.

Some users will access the system and search for images on a very regular basis, while others may only use the database once or twice a year depending on the sort of projects they work on.

In terms of experience, the range of users is extremely wide. Some users will be images specialists, for example those working for newspapers or magazines employed to locate images from a number of sources for their publications. These users, which would include many of Press Association Images' own staff, will spend a large part of their day searching for images and will often have a very clear idea of what they are searching for. At the other extreme may be a user who is asked to find a picture for a company annual report once a year, who has no other involvement with image searching.

Press Association Images employs a large team of account managers and picture researchers to assist all users and advise on search techniques but recognises that some users will not make contact if they are not finding what they want and therefore strives to makes its on–line system as user–friendly as possible.

25.2.2 Types of Search

The system offers users two options for entering their search terms, a simple and an advanced search. Through these search screen users are able to build, if they desire, complex searches to narrow their potential results. Factors such as the date of an image, the photographer's name and the copyright holder can all be used in addition to multiple search terms.

These complex searches are suitable for some of the users and also for some images requirements. A user trying to locate the gas strike picture described earlier knowing the date of the image would be able to locate that specific image very easily by entering 'gas strike' and the specific date.

Analysis of the query logs has been completed by the company and supported by work undertaken by the Department of Information Studies, University of Sheffield (Yang, 2007). From analysing query logs from June 2006 to May 2007, it was found that the top 25 queries were either names, organisations (e,g. sports teams) or broad subject categories (e.g. sports). Analysing zero hits (i.e. when a search returns zero results) showed that the most common reason for this was misspellings of names. The study also found that the majority of queries were very simple, the most common consisting of one or two words.

25.2.3 Challenges

It is clear that though there is the facility to build complex searches, most searches only use simple terms where the number of results returned could be very high. Increased processing speed and a general increase in bandwidth availability does allow users to browse a larger number of images now than they would have been able to do in the past, but with such a large database it is very possible that a user will not find the image they are looking for due to a potentially very high number of results returned. A search on David Beckham, for example, could potentially return in excess of 18,000 pictures. It is possible that a user is just looking for a good recent picture of the footballer which would be returned at the top of search results but they may actually be looking for something more specific and one of the key challenges faced is how to help users find what they are looking for if their starting point is a simple search term.

For such recurring scenarios, an intelligent browsing engine is considered a good solution. Our search engine in its present form does not provide a separate facility for browsing images. To develop such a browsing system is challenging for such a vast image library, as we deal with complex information taxonomy and rich data sets. For example, the number of entities (e.g. people, places, location, organisations, etc.) we are dealing with to supply images can be in millions. An ideal system would allow the annotating of images with descriptive metadata (entities), which could be used for efficient browsing. We have researched in the areas of Semantic Web technologies to address some of these challenges and to develop an intelligent browsing engine.

In the following section we introduce the concept of Semantic Web and outline how we are using semantic technologies to build a browsing engine.

25.3 Semantic Web for Multimedia Applications

25.3.1 Introduction to the Semantic Web

The Web was invented by Tim Berners–Lee among others, a physicist working at the European Organization for Nuclear Research (or CERN). The Semantic Web is seen as an extension of the current World Wide Web (WWW), defined as follows (Berners-Lee et al, 2001):

> "The next generation WWW is a Web in which machines can converse in a meaningful way, rather than a web limited to humans requesting HTML pages."

The fundamental premise of the Semantic Web is to extend the Web's current human–oriented interface to a format that is comprehensible to software programs. Over the years the vision of the Semantic Web has been made realistic by the W3C's[3] standardization process. Based on the concept of autonomous interpretation of machine–understandable metadata, Semantic Web technologies can deliver intelligent management of user–transparent access to an increasingly complex mesh of interrelated information, which makes these technologies especially appealing to organizations with complex information taxonomy and rich data sets such as the Press Association (Moncrieff, 2005), BBC (Kobilarov et al, 2009), Reuters[4] and Yahoo[5]. The practical adoption of the Semantic Web received boost from the emergence of the Linked Data Cloud concept. The Linked Data Cloud refers to the data published on the Web in such a way that it is machine–readable, its meaning is explicitly defined, it is linked to other external data sets, and can in turn be linked to from external data sets (Bizer et al, 2009a). The Linked Data Cloud can be considered as medium for domain experts to come together and share the knowledge about the domains they are expert in by utilizing open Semantic Web standards[6].

25.3.2 Success Stories and Research Areas

The developments in the areas of Linked Data Cloud along with maturing Semantic Web standards such as RDF (Resource description framework), RDFa and SPARQL (Simple Protocol and RDF Query Language) has accelerated the uptake of semantic technologies to address metadata integration, interoperability, search and text–mining problems.

---

[3] `http://www.w3.org/`

[4] `http://www.opencalais.com/`

[5] `http://developer.yahoo.com/searchmonkey/smguide/faq.html`

[6] `http://linkeddata.org/home/`

25.3.2.1 Metadata Integration and Interoperability

The Linked Data Cloud is a distributed architecture where the suppliers of data benefit from each others contributions. However, the pattern is emerging where some of the data sets are more dominant in use and have most number of users within the data sets of cloud and also from outside the cloud. Among these data sets is DBpedia (Bizer et al, 2009b) which is a community effort to extract structured information from Wikipedia and to make this information accessible on the Web. The resulting DBpedia knowledge base currently describes over 2.6 million entities (Bizer et al, 2009b). Similarly, the Freebase[7] data set is generic and is very rich source of entities. The organisations like the British Broadcasting Corporation (BBC) (Kobilarov et al, 2009) and the New York times[8] utilize Linked Data Clouds for enriching their metadata by integrating with the cloud data sets. The developments at Press Association Images is following a similar path as discussed in the next section.

25.3.2.2 Search and Text Mining

The central premise of the Semantic Web is to extend the current World Wide Web that is dominated by free–text search engines such as Google and Yahoo!. The best way to make semantic content searchable for existing search engines is by embedding semantic technologies within existing Web technologies such as XHTML. RDFa[9] is emerging as such a technology that allows adding mark–up to Web pages to make them understandable for machines as well as people. By adding it, browsers, search engines, and other software can understand more of the content of the page, and in so doing offer more services that may end up providing better results for the user. For example, RDFa is adopted by Yahoo!'s search monkey technology[10] (Mika, 2008), Google[11] and other commercial search engines. Within organisations, text–mining systems will come to play a crucial role for building search applications. The majority of established text–mining systems such as the General Architecture for Text Engineering (GATE) (Bontcheva and Cunningham, 2003) and the Thomson Reuter's OpenCalais application [12] increasingly utilize Semantic Web technologies. These complement the role text–mining systems play (extracting important information) by facilitating the management of such information in a highly structured manner.

---

[7] http://blog.freebase.com/2008/10/30/introducing_the_rdf_service/

[8] http://www.beet.tv/2008/06/the-new-york-ti.html

[9] http://www.w3.org/TR/xhtml-rdfa-primer/

[10] http://developer.yahoo.net/blog/archives/2008/09/searchmonkey_support_for_rdfa_enabled.html

[11] http://www.google.com/support/webmasters/bin/answer.py?hl=en&answer=146898

[12] http://www.opencalais.com/

25.3.2.3 Availability of Development Tools

There are many available tools that can assist with the development of Semantic Web applications, ranging from ontology editors to semantic repositories. Semantic repositories are similar to databases in terms of their storage functionalities, but contain additional reasoning capabilities. The work done so far on standardization, especially on the query language SPARQL, has made semantic technologies more accessible for developers and more aligned to existing database technologies Developments in the The Linked Data Cloud have provided a clearer business case for utilizing Semantic Web technologies in a commercial context.

Whilst acknowledging these advances, it is also important to note that there are still certain hindrances for full–scale adoption of Semantic Web technologies. These mainly relate to the level of expertise required in developing Semantic Web applications, the lack of clarity on the use of standards for different types of applications, and as yet being unable to achieve database–like performance using semantic repositories.

25.3.3 The Semantic Web Project at Press Association Images

Press Association Images is looking into the utilization of Semantic Web technologies to improve the image search and browse experience for their customers. The aim is to apply Semantic Web technologies to image retrieval where the semantic annotation of images should allow retrieval engines to make more intelligent decisions about the relevance of an image to a particular user query, especially for complex queries (i.e. those containing entities such as the people names). The use of semantic technologies can significantly improve the computer's understanding of the image objects and their interactions by providing a machine–understandable conceptualization of the various domains that the image represents.

Building Semantic Web browsing systems to achieve this level of intelligence involves development of three components: a rich data set (set of ontologies and knowledge base), a text mining system and a user interface. A screenshot of such an experimental user interface can be seen in Figure 25.3 In the next section we outline our approach in building a PA data set and a text–mining system.

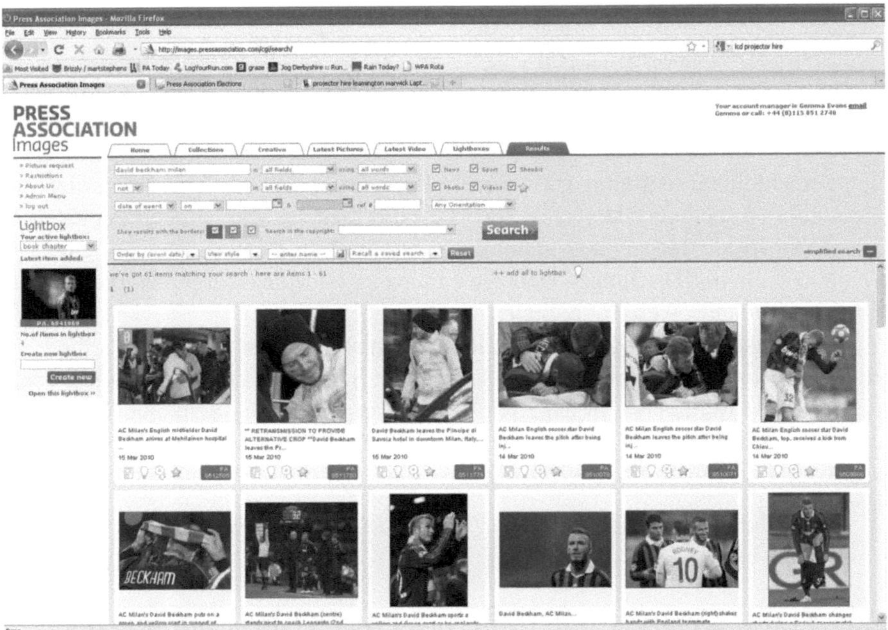

Fig. 25.3: A screenshot of the experimental PA Images search system.

25.4 Utilizing Semantic Web Technologies for Improving User Experience in Image Browsing

25.4.1 PA Data set: Linking to the Linked Data Cloud

1. PA Images Ontology

The first component of the data set is layered OWL (Web Ontology Language) ontologies. These ontologies define entities in news, entertainment and sports domains primarily consisting of people, places, organisations and events. The ontologies also contain hierarchical classification and inter–relationships between these entities such as footballers, sport teams, politicians, stadiums, tournaments, actors, and award events.

2. PA Knowledge Base

The PA Knowledge Base (KB) is the data operating on the PA Images ontology. The number of entities (i.e. people, places, location, organizations, etc.) we are dealing with to supply images can be in millions, hence manual generation of such colossal amounts of data as part of a knowledge base is a daunting task. However, we alleviated the burden of manual compilation of creating such a KB by leveraging the rich amount of structured knowledge publicly available in the Linked Data Cloud, especially DBpedia.

As highlighted in the previous section, the BBC Music beta website and few other websites utilize DBpedia in its original form. As these websites are mainly information providing sites, compared to our browsing application, the main difference in requirements is that our application requires optimal data quality especially in terms of classification and inter–relationships. In terms of data quality, we have found following limitation of the DBpedia knowledge base:

- DBpedia is less formally structured and is governed by the number of ontologies where retrieving a particular class of entity will require joining a number of ontologies. For example, a comprehensive list of footballers can only be retrieved by combining Yago, DBpedia and SKOS ontologies.
- The data quality is inferior (to our expectations) as there are considerable inconsistencies within DBpedia. For example, some of the object properties do not link to other entities and instead link to temporal templates. Another example is the incorrect classification of entities. For example, some of the bands are incorrectly classified as persons.

In addition to the above shortcomings, we have our own view of the world and define them differently in the PA Images ontology. As suggested by the DBpedia authors (Bizer et al, 2009b), an approach to combine the advantages of both worlds is to interlink DBpedia with hand–crafted ontologies, which enables applications to use the formal knowledge from these ontologies together with the instance data from DBpedia.

This brought us to a challenging problem of ontology mapping as the PA Images ontology has a different set of elements compared to the DBpedia ontology. It is also worth noting that there might not be a one–to–one mapping available between the two ontologies; especially the PA Images ontology that contains a smaller number of classes than DBpedia. There are many efforts in ontology mapping research that focuses on the application of automatic ontology mapping (Zhou, 2003; Ding et al, 2005); however, we have preferred to perform the mapping by hand due to the following reasons:

- The accuracy required needs to be close to 100%.
- As mentioned earlier, the coverage of data under DBpedia is richer when using multiple ontologies which require mapping one ontology to many and doing so that the coverage benefits and redundancy is countered.

There is no automatic ontology mapping approach known to us that fulfils the aforementioned criteria. We have successfully used SPARQL CONSTRUCT[13] queries to

[13] http://www.w3.org/TR/rdf-sparql-query/

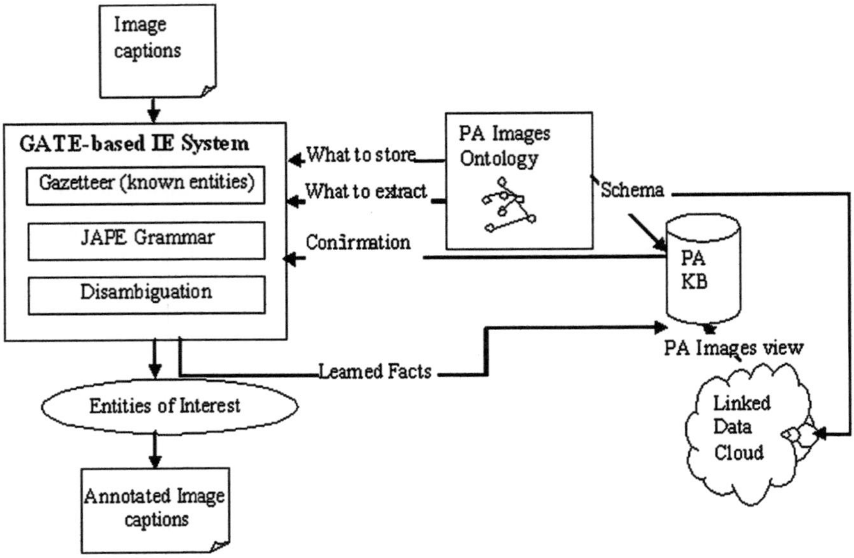

Fig. 25.4: The PA Semantic Annotation System.

achieve an ontology mapping between the PA Images and DBpedia ontologies and to extract the entities from the DBpedia KB and generate a clean, contextualised PA KB.

25.4.2 Information Extraction and Semantic Annotation

It is fundamental to our framework that there is a mechanism to annotate images with descriptive metadata (entities) and to utilize them for efficient browsing. In our framework as illustrated in Figure 25.4, the metadata generation process is handled by a GATE[14]–based text mining system that takes advantage of the rich, domain–specific PA data set. GATE is an integrated development environment for language processing.

The Information Extraction (IE) system utilizes GATE for text–mining and contains three components: gazetteers, the JAPE grammar and a disambiguation module. The gazetteers are lists of known entities that the system makes use of when pre-processing text to perform IE tasks such as Named Entity Recognition (NER). The ontology influences the decision on what information is needed to be stored in these gazetteers. The JAPE grammar rules allow the detection of additional entities while at the same time confirming whether candidate entities detected by the gazetteers are in fact valid. For example a gazetteer containing the location

[14] `http://gate.ac.uk/`

‘Sheffield’ may during initial phases in the language processing pipeline be assigned the named entity type ‘Location’. However, using the JAPE rules, one can remove false hits such as Mr. John Sheffield where Sheffield in this is used as a person name and not placename. Through applying rules such as ‘$< title >. < name >< location >$’, the text string is marked up as a whole as a person. We also encode the context and other heuristics for higher precision and recall using the grammar rules. The disambiguation module deals with any disambiguation generated by previous components. Here, the disambiguation refers to the cases where the same piece of text is either given two class labels (e.g. ‘Liverpool’ as City and Club) or where two or more entity identifiers is assigned to the same piece of text (e.g. ‘Premier League’ as ‘id:Premier_Legue_Darts’ and ‘id: Premier_Legue_Football’). The Knowledge Base plays a crucial role in resolving disambiguation as it has more intelligence embedded in it compared to the IE system itself.

25.5 Conclusions and Future Work

Some of the key challenges for developing a semantic browsing engine are building systems for automatic metadata annotation and utilization of such metadata for improving the end user experience. The systems we outlined in this chapter, namely the PA Data set and the text–mining engine, provide us with an opportunity to build an intelligent browsing system. In the next stages of the project, we are working on building a user interface that exploits this rich data set and semantic annotation process. One of the challenges at this stage is to serve the right balance of metadata through the browsing user interface and not overload the end user with unnecessary information. There are some innovative browsing ideas possible with our framework and we are in the process of evaluating them.

Acknowledgements We would like to thank our project partner Dr Taha Osman from Nottingham Trent University for his work on this project.

References

Berners-Lee T, Hendler J, Lassila O (2001) The semantic web. Scientific American May

Bizer C, Heath T, Berners-Lee T (2009a) Linked data — the story so far. International Journal on Semantic Web and Information Systems 5(3)

Bizer C, Lehmann J, Kobilarov G, Auer S, Becker C, Cyganiak R, Hellmann (2009b) Dbpedia — a crystallization point for the web of data. Journal of Web Semantics: Science, Services and Agents on the World Wide Web 7

Bontcheva K, Cunningham H (2003) The semantic web: A new opportunity and challenge for human language technology. In: Workshop on Human Language Technology for the Semantic Web and Web Services, Florida, USA

Ding Z, Peng Y, Pan R, Yu Y (2005) A bayesian methodology towards automatic ontology mapping. In: Proceedings of the AAAI–05 C&O Workshop on Contexts and Ontologies: Theory, Practice and Applications, Baltimore, MD, USA

Kobilarov G, Scott T, Raimond Y, Oliver S, Sizemore C, Smethurst M, Lee R (2009) Media meets semantic web — how the bbc uses dbpedia and linked data to make connections. In: European Semantic Web Conference, Semantic Web in Use Track, Crete, Greece

Mika P (2008) Anatomy of a searchmonkey. Modalities — the magazine of the semantic web September

Moncrieff C (2005) Living on a Deadline: A History of the Press Association. Virgin Books, Oxford

Yang D (2007) What do people search for in pa photos? Master of Science, University of Sheffield, Sheffield, United Kingdom

Zhou N (2003) A study on automatic ontology mapping of categorical information. In: Proceedings of the 2003 Annual National Conference on Digital Government Research, Boston, MA, USA

Chapter 26
Image Retrieval in a Commercial Setting

Vanessa Murdock, Roelof van Zwol, Lluis Garcia, and Ximena Olivares

Abstract This chapter provides an overview of image retrieval in a commercial setting. It details the types of resources available to commercial systems in conducting image retrieval research, and the challenges in using such resources. In particular the chapter discusses user generated content, click data, and how to evaluate commercial image search systems. It ends with a discussion of the role of benchmark efforts such as ImageCLEF in this type of research.

26.1 Introduction

Image search is becoming increasingly important in commercial search systems with the growth of mobile devices, and an increasing emphasis on visual information. Whether in a Web interface designed for a desktop computer, or a small–screen mobile phone interface, the real–estate is limited for images. Images may be incorporated into an aggregated page as in `http://au.alpha.yahoo.com` (shown in Figure 26.1), or it may be used to enhance Web search results, as in Figure 26.2. For use in Web applications, image retrieval systems must have high precision, so that the two or three images shown are topically appropriate. Since queries to search engines are typically two or three terms, they are often quite vague, and there may be several interpretations of their meaning. For instance a user searching for images of jaguars should be presented with images of animals and cars because we cannot know from the query alone which sense of the term 'jaguar' was intended. Furthermore, even if the user queries unambiguously with 'jaguar animals' the images themselves should be visually distinct. This is especially important with the

Vanessa Murdock, Roelof van Zwol, Lluis Garcia
Yahoo! Research, 177 Diagonal, 08018 Barcelona e-mail: `vmurdock|roelof|lluis@yahoo-inc.com`

Ximena Olivares
Universitat Pompeu Fabra e-mail: `ximena.olivares@upf.edu`

H. Müller et al. (eds.), *ImageCLEF*, The Information Retrieval Series 32,
DOI 10.1007/978-3-642-15181-1_26, © Springer-Verlag Berlin Heidelberg 2010

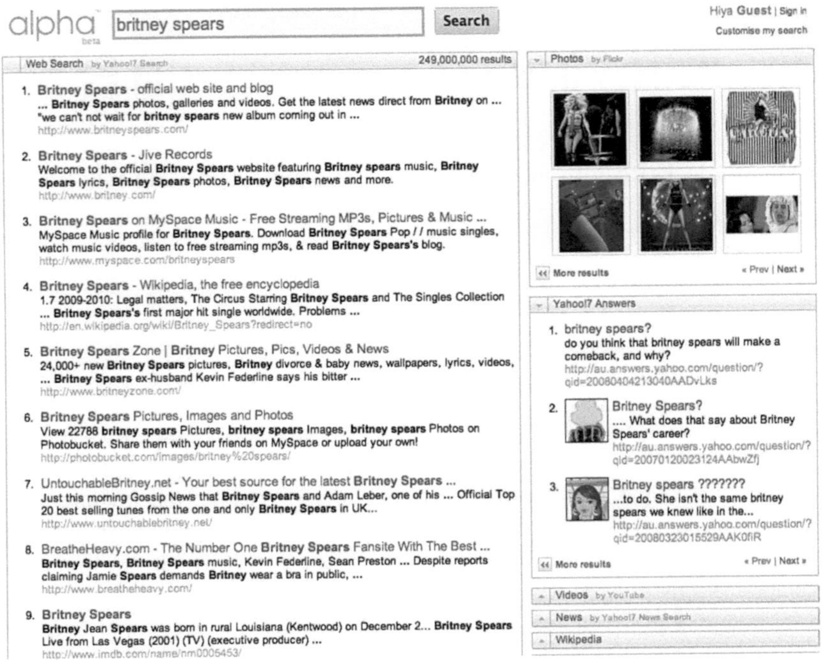

Fig. 26.1: An example of an aggregated search interface, showing results for the query 'Britney Spears'. The aggregated result includes algorithmic search results, images, video, with expandable boxes to show results from multiple verticals (such as Yahoo! Answers).

growing popularity of mobile devices and aggregated search interfaces, where the user is presented with a small number of images.

One limitation of research in image search is that systems rely on a large number of examples in order to generalize sufficiently, but only small amounts of data are available to most researchers. Thanks to Web 2.0 applications, and websites such as Flickr[1], we have vast amounts of data at our disposal — as much as can be crawled in a reasonable amount of time using public APIs (Application Programming Interfaces). This solves the problem of the quantity of data available to us. The data available from these websites is quite rich. It comes with metadata in the form of details about where and when the image was created, textual metadata such as tags, titles, and descriptions attached by the user, and the social context of a user's group memberships, friend relationships, as well as the link structure between documents containing the images.

Access to Web data is not a panacea for image retrieval research, because while we have virtually unlimited quantities of images from the Web, the quality of the data is variable. Images from photo–sharing web sites such as Flickr do not rep-

[1] `http://www.flickr.com/` visited May 2010

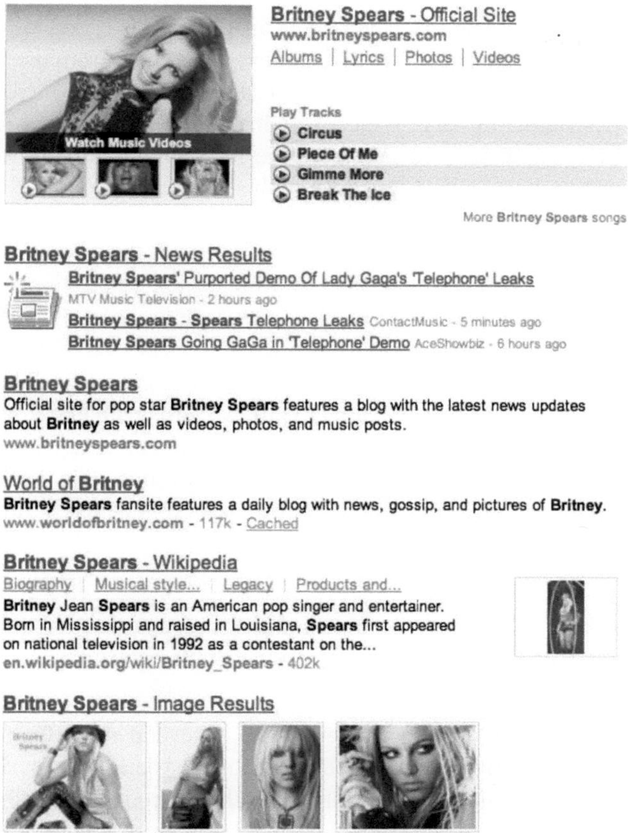

Fig. 26.2: An example of an aggregated result incorporated into the search results on Yahoo.com, resulting from the query 'Britney Spears'. The aggregated result includes images, video, links to song lyrics, news results, official fan websites, and a Wikipedia result. Below this aggregated presentation are the algorithmic search results.

resent images found on the Web as the Flickr photos have user–provided metadata associated with them, whereas Web images are placed in text, may have a caption, but it is unclear how to associate the text surrounding the image with the image itself. In terms of the content of the images, many images uploaded by people to photo–sharing websites are not particularly interesting or informative. They were taken by a person to document their personal experience, and were not intended to be shared with the general public, or to represent iconic concepts. They may be uploaded in bulk, and if they are tagged, the tags may be meaningful only to the photo's owner. Finally, although photo–sharing websites often have images in regular sizes and standard formats, images on the Web in general may be any size or format, complicating automated processing.

An arguably larger issue than the quality of the data is a lack of ground truth for most tasks. Although we can crawl vast amounts of data from the Web, we still must manually assess the data for a specific task. It is this assessment process that prevents us from utilizing large data sets. Efforts such as ImageCLEF are indispensable in this regard. It is far better for the research community to benchmark against a common data set that was built with strict scientific guidelines, than for each research organization to create its own data in an ad hoc manner.

One alternative to manually created data sets is to use the information carried in user clicks. In image search, more than in Web search, when a user clicks on an image it is a clear indication of the image's relevance to the user's search, because the user can actually see the content of the image they click on. Furthermore, when a user clicks on an image, it is often due to the quality of the image. We propose that learning from click data has the potential to kill two birds with one stone: to provide a ranking based on the relevance of the images, and to encourage high quality images to be presented higher in the ranking. Because of this, click data is especially useful for training and evaluating ad hoc image search systems. In this chapter we examine this in more detail.

26.2 Evaluating Large Scale Image Search Systems

Commercial image search engines are evaluated in several ways. One obvious way is to redirect a small portion of the search traffic to the experimental system, and then compare the change in click–through rate. The click–through rate is the percentage of clicks on images for a given query. In general, search engines would like higher click–through rates because this indicates that people are taking the time to interact with the search engine. This is widely accepted as an indication of satisfaction with the search engine. A limitation of this approach is that only experimental systems which have a reasonable chance of success can be evaluated because the search engine is not willing to risk even a small amount of search traffic on a system that might significantly decrease the user experience. Another limitation is the notion that the click–through rate is a surrogate for relevance of the results. Since the user is presented with image thumbnails, it is possible the thumbnail satisfies the search entirely, without the need to click on it. Furthermore, users are likely to click on any thumbnail if the image itself is compelling, even if it is not related to the user's search, such as adult content, or images of popular celebrities.

A second method for evaluating commercial search systems is to sample from the search traffic, and have assessors look at the results for a given set of queries. This type of evaluation provides an assessment of a snapshot of the system, at a given point in time. Since it considers only what was presented to the user, and not what the user clicked on, it gives an independent evaluation of the relevance of the search results. Furthermore, the results can be evaluated to a depth of one hundred or two hundred results, so results are evaluated that the user might never have seen. One limitation of this approach is that the data created by the assessors cannot be used to evaluate other systems. Because the data is not pooled from multiple diverse approaches, any system that is radically different from the existing system is likely to yield poor results, according to the assessment data. The data created by this editorial process is for one–time use only. As such it is extremely expensive to create.

Another method for evaluating a commercial system is to use a small focus group, typically seven to ten people that represent the average user. The users in the focus group interact with the interface, while being observed. The observers note how the interface is used, and the focus group participants have the opportunity to comment on the features of the interface, or the performance of the system. This type of small–scale user study is typically used to assess the human factors in the interface design, rather than as an objective measure of the algorithmic performance of the retrieval system.

In this chapter we discuss the use of click data from the search engine logs both to train and to evaluate image search systems. Click data serves as implicit feedback, and tells us what the user was interested in looking at. As with all methods of evaluating large–scale retrieval systems, it has its pros and cons, which are discussed in more detail below. The main drawback of using click data is that we never know whether the user clicked on an image because it was relevant, or merely because it was eye–catching. For this reason we propose that evaluating on click data be complemented by an evaluation with editorial data. Ideally, the editorial data used would be public data, such as the data provided to ImageCLEF, so that the search engine performance can be compared with state–of–the–art systems.

26.3 Query Logs and Click Data

Though the state–of–the art in image retrieval has progressed significantly over the past few years, large scale retrieval of images on the Web is still a challenge. The textual information associated with an image on a Web page is often sparse and of variable quality. In addition, the extraction of visual features on a large scale requires extensive computing resources and even then it remains unclear how the visual information contributes to the retrieval performance in a domain–unrestricted environment such as the Web. As a consequence Web–based image search engines may not fully incorporate visual content into the ranking strategy. Finally, creating an evaluation set for a Web retrieval system is usually done by manually assessing the ranked lists for a given sample of queries. It is time consuming and costly, and as these sets are extremely small and static compared to the dynamic universe they represent, it is unclear how reliable assessments are made on this type of data. In this chapter we investigate solutions for these limitations. We exploit the vast amount of user–annotated images available from photo sharing sites such as Flickr. We extract light–weight visual features based on the color, texture, and edges in the image. Finally, we take advantage of the enormous amount of click data generated by users of image search engines, both for training and evaluating machine-learned ranking of images.

The success of Flickr and other social media websites has enabled large scale human annotation of media content such as images and video on the Web. Provided with the incentive to make images accessible for friends and family, millions of users upload, share and annotate their collections of personal images and videos through sites like Flickr, and YouTube[2] . User–generated annotations provide useful information about the content and context of an image, necessary for retrieval on the Web.

Image search is unique in that the user searches for images using a query of two or three terms, but the image is deemed relevant based on its visual content. Thus there is a semantic gap between the language used to query (text), and the visual representation of the image itself: the information encapsulated in the visual features cannot be mapped to the textual queries in a straightforward manner. One approach to bridging this gap is to use concept detectors to label regions in the image (Hauptmann et al, 2007; Snoek et al, 2007). Our approach is instead to rely on the machine learned model to bridge this gap by learning from the user clicks how to combine textual and visual information. While the textual features relate the topic of the query to the content and context of an image, the visual features play an important role as an indicator of the quality of an image. Thus visual and textual features work together to ensure that the image is relevant to the query, and of sufficient quality or interest when the user decides to click.

We hypothesize that the click on an image for a given query is a much stronger signal than a click on a snippet in Web search. A user can make a more sophisticated assessment of the relevance of an image based on the image thumbnail, than a user assessing the relevance of a document, given the summary snippet containing one or two lines of text. As the user can often see the entire content of an image by viewing the thumbnail, he may not click on many images, but when he does it can be considered a much more conclusive indicator of the relevance of that image to the query. Thus, while the system can rely on textual similarity between the query and the textual metadata to return relevant images, it can improve the quality of the results by ranking images according to their visual appeal as well. Although visual appeal is subjective, human perception of the attractiveness of photos has been found to be influenced by measurable quantities such as color distribution and coarseness (San Pedro and Siersdorfer, 2009).

We present a learned framework for ranking images that employs click data from image search logs. Click data is generated in virtually limitless amounts by the users of search engines. Previous work in learning to rank Web results (Joachims, 2002; Ciaramita et al, 2008) relies on the list structure of the results to determine which results are relevant and which are not. The assumption is that the user scans the ranked list of results from the top to the bottom until they find a relevant result, which they click. If the user clicks a result at rank three, they are assumed to have rejected the results at ranks one and two. Thus the click at rank three is considered

[2] `http://www.youtube.com/`, visited January 2010

to be a relative preference for that result; it is considered to be *more* relevant than the first two results. This structure of the clicked results is key to the success of learning to rank Web search results.

In this chapter we demonstrate that the block structures developed for list–based representations of the search results can be applied to image results, which have a grid–based presentation. This is a significant difference because in a list–based presentation, the second result always appears below the first result, and is followed by the third result. In image search, the placement of the image on the page is dependent on the browser dimensions, which are established by the user each time they open their browser. There is no guarantee that the third image will be to the right of the second image, and to the left of the fourth image. Furthermore, we do not know whether users scan the page from left to right, or top to bottom, and to what degree they use peripheral vision to reject non-relevant images. We demonstrate that in spite of this, we can predict clicked results with a high degree of accuracy.

We show that a machine–learned model based on either textual or visual features outperforms the standard retrieval baseline, and combining text and visual features significantly improves retrieval over either feature set alone. We propose this is because the textual features relate the content and context of the image to the query, while the visual features represent the aspects of the image that impelled the user to click, such as interestingness.

We investigate whether it is a small subset of features that account for the performance of the classifier. We find that instead, the visual features work in combination to discriminate between the clicked and non–clicked class, and that no single visual feature or class of features accounts for the performance of the classifier. By contrast, the textual features are less democratic, and one or two textual features carry most of the discriminative power.

In this chapter, we propose an efficient framework for ranking images based on the multilayer perceptron, which allows modeling complex nonlinear patterns in the data through hidden layers. The perceptron algorithm is fast and scalable, and allows for the incorporation of any number of features based on any aspect of the data, in our case textual annotations and visual content. We label our data as described below, using the click data from the search engine query logs. Click data is automatically collected in the search logs when users interact with the search engine, thus imposing no further burden on the system. Since the clicks are used as labels, rather than in the feature computation, it is assumed they are not known at test time, thus it is straightforward to produce a ranking for results for a new query. At test time, candidate images can be labeled by the classifier as clicked or non–clicked, and the images predicted to be clicked presented at the top of the ranked list.

26.4 Background Information on Image Search

Machine learning techniques have been used to solve a number of tasks related to image retrieval. For example, Support Vector Machines (SVMs) have been used in a variety of image classification tasks (e.g. Chapelle et al, 1999; Harchaoui and Bach, 2007). Although determining whether the subject matter of the image relates to the topic of the query underlies this process, we do not seek to group images by their subject matter. Rather, our task is to provide a ranking of images most relevant for a user query. As mentioned earlier, San Pedro and Siersdorfer (2009) employ a classifier and low–level visual features as well as textual features that indicate the attractiveness of an image to predict the appeal of an image. They selected as positive examples all photos with at least two favorite assignments, from a large crawl of photos from Flickr, and similarly sized samples randomly drawn from photos that had not been favorited. They find significant improvements from the combination of textual and visual features. We are not attempting to identify attractive images. Rather, we believe that images are clicked because they embody relevance, attractiveness, interestingness, and other indefinable qualities. Because of this our textual features reflect the similarity between the query and the textual metadata, as opposed to the attractiveness of the image.

Central to the problem of image ranking is the problem of relating textual queries to visual image content. Tong et al (2006) propose a propagation method based on a manifold learning algorithm. A small portion of the images are labeled with keywords, and then the labels are propagated to the rest of the collection. Each image is assigned a relevance score for a given keyword by constructing a matrix where each row represents an image, each column represents a keyword, and each cell contains a relevance score. The intuition is to associate textual information with visual content. The experiments were conducted over a collection of 5,000 images extracted from the COREL data set.[3]

In content–based image retrieval the objective is to incorporate the visual characteristics of an image into the search process. Using the Query By Image Content (QBIC) search paradigm similar images are retrieved for a given sample image by extracting visual features from all the images in the collection. The disadvantage of this approach is that the user begins the query process with a sample image, which is not consistent with the current Web search paradigm. Alternatively, high level concepts are derived for the low level features that are extracted from the image content. The problem with this approach is often referred to as the semantic gap problem (Hauptmann et al, 2007), where for each concept a special concept detector is needed to translate the user query into low–level image features. This approach is less suitable for widespread application on the Internet, because no domain restrictions exist on the Web.

Tong and Chang (2001) elicit explicit relevance feedback from users, and then employ active learning with a support vector machine, using features derived from

[3] `http://archive.ics.uci.edu/datasets/Corel+Image+Features` visited January 2010

the color and texture of an image to improve retrieval results. Our work uses similar features based on the color of an image, among others. However they use explicit feedback elicited from the user, and our approach uses implicit feedback in the form of clicks.

Cheng et al (2006) proposed a scalable relevance feedback mechanism using click data for Web image retrieval. Using Rocchio feedback, they add the vector of features representing the query to an 'optimal query,' which is the mean of the vectors of the clicked images. They rank images according to the cosine similarity between the new query vector and the feature vectors representing the images in the collection. Their textual features are based on $tf.idf$ scores of the query and metadata associated with each image. Their visual features are a combination of three color features (color moment, auto–correlogram, and color texture moment). They evaluated their system in a simulated setting, using ten queries, retrieving from a collection of three million images crawled from photo sharing websites.

26.5 Multilayer Perceptron

Learning a ranking from click data was first proposed by Joachims (2002) for document retrieval. In the following sections we adapt his ranking mechanism to image retrieval. Joachims proposed that user clicks are an indicator of relative relevance. That is, a click at rank j indicates that the document at rank j is more relevant than unclicked documents preceding it in the ranked list. Joachims work is the first in a series investigating using click data for learning ranking functions. Elsas et al (2008) extend this idea by learning ranking functions from search results with a committee perceptron using the LETOR data set (Liu et al, 2007).

Ciaramita et al (2008) successfully adapted Joachims' unbiasing model to several learning frameworks: binary classification, ranking, and nonlinear regression and showed positive results on a sponsored search task using commercial query log data. They demonstrate that a multilayer perceptron outperforms both the linear perceptron and a ranking perceptron. In their work the features of an advertisement (ad)–query pair are based entirely on the textual representations of the ad. Their work differs from ours in that the search engine is most interested in generating clicks on ads, thus learning to predict clicks is key to the task of ranking ads. In the case of image search, the search engine would like to encourage people to use the search engine, and thus attempts to maximize the relevance of the search results. Our work is similar to theirs in that we adopt their framework both for training and evaluation, and our images are represented by text in much the same way that ads are. Whereas ads are represented by keywords, titles and a short description, images are represented by tags, titles and a short description. Keywords and tags differ in character, but are similar in their brevity and conciseness. Our work extends their work primarily by showing how this unbiasing framework, which is based on the bias introduced by a linear presentation of the results, can be applied to nonlinear presentations of the results, such as is the case in image search. In addition, we extend their work by considering the visual representation of the data rather than just its textual representation.

In learning from click data we avail ourselves of massive amounts of continually changing data. In principle, several machine learning algorithms could be used. In practice, since we are working with Internet–scale data we require an algorithm that is efficient and scalable. We choose the perceptron because it is efficient and has an on–line formulation so the training data need not be stored in memory all at once. Although the perceptron has been criticized for being limited to modeling linear relationships in the data, we employ a multilayer perceptron with a sigmoidal hidden layer, which allows the modeling of arbitrarily complex patterns (Duda et al, 2000). This can be an important feature in our task where input signals come from different modes (textual and visual) whose combination via latent units can provide a powerful representation. In our data the visual content is represented in a high dimensional feature space.

Because our data is multi–modal, we use multilayer regression for its flexibility in modeling nonlinear relationships. In our setting, an *example* is a vector of features extracted from an image–query pair (a,q), $\mathbf{x} \in R^d$. Each *example* $\mathbf{x}_i$ is labeled with a response value $y_i \in \{-1,+1\}$, where $+1$ indicates a clicked image and -1 indicates a non–clicked image. The learning task is to find a set of weights, $\alpha \in R^d$ which are used to assign a score $F(\mathbf{x}_i;\alpha)$ to examples such that $F(\mathbf{x}_i;\alpha)$ is close to the actual value y_i.

Multilayer networks with sigmoidal nonlinear layers can generate arbitrarily complex contiguous decision boundaries (Bishop, 1995), and have been used successfully in several tasks, including document ranking (Burges et al, 2005). The multilayer perceptron is a fully connected three–layer network with the following structure An input layer of d units, $x_1, x_2, .., x_d$, with $x_0 = 1$ the bias unit; a hidden layer of n_H units, $\mathbf{w} = w_1, w_2, .., w_{n_H}$, plus the bias weight $w_0 = 1$; a one unit y output layer; a weight vector $\alpha^2 \in R^{n_H}$ plus bias unit α_0^2 and finally a weight matrix: $\alpha^1 \in R^{d \times n_H}$ plus bias vector $\alpha_0^1 \in R^{nH}$.

The score $S_{mlp}(\mathbf{x})$ of an *example* $\mathbf{x}$ is computed with a feed-forward pass:

$$S_{mlp}(\mathbf{x}) = y = \sum_{j=1}^{n_H} \alpha_j^2 w_j + \alpha_0^2 = \langle \alpha^2, \mathbf{w} \rangle \tag{26.1}$$

where $w_j = f(net_j)$, and

$$net_j = \sum_{i=1}^{d} \alpha_{ij}^1 x_i + \alpha_0^1 = \langle \alpha_j^1, \mathbf{x} \rangle \tag{26.2}$$

The activation function $f(.)$ of the hidden unit is a sigmoid:

$$f(net) = \frac{1}{1 + \exp^{-a\ net}}. \tag{26.3}$$

Supervised training begins with an untrained network whose parameters are initialized at random. Training is carried out with backpropagation (Rumelhart et al, 1986). An input example $\mathbf{x}_i$ is selected, its score computed with a feed-forward pass and compared to the true value y_i. Then the parameters are adjusted to bring the score closer to the actual value of the input example. The error $\mathbf{E}$ on an example $\mathbf{x}_i$ is the squared difference between the guessed score $S_{mlp}(\mathbf{x}_i)$ and the actual value y_i of $\mathbf{x}_i$, or for brevity $(y_i - s_i)$, $\mathbf{E} = \frac{1}{2}(y_i - s_i)^2$. After each iteration t, α is updated component-wise to α^{t+1} by taking a step in weight space which lowers the error function:

$$\begin{aligned} \alpha^{t+1} &= \alpha^t + \triangle\alpha^t \\ &= \alpha^t + \eta \frac{\partial \mathbf{E}}{\partial \alpha^t} \end{aligned} \tag{26.4}$$

where η is the *learning rate*, which affects the magnitude of the changes in weight space. The weight update for the hidden–to–output weights is:

$$\triangle\alpha_i^2 = \eta\,\delta w_i \tag{26.5}$$

where $\delta = (y_i - z_i)$. The learning rule for the input–to–hidden weights is:

$$\triangle\alpha_{ij}^1 = \eta x_j f'(net_j)\alpha_{ij}^1\delta. \tag{26.6}$$

where f' is the derivative of the nonlinear activation function.

26.6 Click Data

Our data consists of approximately 3.5 million distinct public images from Flickr, and approximately 600,000 unique queries, with their search results collected from the query logs of the Yahoo! image search engine[4]. The actual number of queries is more than 600,000 as many queries will be issued more than once, and the image results presented to the user may be different each time the query is issued, as well as different users issuing the same query may click on different images.

We filter the search results by eliminating images that are not publicly available from Flickr. We construct blocks for each query such that each block contains one clicked image as the positive example, and all unclicked images displayed higher in the ranking as negative examples. Blocks with no negative examples are discarded, for example if the user clicked on the first k results.

Thus, for example, if a user clicked on results at ranks one, three and five in response to a query, two blocks are constructed. In the first block we have a positive example from the image at rank three, and one negative example from the image at rank two. The click at rank one is discarded because we cannot say that the user

[4] `http://images.search.yahoo.com/` visited January 2010

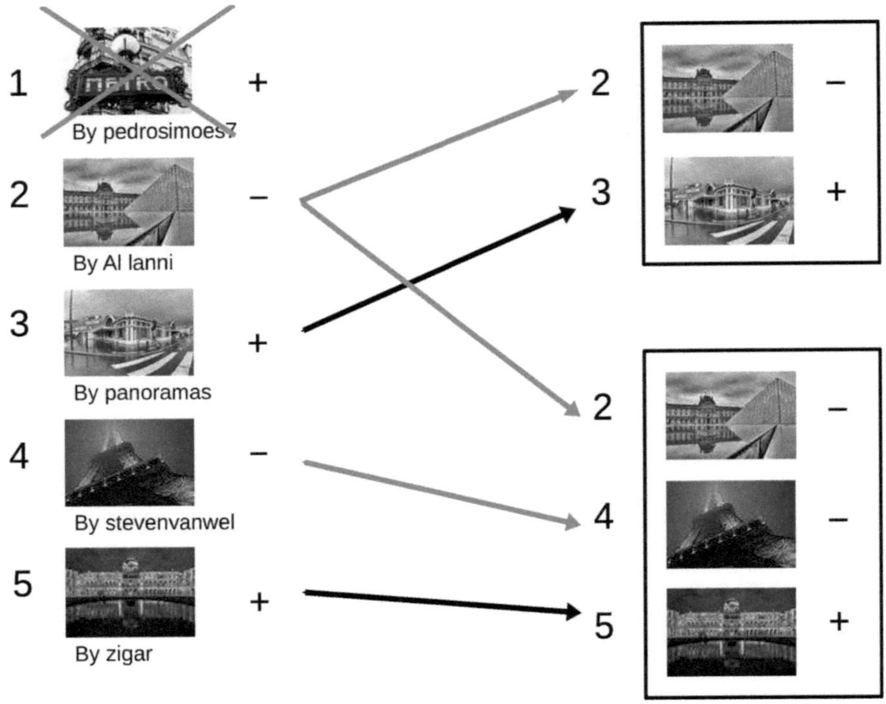

Fig. 26.3: Blocks are constructed from the ranked list of clicked results in response to the query 'Paris' as follows: Clicks at rank one are discarded. For each clicked image, a block consists of the clicked image and all non–clicked images ranked higher. In each block, clicked images are labeled as positive examples, and non–clicked images are labeled as negative examples. All photos shown in this figure are posted on Flickr.

preferred the image at rank one over some other image they saw before. In the second block we have a positive example from the image at rank five, and two negative examples from the images at ranks four and two. This is shown in Figure 26.3. We trained on 1,167,000 blocks, and tested on approximately 250,000 blocks. Parameters were tuned on a held–out set, to maximize the prediction accuracy, using only the textual features. We tuned the number of hidden layers, and the number of training iterations. The optimal performance required fewer than ten training iterations, and one hidden layer.

Each image is represented by two types of information. We collect the textual metadata associated with the image, that is the title, the tags and the description. The tag sets are entered by the owner of the image, as well as by other people who have viewed the image, although in our data the vast majority of tags are entered only by the owner. Tag sets are composed almost entirely of content terms, although some of the terms may not be helpful for the purpose of retrieval. For example, owners of an image might tag the image with the name of the camera, or the length of exposure. In addition the same tag set may be used to tag multiple images uploaded in bulk. In Flickr the tags are lower–cased, spaces are removed, and commas are converted to spaces, thus tags may consist of terms that have been concatenated. As shown in Figure 26.4, the title might contain terms that are not useful for the purposes of text–based retrieval, such as the date the image was taken. The descriptions are often written in natural language, and may be a sentence or two in length. Frequently, the tags, title and description will be written in more than one language, as shown in Figure 26.4.

Each individual picture was downloaded and its visual features computed. Using Hadoop and the Map–Reduce model, textual and visual feature computation was distributed over a grid, allowing fast computation of the features over large amounts of data. Not all images are associated with textual metadata, or the metadata may be incomplete. Also, for some images we were able to collect the metadata, but the image itself was no longer available for download. So for a certain portion of the data, the image was represented by either the text or the visual features.

26.7 Data Representation

We computed twelve textual features over the user–generated content associated with the images, and seven features of the visual content of the images. In addition, a final binary feature, set to one for every example in the data, was intended to reduce the bias in the data. The feature set is normalized by row and by column as described below.

26.7.1 Textual Features

Each image had a set of tags, a title and a description associated with it. We computed text features over each field individually, and then created a fourth 'field' by concatenating the other three. For each of the four, we computed the cosine similarity between the query and the image where the terms were weighted by their tf.idf score. In addition, for each of the four fields we computed the maximum tf.idf score of a query term in the image, and the average tf.idf score of the term in the image.

The tf.idf term weights are given by:

$$w_{q_i,d_j} = \frac{tf_{q_i,d_j}}{\max tf_q} \times \log \frac{N}{n_{q_i}} \tag{26.7}$$

where tf_{q_i,d_j} is the term frequency of the query term q_i in the text associated with the image d_j, $\max tf_q$ is the term frequency of the most frequent query term in the

Paris - Gare du Pont Cardinet - 28-07-2007 - 9h03

Jolie gare des années 1920. Influence angkorienne et ressemblance frappante avec le palais du roi Narai, à Lopburi...Nice, small Paris railway station from the 1920's. Through the Angkorian influence, strong similarities with the King Narai's Palace, in Lopburi....

Tags:
gare
pont
cardinet
paris
batignolles
RER
1920
1925
Angkor
ankorien
Angkorian
Lopburi
Narai
pluie
rain
reflet
reflection
zebra
passage
piétons

Fig. 26.4: An example of the data. The title for the image appears above the image, the description is below, and a sample of the tags appear to the right of the image. The title and description are entered by the person at the time of uploading the image. The tags may be entered by the owner of the image, or by other Flickr users. This image was taken by panoramas and appears with its metadata on the Flickr website.

image text, N is the number of images in the collection, and n_i is the number of images whose text contains term q_i.

Each query and each image are represented as a vector of terms, where each element of the vector is the tf.idf weight of the term. The cosine similarity is the cosine of the angle between the two vectors, normalized to be the unit vector:

$$sim(q_i, d_j) = \frac{\sum_{v=1}^{t} w_{v,d_j} \times w_{v,q_i}}{\sqrt{\sum_{v=1}^{t} w_{v,q_i}^2} \times \sqrt{\sum_{v=1}^{t} w_{v,d_j}^2}} \tag{26.8}$$

26.7.2 Visual Features

For the visual representation of the data we implemented seven global features. These seven global features were chosen because they are standard image descriptors that represent the texture, color, and edges in an image. Furthermore, they are relatively efficient to compute, compared to features based on local regions within the image. Unless otherwise noted, the features correspond to descriptors that are included in MPEG–7. A detailed specification of these descriptors can be found in Salembier and Sikora (2002).

Color histogram (CH). A color histogram describes the color distribution in an image. We discretize the RGB color space into 64 color bins. The color histogram is computed by assigning to each bin the number of pixels that belong to that color range.

Color Autocorrelogram (AC). This descriptor includes the spatial correlation of colors. It is based on the work of Huang et al (1997), and is not included in MPEG–7. The image color is quantized into 64 colors. The correlogram is created from a table indexed by color pairs (i, j), where the kth entry for pair (i, j) specifies the probability of finding a pixel of color j at distance k from a pixel of color i in the image. To reduce the space and improve efficiency, we use the color autocorrelogram, which only captures the spatial correlation between identical colors in the image.

Color layout (CL). The color layout descriptor is a resolution invariant compact descriptor of colors, which is used for high–speed image retrieval (Salembier and Sikora, 2002). This descriptor captures the spatial distribution of the representative colors in an image. The image is divided into 64 blocks. For each block, a representative color is obtained using the average of the pixel colors. Every color component ($YCbCr$) is transformed by an 8 x 8 *Discrete Cosine Transformation* (DCT), obtaining a set of 64 coefficients, which are zigzag-scanned and the first coefficients are nonlinearly quantized.

Scalable color (SC). This descriptor can be interpreted as a Haar–transform applied to a color histogram in the HSV color space (Salembier and Sikora, 2002). The first step is to extract a 256–bin color histogram, normalize it, and nonlinearly map to a 4–bit integer representation. To obtain a smaller descriptor that permits a more scalable representation, the Haar transform is applied across the histogram bins.

CEDD. The color and edge directivity descriptor (CEDD) incorporates both color and texture features in a histogram (Chatzichristofis and Boutalis, 2008). It is limited to 54 bytes per image, making it suitable for large image databases. The image is split into a pre–defined number of blocks, and a color histogram is computed over the HSV color space. For each of the blocks we obtain a 24–bin histogram by applying several rules. A set of five filters is used to extract the texture information associated with the edges that are encountered in the image, and are classified into vertical, horizontal, 45–degree diagonal, 135–degree diagonal, and non–directional edges.

Edge histogram (EH). The edge histogram describes the local edge distribution of the image (Salembier and Sikora, 2002). The image is divided into a 4 x 4 grid. In each of the grid cells the edges are detected and grouped into five categories: vertical, horizontal, 45-degree diagonal, 135-degree diagonal, and non–directional edges. As a result of this processing, we obtain a descriptor with 80 (16×5) coefficients.

Tamura. Tamura et al (1978) identified properties of images that play an important role in describing textures based on human visual perception. In their work they defined six textural features: coarseness, contrast, directionality, line similarity, regularity and roughness. In our work we built a texture histogram using three Tamura features: coarseness, contrast, and directionality.

26.7.2.1 Normalization

The feature vectors were normalized by column and then by row. The mean and the standard deviation were computed for each column, except for the column representing the bias feature, and applied to each element of the matrix using the standard score:

$$SSFV(i,j) = \frac{FV(i,j) - \mu_j}{\sigma_j} \tag{26.9}$$

Where $FV(i,j)$ is the feature value in the row i and column j, μ_j is the mean of the feature values on the column j and σ_j is the standard deviation of the column j. Rows were normalized with the L1 norm:

$$NFV(i,j) = \frac{SSFV(i,j)}{||SSFV(i,j)||} \tag{26.10}$$

26.8 Evaluation and Results

We evaluate the prediction of the clicked event for each block. Each block contains exactly one clicked event. For this reason metrics that give a sense of the overall quality of the ranked list, such as Mean Average Precision (MAP), Precision at k, and Normalized Discounted Cumulative Gain (NDCG), are not meaningful. For example, if we have ten images in a block, precision at rank ten will always be 0.1. As an alternative we could aggregate the block data per query, and evaluate the results that were returned for a given query, independent of users or session. This is also not possible because the results shown to one user, in a given session, might not be the same set of results shown to another user, or even to the same user in a different session. Because of this, we cannot say that an image clicked at rank three in one session was the same image shown at rank three in another session. The images shown to the user in response to the same query might have been completely different, and elicited a completely different response. Thus when evaluating a system using click data, the block structure must be preserved. In practice, to produce a ranked list of images, we compute the features over the query and the candidate images, and use the trained model to produce a prediction for each image. Since the features are computed over a query–image pair, and do not depend on the clicks, the model will produce a prediction for each pair independent of the other pairs. Producing a ranking is then simply a matter of presenting the predicted-clicked images first in the ranked list.

The learning algorithm outputs a score. We rank the images in each block according to their score. We evaluate how well the system predicted the clicked event in a block, using metrics that indicate the rank of the clicked event. Accuracy measures the frequency with which the system predicted the clicked event at the top of the ranked list. Mean Reciprocal Rank (MRR) measures the rank of the clicked event, and is calculated thus:

$$MRR = \frac{1}{N}\sum \frac{1}{rank_i} \qquad (25.11)$$

where there are N blocks and $rank_i$ is the rank of the clicked event in each block.

The vector space model provides a retrieval baseline, with cosine similarity and $tf.idf$ term weights, where the image is represented by the concatenation of its textual annotations. We rank the images for a query within a block by the cosine similarity between the vector of term weights representing the query, and the vector of term weights representing the image.

For the learned baseline, the perceptron was trained with the cosine similarity feature over all fields concatenated, plus the bias feature, with the rows and columns normalized as described above. We would expect the learned baseline to be comparable to the retrieval baseline because both systems are acting on the same information, with the exception of the bias feature, which serves as a prior on the data. We report both baselines for completeness. Table 26.1 shows the results of the retrieval baseline, and the learned baseline.

Incorporating textual features — albeit simple ones — allows us to weight the information carried in the tags, the title, and the description differently. This is important because each field differs substantially in character. The results shown in Table 26.1 seem to confirm our supposition that the ranking benefits from learning different weights for each of the metadata fields.

Using visual features to rank images seems intuitive because we determine whether a photo is relevant based on the visual content of the photo itself. We are unconcerned with the metadata at the moment the image is presented in the ranked list, because it is not visible to us until the image is clicked. The visual features provide an indication of the content of the photo, and our intuition is that photos clicked in response to similar queries would have similar visual characteristics. The results for ranking solely on the visual characteristics of the data are shown in Table 26.1. The MRR results for both the textual features and the visual features are statistically significantly better than for the baseline results, at the $p < 0.001$ level, using a t–test.[5]

[5] The results for accuracy were not tested for significance because accuracy is a binary measure.

Table 26.1: The results for predicting the clicked event in a block. The results indicated with a star are statistically significant compared to the baselines. The results indicated with a dagger are statistically significant compared to the model trained with textual features. All results are significant at the $p < 0.001$ level, using a t–test.

	Accuracy	MRR
Retrieval baseline	0.4198	0.6186
Learned baseline	0.4073	0.6104
Text features	0.5484	0.7034⋆
Visual features	0.5805	0.7233⋆†
Text + Visual features	0.7512	0.8365⋆†

Table 26.2: The ten most discriminative visual features and textual features, ranked by weights produced by models trained on only visual (textual) features.

Rank	Visual Feature	Text Feature
1	CEDD_144	Tags_sim
2	Tamura_3	all_sim
3	CH_64	title_sim
4	EH_19	all_ave_tfidf
5	CEDD_79	all_max_tfidf
6	SC_12	tags_max_tfidf
7	EH_50	title_ave_tfidf
8	SC_35	desc_max_tfidf
9	CEDD_78	title_max_tfidf
10	AC_80	desc_sim

The textual features and the visual features cover completely different aspects of the images. As both features perform well on their own, we would expect the performance of both categories of feature in combination to outperform either category in isolation. The results in Table 26.1 confirm this intuition. The results for MRR for the text and visual features combined are statistically significant at the $p < 0.001$ level, using a t–test, compared to the results for either the textual features or the visual features alone.

26.8.1 Analysis of Features

We would like to know if a subset of the visual features accounts for the results. For example, we can imagine that people find pictures of other people interesting and click on faces even if they are not relevant to the query. To investigate this we examine the weight vector produced by the perceptron (a^1_{ij} in Equation 26.2). In our models, the feature weights range from approximately -2 to 2. Features closer to zero carry less discriminative information in the model than features further from zero. Figure 26.5 shows the distribution of the absolute values of features for two models: one trained on only the visual features, the other trained on all features.

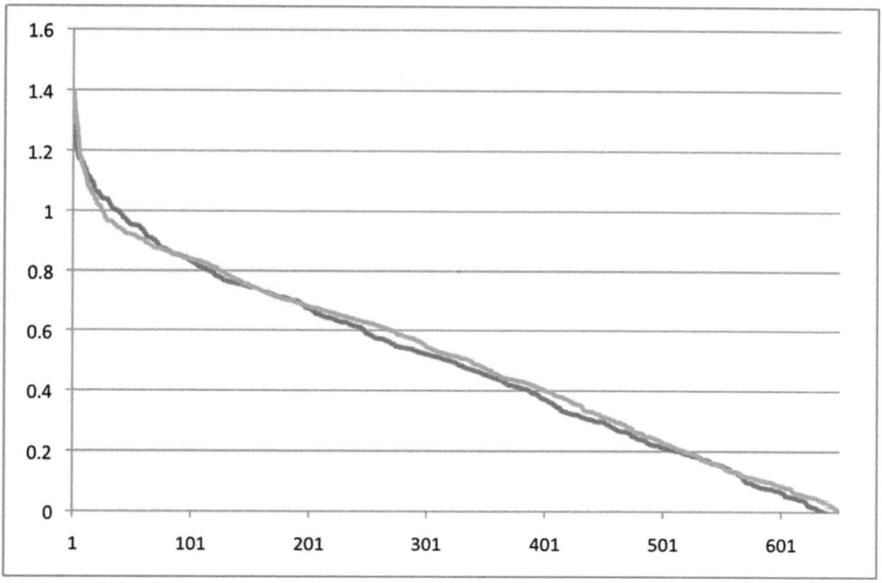

Fig. 26.5: The distribution of feature weights as given in the model trained on visual features, and on all features.

Table 26.2 shows the ten most discriminative visual features, of which the top five are omitted from Figure 26.5 to make it easier to view. We see that no single feature accounts for the discriminative power of the model. The distribution is more–or–less democratic, even with the top features included.

It would be convenient if we could rely on a single class of features, say the color histogram features or the Tamura features, to predict the clicked images. Unfortunately, this did not prove to be the case. We see from Figure 26.6 that the classes of visual features are more or less evenly distributed between highly discriminative features, and features with weights closer to zero. From this we conclude that the features work in combination to determine which images will be clicked.

The textual features are more straightforward. Table 26.2 shows the top ten textual features, ranked by their weights in the model trained only on textual features. The results are unsurprising: once we have computed the textual similarity between the query and all of the metadata, we get little benefit from adding the fields individually. We believe the similarity between the query and the tags to be particularly useful because more images are associated with tags than with the other textual fields, and tags are particularly succinct. They lack stopwords, and often directly indicate the content of the image.

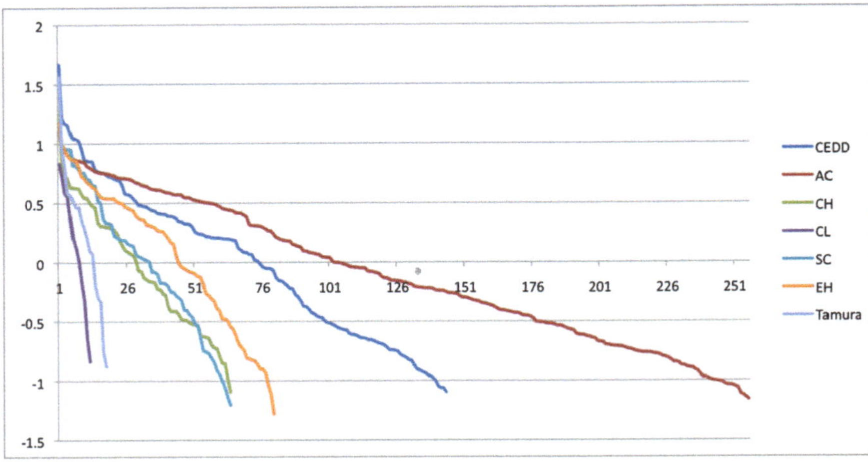

Fig. 26.6: The distribution of feature weights for each class of visual feature.

26.9 Discussion of Results

When a user queries for an image, and is presented with a grid of thumbnails, they often find what they were looking for without the need to click on an image. This is in contrast with traditional document retrieval on the Web, where the user is presented with a list of snippets which are surrogates for the document, and are much less likely to contain the information the user was looking for. For this reason we consider a click on an image a much stronger indicator of the image's relevance or interestingness.

If we take the click to be an indicator of the relevance of the image, we can reduce the reliance on editorial data for developing and evaluating image rankings. Click data is available on a large scale. It is already collected by the system and thus imposes no additional burden to the user or the search engine. In terms of creating an evaluation set, large volumes of queries can be used for training and evaluation, making the sample of data more representative of the population. If the system relies on editorially created data sets, the system will depend on labeling data with human assessors. There is a limit to the number of queries that can be assessed, and the depth in the rankings that can be labeled. Furthermore, it is not clear how to sample from the query stream to create a data set that represents what people are looking for. Click data suffers from none of these restrictions. Finally, what people look for in images, and on the Web in general, changes by season, holidays, current events, movie releases, and so forth. To reflect this, editorial assessments have to be made more often than is practical, while large–scale click data can be sampled at any time.

Unlike Web search, the results of image search are not presented in a ranked list. Therefore, the block construction as used for the research presented in this chapter might not be optimal, as we cannot safely assume that the user favored the clicked image over other images presented higher in the ranked list. It is more likely they favored the clicked image over the surrounding images. However, the layout of the images in the browser is dynamic, and the browser may be resized by the user at any time during the session. We demonstrate that we can predict the clicks based on the ranking of images, without considering the position of the clicked image in relation to the unclicked images.

One of the main findings of this work relates to effective deployment of (low–level) visual features for large scale image retrieval. We have shown how visual features in combination with click data can be deployed effectively by a multilayer perceptron and achieve statistically significant improvement over the machine–learned approach based on textual features. The combination of textual and visual information provides an additional boost in performance. A natural explanation for this is that the different features cover unrelated aspects of the image. Furthermore, no single subset of visual features accounts for the performance of the classifier. Bringing these features together makes the results both textually and visually relevant.

26.10 Looking Ahead

Large–scale image retrieval on the Web poses the challenge of finding relevant images, given a short keyword–based query. The textual information associated with the image is currently the primary source for retrieval. However, when judging images for their relevance to a given query, the assessment is based on the visual characteristics of the image, rather than the text accompanying the image.

We demonstrate how to apply the block structure developed for list–based results presentation to a grid–based image search presentation. Although the assumptions about the bias due to the results presentation in Web search do not hold for image search, the resulting block structure can still be used to accurately predict clicked images. Therefore it is not necessary to know the layout of the image results in order to predict the clicked event.

Furthermore, in this work, we show that the (global) visual features derived from the image content outperform text–based search. This provides evidence for the notion that users decide to click on an image based on the visual information depicted in the thumbnail. We can combine the textual and visual content in a principled and efficient way using a multilayer perceptron. In practice, it is straightforward to use the model to produce a ranking of images, because the features depend on textual and visual properties of the image which are known, and not on its click history, which is unknown at the time of ranking. The perceptron is optimized to be efficient and scalable. It comes with an on–line version, such that training data need not be stored and the training can be updated in a dynamic way.

There are several limitations of this type of approach, however. One is that while we successfully predict what a user will click on, we know nothing of the relevance of the image. It is possible that people click on an image out of curiosity or because the image itself is compelling. Furthermore, since the user will only click on images they see, and they rarely go beyond the first page of results, the system becomes somewhat incestuous, presenting images at the top of the ranking that were presented at the top of the ranking previously.

Another issue with commercial search engines is that they are not replicable. From one month to the next, or even one day to the next, the data may change, and the results presented to two users issuing the same query may be drastically different. We hope that in evaluating on extremely large samples we can mitigate this problem, so that at least the conclusions we draw from experiments on this data are applicable to a different sample of the data. A further issue is that commercial systems are not replicable by people outside of the company, for licensing and legal reasons. Most research organizations do not have access to query logs and click data, thus research in a commercial setting is not directly comparable to research in a non–commercial setting, even when the task setting is similar. Because of this it is difficult, if not impossible, for any result to be independently verified. Since the system cannot be replicated, performance improvements cannot be set as the new standard, and the state–of–the–art cannot advance. However this can be addressed by the existence of public benchmarks which allow us to assess the relative performance of two systems, even when we cannot know the inner workings of the systems themselves. The challenge for ImageCLEF is to design tasks that allow us to benchmark commercial search engines, so that the science produced within these organizations can be exposed, even when the systems themselves cannot be.

Acknowledgements The following person made significant contributions to the work presented in this chapter: Massimiliano Ciaramita wrote the program code for the multilayer perceptron. We would like to thank him for his expertise and hard work in implementing the system described in this chapter.

References

Bishop C (1995) Neural networks for pattern recognition. Oxford University Press, Oxford

Burges C, Shaked T, Renshaw E, Lazier A, Deeds M, Hamilton N, Hullender G (2005) Learning to rank using gradient descent. In: Proceedings of the 22nd International Conference on Machine Learning (ICML), pp 89–96

Chapelle O, Haffner P, Vapnik V (1999) SVMs for histogram–based image classification. IEEE Transactions on Neural Networks 10(5)

Chatzichristofis SA, Boutalis YS (2008) CEDD: Color and edge directivity descriptor: A compact descriptor for image indexing and retrieval. In: Proceedings of the 6th International Conference on Computer Vision Systems, pp 312–322

Cheng E, Jing F, Zhang L, Jin H (2006) Scalable relevance feedback using click-through data for web image retrieval. In: Proceedings of the 14th annual ACM international conference on Multimedia. ACM press, pp 173–176

Ciaramita M, Murdock V, Plachouras V (2008) Online learning from click data for sponsored search. In: Proceedings of the 17th International World Wide Web Conference, Beijing

Duda R, Hart P, Stork D (2000) Pattern classification (2nd ed.) Wiley–Interscience

Elsas J, Carvalho V, Carbonell J (2008) Fast learning of document ranking functions with the committee perceptron. In: Proceedings of the 1st ACM International Conference on Web Search and Data Mining. ACM press

Harchaoui Z, Bach F (2007) Image classification with segmentation graph kernels. In: Proceedings of computer vision and pattern recognition

Hauptmann A, Yan R, Lin WH (2007) How many high–level concepts will fill the semantic gap in news video retrieval? In: Proceedings of the 6th ACM international conference on Image and video retrieval. ACM press, pp 627–634

Huang J, Kumar SR, Mitra M, Zhu WJ, Zabih R (1997) Image indexing using color correlograms. In: Proceedings of the 1997 Conference on Computer Vision and Pattern Recognition. IEEE Computer Society, Washington, DC, USA, p 762

Joachims T (2002) Optimizing search engines using clickthrough data. In: Proceedings of the ACM Conference on Knowledge Discovery and Data Mining. ACM press

Liu TY, Qin T, Xu J, Xiong W, Li H (2007) Letor: Benchmark dataset for research on learning to rank for information retrieval. In: SIGIR Workshop on Learning to Rank for Information Retrieval

Rumelhart D, Hinton G, Williams R (1986) Learning internal representation by backpropagating errors. Nature 323(99):533–536

Salembier P, Sikora T (2002) Introduction to MPEG–7: Multimedia Content Description Interface. John Wiley & Sons, Inc., New York, NY, USA

San Pedro J, Siersdorfer S (2009) Ranking and classifying attractiveness of photos in folksonomies. In: Proceedings of the WWW conference

Snoek CGM, Huurnink B, Hollink L, de Rijke M, Schreiber G, Worring M (2007) Adding semantics to detectors for video retrieval. IEEE Transactions on Multimedia 9(5):975–986

Tamura H, Mori S, Yamawaki T (1978) Texture features corresponding to visual perception. IEEE Transactions on Systems, Man and Cybernetics 8(6)

Tong H, He J, Li M, Ma WY, Zhang HJ, Zhang C (2006) Manifold–ranking–based keyword propagation for image retrieval. EURASIP Journal of Applied Signal Processing 2006(1):190–190

Tong S, Chang E (2001) Support vector machine active learning for image retrieval. In: Proceedings of the 9th annual ACM international conference on Multimedia. ACM press

Chapter 27
An Overview of Evaluation Campaigns in Multimedia Retrieval

Suzanne Little, Ainhoa Llorente, and Stefan Rüger

Abstract This chapter presents an academic and research perspective on the impact and importance of ImageCLEF and similar evaluation workshops in multimedia information retrieval (MIR). Three main themes are examined: the position of ImageCLEF compared with other evaluation conferences; general views on the usefulness of evaluation conferences and possible alternatives, and the impact and real–world meaning of evaluation metrics used within ImageCLEF. We examine the value of ImageCLEF, and related evaluation conferences, for the multimedia IR researcher as providing not only a forum for assessing and comparing outcomes but also serving to promote research aims, provide practical guidance (e.g. standard data sets) and inspire research directions.

27.1 Introduction

This chapter is not an exhaustive review of the impact of ImageCLEF and specific outcomes from ImageCLEF upon research. Rather it gives our multimedia information retrieval (MIR) group's perspective on the importance and usefulness of ImageCLEF in the academic context based on our experience participating in ImageCLEF and similar evaluation conferences and our view of MIR. In this section we outline our experiences participating in ImageCLEF, define key approaches to Information Retrieval (IR) evaluation and present the aims/needs of Multimedia Information Retrieval (MIR) research.

Suzanne Little
KMi, The Open University, Walton Hall, MK7 6AA, UK e-mail: s.little@open.ac.uk

Ainhoa Llorente
KMi, The Open University, Walton Hall, MK7 6AA, UK e-mail: a.llorente@open.ac.uk

Stefan Rüger
KMi, The Open University, Walton Hall, MK7 6AA, UK e-mail: s.rueger@open.ac.uk

H. Müller et al. (eds.), *ImageCLEF*, The Information Retrieval Series 32,
DOI 10.1007/978-3-642-15181-1_27, © Springer-Verlag Berlin Heidelberg 2010

Our Multimedia Information Systems (MMIS) group at the Knowledge Media Institute (KMi) conducts research in the area of multimedia information retrieval including content–based search, automatic image annotation and video shot–boundary detection. MMIS and previously the Multimedia Group at Imperial College London, also led by Stefan Rüger, have participated in both TRECVid and ImageCLEF tasks since 2002 (Pickering and Rüger, 2002; Heesch et al, 2003, 2004; Howarth et al, 2005; Jesus et al, 2005; Magalhães et al, 2006; Overell et al, 2006, 2008; Llorente et al, 2008, 2009; Zagorac et al, 2009).

MMIS participation is generally a team effort with two or more members of the group working together to apply our latest research to the specific tasks for the evaluation campaign. Submission is often a time–consuming task principally due to the need to adapt different input/output formats to work with existing tools and match the required submission format. For example, in 2009 (Zagorac et al, 2009) we used technology developed for the PHAROS project (Bozzon et al, 2009) that used the MPEG–7/MPQF (MPEG Query Format) formats for input/output of the annotation and search tools. Therefore we needed to convert the given media and queries into this format before processing and then convert the output to the required submission style. Processing time for the large volumes of media data also needs to be planned for. Participation is easier when an experienced team member, who has previously submitted runs to TRECVid or ImageCLEF, is available to help.

Section 27.2 describes a number of different evaluation conferences that serve a similar purpose to ImageCLEF but focus on different user tasks or different media types. These evaluations generally conduct performance assessment following what has been termed the 'Cranfield paradigm' (Brookes, 1981; van Rijsbergen, 1989) based on tests performed at Cranfield in the 1960s (Cleverdon et al, 1966). With some variation in the order and which party performs each step, the general process is: a document collection is assembled, a set of test queries is developed, each document is assigned a relevance judgement, each system performs the queries and its output is evaluated using a reserved test set. William Webber has written an extensive blog post[1] discussing how the approach used in the Cranfield tests came to be known as the Cranfield paradigm. Section 27.3 discusses some different viewpoints on the utility of Cranfield based evaluations for driving information retrieval research.

Evaluations based on the Cranfield approach are known as system–based or batch evaluations that compare information retrieval systems primarily on their ability to identify and properly rank documents deemed to be relevant. These evaluations use one or more specific, generally numerical, metrics ranging from straightforward precision and recall calculations to more complex and comprehensive rank–based metrics such as mean average precision, precision at n and cumulative gain (Järvelin and Kekäläinen, 2002) that weight values in favour of returning the most relevant documents first. In contrast, user–based evaluations focus on the performance of the system from a user perspective in fulfilling an information need.

[1] William Webber, 'When did the Cranfield tests become the "Cranfield paradigm"?' `http://blog.codalism.com/?p=817` (accessed 13th May 2010)

These evaluations generally involve direct testing of a system implementation by a user in a situation designed to reflect the real–world. Voorhees (2002) defines the difference as 'user–based evaluation measures the user's satisfaction with the system, while system evaluation focuses on how well the system can rank documents'.

User–oriented evaluation for multimedia often needs to consider extra facets to traditional text search. Cooniss et al (2000, 2003) carried out two widely noted studies of the needs, characteristics, and actions of end users of visual information in the workplace (called VISOR 1 and 2). One of the useful distinctions in their reports is the one between searching for oneself and searching as an intermediary, e g. for a journalist on the other side of the phone. Smeulders et al (2000) identified three types of search, labelled target search, aiming at a specific multimedia object identified by title or other metadata; category search, where the user has no specific object in mind but can describe which features they would like; and search by association, where the user is less specific and happy to browse in order to retrieve multimedia by serendipity.

What are the evaluation needs of a multimedia information retrieval research group? The first is to drive research directions through the exchange of cutting–edge ideas and the establishment of realistic test sets and basic performance benchmarks. The second is to push system performance through open, consistent and comparable evaluation processes that enable clear discussion the strengths, weaknesses and similarities of approaches. The final one is to perform holistic, real–world evaluations of the ability of the system to address user's information needs.

The remainder of this chapter will outline the main evaluation venues for medical image retrieval, discuss the utility of system–based evaluation, focus on the use of metrics to summarise system performance based on relevance judgements and, finally, look at the future requirements for evaluation of multimedia information retrieval systems.

27.2 ImageCLEF in Multimedia IR (MIR)

Since its early conception, information retrieval as a subject has always placed great emphasis on system evaluation (Rüger, 2010). Real user needs are simulated in a laboratory setting with three ingredients: large test collections, information need statements and relevance judgements. The test collection contains a large number of potentially interesting documents from a repository; each information need statement details the type of document that the user would like to retrieve, what the user hopes to see or hear and criteria for how the relevance of documents should be judged. The relevance judgements, also known as ground truth, tell us whether a particular document of the collection is relevant for a particular information need. The value of an evaluation setting like this is that the effectiveness of a particular retrieval method can be measured in a reproducible way. Although this approach has been criticised for its lack of realism and its narrow focus on the pure retrieval aspect of presumably much bigger real tasks, system evaluations are still the main basis on which retrieval algorithms are judged, and on the back of which research flourishes. In this respect evaluation conferences such as INEX for structured XML retrieval, ImageCLEF for image retrieval, MIREX for music retrieval, TRECVid for video retrieval and GeoCLEF for geographic retrieval have a significant and lasting impact on multimedia information retrieval research through reproducibility and comparisons. ImageCLEF is discussed extensively in this book. Here we give a brief summary of INEX, MIREX, TRECVid, GeoCLEF and other evaluation campaigns and compare their structure and aims with ImageCLEF. TRECVid, in particular, is extensively described as its purpose and aims align most closely with those of ImageCLEF.

27.2.1 INEX XML Multimedia Track

In 2002, the INEX[2] Initiative for the Evaluation of XML Retrieval started to provide a test–bed for evaluation of effective access to structured XML content. The organisation of INEX passed from the University of Duisburg to Otago University[3] in the year 2008.

Van Zwol et al (2005) set up an XML multimedia track that was repeated as part of INEX until 2007. It provided a pilot evaluation platform for structured document retrieval systems that combine multiple media types. While in 2005 the collection was made up from Lonely Planet travel guides, the 2006 evaluations used the much larger Wikipedia collection from the INEX main track (Westerveld and van Zwol, 2006). Both collections contain a range of media, including text, image speech, and video — thus modelling real life structured documents. The goal of the multimedia track was to investigate multimedia retrieval from a new perspective, using the structure of documents as the semantic and logical backbone for the retrieval of multimedia fragments.

In contrast to other evaluation fora, INEX's multimedia track was to retrieve relevant document fragments based on an information need with a structured multimedia character, i.e. it focused on the use of document structure to estimate, relate, and combine the relevance of different multimedia fragments. One big challenge for a structured document retrieval system is to combine the relevance of the different media types and XML elements into a single meaningful ranking that can be presented to the user.

[2] `http://inex.is.informatik.uni-duisburg.de/`

[3] `http://www.inex.otago.ac.nz/`

27.2.2 MIREX

The MIREX[4] (Music Information Retrieval Evaluation eXchange) is a Text REtrieval Conference (TREC)–style evaluation effort organised by the International Music Information Retrieval Systems Evaluation Laboratory (IMIRSEL[5]) at the Graduate School of Library and Information Science[6], of the University of Illinois at Urbana-Champaign[7]. It is a community–based evaluation conference for Music Information Retrieval that examines the background, structure, challenges, and contributions of MIREX and provides some insights into the state–of–the–art in Music Information Information Retrieval systems and algorithms. Downie (2008) looks at retrieval research as a whole.

27.2.3 GeoCLEF

The GeoCLEF[8] track was introduced to the Cross Language Evaluation Forum (CLEF) workshop in 2005 as an ad hoc TREC style evaluation for Geographic Information Retrieval (GIR) systems; this provided a uniform evaluation for the growing GIR community and is becoming the de facto standard for evaluating GIR systems. GeoCLEF has moved its home to the University of Hildesheim[9].

The GeoCLEF 2005–08 English corpus consists of approximately 135,000 news articles, taken from the 1995 Glasgow Herald and the 1994 Los Angeles Times; the overall corpus also includes German, Spanish and Portuguese documents. There are 100 GeoCLEF queries from 2005–08 (25 from each year). These topics are generated by hand by the four organising groups. Each query is provided with a title, description and narrative. The title and description contain brief details of the query, while the narrative contains a more detailed description including relevance criteria. The 2005 queries have additional fields for concept, spatial relation and location. However, these fields were discarded in later years as unrealistic. Typical topics of GeoCLEF include *Shark Attacks off Australia and California* (Topic 001) or the rather more difficult *Wine regions around rivers in Europe* (Topic 026). Mandl et al (2008) present an overview of GeoCLEF 2007.

[4] `http://www.music-ir.org/mirex/`

[5] `http://music-ir.org/evaluation/`

[6] `http://www.lis.uiuc.edu/`

[7] `http://www.uiuc.edu/`

[8] `http://ir.shef.ac.uk/geoclef/`

[9] `http://www.uni-hildesheim.de/geoclef/`

27.2.4 TRECVid

The TREC Video Retrieval Evaluation initiative (TRECVid[10]) is an independent evaluation forum devoted to research in automatic segmentation, indexing, and content–based retrieval of digital video. It started out in 2001 as a video track of the TREC[11] conference series and became an independent two–day workshop of is own in 2003. TRECVid is sponsored by the NIST[12] (National Institute of Standards and Technology) with additional support from other US government agencies. Participation in TRECVid has been rising since its early days, and in 2007 54 teams from all over the world took part. Smeaton et al (2006) give an overview of the TREC Video Retrieval Evaluation initiative.

The information need of an example topic for the 2003 TRECVid Interactive Track is described as "Find shots from behind the pitcher in a baseball game as he throws a ball that the batter swings at". Other search topics may be exemplified by short video clips or a combination of video clips and images. The 2003 TRECVid test collection repository consists of video shots from mainly US news programmes.

Every year 25 topics are released to all participating groups, who would have pre–processed and indexed the test collection prior to this. The rules for the interactive task of the search track allow searchers to spend 15 minutes per topic to find as many relevant shots as possible; they are free to create a search engine query from the given topic in any way they see fit, modify their query, and collect shots that they deem relevant. Each participating group returns the results of their searches to NIST, who are then responsible for assessing the returned shots from all the participating groups. The assessors, often retired intelligence workers, would look at a pool of results for each topic and assess the relevance of each shot in the pool for a topic. In order to make the best use of the assessors' time, only the union of the top *n*, say 100, of all the results from different groups for a particular topic is put into this pool. The explicitly assessed shots for each topic form the relevance judgements. Shots that were not assessed during this procedure are those that none of the many participating systems reported in their respective top *n* results, and the implicit assumption is that these unassessed shots are *not* relevant. The reason for this is the prohibitive cost of assessing all shots against all topics.

This ground truth is then the basis on which participating retrieval systems can be compared. It is possible to use this setting for later evaluation outside the TRECVid programme: the only slight disadvantage is that the assessed algorithm would not have contributed to the pooling process; hence, if the new algorithm uncovered many relevant shots that no other algorithm of the participating groups has reported in their top *n* results, then these would be treated as irrelevant.

The interactive task is only one task among many. There are *manual tasks* where the searchers are allowed to formulate and submit a query *once* for each topics without further modification; there is an *automated task* where the generation of the

---

[10] `http://trecvid.nist.gov/`

[11] `http://trec.nist.gov/`

[12] `http://www.nist.gov/`

computer query from a search topic is fully automated without any human intervention. These three tasks form the *search track* of the TRECVid evaluation conference, which is one of typically three to five tracks, each year. Over the years other tracks have included:

Shot segmentation, i.e. the sectioning of a video into units that result from a single operation of the camera, is a basic but essential task that any video processing unit has to carry out. Hard cuts, where adjacent shots are basically edited by simply concatenating the shots, are relatively easy to detect as the frames of a video change abruptly. Modern editing techniques deploy gradual transmissions, though, e.g. fade out/in, which provide continuity between shots and thus are harder to detect. Shot segmentation algorithms vary widely in their efficiency, i.e. how much faster (or slower) they are than playing the video. Generally, algorithms that need to decode the video stream into frames tend to be slower than algorithms that operate on the compressed video format.

The *story segmentation* track is meant to identify the (news) story boundaries with their time. A news story is defined as a segment of news broadcast with a coherent focus. While a story can be composed of multiple shots (e.g. an anchorperson introduces a reporter, who interviews someone in the field and uses archive material to explain the background), a single shot can contain story boundaries, e.g. an anchorperson switching to the next news topic. Although this track is non–trivial, it has only been part of TRECVid for a couple of years.

In 2007 TRECVid introduced new video genres taken from a real archive in addition to its previous focus on news: news magazine, science news, news reports, documentaries, educational programming and archival video. The idea was to see how well the video retrieval and processing technologies apply to new sorts of data.

In addition to that, the BBC Archive has provided about 100 hours of unedited material (also known as *rushes*) from five dramatic series to support an exploratory track of *rushes summarisation*: systems should construct a very short video clip that includes the major objects and events of the original video. At a dedicated workshop at ACM Multimedia, Over and Smeaton (2007) presented the results of these efforts.

The *surveillance event detection* track is a more recent addition to TRECVid that operates on around 150 hours of UK Home Office surveillance data at London Gatwick International Airport.

The *content–based copy detection* track tries to identify modified segments of a video under a variety of transformations such as a change of aspect ratio, colour, contrast, encoding, bit rate, addition of material, deletion of material, picture in picture in the video part or bandwidth limitation and variate mixing with other audio content in the audio part. Real world applications would be copyright control, de–duplication in large data repositories, grouping of video results in large video repositories or advertisement tracking.

Feature extraction tracks have played an important role throughout the lifetime of TRECVid. Many requests for archival video contain requests for specific features (see above discussion in this section). One of the frequently required aspects is that of a specific camera motion. In the low–level feature extraction version, camera motions such as *pan (left or right)* or *tilt (up or down)* had to be detected. Generally, owing to the semantic gap, high level feature extraction tasks are more difficult. They concern semantic concepts such as *indoor*, *outdoor*, *people*, *face*, *text overlay*, *speech*, etc. These concepts can be very useful additional search criteria to home in on many real–world requests. Smeaton et al (2009) have summarised the work done on the TRECVid high-level feature task and show the progress made across the spectrum of various approaches.

27.2.5 VideOlympics

The VideOlympics[13] (Snoek et al, 2008), held most recently at CIVR 2009, is not an evaluation campaign in the traditional sense but rather an opportunity for researchers with video retrieval systems to demonstrate their work through real–time user evaluations in a demonstration session format. It uses data from TRECVid and is therefore aimed principally at TRECVid participants. It is not intended to produce comparative results for publication but to inform the audience about the state–of–the–art in video retrieval and promote discussion about user interfaces for video search.

27.2.6 PASCAL Visual Object Classes (VOC) Challenge

In 2005, the PASCAL Visual Object Classes challenge[14] appeared supported by the EU–funded PASCAL2 Network of Excellence on Pattern Analysis, Statistical Modelling and Computational Learning[15]. The goal of this challenge is to recognise objects from a number of visual object classes in realistic scenes. It is fundamentally a supervised learning problem in that a training set of labelled images is provided. The twenty object classes that were selected belonged to the following categories: person, animal, vehicle, and objects typically found in an indoor scene. The challenge is divided into three main tasks: the classification task, which predicts the presence or absence of an instance of the class in the test image; the detection task, which determines the bounding box and label of each object in the test image; and the segmentation task, which generates pixel–wise segmentations giving the class of the object visible at each pixel.

The workshop where participants are invited to show their results is co–located with a relevant conference in computer vision such as the International Conference on Computer Vision or the European Conference on Computer Vision. The 2010 edition added new tasks such as still image action classification and large scale visual recognition.

---

[13] http://www.videolympics.org/

[14] http://pascallin.ecs.soton.ac.uk/challenges/VOC/

[15] http://www.pascal-network.org/

27.2.7 MediaEval and VideoCLEF

MediaEval[16] is a benchmarking initiative launched by the PetaMedia Network of Excellence in late 2009 to serve as an umbrella organization to run multimedia benchmarking evaluations. It is a continuation and extension of VideoCLEF, which ran as a track in the CLEF campaign in 2008 and 2009.

The initiative is divided into several tasks. In the 2010 edition, there are two annotation tasks called the tagging task but designed with two variations, the professional version and the wild wild web version. The professional version tagging task requires participants to assign semantic theme labels from a fixed list of subject labels to videos. The task uses the TRECVid data collection from the Netherlands Institute for Sound and Vision. However, the tagging task is completely different than the original TRECVid task since the relevance of the tags to the videos is not necessarily dependent on what is depicted in the visual channel. The wild wild web version task requires participants to automatically assign tags to videos using features derived from speech, audio, visual content or associated textual or social information. Participants can chose which features they wish to use and are not obliged to use all features. The data set provided is a collection of Internet videos.

Additional tasks for the 2010 initiative are the placing task or geotagging where participants are required to automatically guess the location of the video by assigning geo–coordinates (latitude and longitude) to videos using one or more of: video metadata such as tags or titles, visual content, audio content, social information. Any use of open resources, such as gazetteers, or geo–tagged articles in Wikipedia is encouraged. The goal of the task is to come as close to possible to the geo–coordinates of the videos as provided by users or their GPS devices. Other tasks are the affect task whose main goal is to detect videos with high and low levels of dramatic tension; the passage task where, given a set of queries and a video collection, participants are required to automatically identify relevant jump–in points into the video based on the combination of modalities such as audio, speech, visual, or metadata; and the linking task, where participants are asked to link the multimedia anchor of a video to a relevant article from the English language Wikipedia.

One of the strongest points of this competition is that it attempts to complement rather than duplicate the tasks assigned to in the TRECVid evaluation campaign. Traditionally, TRECVid tasks are mainly focused on finding objects and entities depicted in the visual channel whereas MediaEval concentrates on what a video is about as a whole.

[16] `http://www.multimediaeval.org`

27.2.8 Past Benchmarking Evaluation Campaigns

This section summarises other relevant benchmarking evaluation campaigns that have previously been operative in this area. They are worth mentioning as their research questions, objectives, results, and the used data sets persist online.

The *Face Recognition Vendor Test* (FRVT)[17] 2006 was the latest in a series of large scale independent evaluations for face recognition systems organised by the U.S. National Institute of Standards and Technology. Previous evaluations in the series were the FERET, FRVT 2000, and FRVT 2002. The primary goal of the FRVT 2006 was to measure progress of prototype systems and commercial face recognition systems since FRVT 2002. Additionally, FRVT 2006 evaluated the performance on high resolution still imagery, 3–D facial scans, multi–sample still facial imagery, and re–processing algorithms that compensate for pose and illumination.

A short–lived evaluation campaign that only ran for one year in 2006, *ImagEVAL*[18] was a French initiative that tried to bring some answers to the question posed by Carol Peters, in the CLEF workshop in 2005, where she wondered why systems that show very good results in the CLEF campaigns have not achieved commercial success. The point of view of ImagEVAL was that the evaluation criteria 'do not reflect the real use of the systems'. Thus, this initiative was launched in France in 2006, mainly concentrated on the French research domain, although it was accessible to other researchers as well. The campaign was divided into several tasks relating to image analysis including object detection, querying, text detection and recognising transformed images. The focus of this evaluation campaign was certainly closer to the user–oriented perspective and it hoped to improve methods of technological evaluation so that end–user criteria could also be included.

During the 2000 Internet Imaging Conference, a suggestion was made to hold a public contest to assess the merits of various image retrieval algorithms. Since the contest would require a uniform treatment of image retrieval systems, the concept of a benchmark quickly entered into the scenario. This contest became known as the *Benchathlon*[19] and was held at the Internet Imaging Conference in January 2001. Despite their initial objectives no real evaluation ever took place although many papers were published in this context and a reference database created.

The *Classification of Events, Events, Activities and Relationships* (CLEAR)[20] evaluation conference was an international effort to evaluate systems that are designed to recognise events, activities, and their relationships in interaction scenarios. Its main goal was to bring together projects and researchers working on related technologies in order to establish a common international evaluation in this field. It was divided into the following tasks: person tracking (2–D and 3–D, audio–only, video–only, multimodal); face tracking; vehicle tracking; person identification (audio–only, video–only, multimodal); head pose estimation (2–D and 3–D); and

[17] `http://www.frvt.org/`

[18] `http://www.imageval.org/e_presentation.html`

[19] `http://www.benchathlon.net/`

[20] `http://clear-evaluation.org/`

acoustic event detection and classification. The latest edition, held in 2007, was supported by the European Integrated project 'Computers In the Human Interaction Loop' (CHIL) and the U.S. National Institute of Standards and Technology (NIST).

27.2.9 Comparison with ImageCLEF

The principal difference between these evaluation conferences is in the document and query types that they focus on. This necessarily leads to differences in the way in which the tasks are structured and the evaluation metrics used. Core similarities remain — evaluating the state of the art in information retrieval for structured text, music, image, video or geographical queries. While the specific metrics may vary, they remain based on the notion of document relevance and ranking a retrieved list of documents. This is also true for annotation tasks where the confidence of a label is used.

A common feature among many evaluation campaigns is the regular changes to the tasks or strands of the challenge. Tasks not only get new content each year but may change their focus, evolving to meet the needs of the research community and the latest challenges. New tasks are constantly proposed and old tasks retired. The ongoing evolution and diversity of the challenges helps to keep evaluation campaigns relevant.

Multimedia objects are a highly multidimensional and collections often also include transcripts or other text–based metadata that is useful for information retrieval purposes. Until the 2009 TRECVid the video retrieval task required a run to be submitted that only used the text data to demonstrate that an improvement was achieved using the media content over that of using text alone. In the early days of TRECVid it was often found that content–based or visual methods displayed little or no improvement over using the video transcription to retrieve video segments. Recent results have consistently demonstrated that using content–based methods has improved system performance and hence this requirement has been dropped.

ImageCLEF, VideoCLEF and GeoCLEF obviously have roots in the CLEF multi–lingual text retrieval evaluation conference and thus also have a focus on cross–language retrieval. By necessity, multi–linguality requires the inclusion of text data in the document collection — images are generally language independent.

TRECVid has an option in the retrieval track which allows searches to be performed by a user interacting with the search system to submit and refine queries. The resulting evaluation is only conducted on the ranked results list and, officially at least, does not include capturing user feedback on the system usability for comparison. VideOlympics, which is based on the interactive track of TRECVid, starts to move towards user–based evaluations but does not produce comparative evaluations, only demonstrations.

ImageCLEF has also increased its focus in recent years on the use of ontologies or knowledge models (e.g. Wikipedia) to improve performance. This allows better inclusion of contextual information in the queries and a potentially better user experience — although this has yet to be exhaustively tested within ImageCLEF. Newly proposed evaluation metrics aim to judge the importance and effectiveness of structured knowledge in MIR.

27.3 Utility of Evaluation Conferences

A great deal has been written from the 1960s to the present time regarding the utility of batch system analysis of test collections for evaluating information retrieval approaches as compared with the real needs of users in their information environment. Apart from operational criticisms relating to the methods of generating 'enough' data, the difficulty in defining relevance and determining appropriate queries, the major criticism is that system–based approaches are too far removed from the reality of user interactions and information requirements.

In defence of applying the Cranfield approach for system evaluation, Voorhees (2002) discussed the philosophical implications and concluded that, within limits, laboratory tests are a valid tool for performing this type of evaluation. This was based on analysis of a series of experiments run on TREC collections that demonstrated that comparative evaluations remain stable despite changes in the relevance judgements. Salton (1992) examines a number of key criticisms regarding laboratory based retrieval system evaluation and finally concludes that "there should be no question at this point about the usefulness and effectiveness of properly designed automatic text retrieval systems". Harman has, in general, been supportive of Cranfield based evaluation particularly in her role at TREC (Harman, 2005). We await with interest her keynote talk at SIGIR 2010 titled 'Is the Cranfield Paradigm Outdated?'[21].

In contrast, Järvelin (2009) argues strongly that Cranfield–style approaches are limited and insufficient to explain searcher behaviour principally on the basis that the resulting comparison and analysis lacks inclusion of the user contexts. Hersh et al (2000); Turpin and Hersh (2001) present results that argue that end users perceive little or no difference between the performance of a baseline system and one shown to be 'significantly' better in relevance–based evaluations. Almost 20 years ago Brookes (1981) questioned the continuing usefulness of applying the 'Cranfield paradigm' for information retrieval evaluation stating that it was an evaluation from the 'computer science' side and did not reach out to fulfil the needs of 'information science'. More recently, Järvelin (2009) stated "there is mounting evidence that we may not be able to improve human performance by further improving traditional research effectiveness".

In spite of appearances, the conclusions from the literature summarised here are not incompatible. Rather it is clear that evaluation of information retrieval systems requires consideration of the system *in situ* rather than solely *in vitro* or *in silico*. Certain tracks and tasks of the various evaluation conferences do consider the user.

[21] `http://www.sigir2010.org/doku.php?id=program:keynotes`

For example, both TRECVid and ImageCLEF have had interactive strands for retrieval tasks that involve user participation. TRECVid allows iterative querying of the system to develop the results list, mimicking user behaviour in real life where queries are refined based on the results list. ImageCLEF 2003, 2004 and 2005 had an interactive image retrieval task that used user questionnaires to explore variations of retrieval systems within a submission. That is, results from participants were not compared. From 2006 this task was merged with the main interactive CLEF track (iCLEF[22]).

Many of the papers written on the topic of information retrieval evaluation and referenced here predate the Internet and are heavily focused on traditional text–based IR. What is the implication for multimedia? Batch system analysis is dependent on the creation of sufficiently large and well–annotated test sets. Building such sets of multimedia documents is time consuming and expensive. Evaluation campaigns are invaluable in providing standard data sets and a forum to conduct such experiments. However, in many ways user–based evaluation is even more critical for multimedia document search and retrieval. The high density of information contained in images, audio or video and the often subjective interpretation create more complications for determining relevance and increase the importance of personal context.

Saracevic (1995) stated that IR was increasingly being embedded in other systems — e.g. the Internet, digital libraries — and noted that new evaluations in this context needed to be incorporated. Since this paper was published, both ImageCLEF and TRECVid have embraced the knowledge context found on the Internet with tasks that incorporate contextual knowledge about images from Web pages, use of Wikipedia and inclusion of social networking information.

Finally, there is of course more to evaluation conferences than simply the chance to execute a batch system evaluation with a suitably large and well documented test set. The collaborative nature, the focus and time–pressure of producing a submission, and the opportunity to openly share and explore approaches, are in many ways the more valuable result of participation.

27.4 Impact and Evolution of Metrics

In the previous section we described a number of different evaluation campaigns and looked at some of the criticisms of evaluation conferences for information retrieval research. We found that while a holistic approach to evaluation is required there is benefit in applying batch system–based evaluations. Here we look at the use of metrics to assess information retrieval performance. The definitions and explanations of the metrics used in ImageCLEF have been given elsewhere in this book. In this section we aim to discuss the real–world impact of using these metrics to assess information retrieval research and some of the problems that can be caused

[22] `http://nlp.uned.es/iCLEF/`

by over reliance on judging performance using only a narrow selection of numerical metrics. We also suggest some newly proposed metrics for evaluation that may help.

Perhaps the biggest challenge to conducting system–based evaluations is the need to assign a relevance judgement to each document in the data set for each query. This raises significant practical problems as manual annotation is time–consuming, expensive and subjective — all motivations behind research into automatic image annotation or content–based search. Evaluation campaigns often go some way towards sharing this cost among research groups. TRECVid, for example, conducts a shared annotation phase where participants manually annotate subsets of the training data to share and use in all systems. Pooling of results from all submissions is also used to reduce the volume of documents that need to be assessed in the testing phase. This may result in some relevant documents being missed but is generally accepted to provide a more efficient cost–effective method of using very large data sets.

Philosophically a larger question is how to define 'relevance' for IR. Borlund (2003) provides an extensive review of the concept of relevance in IR and its importance in evaluation.Cosijn and Ingwersen (2000) also discuss the difficulties of consistently and accurately defining relevance and propose a model based on the notion of socio–cognitive relevance. Saracevic (1997) in his acceptance speech for the 1997 ACM SIGIR Gerald Salton Award talks about the impossibility of separating users from the notion of relevance — by its very definition it requires user involvement and user judgement.

Ellis (1996) describes the 'dilemma of measurement' in information science that seeks to perform exact measurements in the scientific style but uses human judgement of relevance and concluded that the Cranfield tests "oversimplified the inherent complexity of the retrieval interaction in the pursuit of quantification". This tension between the desire for a clear, quantitative method for comparing and defining improvements in IR and the fundamental variations that occur when using human judgements about document relevance continues to drive research into IR evaluation methodologies.

Soboroff et al (2001) conducted interesting experiments that extend those by Voorhees (2000) into the impact of differing relevance judgements on comparative system performance. Using data from TREC, the hypothesis that variations in relevance had minimal impact on the relative ranking of systems was assessed. Interestingly, they found that even random assignments of relevance based on the pooled TRECVid results produced rankings that correlated positively with the official TREC result, although it was not possible to predict system performance. This reinforces the view that evaluation campaign results should be used carefully outside of the context of the comparative workshop. Earlier work by Zobel (1998), who proposed a new method for pooling system results, reached similar conclusions.

Buckley and Voorhees (2000) ask two questions regarding evaluation methods for IR: "how to build and validate good test collections" and "what measures should be used to evaluate retrieval effectiveness". They examine a number of common metrics based on precision and recall of relevant documents, discuss the general rules of thumb (e.g. for size of the data or query set) and look at a method for quantifying the confidence that can be placed in the experimental conclusions. As a result, they suggest that IR evaluation papers should include results from several collections.

The precision and recall metrics used by Buckley and Voorhees (2000) are fairly standard within IR. Other terms that are commonly used include those that aim to quantitatively measure the overall improvement or gain achieved by one ranked list over another (Järvelin and Kekäläinen, 2002). That is, is the order of results provided by system A better than that provided by system B? Sakai (2007) reviews a number of graded–relevance retrieval metrics and concludes that they are "at least as stable and sensitive as Average Precision, and are fairly robust to the choice of gain values". Researchers are also exploring areas such as novelty and diversity to compare the performance of systems from a more user–friendly perspective (Clarke et al, 2008).

Newer metrics that aim to address limitations with measures that use recall have also been proposed. Moffat and Zobel (2008) suggest *rank–biased precision* derived from a model of user behaviour as a replacement for average precision that measures the behaviour as observed by the user. This publication also lays out a clear case for and against the use of measures such as average precision and the benefits of metrics that consider a broader range of the user's perspective.

The Photo Annotation task at ImageCLEF 2009 calculated a new evaluation metric based on ontology scoring from Nowak et al (2010) that aimed to measure how information obtained from ontologies improved system performance. This metric supported the focus on multi–modal approaches to photo annotation. Measures such as this, while not necessarily improving the user focus of evaluation, do extend the scope of evaluation campaigns and enrich the discussion surrounding the system performance beyond that of single quantitative rankings.

Finally, too narrow a focus on single quantitative evaluation metrics can lead to over–fitting of the system to produce optimal results for one or more evaluation campaigns that cannot be transferred to real world performance. Both TRECVid and ImageCLEF aim to mitigate this by providing multiple metrics for judging performance and comparing systems internally. There is also a clear expectation that precludes participants making exaggerated claims about the system performance outside of the evaluation workshop — particularly in a commercial setting. In TREC and TRECVid this takes the form of an explicit user agreement signed by participants.

27.5 Conclusions

In this chapter we have described the ImageCLEF/TRECVid participation experience of a multimedia IR research group and examined the issues surrounding evaluation campaigns in IR. The main issues in IR evaluation focus on the weaknesses of system–based evaluation in isolation, the problems in assessing IR system performance outside of the real–world context and without user input, and the seductive difficulty in finding a quantitative measure of IR success. We have not presented an exhaustive analysis of all of the available literature on evaluation in information retrieval — there is an extensive volume of literature stretching back over almost 50 years of research in information science.

It is unfortunate that parties external to the evaluation communities can sometimes view them as a 'competition' and judge system performance in isolation based only on the numbers. Research has shown that the use of system evaluation based on the Cranfield approach can be a valid and useful approach to drive the development of information retrieval. It is also clear that these judgements cannot necessarily be applied outside of the evaluation workshop context to determine the absolute, real–world usefulness or effectiveness of one system over another.

User–based evaluation of multimedia information retrieval systems is challenging due to difficulties in finding appropriate users, setting up systems with consistent and functionally complete user interfaces (that do not impact excessively on the perception of performance), and running experiments that are comparable and reproducible. Evaluation campaigns do a fantastic job of helping researchers conduct quality, large–scale system–based evaluations that drive research and improve technology. As multimedia information retrieval moves forward, we believe that holistic user focused evaluations will become increasingly important. How can the communities of researchers and users mobilised by evaluation campaigns contribute to this process?

ImageCLEF is useful and fulfils a significant need in multimedia IR. It is important to also consider a holistic view of information retrieval systems and not to focus solely on the ranking or single performance values. Evolving tasks that incorporate external context and begin to include users and interactivity are improving the outcomes for multimedia IR. Emerging metrics that focus on other aspects are providing new insights into IR system performance and will be beneficial to incorporate into future evaluation campaigns. The diversity of tasks in evaluation campaigns helps to approximate user needs in small, specific situations. We believe this fragmentation helps to simulate the variety of contexts and situations that occur in MIR and contributes to improving real-world information retrieval systems. The diversity found within evaluation campaigns will continue to drive multimedia IR research and play a vital role in future developments.

Acknowledgements The authors would like to acknowledge the contributions of many current and past members of the MMIS group in participating in ImageCLEF and TRECVid — Paul Browne, Shyamala Doraisamy, Daniel Heesch, Peter Howarth, Rui Hu, Haiming Liu, João Magalhães, Simon Overell, Marcus Pickering, Adam Rae and Alexei Yavlinsky, among many others.

References

Borlund P (2003) The concept of relevance in IR. Journal of the American Society for information Science and Technology 54(10):913–925

Bozzon A, Brambilla M, Fraternali P, Nucci F, Debald S, Moore E, Neidl W, Plu M, Aichroth P, Pihlajamaa O, Laurier C, Zagorac S, Backfried G, Weinland D, Croce V (2009) PHAROS: an audiovisual search platform. In: ACM International conference on research and development in information retrieval. ACM press, p 841

Brookes B (1981) Information technology and the science of information. Information retrieval research. London: Butterworths 1–8

Buckley C, Voorhees E (2000) Evaluating evaluation measure stability. In: Proceedings of the 23rd annual international ACM SIGIR conference on research and development in information retrieval, p 40. ACM press

Clarke C, Kolla M, Cormack G, Vechtomova O, Ashkan A, Büttcher S, MacKinnon I (2008) Novelty and diversity in information retrieval evaluation. In: Proceedings of the 31st annual international ACM SIGIR conference on research and development in information retrieval. ACM press, pp 659–666

Cleverdon C, Mills J, Keen E (1966) Factors determining the performance of indexing systems,(Volume 1: Design). Cranfield: College of Aeronautics

Cooniss L, Ashford A, Graham M (2000) Information seeking behaviour in image retrieval. VISOR 1 final report. Technical report, Library and Information Commission Research Report, British Library

Cooniss L, Davis J, Graham M (2003) A user–oriented evaluation framework for the development of electronic image retrieval systems in the workplace: VISOR 2 final report. Technical report, Library and Information Commission Research Report, British Library

Cosijn E, Ingwersen P (2000) Dimensions of relevance. Information Processing & Management 36(4):533–550

Downie JS (2008) The music information retrieval evaluation exchange (2005–2007): a window into music information retrieval research. Acoustical Science and Technology 29(4):247–255

Ellis D (1996) The dilemma of measurement in information retrieval research. Journal of the American Society for Information Science 47(1):23–36

Harman D (2005) The importance of focused evaluations: A Case Study of TREC and DUC. Charting a New Course: Natural Language Processing and Information Retrieval 16:175–194

Heesch D, Pickering M, Rüger S, Yavlinsky A (2003) Video retrieval using search and browsing with key frames. In: TREC Video Retrieval Evaluation

Heesch D, Howarth P, Magalhães J, May A, Pickering M, Yavlinsky A, Rüger S (2004) Video retrieval using search and browsing. In: TREC Video Retrieval Evaluation

Hersh W, Turpin A, Price S, Chan B, Kramer D, Sacherek L, Olson D (2000) Do batch and user evaluations give the same results? In: Proceedings of the 23rd annual international ACM SIGIR conference on research and development in information retrieval, p 24. ACM press

Howarth P, Yavlinsky A, Heesch D, Rüger S (2005) Medical image retrieval using texture, locality and colour. In: Proceedings of the Cross Language Evaluation Forum 2004. Lecture Notes in Computer Science (LNCS), vol 3491. Springer, pp 740–749

Järvelin K (2009) Explaining user performance in information retrieval: Challenges to IR evaluation. Advances in Information Retrieval Theory 289–296

Järvelin K, Kekäläinen J (2002) Cumulated gain–based evaluation of ir techniques. ACM Transactions on Information Systems 20(4):422–446

Jesus R, Magalhães J, Yavlinsky A, Rüger S (2005) Imperial college at trecvid. In: TREC Video Retrieval Evaluation, Gaithersburg, MD

Llorente A, Zagorac S, Little S, Hu R, Kumar A, Shaik S, Ma X, Rüger S (2008) Semantic video annotation using background knowledge and similarity–based video retrieval. In: TREC Video Retrieval Evaluation (TRECVid, Gaithersburg, MD)

Llorente A, Little S, Rüger S (2009) MMIS at ImageCLEF 2009: Non–parametric density estimation algorithms. In: Working notes of CLEF 2009, Corfu, Greece

Magalhães J, Overell S, Yavlinsky A, Rüger S (2006) Imperial college at TRECVID. In: TREC Video Retrieval Evaluation, Gaithersburg, MD

Mandl T, Gey F, Di Nunzio G, Ferro N, Larson R, Sanderson M, Santos D, Womser-Hacker C, Xie X (2008) GeoCLEF 2007: the CLEF 2007 cross–language geographic information retrieval track overview. In: Advances in Multilingual and Multimodal Information Retrieval. Lecture Notes in Computer Science (LNCS), vol 5152. Springer, pp 745–772

Moffat A, Zobel J (2008) Rank–biased precision for measurement of retrieval effectiveness. ACM Transactions on Information Systems 27(1):1–27

Nowak S, Lukashevich H, Dunker P, Rüger S (2010) Performance measures for multilabel classification — a case study in the area of image classification. In: ACM SIGMM International Conference on Multimedia Information Retrieval, Philadelphia, Pennsylvania

Over P, Smeaton A (eds) (2007) TVS 2007: proceedings of the international workshop on TRECVid video summarization. ACM press

Overell S, Magalhães J, Rüger S (2006) Place disambiguation with co–occurrence models. In: Working notes CLEF 2006, Alicante, Spain

Overell S, Llorente A, Liu HM, Hu R, Rae A, Zhu J, Song D, Rüger S (2008) MMIS at ImageCLEF 2008: Experiments combining different evidence sources. In: Working notes of CLEF 2008, Aarhus, Denmark

Pickering M, Rüger S (2002) Multi–timescale video shot–change detection. In: Text Retrieval Conf, NIST (Trec, Gaithersburg, MD, Nov 2001), NIST Special Publication 500–250, pp 275–278

van Rijsbergen CJ (1989) Towards an information logic. In: Proceedings of the 12th annual international ACM SIGIR conference on research and development in information retrieval. ACM press, pp 77–86

Rüger S (2010) Multimedia information retrieval. Lecture notes in the series Synthesis Lectures on Information Concepts, Retrieval, and Services. Morgan–Claypool

Sakai T (2007) On the reliability of information retrieval metrics based on graded relevance. Information Processing & Management 43(2):531–548

Salton G (1992) The state of retrieval system evaluation. Information Processing and Management 28(4):441–449. Special Issue: Evaluation Issues in Information Retrieval

Saracevic T (1995) Evaluation of evaluation in information retrieval. In: Proceedings of the 18th annual international ACM SIGIR conference on research and development in information retrieval, pp 138–146. ACM press

Saracevic T (1997) Users lost: Reflections on the past, future, and limits of information science. ACM SIGIR Forum 31(2):16–27

Smeaton A, Over P, Kraaij W (2006) Evaluation campaigns and TRECVid. In: ACM International Workshop on Multimedia Information Retrieval. ACM press

Smeaton A, Over P, Kraaij W (2009) High–level feature detection from video in TRECVid: a 5–year retrospective of achievements. In: Divakaran A (ed) Multimedia Content Analysis: Theory and Applications. Springer, pp 151–174

Smeulders A, Worring M, Santini S, Gupta A, Jain R (2000) Content–based image retrieval at the end of the early years. IEEE Transactions on Pattern Analysis and Machine Intelligence 22(12):1349–1380

Snoek CGM, Worring M, de Rooij O, van de Sande KEA, Yan R, Hauptmann AG (2008) VideOlympics: Real–time evaluation of multimedia retrieval systems. IEEE MultiMedia 15(1):86–91

Soboroff I, Nicholas C, Cahan P (2001) Ranking retrieval systems without relevance judgments. In: Proceedings of the 24th annual international ACM SIGIR conference on research and development in information retrieval, pp 66–73. ACM press

Turpin A, Hersh W (2001) Why batch and user evaluations do not give the same results. In: Proceedings of the 24th annual international ACM SIGIR conference on research and development in information retrieval, pp 225–231. ACM press

Voorhees EM (2000) Variations in relevance judgments and the measurement of retrieval effectiveness. Information Processing & Management 36(5):697–716

Voorhees EM (2002) The philosophy of information retrieval evaluation. In: Revised Papers from the Second Workshop of the Cross–Language Evaluation Forum on Evaluation of Cross–Language Information Retrieval Systems — CLEF 2001. Lecture Notes in Computer Science (LNCS). Springer, London, UK, pp 355–370

Westerveld T, van Zwol R (2006) The INEX 2006 multimedia track. In: Comparative Evaluation of XML Information Retrieval Systems, International Workshop of the Initiative for the Evaluation of XML Retrieval. Lecture Notes in Computer Science (LNCS), vol 4518. Springer, pp 331–344

Zagorac S, Llorente A, Little S, Liu HM, Rüger S (2009) Automated content based video retrieval. In: TREC Video Retrieval Evaluation (TRECVid, Gaithersburg, MD)

Zobel J (1998) How reliable are the results of large–scale information retrieval experiments? In: Proceedings of the 21st annual international ACM SIGIR conference on research and development in information retrieval. ACM press, pp 307–314

van Zwol R, Kazai G, Lalmas M (2005) INEX 2005 multimedia track. In: Advances in XML Information Retrieval and Evaluation, International Workshop of the Initiative for the Evaluation of XML Retrieval. Lecture Notes in Computer Science (LNCS), vol 3977. Springer, pp 497–510

Glossary

ABIR Annotation–Based Image Retrieval

AP Average Precision

API Application Programming Interfaces

ARRS American Roentgen Ray Society

AUC Area Under Curve

BoW Bag–of–Words

BPref Binary Preference

CAD Computer–Assisted Diagnosis

CC Cover Coefficient

CBIR Content–Based Image Retrieval

CEDD Colour and Edge Directivity Descriptor

CLEF Cross–Language Evaluation Forum

CLIR Cross–Language Information Retrieval

CORI Clinical Outcomes Research Initiative

CR Cluster Recall

CRFH Composed Receptive Field Histograms

CT Computed Tomography

C^3M Cover–Coefficient–based Clustering Methodology

DAS Discriminative Accumulation Scheme

DBSCAN Density–based Spatial Clustering of Applications with Noise

DCG Discounted Cumulative Gain

H. Müller et al. (eds.), *ImageCLEF*, The Information Retrieval Series 32,
DOI 10.1007/978-3-642-15181-1, © Springer-Verlag Berlin Heidelberg 2010

DCT Discrete Cosine Transform

DE Document Expansion

DFR Divergence From Randomness

DGFR Diverging Gradient Field Response

DICOM Digital Imaging and Communications in Medicine

DIRECT Distributed Information Retrieval Evaluation Campaign Tool

EXIF Exchangeable Image File Format

EER Equal Error Rate

EM Expectation Maximization

FDA Food and Drug Administration

FlexiTile Flexible Tiling

GATE General Architecture for Text Engineering

GGN Ground Glass Nodules

GMAP Geometric Mean Average Precision

GMM Gaussian Mixture Model

HEAL Health Education Assets Library

HME Hierarchical Mixture of Experts

HSV Hue and Saturation Values

IAPR International Association of Pattern Recognition

ICPR International Conference on Pattern Recognition

IDOL Image Database for rObot Localization

IG Information Gain

INEX INitiative for the Evaluation of XML retrieval

infAP inferred Average Precision

IR Information Retrieval

IRM Integrated Retrieval Model

IRS Information Retrieval System

IRMA Image Retrieval in Medical Applications

ISJ Interactive Search and Judge

JD Jeffrey Divergence

JIRS Java Information Retrieval System

KB Knowledge Base

KNN K–Nearest Neighbor algorithm

LBP Local Binary Patterns

LCA Local Context Analysis

LDA Latent Dirichlet Allocation

LEP Large Electron Positron collider

LIDC Lung Image Database Consortium

LM Language Model

LS–SVM Least Squares Support Vector Machines

LTU LookThatUp

MAP Mean Average Precision

MDCT Multi Detector Computer Tomography

MeSH Medical Subject Heading

MIL Multiple Instance Learning

MIR Mallinckrodt Institute of Radiology/Multimedia Image Retrieval

MMR Maximum Margin Relevance

MPQF MPEG QUery Format

MR Magnetic Resonance

MRR Mean Reciprocal Rank

NCI National Cancer Institute

NDCG Normalized Discounted Cumulative Gain

NIST National Institute of Standards and Technology

NLM National Library of Medicine

NLP Natural Language Processing

NRBP Novelty and Rank–Based Precision

NTCIR NII Test Collection for IR Systems

OS Ontology Score

OWL Web Ontology Language

PA Press Association

PACS Picture Archiving and Communication System

PACT PCA Census Transform Histograms

PCA Principal Component Analysis

PDQ Pretty Damned Quick cancer information retrieval system of the National Cancer Institute, USA

PEIR Pathology Educational Instructional Resource

PLSA Probabilistic Latent Semantic Analysis

PLSI Probabilistic Latent Semantic Indexing

PMID PubMed Identifier

POS Part Of Speech/Programmable Option Select

PRP Probability Ranking Principle

PRF Pseudo–Relevance Feedback

PSN Part Solid Nodule

PT Portfolio Theory

QA Question Answering

QBIC Query By Image Content

QBSI Query By Spatial Icons

QE Query Expansion

QPRP Quality Performance Reporting Program

R/P Recall/Precision

RA Relational Analysis

RBF Radial Basis Function

RBP Rank–Biased Precision

RDF Resource Description Framework

RDM Reconciled Detection Map

RF Relevance Feedback

RIS Remote Installation Services

ROC Receiver Operating Curve

ROI Region Of Interest

RSNA Radiological Society of North America

RSV Relevance Status Value

SAC St. Andrews Collection

SAM Spatial Aggregation Map

SGML Standard Generalized Markup Language

SIFT Scale Invariant Feature Transform

SINTRAM SINai TRAnslation Module

SLAM Simultaneous Localization and Mapping

SLT Statistical Learning Theory

SPARQL Simple Protocol and RDF Query Language

SPEC Standard Performance Evaluation Corporation

S–recall Sub–topic recall

SVD Singolar Value Decomposition

SVM Support Vector Machines

TBIR Text–Based Image Retrieval

TF–IDF Term Frequency–Inverse Document Frequency

TPC Transaction Processing Performance Council

TPS Term Phrase Selection

TREC Text REtrieval Conference

UMLS Unified Medical Language System

VIR Visual Information Retrieval

VOI Volume Of Interest

VOC Visual Object Classes

VQT Visual Query Term

VSM Vector Space Model

WSD Word Sense Disambiguation

Index

H. Müller et al. (eds.), *ImageCLEF*, The Information Retrieval Series 32,
DOI 10.1007/978-3-642-15181-1, © Springer-Verlag Berlin Heidelberg 2010

MIX
Papier aus verantwortungsvollen Quellen
Paper from responsible sources
FSC® C105338

FSC
www.fsc.org

If you have any concerns about our products,
you can contact us on
ProductSafety@springernature.com

In case Publisher is established outside the EU,
the EU authorized representative is:
Springer Nature Customer Service Center GmbH
Europaplatz 3, 69115 Heidelberg, Germany

Printed by Libri Plureos GmbH
in Hamburg, Germany